美國專利申請
及審查實務

冠群國際專利商標聯合事務所

蕭財政、吳珮琪　編著

美國專利申請及審查實務
總目錄

第 1 章　專利申請案之種類 及申請程序

目　錄

§1.1 美國專利申請案之種類

美國專利申請案有許多不同之名稱，依分類之不同有以下數種：

(A) 依請求項之標的分類

① 實用申請案（utility application）：標的為裝置、產品、方法及組合物。

② 新式樣申請案（design application）：標的為物品之外觀設計。

③ 植物申請案（plant application）：標的為以無性繁殖方式複製出的新且獨特的植物新品種。

(B) 依發明人之人數分類

① 單獨申請案（sole application）：發明人只有一人之專利申請案。

② 共同申請案（joint application）：發明人有二人以上之專利申請案。

(C) 依所申請之國家分類

① 國際申請案（international application）：依照國際合作條約（Patent Corporation Treaty, PCT）申請而再進入各國國家階段前之申請案。

② 國家申請案（national application）：包括向美國專利商標局提出申請之專利案件及國際申請案進入美國階段之申請案（national stage application）。

(D) 依是否可取得專利權分類

① 暫時申請案（provisional application）：發明人為了及早取得申請日之申請案，不進行實體審查，不會核發專利權。

② 非暫時申請案（non-provisional application）：實用專利申請案，新式樣專利申請案，植物專利申請案等正規（regular）的專利申請案。

(E) 依與其他申請案之關係分類

① 母申請案（parent application）：指發明人對一發明第一次提出申請之申請案，不包括暫時申請案。

② 延續申請案（continuing application）：指在母申請案提出之後再提出申請而延續母申請案程序之申請案。

③ 一般延續案（continuation application）：請求母申請案所揭露之發明，而發明之揭露與母案相同之申請案。

④ 分割申請案（divisional application）：請求從母申請案分割出來之標的且發明之揭露與母案相同之申請案。

⑤ 部份延續案（continuation-in-part application）：發明之揭露中加入新事項之專利申請案。

⑥ 再發證申請案（reissue application）：已發證後修改請求項再提出之申請案。

⑦ 原申請案（original application）：相對於再發證申請案之原已發證之申請案。

⑧ 再審查申請案（re-examination application）：已發證之專利申請案有專利性之實質新問題時所提出之申請案。

⑨ 取代申請案（substitute application）：為同一申請人在放棄一專利申請案之後，重新提出實質上相同之申請案，亦稱作重複申請案（duplicate application）或重新申請申請案（refiled application）。

⑩ 法定發明註冊申請案（Statutory Invention Registration, SIR）：為一種防衛申請案，申請人如果不想取得專利權，可將非暫時申請案在審查期間變更為法定註冊申請案並公告而達到防衛之目的。

37CFR 1.9 Definitions.

(a)(1) A national application as used in this chapter means a U.S. application for patent which was either filed in the Office under 35 U.S.C. 111, or which entered the national stage from an international application after compliance with 35 U.S.C. 371.

(2) A provisional application as used in this chapter means a U.S. national application for patent filed in the Office under 35 U.S.C. 111(b).

(3) A nonprovisional application as used in this chapter means a U.S. national application for patent which was either filed in the Office under 35 U.S.C. 111(a), or which entered the national stage from an international application after compliance with 35 U.S.C. 371.

(b) An international application as used in this chapter means an international application for patent filed under the Patent Cooperation Treaty prior to entering national processing at the Designated Office stage.

MPEP 201.01 Sole

An application wherein the invention is presented as that of a single person is termed a sole application.

MPEP 201.02 Joint

A joint application is one in which the invention is presented as that of two or more persons. See MPEP § 605.07.

MPEP 201.04 Parent Application

The term "parent" is applied to an earlier application of an inventor disclosing a given invention. Such invention may or may not be claimed in the first application. Benefit of the filing date of copending parent application may be claimed under 35 U.S.C. 120. The term parent will not be used to describe a provisional application.

MPEP 201.04(a) Original Application

"Original" is used in the patent statute and rules to refer to an application which is not a reissue application. An original application may be a first filing or a continuing application.

MPEP 201.09 Substitute Application [R-3]

The use of the term "Substitute" to designate any application which is in essence the duplicate of an application by the same applicant abandoned before the filing of the later application, finds official recognition in the decision *Ex parte Komenak*, 45 USPQ 186, 1940 C.D. 1, 512 O.G. 739 (Comm'r Pat. 1940). Current practice does not require applicant to insert in the specification reference to the earlier application; however, attention should be called to the earlier application. The notation in the file history (see MPEP § 202.02) that one application is a "Substitute" for another is printed in the heading of the patent copies. See MPEP § 202.02.

MPEP 201.10 Refile [R-2]

No official definition has been given the term "Refile," though it is sometimes used as an alternative for the term "Substitute."

If the applicant designates his or her application as "Refile" and the examiner finds that the application is in fact a duplicate of a former application by the same party which was abandoned prior to the filing of the second application, the examiner should require the substitution of the word "substitute" for "refile", since the former term has official recognition.

§1.2 美國境內完成之發明之申請程序

要申請美國專利之發明如果是在美國境內完成（付諸實施是在美國），則依照 35USC§184 之規定，需先在美國提出申請。例如一台灣公司在美國的研究機構有新的發明要申請專利時，需在美國先提出專利之申請，經美國專利局授權核准後才能在外國（例如台灣）提出專利申請。

---

35 U.S.C. 184 Filing of application in foreign country.

Except when authorized by a license obtained from the Commissioner of Patents a person shall not file or cause or authorize to be filed in any foreign country prior to six months after filing in the United States an application for patent or for the registration of a utility model, industrial design, or model in respect of an invention made in this country. A license shall not be granted with respect to an invention subject to an order issued by the

Commissioner of Patents pursuant to section 181 of this title without the concurrence of the head of the departments and the chief officers of the agencies who caused the order to be issued. The license may be granted retroactively where an application has been filed abroad through error and without deceptive intent and the application does not disclose an invention within the scope of section 181 of this title.

＊＊＊＊＊＊＊＊＊＊＊

§1.3 美國專利法 35USC§184 之國外申請授權之規定

依照 35USC§184，除非獲得美國專利局局長之授權（核准），任何人在美國境內所完成之發明在美國提出專利申請之後的六個月之前均不得對此發明在外國提出發明專利申請，實用新型之註冊或工業新式樣之註冊。對於依照 35USC§181 受專利局長的保密命令的發明，非經提出保密命令之相關部門主管之同意，將不會給予國外申請之授權。當一申請案在國外申請是由於錯誤且無欺騙之意圖，而且申請案並未揭露屬於 35USC§181 之範圍內的發明時，此國外申請之授權可追溯核准。申請案包括其改良、修正或補充或分割之申請案。提出美國專利之申請就被視為是自動提出請願，請求美國專利局准予國外申請之授權。在美國專利局之申請收據上會指出是否已授權可向國外提出申請。如果此自動申請未核准，則申請人必需另外提出請願，請求准予國外申請之授權。提出請願時必需註明申請案號、申請日、發明人、發明名稱。如果有二個以上專利申請案要合併在國外申請時則需各別註明。而且如果要在國外申請之案件中包括了美國專利申請案未揭露的部份，則此未揭露的部份需在請願書中敘明。申請人若未獲得美國專利局之授權去申請國外專利，則將無法獲得專利或之後獲得之專利將無效，除非申請人未經授權之國外申請專利是因錯誤且未有欺騙之意圖而且申請專利之標的不會對國家安全有害，而且之後有追溯提出國外申請授權之許可。如果是惡意地不提出國外申請之授權，將另被科以最多美金 1 萬元之罰金或併科最多二年之監禁。

37CFR§5.12 Petition for license.

(a) Filing of an application for patent for inventions made in the United States will be considered to include a petition for license under 35 U.S.C. 184 for the subject matter of the application. The filing receipt will indicate if a license is granted. If the initial automatic petition is not granted, a subsequent petition may be filed under paragraph (b) of this section.

(b) A petition for license must include the fee set forth in § 1.17(g) of this chapter, the petitioner's address, and full instructions for delivery of the requested license when it is to be delivered to other than the petitioner. The petition should be presented in letter form.

37 CFR 5.14 Petition for license; corresponding U.S. application.

(a) When there is a corresponding United States application on file, a petition for license under § 5.12(b) must also identify this application by application number, filing date, inventor, and title, but a copy of the material upon which the license is desired is not required. The subject matter licensed will be measured by the disclosure of the United States application.

(b) Two or more United States applications should not be referred to in the same petition for license unless they are to be combined in the foreign or international application, in which event the petition should so state and the identification of each United States application should be in separate paragraphs.

(c) Where the application to be filed or exported abroad contains matter not disclosed in the United States application or applications, including the case where the combining of two or more United States applications introduces subject matter not disclosed in any of them, a copy of the application as it is to be filed in the foreign country or international application which is to be transmitted to a foreign international or national agency for filing in the Receiving Office, must be furnished with the petition. If however, all new matter in the foreign or international application to be filed is readily identifiable, the new matter may be submitted in detail and the remainder by reference to the pertinent United States application or applications.

35 U.S.C. 185 Patent barred for filing without license.

Notwithstanding any other provisions of law any person, and his successors, assigns, or legal representatives, shall not receive a United States patent for an invention if that person, or his successors, assigns, or legal representatives shall, without procuring the license prescribed in section 184 of this title, have made, or consented to or assisted another's making, application in a foreign country for a patent or for the registration of a utility model, industrial design, or model in respect of the invention. A United States patent issued to such person, his successors, assigns, or legal representatives shall be invalid, unless the failure to procure such license was through error and without deceptive intent, and the patent does not disclose subject matter within the scope of section 181 of this title.

35 U.S.C. 186 Penalty.

Whoever, during the period or periods of time an invention has been ordered to be kept secret and the grant of a patent thereon withheld pursuant to section 181 of this title, shall, with knowledge of such order and without due authorization, willfully publish or disclose or authorize or cause to be published or disclosed the invention, or material information with respect thereto, or whoever willfully, in violation of the provisions of section 184 of this title, shall file or cause or authorize to be filed in any foreign country an application for patent or for the registration of a utility model, industrial design, or model in respect of any invention made in the United States, shall, upon conviction, be fined not more than $10,000 or imprisoned for not more than two years, or both.

§1.4　追溯請求國外申請授權之規定及程序

如果要追溯請求國外申請授權則需以請願（petition）方式進行，請願書要包括以下事項：

(1)　未經授權之專利申請案已申請之國家一覽表，

(2)　在各國申請之申請日期，

(3)　一認證聲明（宣誓書）包括以下各點：

　　(i) 聲明在國外申請時及目前所申請之標的並非受秘密命令之案件；

　　(ii) 聲明在發現有國外申請授權規定後，積極地提出國外申請之授權；

　　(iii)解釋未經授權是出於錯誤且無欺騙之意圖。

(4)　相關規費。

上述之聲明（宣誓書）之中必需包括相關的事實證據而不能只聲明是出於錯誤並無欺騙之意圖。相關的事實證據包括知曉此事之人士之陳述及相關文件之影本。

提出追溯申請如果不被接受則會有 30 天的時間可重新提出請願。如未重新提出，則會發下最終決定，發下最終決定後二個月內申請人可再提請願，如未提請願則最終決定確認，申請人將無法獲得專利。

37CFR §5.25 Petition for retroactive license.

(a) A petition for retroactive license under 35 U.S.C. 184 shall be presented in accordance with § 5.13 or § 5.14(a), and shall include:

(1) A listing of each of the foreign countries in which the unlicensed patent application material was filed,

(2) The dates on which the material was filed in each country,

(3) A verified statement (oath or declaration) containing:

(i) An averment that the subject matter in question was not under a secrecy order at the time it was filed abroad, and that it is not currently under a secrecy order,

(ii) A showing that the license has been diligently sought after discovery of the proscribed foreign filing, and

(iii) An explanation of why the material was filed abroad through error and without deceptive intent without the required license under § 5.11 first having been obtained, and

(4) The required fee (§ 1.17(g) of this chapter).

(b) The explanation in paragraph (a) of this section must include a showing of facts rather than a mere allegation of action through error and without deceptive intent. The showing of facts as to the nature of the error should include statements by those persons having personal knowledge of the acts regarding filing in a foreign country and should be accompanied by copies of any necessary supporting documents such as letters of transmittal or instructions for filing. The acts which are alleged to constitute error without deceptive intent should cover the period leading up to and including each of the proscribed foreign filings.

(c) If a petition for a retroactive license is denied, a time period of not less than thirty days shall be set, during which the petition may be renewed. Failure to renew the petition within the set time period will result in a final denial of the petition. A final denial of a petition stands unless a petition is filed under § 1.181 within two months of the date of the denial. If the petition for a retroactive license is denied with respect to the

invention of a pending application and no petition under § 1.181 has been filed, a final
rejection of the application under 35 U.S.C. 15 will be made.

實例

　　申請人於美國境內發明了一開罐器，並於 2002 年 10 月 10 日提出美國專利
申請案，之後美國專利局發下的申請收據上並未註明已授權申請國外專利。申請
人於 2003 年 6 月 10 日對該開罐器之改良提出了一部份延續案之申請，同時擬向
國外申請該開罐器及改良之開罐器，請問申請人如果提出國外之申請是否違反
35USC§184 之規定？

　　申請人如果在國外申請原母案之開罐器並未違反 35USC §184 之規定，因為
在申請收據上雖未授權可申請國外專利，但自申請日起已超過 6 個月，故可提出
國外專利之申請。但如果在 2003 年 6 月 10 日起之半年內提出該部份延續案（CIP）
案之改良的開罐器的國外申請，則違反 35USC§184 之規定。

§1.5　取得非暫時申請案之申請日需具備的文件及補件期限

　　將符合 35USC§112 之說明書，至少一請求項及圖式送入美國專利商標局之
日期即為一非暫時申請案之申請日。亦即，要取得一有效的非暫時申請案之申請
日只需要說明書，至少一請求項及圖式（如需要）即可。發明人名稱、受讓人名
稱、宣誓書、讓渡書、委任書及申請規費等均可以後補。

　　若一非暫時申請案送入之文件不符合上述申請日要件（filing date
requirements）時，美國專利局會發下申請案不完整之通知（notice of incomplete
application）通知申請人於二個月內補正（可延 5 個月），如果申請人要美國專
利局重新檢閱所送之文件，則可提出請願。如未提出請願，則該非暫時申請案之
申請日即為上述文件補正完成之日期。如果申請人在期限內未作補正，則此申請
案將被放棄。

　　一非暫時申請案如果符合上述申請日要件並已賦予一申請日，申請案號，但
仍缺某些文件或規費，例如宣誓書、委任書、申請規費、檢索規費、審查規費、

超項費，多項附屬項規費，英文說明書，所主張之暫時申請案之英文翻譯等時，美國專利局會發下補件通知（notice to file missing parts），並給予申請人二個月的時間補正（可再延 5 個月）。

　　一非暫時申請案如果符合申請日要件，但送入之文件並不正確，例如圖式不符規定，說明書格式不符規定或宣誓書有錯誤等時，美國專利局則會發下補送正確文件通知（notice to file corrected application），並給申請人二個月的時間補正（可再延 5 個月）。

37CFR §1.53 Application number, filing date, and completion of application.

(a) Application number. Any papers received in the Patent and Trademark Office which purport to be an application for a patent will be assigned an application number for identification purposes.

(b)Application filing requirements - Nonprovisional application. The filing date of an application for patent filed under this section, except for a provisional application under paragraph (c) of this section or a continued prosecution application under paragraph (d) of this section, is the date on which a specification as prescribed by 35 U.S.C. 112 containing a description pursuant to § 1.71 and at least one claim pursuant to § 1.75, and any drawing required by § 1.81(a) are filed in the Patent and Trademark Office.

＊＊＊＊＊＊＊＊＊＊＊

(e) Failure to meet filing date requirements.

(1) If an application deposited under paragraph (b), (c), or (d) of this section does not meet the requirements of such paragraph to be entitled to a filing date, applicant will be so notified, if a correspondence address has been provided, and given a period of time within which to correct the filing error.

＊＊＊＊＊＊＊＊＊＊＊

(2) Any request for review of a notification pursuant to paragraph (e)(1) of this section, or a notification that the original application papers lack a portion of the specification or drawing(s), must be by way of a petition pursuant to this paragraph accompanied by the fee set forth in § 1.17(f). In the absence of a timely (§ 1.181(f)) petition pursuant to this paragraph, the filing date of an application in which the applicant was notified of a filing error pursuant to paragraph (e)(1) of this section will be the date the filing error is corrected.

(3) If an applicant is notified of a filing error pursuant to paragraph (e)(1) of this section, but fails to correct the filing error within the given time period or otherwise timely (§ 1.181(f)) take action pursuant to this paragraph, proceedings in the application will be considered terminated. Where proceedings in an application are terminated pursuant to this paragraph, the application may be disposed of, and any filing fees, less the handling fee set forth in § 1.21(n), will be refunded.

(f) *Completion of application subsequent to filing-Nonprovisional (including continued prosecution or reissue) application.*

(1) If an application which has been accorded a filing date pursuant to paragraph (b) or (d) of this section does not include the basic filing fee, the search fee, or the examination fee, or if an application which has been accorded a filing date pursuant to paragraph (b) of this section does not include an oath or declaration by the applicant pursuant to §§ 1.63, 1.162 or § 1.175, and applicant has provided a correspondence address (§1.33(a)), applicant will be notified and given a period of time within which to pay the basic filing fee, search fee, and examination fee, file an oath or declaration in an application under paragraph (b) of this section, and pay the surcharge if required by § 1.16(f) to avoid abandonment.

(2) If an application which has been accorded a filing date pursuant to paragraph (b) of this section does not include the basic filing fee, the search fee, the examination fee, or an oath or declaration by the applicant pursuant to §§ 1.63, 1.162 or § 1.175, and applicant has not provided a correspondence address (§ 1.33(a)), applicant has two months from the filing date of the application within which to pay the basic filing fee, search fee, and examination fee, file an oath or declaration, and pay the surcharge required by § 1.16(f) to avoid abandonment.

(3) If the excess claims fees required by §§ 1.16 (h) and (i) and multiple dependent claim fee required by § 1.16 (j) are not paid on filing or on later presentation of the claims for which the excess claims or multiple dependent claim fees are due, the fees required by §§ 1.16 (h), (i) and (j) must be paid or the claims canceled by amendment prior to the expiration of the time period set for reply by the Office in any notice of fee deficiency. If the application size fee required by § 1.16 (s) (if any) is not paid on filing or on later presentation of the amendment necessitating a fee or additional fee under § 1.16(s), the fee required by § 1.16 (s) must be paid prior to the expiration of the time period set for reply by the Office in any notice of fee deficiency in order to avoid abandonment.

＊＊＊＊＊＊＊＊＊＊＊

§1.6　取得申請日之要件：說明書缺頁，但包含有至少一請求項

專利申請案送入美國專利局後，如果經初步審查發現說明書有缺頁，但有請求項及圖式時，會發下缺文件通知（notice of omitted items）通知該申請人於二個月內答覆。

如為下列情形該專利申請案可取得申請日：

(A) 申請人主張缺頁的部份確實有送入美國專利局，此時需附上證據（例如由美國專利局確收之回郵明信片，上面註明說明書幾頁）提出請願。如果美國專利局經查證事實上缺頁部份有送入專利局，則申請日即為原申請日且請願之費用可退費。此二個月的答覆期限不能延期。

(B) 如果缺頁部份確實未送入美國專利局，則申請人需補送缺頁部份及補充的宣誓書（需提到此缺頁部份）並提出請願，則申請日為補送缺頁之日期。

(C) 如果申請人認為缺頁部份對發明之揭露並不會影響，則可提出先前修正（preliminary amendment）將說明書之頁碼重新編，並刪除不完整之句子，則申請日可維持原來之申請日。此外，如果申請案有主張國外優先權，或母案或暫時案之優先權，而此缺頁部份是因疏忽所造成時，可依照 37CFR§1.57(a)之規定以先前修正的方式將缺頁部份之敘述補入，就可維持原來的申請日。

MPEP 601.01(d) Application Filed Without All Pages of Specification [R-7]

The Office of **>Patent Application Processing (OPAP)< reviews application papers to determine whether all of the pages of specification are present in the application. If the application is filed without all of the page(s) of the specification, but containing something that can be construed as a written description, at least one drawing figure, if necessary under 35 U.S.C. 113 (first sentence), and, in a nonprovisional application, at least one claim, *>OPAP< will mail a "Notice of Omitted Items" indicating that the application papers so deposited have been accorded a filing date, but are lacking some page(s) of the specification.

If the application does not contain anything that can be construed as a written description, *>OPAP< will mail a Notice of Incomplete Application indicating that the application lacks the specification required by 35 U.S.C. 112 and no filing date is granted.

I.APPLICATION ENTITLED TO FILING DATE

The mailing of a "Notice of Omitted Item(s)" will permit the applicant to:

(A) promptly establish prior receipt in the USPTO of the page(s) at issue. An applicant asserting that the page(s) was in fact received by the USPTO with the application papers must, within 2 months from the date of the "Notice of Omitted Item(s)," file a petition under 37 CFR 1.53(e) with the petition fee set forth in 37 CFR 1.17(f), along with evidence of such deposit (37 CFR 1.181(f)). The petition fee will be refunded if it is determined that the page(s) was in fact received by the USPTO with the application papers deposited on filing. The 2-month period is not extendable under 37 CFR 1.136;

(B) promptly submit the omitted page(s) in a nonprovisional application and accept the date of such submission as the application filing date. An applicant desiring to submit the omitted page(s) in a nonprovisional application and accept the date of such submission as the application filing date must, within 2 months from the date of the "Notice of Omitted Item(s)," file any omitted page(s) with an oath or declaration in compliance with 37 CFR 1.63 and 37 CFR 1.64 referring to such page(s) and a petition under 37 CFR 1.182 with the petition fee set forth in 37 CFR 1.17(f), requesting the later filing date (37 CFR 1.181(f)). The 2-month period is not extendable under 37 CFR 1.136; or

(C) accept the application as deposited in the USPTO. Applicant may accept the application as deposited in the USPTO by either:

(1) not filing a petition under 37 CFR 1.53(e) or 37 CFR 1.182 (and the required petition fee) as discussed above within 2 months of the date of the "Notice of Omitted Item(s)". The failure to file a petition under 37 CFR 1.53(e) or 37 CFR 1.182 will be treated as constructive acceptance by the applicant of the application as deposited in the USPTO. The application will maintain the filing date as of the date of deposit of the application papers in the USPTO, and the original application papers (i.e., the original disclosure of the invention) will include only those application papers present in the USPTO on the date of deposit. Amendment of the specification is required in a

nonprovisional application to renumber the pages consecutively and cancel any incomplete sentences caused by the absence of the omitted page(s). Such amendment should be by way of preliminary amendment submitted prior to the first Office action to avoid delays in the prosecution of the application, or

(2) filing an amendment under 37 CFR 1.57(a). If an application was filed on or after September 21, 2004, and contains a claim under 37 CFR 1.55 for priority of a prior-filed foreign application, or a claim under 37 CFR 1.78 for the benefit of a prior-filed provisional, nonprovisional, or international application that was present on the filing date of the application, and the omitted portion of the specification was inadvertently omitted from the application and is completely contained in the prior-filed application, applicant may submit an amendment to include the inadvertently omitted portion of the specification pursuant to 37 CFR 1.57 (a). Such amendment should be by way of a preliminary amendment and the preliminary amendment must be submitted within 2 months from the date of the "Notice of Omitted Item(s)." The amendment should be identified as an amendment pursuant to 37 CFR 1.57(a) and must comply with the requirements of 37 CFR 1.57(a) and 37 CFR 1.121. See MPEP § 201.17. The application will maintain the filing date as of the date of deposit of the original application papers in the USPTO. The original application papers (i.e., the original disclosure of the invention) will include only those application papers present in the USPTO on the original date of deposit. The 2-month period is not extendable under 37 CFR 1.136.

＊＊＊＊＊＊＊＊＊＊＊

§1.7　取得申請日之要件：有完整之說明書，請求項但無圖式

依照 35USC§113，如果圖式是要了解所請求專利的標的所必需的，申請人就需在申請時提供圖式。在提出申請後所補送之圖式不能用來克服(i)因缺乏據以實施之揭露或其他不足之揭露所造成之說明書之揭露不充份，及(ii)不能用來解釋請求項之範圍而補充原始之揭露。

因而，當一申請案提申後，如果美國專利局之初步審查辦公室（OIPE）發現沒有圖式，就會先檢查說明書本文中是否有圖式之簡單說明，如果沒有圖式之簡單說明，就再判斷是否圖式是要了解請求標的所必需的。通常有至少一方法請

求項或組合物請求項時，OIPE 就會認為圖式是非必需的。圖式被認為不是必需用來了解請求的標的的情形還有下列數種：

(A) 經塗覆之物品或產品：發明乃在於塗覆或沈浸一習知的片狀物、紙、布或習知特定組成之物品，除非物品請求項包含了結構或構造上之特徵；

(B) 由特別的材料或組成製成之物品：發明乃為製造一特別的材料或組成之物品的方法，除非物品請求項包合結構或構造之特徵；

(C) 層積之物品；所請求發明僅為片狀物之層積或塗覆，除非物品請求項包括結構或構造之特徵（非層積之順序）；

(D) 物品、裝置或系統，其特徵僅為具有特定之材料，所請求之發明僅是在習知之物品，裝置或系統上使用特定之材料，例如(1)液壓系統，其特徵僅在於使用一特殊之液壓流體；(2)包裝的綑線，其包裝之結構為習知的，只是使用一特定的材料；

實用專利申請案所請求之標的為上述者，且其說明書本文中並無圖式之說明，如果並未附上圖式則美國專利局就會認為圖式是不需要的，而給予一申請日。

實用專利申請案所請求之標的為上述者，但說明書本文中有圖式之說明，如果並未附上圖式，則美國專利局就會認為圖式是必需的，就會發下申請不完整之通知，而不賦予申請日。申請人如果確實有將圖式送入但 USPTO 未收到，則可檢附證據（如申請時回執之明信片上有註明圖式幾張）提出請願，或者亦可提出請願主張圖式並非了解所請求之發明的必需者。而如果專利申請有主張國外優先權或暫時申請案或母案之優先權時可提出請願，主張圖式是因疏忽未送入，可依 37CFR §1.157(a)之規定將圖式以先前修正之方式補入，如請願被核准，則可取得申請日。

---

35 U.S.C. 113 Drawings.

The applicant shall furnish a drawing where necessary for the understanding of the subject matter sought to be patented. When the nature of such subject matter admits of illustration by a drawing and the applicant has not furnished such a drawing, the Director may require its submission within a time period of not less than two months from the sending of a notice thereof. Drawings submitted after the filing date of the application may not be used (i) to overcome any insufficiency of the specification due to lack of an enabling disclosure or otherwise inadequate disclosure therein, or (ii) to supplement the original disclosure thereof for the purpose of interpretation of the scope of any claim.

MPEP 601.01(f) Application Filed Without Drawings [R-7]

35 U.S.C. 111(a)(2)(B) and 35 U.S.C. 111(b)(1)(B) each provide, in part, that an "application shall include . . . a drawing as prescribed by section 113 of this title" and 35 U.S.C. 111(a)(4) and 35 U.S.C. 111(b)(4) each provide, in part, that the "filing date. . . shall be the date on which . . . any required drawing are received in the Patent and Trademark Office." 35 U.S.C. 113 (first sentence) in turn provides that an "applicant shall furnish a drawing where necessary for the understanding of the subject matter sought to be patented."

Applications filed without drawings are initially inspected to determine whether a drawing is referred to in the specification, and if not, whether a drawing is necessary for the understanding of the invention. 35 U.S.C. 113 (first sentence).

It has been USPTO practice to treat an application that contains at least one process or method claim as an application for which a drawing is not necessary for an understanding of the invention under 35 U.S.C. 113 (first sentence). The same practice has been followed in composition applications. Other situations in which drawings are usually not considered necessary for the understanding of the invention under 35 U.S.C. 113 (first sentence) are:

(A) Coated articles or products: where the invention resides solely in coating or impregnating a conventional sheet (e.g., paper or cloth, or an article of known and conventional character with a particular composition), unless significant details of structure or arrangement are involved in the article claims;

(B) Articles made from a particular material or composition: where the invention consists in making an article of a particular material or composition, unless significant details of structure or arrangement are involved in the article claims;

(C) Laminated structures: where the claimed invention involves only laminations of sheets (and coatings) of specified material unless significant details of structure or arrangement (other than the mere order of the layers) are involved in the article claims; or

(D) Articles, apparatus, or systems where sole distinguishing feature is presence of a particular material: where the invention resides solely in the use of a particular material in an otherwise old article, apparatus or system recited broadly in the claims, for example:

(1) A hydraulic system distinguished solely by the use therein of a particular hydraulic fluid;

(2) Packaged sutures wherein the structure and arrangement of the package are conventional and the only distinguishing feature is the use of a particular material.

A nonprovisional application having at least one claim, or a provisional application having at least some disclosure, directed to the subject matter discussed above for which a drawing is usually not considered essential for a filing date, not describing drawing figures in the specification, and filed without drawings will simply be processed, so long as the application contains something that can be construed as a written description. A nonprovisional application having at least one claim, or a provisional application having at least some disclosure, directed to the subject matter discussed above for which a drawing is usually not considered essential for a filing date, describing drawing figure(s) in the specification, but filed without drawings will be treated as an application filed without all of the drawing figures referred to in the specification as discussed in MPEP § 601.01(g), so long as the application contains something that can be construed as a written description. In a situation in which the appropriate Technology Center (TC) determines that drawings are necessary under 35 U.S.C. 113 (first sentence) the filing date issue will be reconsidered by the USPTO. The application will be returned to the Office of **>Patent Application Processing (OPAP)< for mailing of a "Notice of Incomplete Application."

Applicant may file a petition under 37 CFR 1.53(e) with the petition fee set forth in 37 CFR 1.17(f), asserting that (A) the drawing(s) at issue was submitted, or (B) the drawing(s) is not necessary under 35 U.S.C. 113 (first sentence) for a filing date. The petition must be accompanied by sufficient evidence to establish applicant's entitlement to the requested filing date (e.g., a date-stamped postcard receipt (MPEP § 503) to establish prior receipt in the USPTO of the drawing(s) at issue). Alternatively, applicant may submit drawing(s) accompanied by an oath or declaration in compliance with 37 CFR 1.63 and 1.64 referring to the drawing(s) being submitted and accept the date of such submission as the application filing date.

As an alternative to a petition under 37 CFR 1.53(e), if the drawing(s) was inadvertently omitted from an application filed on or after September 21, 2004, and the application contains a claim under 37 CFR 1.55 for priority of a prior-filed foreign application, or a claim under 37 CFR 1.78 for the benefit of a prior-filed provisional, nonprovisional, or international application, that was present on the filing date of the application, and the inadvertently omitted drawing(s) is completely contained in the prior-filed application, the applicant may submit the omitted drawing(s) by way of an amendment in compliance with 37 CFR 1.57(a). The amendment must be by way of a petition under 37 CFR 1.57(a)(3) accompanied by the petition fee set forth in 37 CFR 1.17(f). See MPEP § 201.17.

§1.8 取得申請日之要件：有完整之說明書，請求項，但圖式有缺頁

　　美國專利之專利申請案處理辦公室（Office of Patent Application Processing, OPAP）會先檢查在說明書中所提到的各圖式是否有附在申請書之中。如果發現圖式有缺少時會發下缺文件通知，請申請人於二個月內回覆並同時給予申請案一申請日。

　　如為下列情形，該專利申請案可取得申請日：

(A) 申請人主張缺少的圖確實有送入美國專利局，此時需附上證據（例如由美國專利局確收之回郵明信片，上面註明圖式幾頁）提出請願，如果經美國專利局查證事實上缺的圖有送入專利局，則申請日即為原申請日。此二個月答覆期限不能延期。

(B) 如果缺的圖並未送入美國專利局，則申請人需補送所缺的圖及補充的宣誓書（需提到此缺的圖）並提出請願，則申請日為補送圖式之日。

(C) 如果申請人認為缺圖對發明之揭露並不會影響，則可提出先前修正將說明書中該缺少圖的說明刪除，並修正圖號順序，則申請日可維持原來之申請日。此外，如果申請案有主張國外優先權或母案或暫時案之優先權，而此缺圖是因疏忽所造成時可依照 37CFR§1.57(a)之規定以先前修正的方式將所缺之圖補入，就可維持原來的申請日。

MPEP 601.01(g) Applications Filed Without All Figures of Drawings [R-7]

　　The Office of **>Patent Application Processing (OPAP)< reviews application papers to determine whether all of the figures of the drawings that are mentioned in the specification are present in the application. If the application is filed without all of the drawing figure(s) referred to in the specification, and the application contains something that can be construed as a written description, at least one drawing, if necessary under 35 U.S.C. 113(first sentence), and, in a nonprovisional application, at least one claim,

>OPAP< will mail a "Notice of Omitted Item(s)" indicating that the application papers so deposited have been accorded a filing date, but are lacking some of the figures of drawings described in the specification.

The mailing of a "Notice of Omitted Item(s)" will permit the applicant to:

(A) promptly establish prior receipt in the USPTO of the drawing(s) at issue. An applicant asserting that the drawing(s) was in fact received by the USPTO with the application papers must, within 2 months from the date of the "Notice of Omitted Item(s)," file a petition under 37 CFR 1.53(e) with the petition fee set forth in 37 CFR 1.17(f), along with evidence of such deposit (37 CFR 1.181(f)). The petition fee will be refunded if it is determined that the drawing(s) was in fact received by the USPTO with the application papers deposited on filing. The 2-month period is not extendable under 37 CFR 1.136;

(B) promptly submit the omitted drawing(s) in a nonprovisional application and accept the date of such submission as the application filing date. An applicant desiring to submit the omitted drawing(s) in a nonprovisional application and accept the date of such submission as the application filing date must, within 2 months from the date of the "Notice of Omitted Item(s)," file any omitted drawing(s) with an oath or declaration in compliance with 37 CFR 1.63 and 37 CFR 1.64 referring to such drawing(s) and a petition under 37 CFR 1.182 with the petition fee set forth in 37 CFR 1.17(f), requesting the later filing date (37 CFR 1.181(f)). The 2-month period is not extendable under 37 CFR 1.136; or

(C) accept the application as deposited in the USPTO. Applicant may accept the application as deposited in the USPTO by either:

(1) not filing a petition under 37 CFR 1.53(e) or 37 CFR 1.182 (and the required petition fee) as discussed above within 2 months of the date of the "Notice of Omitted Item(s)." The failure to file a petition under 37 CFR 1.53(e) or 37 CFR 1.182 will be treated as constructive acceptance by the applicant of the application as deposited in the USPTO. The application will maintain the filing date as of the date of deposit of the original application papers in the USPTO. The original application papers (i.e., the original disclosure of the invention) will include only those application papers present in the USPTO on the original date of deposit. Amendment of the specification is required in a nonprovisional application to cancel all references to the omitted drawing, both in the brief and detailed descriptions of the drawings and including any reference numerals shown only in the omitted drawings. In addition, an amendment with replacement sheets of drawings in compliance with 37 CFR 1.121(d) is required in a nonprovisional application to renumber the drawing figures consecutively, if necessary, and amendment of the specification is required to correct the references to the drawing figures to correspond with any relabeled drawing figures, both in the brief and detailed descriptions

of the drawings. Such amendment should be by way of preliminary amendment submitted prior to the first Office action to avoid delays in the prosecution of the application, or

(2) filing an amendment under 37 CFR 1.57 (a). If an application was filed on or after September 21, 2004, and contains a claim under 37 CFR 1.55 for priority of a prior-filed foreign application, or a claim under 37 CFR 1.78 for the benefit of a prior-filed provisional, nonprovisional, or international application that was present on the filing date of the application, and the omitted portion of the drawings was inadvertently omitted from the application and is completely contained in the prior-filed application, applicant may submit an amendment to include the inadvertently omitted portion of the drawings pursuant to 37 CFR 1.57(a). Such amendment should be by way of a preliminary amendment and the preliminary amendment must be submitted within 2 months from the date of the "Notice of Omitted Item(s)." The amendment should be identified as an amendment pursuant to 37 CFR 1.57 (a) and must comply with the requirements of 37 CFR 1.57 (a) and 37 CFR 1.121. See MPEP § 201.17. The application will maintain the filing date as of the date of deposit of the original application papers in the USPTO. The original application papers (i.e., the original disclosure of the invention) will include only those application papers present in the USPTO on the original date of deposit.

The 2-month period is not extendable under 37 CFR 1.136.

實例 1

一專利申請案提出申請之後，發明人發現圖式中有一張圖並未送入，但是在說明書之圖式簡單說明中及詳細敘述中有提到此圖式，發明人在申請時已簽署了宣誓書，且宣誓書已提交，發明人認為即使無該張圖式，說明書已充分揭露。請問如果要保有原申請日需提供哪些文件？進行哪些程序？

提出修正將說明書中有關該圖之敘述刪除並提出請願（petition）及繳費請求維持原申請日，同時需提出發明人之補充宣誓書。

實例 2

一台灣專利事務所於 2006 年 5 月 23 日收到客戶指示需於 2006 年 5 月 23 日當天提出一美國專利的申請以主張台灣發明專利的優先權。台灣發明專利申請案乃於 2005 年 5 月 23 日提申，由於尚未撰寫英文說明書且發明人尚未簽署任何表格，故該台灣專利事務所以中文說明書於當日傳送至合作之美國律師處，請求當日以中文說明書提出申請。此美國專利申請案確實於 2006 年 5 月 23 日提出申請。之後，美國律師收到美國專利局之通知需於 2006 年 8 月 1 日前補送英文說明書及宣誓書（讓渡書）委任書等文件。此台灣專利事務所於 2006 年 7 月 20 日完成英文說明書並交發明人簽署相關表格，英文說書及表格於 2006 年 7 月 23 日以電子郵件送交美國律師處。美國律師收到英文說明書及表格時發現於 2006 年 5 月 23 日送入之中文說明書並無圖式，而此圖式對於所請求之發明乃是必需的。美國律師於 2006 年 7 月 24 日將英文說書及圖式及相關表格送入美國專利局，請問美國專利申請案之申請日為哪一天？是否可主張台灣優先權？

此美國專利申請案之申請日為 2006 年 7 月 24 日，即補入圖式之日期。由於 2006 年 7 月 24 日至 2005 年 5 月 23 日已超過一年，故無法主張台灣申請案之優先權。

§1.9 提出專利申請之方法

有 4 種方法可提出專利的申請以取得申請日：(1)親至美國專利局送件(2)以美國郵局（United States Postal Service USPS）的一般郵件送件(3)以美國郵局之郵局對收件人之快速郵件（express mail）送件(4)以電子申請送件。

如果親至美國專利局投件或由私人快遞投件，則需直接送至 Customer Service Window。美國專利局會在收到的文件上蓋上收件章（stamp of receipt）以確認收到日期。申請日為申請文件確實被美國專利局收到之日期（如果申請文件符合申請日要件）。如果以美國郵局一般郵件將申請文件寄至美國專利局，則申請日為美國專利局收到申請文件之日期。如果以美國郵局（USPS）之郵局對收件人之快速郵件（Express Mail Post Office to Addressee）送件，則其申請日乃為向美國郵局投件之日期。

當以上述前三種方式提出申請，而且申請人同時呈送具有回郵及住址之明信片，且此明信片上註明了所呈送之申請文件的細目：①申請人姓名，②發明之名稱，③說明書及請求項之頁數，圖之張數，④是否包括宣誓書，其他表格⑤如申請資料表、申請文件等，⑥規費之金額或付費方式的話，美國專利局就會在明信片上蓋上收文章並給予一申請案號，並將明信片寄回給申請人（代理人）。但需注意的是美國專利局在明信片上蓋章並不一定表示該申請文件已齊備，而所給予之申請案號亦可能有所變更。

　　除了上述三種方法提出專利申請以外，申請人亦可以電子申請方式來送件。以電子申請除了迅速方便之外，還可立即得到申請案號。此外，對於小企業或個人之申請人，可減免申請之規費，故電子申請已漸漸被專利代理人使用。

　　上述四種方式可適用非暫時專利申請案，暫時專利申請案、延續案、分割案等申請案之申請。而新式樣之 CPA 申請案，除了可以上述四種方法申請外，還可以傳真方式送件以取得申請日。

　　美國專利局之 OIPE（程序審查辦公室）在審查申請人所送入的文件判斷符合取得申請日之最少文件之後，就會發給申請人（代理人）官方之申請收據。此申請收據會記載申請日，申請號碼，確認號碼（confirmation number），專利分類，技術單位（art unit），延續案資料，國外優先權資料，大小實體狀態等。此申請收據上之資料乃為美國專利局之官方正式資料，因此在回郵明信片上之申請案號如與申請收據上號碼不符時，乃以申請收據上之資料為準。因而，當發現申請收據上之資訊不正確時，申請人應向美國專利局提出修正。

MPEP 502

　　All applications (provisional and nonprovisional) may be sent to the U.S. Patent and Trademark Office by mail (see MPEP § 501), or they may be hand-carried to the Customer Service Window. New utility patent applications and provisional applications can also be filed via the Office's Electronic Filing System (EFS). See MPEP §1730, subsection II.B. A continued prosecution application (CPA) filed under 37 CFR 1.53(d) (available for design applications only), amendments, and other papers may be sent to the U.S. Patent and Trademark Office by mail (see MPEP § 501), by facsimile (see MPEP § 502.01) or hand-carried to the Customer Service Window. Any correspondence sent to the U.S. Patent and Trademark Office should include the sender's return address and ZIP Code designation.

　　All correspondence with the Office, except for communications elating to pending litigation and certain disciplinary proceedings as specified in 37 CFR 1.1(a)(3), may be filed directly at the Customer Service Window at the following address, or delivered by private courier to:

　　Assistant Commissioner for Patents [or Commissioner of Patents and Trademarks or Assistant Commissioner for Trademarks]

USPTO Office of Initial Patent Examination

Crystal Plaza Building 2, Room 1B03

2011 South Clark Place

Arlington, VA 22202

＊ ＊ ＊ ＊ ＊ ＊ ＊ ＊ ＊ ＊ ＊

MPEP 503

＊ ＊ ＊ ＊ ＊ ＊ ＊ ＊ ＊ ＊ ＊

If a self-addressed postcard is submitted with a patent application, that postcard will be provided with both the receipt date and application number prior to returning it to the addressee. The application number identified on such a postcard receipt is merely the preliminary assignment of an application number to the application, and should not be relied upon (e.g., with respect to foreign filings) as necessarily representing the application number assigned to such application. See 37 CFR 1.53(b).

The identifying data on the postcard should include:

(A) applicant's name(s);

(B) title of invention;

(C) number of pages of specification, claims (for nonprovisional applications), and sheets of drawing;

(D) whether oath or declaration is included;

(E) a list of any additional forms included with the application (e.g., application transmittal form, application data sheet, fee transmittal form, and/or provisional application cover sheet); and

(F amount and manner of paying the fee.

A return postcard should be attached to each patent application for which a receipt is desired.

It is important that the return postcard itemize all of the components of the application. If the postcard does not itemize each of the components of the application, it will not serve as evidence that any component which was not itemized was received by the United States Patent and Trademark Office (USPTO).

It should be recognized that the identification of an application by application number does not necessarily signify that the USPTO has accepted the application as complete (37 CFR 1.53(a)).

OIPE mails a filing receipt to the attorney or agent, if any, otherwise to the applicant, for each application filed which meets the minimum requirements to receive a filing date. The filing receipt includes the application number, filing date, a confirmation number, a suggested class in the U.S. Patent Classification System (see MPEP § 902.01), and the number of an art unit where the application is likely to be examined. The filing receipt also includes other information about the application as applicable, such as continuing data, national stage data, foreign priority data, foreign filing license data, entity status information, and the date the Office anticipates publishing the application under 35 U.S.C. 122(b). The filing receipt represents the official assignment by the USPTO of a specific application number and confirmation number to a particular application. See 37 CFR 1.54 (b). The application number officially assigned to an application on the filing receipt may differ from the application number identified on a postcard receipt submitted with such application, and, as between inconsistent filing receipts and postcard receipts, the application number on the filing receipt is controlling.

§1.10　以郵局對收件人之快速郵件提出專利申請

自 1996 年 12 月 2 日起申請人提出專利申請可使用美國郵局（USPS）之郵局對收件人之快速郵件（Express Mail Post Office to Addressee）送件。以此方法提出專利申請的好處是向美國郵局投遞之日期即為專利之申請日，即使是在假日，如週日向美國郵局投遞，該週日即為專利申請日。

向美國郵局投遞之日期（date in）乃顯示在快速郵件之郵寄標籤（mailing label）或其他美國郵局之正式單據上。而美國專利局就會以郵寄標籤上之投遞日期作為申請日。因而以此方式送件時，需確認美國郵局之人員正確且清楚地將投遞日期印在郵寄標籤之上。如果美國專利局收到快速郵件時投遞日期不清楚的話，美國專利局將以收到快速郵件之日期為申請日。另外，如果去投遞快速郵件時，美國郵局人員已下班而將快速郵件投入郵件箱（drop box）時，有可能郵寄標籤上顯示的投遞日（date in）會與真正的投遞日不一致。

37CFR§ 1.10 Filing of correspondence by "Express Mail."

(a)(1) Any correspondence received by the U.S. Patent and Trademark Office (USPTO) that was delivered by the "Express Mail Post Office to Addressee" service of the United States Postal Service (USPS) will be considered filed with the USPTO on the date of deposit with the USPS.

(2) The date of deposit with USPS is shown by the "date in" on the "Express Mail" label or other official USPS notation. If the USPS deposit date cannot be determined, the correspondence will be accorded the USPTO receipt date as the filing date. See § 1.6(a).

(b) Correspondence should be deposited directly with an employee of the USPS to ensure that the person depositing the correspondence receives a legible copy of the "Express Mail" mailing label with the "date-in" clearly marked. Persons dealing indirectly with the employees of the USPS (such as by deposit in an "Express Mail" drop box) do so at the risk of not receiving a copy of the "Express Mail" mailing label with the desired "date-in" clearly marked. The paper(s) or fee(s) that constitute the correspondence should also include the "Express Mail" mailing label number thereon. See paragraphs (c), (d) and (e) of this section.

＊＊＊＊＊＊＊＊＊＊＊＊

§1.11　非暫時專利申請案之形式審查對圖式之要求

由於 2000 年 11 月 29 日以後提出申請之實用申請案及植物申請案會作核准前公告（自申請日或優先權日起 18 個月後公告），因而美國專利局的初步審查辦公室（OIPE）會對 2000 年 11 月 29 日以後申請之實用申請案及植物申請案作圖式之初步審查以確認圖式是否適合被掃描而用作公告之用。新式樣之專利申請案雖不作核准前公告但 OIPE 仍會對圖式作初步的審查。此 OIPE 之圖式初步審查所採取的標準並不像 37 CFR 1.84 所規定之正式圖式之標準那樣嚴格。因此有時申請人自認為是非正式的圖式亦可被接受。如果 OIPE 認為圖式不能被接受（不符合掃描之標準），就會通知申請人並給二個月的期間補送可接受之圖式。補送 OIPE 要求之圖式不需繳規費。

圖式有下列情形之一時，OIPE 會通知申請人補送可接受之圖式：

(A) 圖式之線太淡無法清楚複製，或參考數字（文字）、圖號不清楚（圖式中之參考文字需至少 1/8 英吋高）；

(B) 圖式之引線（圖式之部份與參考數字之連線）漏掉；

(C) 包含多餘之參考文字或圖號或參考文字非英文，如流程圖中包含非英文之文字；

(D) 紙張太小及邊線空白不符標準，紙張需為 21.0cm×29.7cm（A4）或 21.6cm×27.9cm，且上方邊緣需留至少 2.5cm 空白，右方邊緣需留至少 1.5cm 空白，左方邊緣需留至少 2.5cm 空白，下方邊緣需留至少 1.0cm 空白；

(E) 圖式之編號應標示為 Fig.1，Fig.2……或 Fig.1A，Fig 1B，Fig 2A，Fig 2B；

(F) 包括有相片，而此相片可用線圖代替，或相片不清楚；

(G) 包含彩色圖式或彩色相片，但未同時提出請願。

MPEP 507 Drawing Review in the Office of Initial Patent Examination [R-3]

The Office has revised the drawing review process to implement the eighteen-month publication of patent applications. Under the revised drawing review process, the Office of Initial Patent Examination (OIPE) performs an initial review of drawings in new utility and plant patent applications filed on or after November 29, 2000 to see if the drawings can be effectively scanned for publication purposes. Design applications are not published. Therefore, drawings filed in design patent applications (whether filed before, on or after November 29, 2000) will be reviewed but not for publication purposes. The standard of review employed by OIPE is such that most drawings, including those that have been indicated by applicant to be informal drawings, will be accepted.

OIPE inspects the drawings to see if they can be effectively scanned and adequately reproduced. If the drawings are not acceptable, OIPE will object to the drawings and notify applicant that a timely submission of acceptable drawings (e.g., drawings which can be scanned) is required. This initial review process in OIPE is necessary in order to ensure that applications can be timely published.

Under the OIPE review process, OIPE may object to and require corrected drawings within a set time period, if the drawings:

(A)　have a line quality that is too light to be reproduced (weight of all lines and letters must be heavy enough to permit adequate reproduction) or text that is illegible (reference characters, sheet numbers, and view numbers must be plain and legible). See 37 CFR 1.84(l) and (p)(1);

(B)　have missing lead lines. See 37 CFR 1.84(q). Lead lines are those lines between the reference characters and the details referred to;

(C)　contain excessive text or text that is not in English (including, for example, a flow chart that was originally not in English that has been marked up to include the English text). See 37 CFR 1.84(o) and (p)(2) and 37 CFR 1.52(d)(1);

(D)　do not have the appropriate margin or are not on the correct size paper. See 37 CFR 1.84(f) and (g). Each sheet must include a top margin of at least 2.5 cm. (1 inch), a left size margin of at least 2.5 cm. (1 inch), a right size margin of at least 1.5 cm. (5/8 inch), and a bottom margin of at least 1.0 cm. (3/8 inch). The size of the sheets on which drawings are made must be either 21.0 cm. by 29.7 cm. (DIN size A4) or 21.6 cm. by 27.9 cm. (8-1/2 by 11 inches);

(E)　have more than one figure and each figure is not labeled "Fig." With a consecutive Arabic numeral (1, 2, etc.) or an Arabic numeral and capital letter in the English alphabet (A, B, etc.). See 37 CFR 1.84(u)(1);

(F)　include photographs of the claimed invention which are capable of illustration by other medium such as ink drawings, and which are illegible after scanning. See 37 CFR 1.84(b); and

(G)　contain color drawings or color photographs, but not a petition to accept color drawings/photographs. Note that the requirement ** for a black and white photocopy of any color drawings/photographs has been *>eliminated<. **

If OIPE objects to the drawings and sends applicant a Notice requiring submission of corrected drawings within a set time period (usually two months), corrected drawings must be filed, in paper, to the mailing address set forth in the Notice, along with any other items required by OIPE, to avoid abandonment of the application. No fee will be necessary for filing corrected drawings which are required by OIPE. Otherwise, in most situations, patent application publications and patents will reflect the quality of the drawings that are included with a patent application on filing unless applicant voluntarily submits better quality drawings as set forth *>in MPEP § 1121<.

§1.12　申請及審查過程中主張申請人為小實體之好處，小實體之定義

　　美國專利申請的申請人如為獨立的發明人，或受讓人為個人，非營利機構或小企業時，某些申請，審查及法律程序之費用可予以減少。例如基本申請費、檢索費、審查費、申請案超頁費、超項費、延期費、復活費、訴願費、領證費、法定棄權費（statutory disclaimer fee）及專利維持費可減少 50%，此外，如果是實用專利申請案在 2004 年 12 月 8 日以後以電子申請的話，申請費可減少 75%，此 75%費用之減少不適用於新式樣專利申請案、植物專利申請案、再發證專利申請案及暫時申請案。

　　其餘的費用，如請願費、處理費（不包括復活費）、提供文件費、證書修正費、請求再審查費用、國際專利申請費及其他費用，即使是個人或小企業或非營利機構均無減少。

　　37CFR§1.27 所定義的小實體包括三種：(i)個人(ii)小企業(iii)非營利機構。所謂個人乃指發明人或接受發明人權利轉讓的個人，此發明人或個人未將其發明之權利轉讓、授權或有義務將其權利轉讓、授權給任何人者。如果此發明人或個人有將其發明之權利轉讓或授權給他人，而接受轉讓者亦為小實體時，此發明人或個人仍為小實體。所謂小企業（small business concern）乃是符合 37CFR 121.801-121.805 之標準者，亦即員工人數（包括非正式員工及相關企業之員工）不超過 500 人者。所謂非營利機構乃指大學或以上之教育機構或研究組織，其為非營利者。上述三種小實體不論其位於哪一國家，亦即外國之小實體亦可適用費用減少之規定。政府機關，不論是美國或外國的政府機關均不是非營利機構。政府擁有之公司（國營企業）亦不是非營利機關。但是如果依照 35USC202(c)(4) 接受政府機關之金錢補助作研發，而需將專利授權（license）給該政府機關的私人的研究機構則仍是非營利機構。

37 CFR 1.27 Definition of small entities and establishing status as a small entity to permit payment of small entity fees; when a determination of entitlement to small entity status and notification of loss of entitlement to small entity status are required; fraud on the Office.

**>

(a) *Definition of small entities.* A small entity as used in this chapter means any party (person, small business concern, or nonprofit organization) under paragraphs (a)(1) through (a)(3) of this section.

(1) *Person.* A person, as used in paragraph (c) of this section, means any inventor or other individual (*e.g.*, an individual to whom an inventor has transferred some rights in the invention) who has not assigned, granted, conveyed, or licensed, and is under no obligation under contract or law to assign, grant, convey, or license, any rights in the invention. An inventor or other individual who has transferred some rights in the invention to one or more parties, or is under an obligation to transfer some rights in the invention to one or more parties, can also qualify for small entity status if all the parties who have had rights in the invention transferred to them also qualify for small entity status either as a person, small business concern, or nonprofit organization under this section.

(2) *Small business concern.* A small business concern, as used in paragraph (c) of this section, means any business concern that:

(i) Has not assigned, granted, conveyed, or licensed, and is under no obligation under contract or law to assign, grant, convey, or license, any rights in the invention to any person, concern, or organization which would not qualify for small entity status as a person, small business concern, or nonprofit organization; and

(ii) Meets the size standards set forth in 13 CFR 121.801 through 121.805 to be eligible for reduced patent fees. Questions related to standards for a small business concern may be directed to: Small Business Administration, Size Standards Staff, 409 Third Street, SW., Washington, DC 20416.

(3) *Nonprofit Organization.* A nonprofit organization, as used in paragraph (c) of this section, means any nonprofit organization that:

(i) Has not assigned, granted, conveyed, or licensed, and is under no obligation under contract or law to assign, grant, convey, or license, any rights in the invention to any person, concern, or organization which would not qualify as a person, small business concern, or a nonprofit organization; and

(ii) Is either:

(A)   A university or other institution of higher education located in any country;

(B)   An organization of the type described in section 501(c)(3) of the Internal Revenue Code of 19 86 (26 U.S.C. 501(c)(3)) and exempt from taxation under section 501(a) of the Internal Revenue Code (26 U.S.C. 501(a));

(C)   Any nonprofit scientific or educational organization qualified under a nonprofit organization statute of a state of this country §35 U.S.C. 201 (i)); or

(D)   Any nonprofit organization located in a foreign country which would qualify as a nonprofit organization under paragraphs (a)(3)(ii)(B) of this section or (a)(3)(ii)(C) of this section if it were located in this country.

(4) *License to a Federal agency*. (i) For persons under paragraph (a)(1) of this section, a license to the Government resulting from a rights determination under Executive Order 10096 does not constitute a license so as to prohibit claiming small entity status.

(ii) For small business concerns and nonprofit organizations under paragraphs (a)(2) and (a)(3) of this section, a license to a Federal agency resulting from a funding agreement with that agency pursuant to 35 U.S.C. 202 (c)(4) does not constitute a license for the purposes of paragraphs (a)(2)(i) and (a)(3)(i) of this section.

(5) *Security Interest*. A security interest does not involve an obligation to transfer rights in the invention for the purposes of paragraphs (a)(1) through (a)(3) of this section unless the security interest is defaulted upon.

＊＊＊＊＊＊＊＊＊＊＊

§1.13 如何主張為小實體，在審查中及發證後小實體狀態改變之修改

申請人要主張為小實體時，需向美國專利局提出書面的聲明（verified statement），此聲明書並無格式之規定，聲明書中需清楚地敘述其為小實體並簽名。在一專利申請案中一旦已主張了小實體而成為小實體之狀態（例如在申請時主張為小實體），則一直到繳領證費或專利維持費之前，即使小實體之狀態有改變，均可繳納小實體之費用。亦即，申請人在繳領證費或維持費時才需向美國專利局再確定是否仍為小實體，如已非小實體則需聲明，並繳納大實體之費用。

　　主張小實體必需逐案提出，即使是延續案、分割案、部份延續案、CPA 案或再發證案均需重新主張小實體。但是提出 RCE 則不必重新主張小實體，因為 RCE 不被視為一新的申請案。

　　主張小實體除了提出書面的聲明外，實務上，如果申請人在提出專利申請時繳了正確金額之小實體之基本申請費（exact small entity basic filing fee）就被認為是主張了小實體。而且即使申請人繳了不同種類的小實體基本申請費，例如申請實用專利申請案卻繳了新式樣的小實體基本申請費，則仍會被認為是已主張了小實體。如果在申請時未提出書面聲明，主張小實體，亦未繳小實體之基本申請費，之後要主張小實體時就只能提出書面的聲明。

　　主張小實體可由發明人（如無讓渡）或受讓人或專利代理人提出，且只需其中一人提出就可。例如一專利申請案有三個發明人（未讓渡）只要由其中一人提出聲明並簽名即可。又，例如一專利申請案有二位發明人，其中一發明人將權利讓渡一非營利機構，則只要該未讓渡之發明人或該非營利機構之代表人之一簽署小實體之聲明即可。簽署小實體之人士需經過完整及徹底的調查，確定是小實體，且是知道事實者，例如總經理。

　　當提出申請後如果變成大實體，一直到繳領證費或繳維持費前仍可繳小實體之費用。但繳領證費或維持費時則需繳大實體之費用，同時需作聲明已變成大實體，不能只繳大實體之費用而不另作書面聲明。

　　如果申請人為小實體但是在提出申請時卻繳了大實體之費用，則可在繳費之後的三個月內請求退還多繳之費用，同時主張為小實體。

　　如果申請人誤認自己是小實體而繳了小實體之費用，則當發現此錯誤時就需補繳所有已繳費用之差額。差額之計算基準為補繳費用時大實體的規費減去當時實際所繳之小實體的規費。

MPEP 509.03

　　In order to establish small entity status for the purpose of paying small entity fees, any party (person, small business concern or nonprofit organization) must make an assertion of entitlement to small entity status in the manner set forth in 37 CFR 1.27(c)(1) or (c)(3), in the application or patent in which such small entity fees are to be paid. Under

37 CFR 1.27, as long as all of the rights remain in small entities, the fees established for a small entity can be paid. This includes circumstances where the rights were divided between a person, a small business concern, and a nonprofit organization, or any combination thereof.

Under 37 CFR 1.4(d)(*>4<), an assertion of entitlement to small entity status, including the mere payment of an exact small entity basic filing fee, inherently contains a certification under 37 CFR 10.18(b). It is not required that an assertion of entitlement to small entity status be filed with each fee paid. Rather, once status as a small entity has been established in an application or patent, fees as a small entity may thereafter be paid in that application or patent without regard to a change in status until the issue fee is due or any maintenance fee is due. 37 CFR 1.27(g)(1). Notification of a loss of entitlement to small entity status must be filed in the application or patent prior to paying, or at the time of paying, the earliest of the issue fee or any maintenance fee due after the date on which status as a small entity is no longer appropriate. 37 CFR 1.27(g)(2).

Status as a small entity may be established in a provisional application by complying with 37 CFR 1.27.

Status as a small entity must be specifically established in each application or patent in which the status is available and desired. Status as a small entity in one application or patent does not affect any other application or patent, including applications or patents which are directly or indirectly dependent upon the application or patent in which the status has been established. The filing of an application under 37 CFR 1.53 as a continuation-in-part, continuation or division (including a continued prosecution application under 37 CFR 1.53(d)), or the filing of a reissue application requires a new assertion as to continued entitlement to small entity status for the continuing or reissue application. Submission of a request for continued examination (RCE) under 37 CFR 1.114 does not require a new determination or assertion of entitlement to small entity status since it is not a new application.

＊＊＊＊＊＊＊＊＊＊＊

I.ASSERTION BY WRITING

Small entity status may be established by the submission of a simple written assertion of entitlement to small entity status. The assertion must be signed, clearly identifiable, and convey the concept of entitlement to small entity status. 37 CFR 1.27(c)(1). The written assertion is not required to be presented in any particular form. Written assertions of small entity status or references to small entity fees will be liberally interpreted to represent the required assertion. The written assertion can be made in any paper filed in or with the application and need be no more than a simple sentence or a box checked on an application transmittal letter.

Practitioners may continue to use former USPTO forms or similar forms if they believe such small entity forms serve an educational purpose for their clients.

II.PARTIES WHO CAN ASSERT AND SIGN AN ENTITLEMENT TO SMALL ENTITY STATUS BY WRITING

The parties who can assert entitlement to small entity status by writing include all parties permitted by 37 CFR 1.33(b) to file a paper in an application, including a registered practitioner. 37 CFR 1.27(c)(2)(i). Additionally, one of the individuals identified as an inventor, or a partial assignee, can also sign the written assertion. 37 CFR 1.27(c)(2)(ii) and (iii). By way of example, in the case of three *pro se* inventors for a particular application, one of the three inventors upon filing the application can submit a written assertion of entitlement to small entity status. and thereby establish small entity status for the application, (but see paragraph VI. below). Where rights are divided between a person, small business concern, and nonprofit organization, or any combination thereof, only one party is required to assert small entity status. For example, where one of two inventors has assigned his or her rights in the invention, it is sufficient if either of the two inventors or the assignee asserts entitlement to small entity status.

IV. ASSERTION BY PAYMENT OF SMALL ENTITY BASIC FILING OR BASIC NATIONAL FEE

The payment of an exact small entity basic filing (37 CFR 1.16(a), (*>b<), (*>c<), (*>d<), or (*>e<)) or basic national fee (37 CFR 1.492(a)**) is also considered to be a sufficient assertion of entitlement to small entity status. 37 CFR 1.27(c)(3). An applicant filing a patent application and paying an exact small entity basic filing or basic national fee automatically establishes small entity status for the application even without any other assertion of small entity status. This is so even if an applicant inadvertently selects the wrong type of small entity basic filing or basic national fee for the application being filed (e.g., the exact small entity basic filing fee for a design application is selected but the application is a utility application). If small entity status was not established when the basic filing or basic national fee was paid, such as by payment of a *>non-small< entity basic filing or basic national fee, a later claim to small entity status requires a written assertion under 37 CFR 1.27(c)(1). Payment of a small entity fee other than a small entity basic filing or basic national fee (e.g., extension of time fee, or issue fee) without inclusion of a written assertion is not sufficient.

Even though applicants can assert small entity status only by payment of an exact small entity basic filing or basic national fee, the Office encourages applicants to also file a written assertion of small entity status as well as to pay the exact amount of the small entity basic filing or basic national fee. The Office's application transmittal forms include a check box that can be used to submit a written assertion of small entity status. A written assertion will provide small entity status should applicant fail to pay the exact small entity basic filing or basic national fee. The provision providing for small entity status by payment of an exact small entity basic filing or basic national fee is intended to act as a safety net to avoid possible financial loss to inventors or small businesses that qualify for small entity status.

＊＊＊＊＊＊＊＊＊＊＊

VI. CONTINUED OBLIGATIONS FOR THOROUGH INVESTIGATION OF SMALL ENTITY STATUS

While small entity status is not difficult to obtain, it should be clearly understood that applicants need to do a complete and thorough investigation of all facts and circumstances before making a determination of actual entitlement to small entity status. 37 CFR 1.27(f). Where entitlement to small entity status is uncertain, it should not be claimed.

The assertion of small entity status (even by mere payment of the exact small entity basic filing fee) is not appropriate until such an investigation has been completed. For example, where there are three *pro se* inventors, before one of the inventors pays the small entity basic filing or basic national fee to establish small entity status, the single inventor asserting entitlement to small entity status should check with the other two inventors to determine whether small entity status is appropriate.

Furthermore, once status as a small entity has been established in an application, a new determination of entitlement to small entity status is needed (1) when the issue fee is due and (2) when any maintenance fee is due. It should be appreciated that the costs incurred in appropriately conducting the initial and subsequent investigations may outweigh the benefit of claiming small entity status. For some applicants it may be desirable to file as a *>non-small< entity (by not filing a written assertion of small entity status and by submitting *>non-small< entity fees) rather than undertaking the appropriate investigations which may be both difficult and time-consuming and which may be cost effectiveonly where several applications are involved.

The intent of 37 CFR 1.27 is that the person making the assertion of entitlement to small entity status is the person in a position to know the facts about whether or not status as a small entity can be properly established. That person, thus, has a duty to investigate the circumstances surrounding entitlement to small entity status to the fullest extent. It is important to note that small entity status must not be claimed unless the person or persons can unequivocally make the required self-certification.

＊＊＊＊＊＊＊＊＊＊＊

VII. REMOVAL OF STATUS

Once small entity status is established in an application or patent, fees as a small entity may thereafter be paid in that application or patent without regard to a change in status until the issue fee is due or any maintenance fee is due. 37 CFR 1.27(g)(1). 37 CFR 1.27(g)(2) requires that notification of any change in status resulting in loss of entitlement to small entity status be filed in the application or patent prior to paying, or at the time of paying, the earliest of the issue fee or any maintenance fee due after the date on which status as a small entity is no longer appropriate. 37 CFR 1.27(g)(2) also requires that the notification of loss of entitlement to small entity status be in the form of a specific written assertion to that extent, rather than only payment of a *>non-small< entity fee. For example, when paying the issue fee in an application that has previously been accorded small entity status and the required new determination of continued entitlement to small entity status reveals that status has been lost, applicant should not just simply pay the *>non-small entity< issue fee or cross out the recitation of small entity status on Part B of the Notice of Allowance and Fee(s) Due (PTOL-85), but should **>(A) check the appropriate box on Part B of the PTOL-85 form to indicate that there has been a change in entity status and applicant is no longer claiming small entity status, and (B) pay the fee amount for a non-small entity.<

IX. REFUNDS BASED ON LATER ESTABLISHMENT OF SMALL ENTITY STATUS

37 CFR 1.28(a) provides a three-month time period for requesting a refund of a portion of a *>non-small< entity fee based on later establishment of small entity status. The start date of the three-month refund period of 37 CFR 1.28(a) is the date the full fee has been paid.

＊＊＊＊＊＊＊＊＊＊＊

X. CORRECTING ERRORS IN SMALL ENTITY STATUS

＊＊＊＊＊＊＊＊＊＊＊

37 CFR 1.28(c) provides that if small entity status is established in good faith and the small entity fees are paid in good faith, and it is later discovered that such status as a small entity was established in error or through error the Office was not notified of a change of status, the error will be excused upon compliance with the separate submission and itemization requirements of 37 CFR 1.28(c)(1) and (c)(2), and the deficiency payment requirement of 37 CFR 1.28(c)(2). The deficiency amount owed under 37 CFR 1.28(c) is calculated using the date on which the deficiency was paid in full. See 37 CFR 1.28(b)(2).

實例 1

一 X 公司總裁 A 有一新的發明想申請美國的實用專利，此 X 公司雇用了 450 位正式員工，20 位計時員工及 40 位臨時工。X 公司並無任何相關之子公司。發明人 A 亦未將其發明讓渡給任何人或公司，亦無任何義務將其發明讓渡給 X 公司或其他任何人或任何機構。請問發明人 A 如以個人提出美國專利申請是否可主張為小實體？

可以主張為小實體。雖然發明人 A 為 X 公司之總裁，且 X 公司之員工多於 500 人，但發明人 A 並無義務將其發明讓給 X 公司或其他任何人或公司或機構，故可主張為小實體。

實例 2

一專利申請案於 2008 年 7 月 1 日提出申請，申請時美國代理人主張小實體且共繳了申請費 USD 150，檢索費 USD 250，審查費 USD 100，讓渡費 USD 40，獨立項超項費 USD 200（2 項），總項數超項費 USD 100（超 4 項），共 USD 840 之規費。之後在 2009 年 1 月 1 日美國代理人被此專利申請案之受讓人告知在 2008 年 6 月 1 日時已將本申請案之發明的所有權利讓渡給一台灣的政府機關。請問美國代理人是否需向美國專利局提出修正小實體為大實體？如需補繳規費需補繳多少的規費？

由於本專利申請案在提申之前就已讓渡給台灣的政府機關（大實體）故依 37CFR 1.28 需向美國專利局提出修正並補繳規費。由於美國專利局之規費乃在 2008 年 10 月 20 日起就漲價，故需依照 2008 年 10 月 20 日時之規費補繳差額 USD 938。差額之計算如下：

已繳金額（小企業）	2008/10/20 之後之規費（大企業）	差額
申請費：USD 150	USD 330	USD 180
檢索費：USD 250	USD 540	USD 290
審查費：USD 100	USD 220	USD 120
讓渡費：USD 40	USD 40	-
獨立項超項費 USD 200	USD 440	USD 240
總項超項費 USD 100	USD 208	USD 108
USD 840	USD 1778	USD 938

第 2 章　發明人

目　錄

§2.1　發明人（inventor），共同發明人（joint inventors）

　　發明人乃指構思了所請求之發明的自然人。例如一專利申請案有數項請求項（claims）則每一請求項可能有不同之發明人。如果僅僅將構思付諸實施之人並非發明人。發明人並非一法人，亦非發明之受讓人。如果有二人以上共同構思了一請求之發明（claimed invention），例如二人共同合作，研發方向相同或參考另一人之報告或接受另一人之建議而共同構思了所請求之發明則為共同發明人。共同發明人不必實際在一起研發也不必在同一時間研發，也不必有相同量或相同型式之貢獻，或對每一請求項（claim）之標的有貢獻。

MPEP 2137.01

　　The definition for inventorship can be simply stated: "The threshold question in determining inventorship is who conceived the invention. Unless a person contributes to the conception of the invention, he is not an inventor. Insofar as defining an inventor is concerned, reduction to practice, per se, is irrelevant

35 U.S.C. 116 Inventors.

　　When an invention is made by two or more persons jointly, they shall apply for patent jointly and each make the required oath, except as otherwise provided in this title. Inventors may apply for a patent jointly even though (1) they did not physically work together or at the same time, (2) each did not make the same type or amount of contribution, or (3) each did not make a contribution to the subject matter of every claim of the patent.

實例 1

　　X 公司的鋸片部門技術經理 A 指示其部門成員 B 研發能夠較現有的鋸片有更好的切割效果的鋸片。B 參考了各種鋸片之形狀，最後想到具有對稱鋸齒之鋸片可能有較好的切割效果。接著，B 就指示其助理 C 製造出此一具有對稱鋸齒之鋸片並進行標準測試，結果確實切割效果較佳。X 公司欲提出專利申請，請問發明人為何人？

　　發明人為 B。決定發明人之關鍵是誰構思了該發明，除非其對發明之構思有貢獻，否則不是發明人，付諸實施之人非發明人。

實例 2

　　A 與 B 在 C 公司之不同部門工作。A 研發出一種影印紙表面的絕緣塗料。B 知曉 A 之發明之後，研發出一金屬層可作為影印紙置於基材與 A 發明之絕緣紙之中間層。C 公司提出專利申請，請求項為：一影印紙包括一基材，一金屬置於該基紙之上，以及一絕緣塗料塗覆於該金屬層之上。請問此申請案之發明人為何人？

　　A 及 B 均為發明人（共同發明人）。一專利申請案之共同發明人必須對所請求之發明（claimed invention）有共同之研發行為，例如(1)合作(2)研發方向相同或(3)參考另一人之相關報告或接受另一人之建議。

實例 3

　　A 為 X 公司之總公司研發部之員工，A 發明了一新穎的人工心臟泵浦，此泵浦在抽吸血液時對血液細胞之作用力極溫和。A 將此發明付諸實施不久，就將此發明之內容交由 X 公司內之專利律師準備專利說明書。同時 X 公司之分公司的另一員工 B 將 A 研發之人工心臟泵浦加上一個自己研發之閥以將此泵浦小型化。B 未與 A 見過面且未曾與 A 討論自己的發明，但經由內部文件獲知 A 之發明內容。專利律師收到 B 之發明提案書後就修正其專利說明書及申請專利範圍，同時請求及揭露了 A 之泵浦及 B 之泵浦。在 A 及 B 各自發明其泵浦時，A 及 B 均有義務將其發明讓渡給 X 公司。請問此專利申請之發明人為何人？

　　A 及 B 均為發明人。發明人不必實際上在一起研發，也不必在同時間研發，也不必有相同量或相同型式之貢獻或對每請求項之標的有貢獻。

§2.2　美國專利申請案申請人應簽署之文件

　　美國專利申請案應由發明人提出申請，亦即美國專利申請案所謂之申請人乃指發明人而言。非暫時專利申請案（實用專利申請案、植物專利申請案及新式樣專利申請案）之申請人應簽署宣誓書（Oath 或 declaration），於宣誓中發明人必需宣誓相信自己是專利申請案之發明的原始（original）及第 1 個（first）發明人。所謂原始的發明人乃指發明的構思是自己創造出而非完全從他人之處得到構想。而所謂第 1

個發明人乃指第 1 個構思該發明且第 1 個將之付諸實施，或者是第 1 個構思了該發明且自別人構思該發明至自己將之付諸實施之期間有勤勉度而言。

　　暫時專利申請之發明人不必簽署宣誓書。暫時申請案申請時首頁（cover sheet）上所載明之發明人就被認為是暫時申請案之發明人。非暫時申請案之發明人如果有將申請權讓渡或有委托代理人則需另簽署讓渡書及委任書。暫時申請案之發明人即使有讓渡或委任代理人亦不必簽署任何讓渡書或委任書。

---

37 CFR 1.41 Applicant for patent.

(a) A patent is applied for in the name or names of the actual inventor or inventors.

(1) The inventorship of a nonprovisional application is that inventorship set forth in the oath or declaration as prescribed by § 1.63, except as provided for in §§ 1.53(d)(4) and 1.63(d). If an oath or declaration as prescribed by § 1.63 is not filed during the pendency of a nonprovisional application, the inventorship is that inventorship set forth in the application papers filed pursuant to § 1.53(b), unless applicant files a paper, including the processing fee set forth in § 1.17(i), supplying or changing the name or names of the inventor or inventors.

(2) The inventorship of a provisional application is that inventorship set forth in the cover sheet as prescribed by § 1.51(c)(1). If a cover sheet as prescribed by § 1.51(c)(1) is not filed during the pendency of a provisional application, the inventorship is that inventorship set forth in the application papers filed pursuant to § 1.53(c), unless applicant files a paper including the processing fee set forth in § 1.17(q), supplying or changing the name or names of the inventor or inventors.

＊＊＊＊＊＊＊＊＊＊＊

(b) Unless the contrary is indicated the word "applicant" when used in these sections refers to the inventor or joint inventors who are applying for a patent, or to the person mentioned in §§ 1.42, 1.43 or 1.47 who is applying for a patent in place of the inventor.

＊＊＊＊＊＊＊＊＊＊＊

35 U.S.C. 115 Oath of applicant.

The applicant shall make oath that he believes himself to be the original and first inventor of the process, machine, manufacture, or composition of matter, or improvement thereof, for which he solicits a patent; and shall state of what country he is a citizen. Such oath may be made before any person within the United States authorized by law to administer oaths, or, when made in a foreign country, before any diplomatic or consular officer of the United States authorized to administer oaths, or before any officer having an official seal and authorized to administer oaths in the foreign country in which the applicant may be, whose authority is proved by certificate of a diplomatic or consular officer of the United States, or apostille of an official designated by a foreign country which, by treaty or convention, accords like effect to apostilles of designated officials in the United States. Such oath is valid if it complies with the laws of the state or country where made. When the application is made as provided in this title by a person other than the inventor, the oath may be so varied in form that it can be made by him. For purposes of this section, a consular officer shall include any United States citizen serving overseas, authorized to perform notarial functions pursuant to section 1750 of the Revised Statutes, as amended (22 U.S.C. 4221).

§2.3　發明人死亡，失能時，專利申請案之審查

　　在申請一非暫時專利申請案之前或要簽署宣誓書之前，如果發明人已死亡，則應由其法律上之代表人（執行遺囑者（executor），遺產管理人（administrator））簽署宣誓書及提出專利申請。如果在提出專利申請之後但在核准之前死亡則專利將發給其法定代表人。而在申請一非暫時專利申請案之前或要簽署宣誓書之前或提出專利申請之後，發明人喪失心智或在法律上失能，則應由其法律上之代表人（監護人（guardian），管理人（conservator））簽署宣誓書，提出專利申請及取得專利。在上述兩種情形，法律上的代表人在簽署宣誓書時，宣誓書需註明其為法律上之代表人同時亦要註明其國籍，永久住址及郵寄住址。美國專利局現已不要求法定代表人提出證據證明其具有法律上代表人之權力。如果一專利申請案只有單獨一個發明人（sole inventor）而在申請之後，發明人死亡，則此專利申請案之受讓人或發明人之法定代理人可介入（intervene）專利申請之後續審查。如果不介入，則專利申請案就被放棄。但是如果發明人死亡時專利申請案已可核准，則美國專利局仍會將專利案件發證。

　　如果專利申請案有一個以上的發明人，而其中一個在提出申請後死亡，則存活之發明人必需呈送美國專利局文件，證明該發明人已死亡，則以後的審查程序只由存活的發明人簽署相關文件。如果死亡的發明人的代表人介入，則該代表人需簽署宣誓書，則後續的審查程序由存活的發明人及該代表人共同簽署及進行。

　　如果發明人死亡或失能，但並無法律上的代表人，則應由法院指定代表人來簽署宣誓書及進行後續審查程序。

　　另外，如果發明人為未成年人（未滿 18 歲），只要其能了解相關文件的內容且可簽名，就應由其簽署宣誓書而不必由其法定代表人簽署。

---

37 CFR §1.42 When the inventor is dead.

In case of the death of the inventor, the legal representative (executor, administrator, etc.) of the deceased inventor may make the necessary oath or declaration, and apply for and obtain the patent. Where the inventor dies during the time intervening between the filing of the application and the granting of a patent thereon, the letters patent may be issued to the legal representative upon proper intervention.

37 CFR § 1.43 When the inventor is insane or legally incapacitated.

In case an inventor is insane or otherwise legally incapacitated, the legal representative (guardian, conservator, etc.) of such inventor may make the necessary oath or declaration, and apply for and obtain the patent.

37 CFR§ 1.64 Person making oath or declaration.

(a) The oath or declaration (§ 1.63), including any supplemental oath or declaration (§ 1.67), must be made by all of the actual inventors except as provided for in §§ 1.42, 1.43, 1.47, or § 1.67.

(b) If the person making the oath or declaration or any supplemental oath or declaration is not the inventor (§§ 1.42, 1.43,1.47, or §1.67), the oath or declaration shall state the relationship of the person to the inventor, and, upon information and belief, the facts which the inventor is required to state. If the person signing the oath or declaration is the legal representative of a deceased inventor, the oath or declaration shall also state that the person is a legal representative and the citizenship, residence, and mailing address of the legal representative.

MPEP 409 Death, Legal Incapacity, or Unavailability of Inventor [R-5]

If the inventor is dead, insane, or otherwise legally incapacitated, refuses to execute an application, or cannot be found, an application may be made by someone other than the inventor, as specified in 37 CFR **>1.42,1.43 and 1.47<, and 37 CFR 1.423, MPEP § 409.01 - § 409.031(j).

A minor (under age 18) inventor may execute an oath or declaration under 37 CFR 1.63 as long as the *>minor< is competent to sign (i.e., understands the document that he or she is signing); a legal representative is not required to execute an oath or declaration on the minor's behalf. See 37 CFR 1.63(a)(1).

Employees of the United States Patent and Trademark Office (Office) who were inventors are not permitted to sign an oath or declaration for patent application (37 CFR 1.63) during the period of their employment with the Office and one year thereafter. 35 U.S.C. 4. These employees (inventors) will be treated as being unavailable to sign the oath or declaration pursuant to 37 CFR 1.47.

MPEP 409.01 Death of Inventor [R-5]

Unless a power of attorney is coupled with an interest (i.e., **>a patent practitioner< is assignee or part-assignee), the death of the inventor (or one of the joint inventors) terminates the power of attorney given by the deceased inventor. A new power from the heirs, administrators, executors, or assignees is necessary if the deceased inventor is the sole inventor or all powers of attorney in the application have been terminated (but see MPEP § 409.01(f)). See also 37 CFR 1.422.

MPEP 409.01(a) Prosecution by Administrator or Executor

One who has reason to believe that he or she will be appointed legal representative of a deceased inventor may apply for a patent as legal representative in accordance with 37 CFR 1.42.

Application may be made by the heirs of the inventor, as such, if there is no will or the will did not appoint an executor and the estate was under the sum required by state law for the appointment of an administrator. The heirs should identify themselves as the legal representative of the deceased inventor in the oath or declaration submitted pursuant to 37 CFR 1.63 and 1.64.

MPEP 409.01(b) Proof of Authority of Administrator or Executor

The Office no longer requires proof of authority of the legal representative of a deceased or incapacitated inventor. Although the Office does not require proof of authority to be filed, any person acting as a legal representative of a deceased or incapacitated inventor should ensure that he or she is properly acting in such a capacity

MPEP 409.01(f) Intervention of Executor Not Compulsory

When an inventor dies after filing an application and executing the oath or declaration required by 37 CFR 1.63, the executor or administrator should intervene, but the allowance of the application will not be withheld nor the application withdrawn from issue if the executor or administrator does not intervene.

This practice is applicable to an application which has been placed in condition for allowance or passed to issue prior to notification of the death of the inventor. See MPEP § 409.01.

When a joint inventor of a pro se application dies after filing the application, the living joint inventor(s) must submit proof that the other joint inventor is dead. Upon submission of such proof, only the signatures of the living joint inventors are required on the papers filed with the USPTO if the legal representative of the deceased inventor does not intervene. If the legal representative of the deceased inventor wishes to intervene, the legal representative must submit an oath or declaration in compliance with 37 CFR 1.63 and 1.64 (e.g., stating that he or she is the legal representative of the deceased inventor and his or her residence, citizenship and post office address). Once the legal representative of the deceased inventor intervenes in the pro se application, the signatures of the living joint inventors and the legal representative are required on the papers filed with the USPTO.

MPEP 409.02 Insanity or Other Legal Incapacity [R-3]

37 CFR 1.43 When the inventor is insane or legally incapacitated.

In case an inventor is insane or otherwise legally incapacitated, the legal representative (guardian, conservator, etc.) of such inventor may make the necessary oath or declaration, and apply for and obtain the patent.

When an inventor *>becomes legally incapacitated< prior to the filing of an application and prior to *>executing< the oath or declaration required by 37 CFR 1.63 and no legal representative has been appointed, one must be appointed by a court of competent jurisdiction for the purpose of execution of the oath or declaration of the application.

§2.4　共同發明人之一找不到或拒絕簽署宣誓書

當共同發明人之一找不到或拒絕簽署宣誓書時，應提出請願，並由其他共同發明人簽署宣誓書以提出專利申請。此專利申請應符合下列要件：

(A) 所有其他的發明人簽署 37CFR 1.63 之宣誓書，同時代替不能簽署之共同發明人簽署 37 CFR 1.64 之宣誓書，並將不能簽署之發明人之簽名欄留空白表示代簽。

(B) 需提出事實證據證明經過相當的努力仍不能找到該發明人或該發明人仍拒絕簽署宣誓書。

(C) 陳述該不能簽名之發明人的最後住址。

---

35 U.S.C. 116 Inventors.

＊＊＊＊＊＊＊＊＊＊＊

If a joint inventor refuses to join in an application for patent or cannot be found or reached after diligent effort, the application may be made by the other inventor on behalf of himself and the omitted inventor. The Director, on proof of the pertinent facts and after such notice to the omitted inventor as he prescribes, may grant a patent to the inventor making the application, subject to the same rights which the omitted inventor would have had if he had been joined. The omitted inventor may subsequently join in the application.

Whenever through error a person is named in an application for patent as the inventor, or through an error an inventor is not named in an application, and such error arose without any deceptive intention on his part, the Director may permit the application to be amended accordingly, under such terms as he prescribes

37CFR § 1.47 Filing when an inventor refuses to sign or cannot be reached.

(a) If a joint inventor refuses to join in an application for patent or cannot be found or reached after diligent effort, the application may be made by the other inventor on behalf of himself or herself and the nonsigning inventor. The oath or declaration in such an application must be accompanied by a petition including proof of the pertinent facts, the fee set forth in §1.17(g), and the last known address of the nonsigning inventor. The nonsigning inventor may subsequently join in the application by filing an oath or declaration complying with § 1.63.

MPEP 409.03(a) At Least One Joint Inventor Available

37 CFR 1.47(a) and 35 U.S.C. 116, second paragraph, requires all available joint inventors to file an application "on behalf of" themselves and on behalf of a joint inventor who "cannot be found or reached after diligent effort" or who refuses to "join in an application."

In addition to other requirements of law (35 U.S.C. 111(a) and 115), an application deposited in the U.S. Patent and Trademark Office pursuant to 37 CFR 1.47(a) must meet the following requirements:

(A) All the available joint inventors must (1) make oath or declaration on their own behalf as required by 37 CFR 1.63 or 1.175 (see MPEP § 602, § 605.01, and § 1414) and (2) make oath or declaration on behalf of the nonsigning joint inventor as required by 37 CFR 1.64. An oath or declaration signed by all the available joint inventors with the signature block of the nonsigning inventor(s) left blank may be treated as having been signed by all the available joint inventors on behalf of the nonsigning inventor(s), unless otherwise indicated.

(B) The application must be accompanied by proof that the nonsigning inventor (1) cannot be found or reached after diligent effort or (2) refuses to execute the application papers. See MPEP § 409.03(d).

(C) The last known address of the nonsigning joint inventor must be stated. See MPEP § 409.03(e).

§2.5　發明人都找不到或均拒絕簽署宣誓書

當經過相當的努力仍找不到任何一個發明人或發明人都拒絕簽署宣誓書時，可由發明人所讓渡或以書面同意讓渡之人士或者由具有所有權之人士簽署 37 CFR 1.64 之宣誓書，同時提出請願。請願書需包括下列事實證據：(1)與發明人之關係(2)發明人找不到或拒絕簽署之事實(3)發明人之最後地址(4)需要保有此申請的權利以避免不可補償之損失，(5)規費，以提出專利之申請。

---

35 U.S.C. 118 Filing by other than inventor.

Whenever an inventor refuses to execute an application for patent, or cannot be found or reached after diligent effort, a person to whom the inventor has assigned or agreed in writing to assign the invention or who otherwise shows sufficient proprietary interest in the matter justifying such action, may make application for patent on behalf of and as agent for the inventor on proof of the pertinent facts and a showing that such action is necessary to preserve the rights of the parties or to prevent irreparable damage; and the Director may grant a patent to such inventor upon such notice to him as the Director deems sufficient, and on compliance with such regulations as he prescribes.

37 CFR 1.47(b)

＊＊＊＊＊＊＊＊＊＊＊

b)Whenever all of the inventors refuse to execute an application for patent, or cannot be found or reached after diligent effort, a person to whom an inventor has assigned or agreed in writing to assign the invention, or who otherwise shows sufficient proprietary interest in the matter justifying such action, may make application for patent on behalf of and as agent for all the inventors. The oath or declaration in such an application must be accompanied by a petition including proof of the pertinent facts, a showing that such action is necessary to preserve the rights of the parties or to prevent irreparable damage, the fee set forth in § 1.17(g), and the last known address of all of the inventors. An inventor may subsequently join in the application by filing an oath or declaration complying with § 1.63.

* * * * * * * * * * * *

MPEP 409.03(b) No Inventor Available

Filing under 37 CFR 1.47(b) and 35 U.S.C. 118 is permitted only when no inventor is available to make application. These provisions allow a "person" with a demonstrated proprietary interest to make application "on behalf of and as agent for" an inventor who "cannot be found or reached after diligent effort" or who refuses to sign the application oath or declaration. The word "person" has been construed by the U.S. Patent and Trademark Office to include juristic entities, such as a corporation. Where 37 CFR 1.47(a) is available, application cannot be made under 37 CFR 1.47(b).

In addition to other requirements of law (35 U.S.C. 111(a) and 115), an application deposited pursuant to 37 CFR 1.47(b) must meet the following requirements:

(A) The 37 CFR 1.47(b) applicant must make the oath required by 37 CFR 1.63 and 1.64 or 1.175. Where a corporation is the 37 CFR 1.47(b) applicant, an officer (President, Vice-President, Secretary, Treasurer, or Chief Executive Officer) thereof should normally sign the necessary oath or declaration. A corporation may authorize any person, including an attorney or agent registered to practice before the U.S. Patent and Trademark Office, to sign the application oath or declaration on its behalf. Where an oath or declaration is signed by a registered attorney or agent on behalf of a corporation, either proof of the attorney"s or agent"s authority in the form of a statement signed by an appropriate corporate officer must be submitted, or the attorney or agent may simply state that he or she is authorized to sign on behalf of the corporation. Where the oath or declaration is being signed on behalf of an assignee, see MPEP § 324. An inventor may not authorize another individual to act as his or her agent to sign the application oath or declaration on his or her behalf. Staeger v. Commissioner, 189 USPQ 272 (D.D.C. 1976), In re Striker, 182 USPQ 507 (Comm"r Pat. 1973). Where an application is executed by one other than the inventor, the declaration required by 37 CFR 1.63 must state the full name, residence, post office address, and citizenship of the nonsigning inventor. Also, the title or position of the person signing must be stated if signing on behalf of a corporation under 37 CFR 1.47(b).

(B) The 37 CFR 1.47(b) applicant must state his or her relationship to the inventor as required by 37 CFR 1.64.

(C) The application must be accompanied by proof that the inventor (1) cannot be found or reached after a diligent effort or (2) refuses to execute the application papers. See MPEP § 409.03(d).

(D) The last known address of the inventor must be stated. See MPEP § 409.03(e).

(E) The 37 CFR 1.47(b) applicant must make out a prima facie case (1) that the invention has been assigned to him or her or (2) that the inventor has agreed in writing to assign the invention to him or her or (3) otherwise demonstrate a proprietary interest in the subject matter of the application. See MPEP § 409.03(f).

(F) The 37 CFR 1.47(b) applicant must prove that the filing of the application is necessary (1) to preserve the rights of the parties or (2) to prevent irreparable damage. See MPEP § 409.03(g).

§2.6　提出事實證據證明發明人無法找到或拒絕簽署

(Ⅰ)發明人無法找到

依照 37CFR§1.47 需經過相當的努力找不到發明人才符合要件，因而，需提出一聲明，記載已經過相當的努力仍找不到發明人的事實。例如發明人正在渡假或不在，故暫時無法簽署宣誓書，是不會被接受的理由，或發明人生病住院，亦不是可以接受之理由。通常會被接受的事實是：將專利申請書及空白的宣誓書以掛號郵寄發明人請其簽署，但被退回，加上以電話數次連絡，及 e-mail 連絡，均無法連絡上，或者透過其他人試圖連絡均找不到，附上被退回之掛號信函，被退回之 e-mail 或某年某月某日電話連絡等書面的事實證據才是可接受的。

(Ⅱ)發明人拒絕簽署

發明人拒絕簽署宣誓書之事實證據只有當確實寄送（送交）該專利申請之文件（說明書、圖式、宣誓書）給發明人而且發明人確實了解要簽署的情況下仍拒絕簽署才符合要件。通常是將發明人要簽署的文件（宣誓書）及專利申請書以掛號寄至發明人或其代理人之最後地址，但發明人拒收此文件或者表明不必將文件寄給他，就可表示發明人拒絕簽署。如果發明只以口頭表示拒絕簽署，則請願書中需註明時間及地點。如果發明人以書面拒絕簽署，則需附上該書面之影本。另外，在請願書中需詳述在何種情況下，發明人之何種行為構成拒絕簽署之理由，如果發明人有告知拒絕簽署之理由，此理由亦需記載在請願書中。

MPEP409.03(d)

＊＊＊＊＊＊＊＊＊＊＊

II.REFUSAL TO JOIN

A refusal by an inventor to sign an oath or declaration when the inventor has not been presented with the application papers does not itself suggest that the inventor is refusing to join the application unless it is clear that the inventor understands exactly what he or she is being asked to sign and refuses to accept the application papers. A copy of the application papers should be sent to the last known address of the nonsigning inventor, or, if the nonsigning inventor is represented by counsel, to the address of the nonsigning inventor's attorney. The fact that an application may contain proprietary information does not relieve the 37 CFR 1.47 applicant of the responsibility to present the application papers to the inventor if the inventor is willing to receive the papers in order to sign the oath or declaration. It is noted that the inventor may obtain a complete copy of the application, unless the inventor has assigned his or her interest in the application, and the assignee has requested that the inventor not be permitted access. See MPEP § 106. It is reasonable to require that the inventor be presented with the application papers before a petition under 37 CFR 1.47 is granted since such a procedure ensures that the inventor is apprised of the application to which the oath or declaration is directed. In re Gray, 115 USPQ 80 (Comm'r Pat. 1956).

Where a refusal of the inventor to sign the application papers is alleged, the circumstances of the presentation of the application papers and of the refusal must be specified in a statement of facts by the person who presented the inventor with the application papers and/or to whom the refusal was made. Statements by a party not present when an oral refusal is made will not be accepted.

Proof that a bona fide attempt was made to present a copy of the application papers (specification, including claims, drawings, and oath or declaration) to the nonsigning inventor for signature, but the inventor refused to accept delivery of the papers or expressly stated that the application papers should not be sent, may be sufficient. When there is an express oral refusal, that fact along with the time and place of the refusal must be stated in the statement of facts. When there is an express written refusal, a copy of the document evidencing that refusal must be made part of the statement of facts. The document may be redacted to remove material not related to the inventor's reasons for refusal.

When it is concluded by the 37 CFR 1.47 applicant that a nonsigning inventor's conduct constitutes a refusal, all facts upon which that conclusion is based should be stated in the statement of facts in support of the petition or directly in the petition. If there is documentary evidence to support facts alleged in the petition or in any statement of facts, such evidence should be submitted. Whenever a nonsigning inventor gives a reason for refusing to sign the application oath or declaration, that reason should be stated in the petition.

§2.7　未簽名的發明人之權利

未簽名的發明人可抗議其在某一申請案中被指定為發明人。未簽名的發明人也有權利檢視、影印申請案中所有的文件，可委任專利律師（專利代理人）進行任何相關之程序。未簽名之發明人可在之後加入成為發明人，此時需簽署 37CFR 1.63 之宣誓書作為發明人。即使此未簽名之發明人後來加入了，但在無原申請人之同意之下不能撤銷原申請人所委任的代理人或新委任代理人。另外，如果未簽名的發明人覺得自已是該專利申請案之唯一發明人時，可自行提出一專利申請案，以讓自已的申請案與原來之申請案進入抵觸程序（interference）。雖然美國專利局允許未簽名的發明人檢視、影印申請案之所有文件及參加相關程序，但是對於 inter partes 程序則不准參加案件之審查。

MPEP 1.409.03(i) Rights of the Nonsigning Inventor [R-3]

The nonsigning inventor (also referred to as an "inventor designee") may protest his or her designation as an inventor. The nonsigning inventor is entitled to inspect any paper in the application, order copies thereof at the price set forth in 37 CFR 1.19, and make his or her position of record in the file wrapper of the application. Alternatively, the nonsigning inventor may arrange to do any of the preceding through a registered patent attorney or agent.

While the U.S. Patent and Trademark Office will grant the nonsigning inventor access to the application, inter partes proceedings will not be instituted in 37 CFR 1.47 case. In re Hough, 108

USPQ 89 (Comm"r Pat. 1955). A nonsigning inventor is not entitled to a hearing (Cogar v. Schuyler, 464 F.2d 747, 173 USPQ 389 (D.C. Cir. 1972)), and is not entitled to prosecute the application if status under 37 CFR 1.47 has been accorded, or if proprietary interest of the 37 CFR 1.47(b) applicant has been shown to the satisfaction of the U.S. Patent and Trademark Office.

A nonsigning inventor may join in a 37 CFR 1.47 application. To join in the application, the nonsigning inventor must file an appropriate 37 CFR 1.63 oath or declaration. Even if the nonsigning inventor joins in the application, he or she cannot revoke or give a power of attorney without agreement of the 37 CFR 1.47 applicant. >See MPEP § 402.10.<

The rights of a nonsigning inventor are protected by the fact that the patent resulting from an application filed under 37 CFR 1.47(b) and 35 U.S.C. 118 must issue to the inventor, and in an application filed under 37 CFR 1.47(a) and 35 U.S.C. 116, the inventor has the same rights that he or she would have if he or she had joined in the application. In re Hough, 108 USPQ 89 (Comm"r Pat. 1955).

If a nonsigning inventor feels that he or she is the sole inventor of an invention claimed in a 37 CFR 1.47 application naming him or her as a joint inventor, the nonsigning inventor may file his or her own application and request that his or her application be placed in interference with the 37 CFR 1.47 application. If the claims in both the nonsigning inventor"s application and the 37 CFR 1.47 application are otherwise found allowable, an interference may be declared.

§2.8　非暫時專利申請案之宣誓書之簽署

宣誓書上需記載發明人之姓名、國籍（citizenship）、永久住址（residence）、郵寄地址（mailing or post office address）。永久住址只要記載城市名及州名(或外國國家名稱即可)發明人之姓名必需是全名，包括姓（family name）及至少一名（given name），例如全名如果是 John Paul Doe，必需註明為 John P. Doe 或 J. Paul Doe，不可僅記載為 J. P Doe。發明人之郵寄地址乃指其通常用來收郵件之住址，不能以郵寄住址作為永久地址。

發明人之簽名可以為任何形式，並不一定要用英文，只要確實是由發明人簽名即可。發明人簽名時需確實已閱讀了所附的說明書、請求項及圖式，所謂所附的說明書、圖式及請求項並不是說明書、圖式及請求項需與宣誓書訂在一起，只要發明人在簽名時說明書、圖式、請求項在一起，且已閱讀即可。發明人在簽署了宣誓書之後，就不能再對說明書（圖式、請求項）作修改、塗改、刪除、插入文字等，如有修改、塗改、刪除、插入文字，則需再重簽一份宣誓書。如果發明人在簽署宣誓書之前，要直接在說明書（圖式、請求項）上作修改，則需在修改處簽上名字之縮寫及註明日期。

發明人簽名時亦可將其頭銜，例如 Dr.簽在其名字之前，但此頭銜不會印在專利證書上。

當發明人死亡或失能而無法簽署宣誓書時，可由其法定代表人簽名，此時法定代表人必需在宣誓書上註明為法定代表人，且記載其國籍、永久地址、郵寄地址。如果發明人找不到或拒絕簽署時，則可由具有權利之人（如受讓人）簽署，但不能由受委任之專利代理人（專利律師）簽署。

如有共同發明人時，則獲准專利之後、專利證書上之發明人順序將依照申請時宣誓書所打上的發明人順序列印，由於發明人之順序與法律上權利無關，通常不能修改。如要修改發明人順序則需提出請願並繳交規費。有共同發明人時，各發明人之簽名亦可不必簽在同一張宣誓書上，但每一張宣誓書上均需列上所有發明人之名字以讓各別簽署的發明人知道其他共同發明人。

發明人在簽署宣誓書時應同時簽上日期，簽署之日期亦可事先打字好交由發明人於該日期簽署。即使發明人簽署之日期與申請之日期相隔很久（例如超過二年），現在亦可被美國專利局接受。

美國專利局現可接受宣誓書之影本，正本可不必送入美國專利局。

MPEP 605.01 Applicant's Citizenship

The statute (35 U.S.C. 115) requires an applicant, in a nonprovisional application, to state his or her citizenship. Where an applicant is not a citizen of any country, a statement to this effect is accepted as satisfying the statutory requirement, but a statement as to citizenship applied for or first papers taken out looking to future citizenship in this (or any other) country does not meet the requirement.

＊＊＊＊＊＊＊＊＊＊＊

MPEP 605.02 Applicant's Residence

Applicant's place of residence, that is, the city and either state or foreign country, is required to be included in the oath or declaration in a nonprovisional application for compliance with 37 CFR 1.63 unless it is included in an application data sheet (37 CFR 1.76).

＊＊＊＊＊＊＊＊＊＊＊

EXECUTION OF OATH OR DECLARATION ON BEHALF OF INVENTOR

The oath or declaration required by 35 U.S.C. 115 must be signed by all of the actual inventors, except under limited circumstances.35 U.S.C. 116 provides that joint inventors can sign on behalf of an inventor who cannot be reached or refuses to join. See MPEP § 409.03(a). 35 U.S.C. 117 provides that the legal representative of a deceased or incapacitated inventor can sign on behalf of the inventor. If a legal representative executes an oath or declaration on behalf of a deceased inventor, the legal representative must state that the person is a legal representative and provide the citizenship, residence, and mailing address of the legal representative. See 37 CFR 1.64, MPEP § 409.01 and § 409.02. 35 U.S.C. 118 provides that a party with proprietary interest in the invention claimed in an application can sign on behalf of the inventor, if the inventor cannot be reached or refuses to join in the filing of the application. See MPEP § 409.03(b) and § 409.03(f). The oath or declaration may not be signed by an attorney on behalf of the inventor, even if the attorney has been given a power of attorney to do so. Opinion of Hon. Edward Bates, 10 Op. Atty. Gen. 137 (1861). See also Staeger v.

MPEP 605.04(b) One Full Given Name Required

37 CFR 1.63(a)(2) requires that each inventor be identified by full name, including the family name, and at least one given name without abbreviation together with any other given name or initial in the oath or declaration. For example, if the applicant"s full name is "John Paul Doe," either "John P. Doe" or "J. Paul Doe" is acceptable.

Form paragraphs 6.05 (reproduced in MPEP § 602.03) and 6.05.18 may be used to notify applicant that the oath or declaration is defective because the full given name of each inventor has not been adequately stated.

MPEP 605.04(e) May Use Title With Signature

It is permissible for an applicant to use a title of nobility or other title, such as "Dr.", in connection with his or her signature. The title will not appear in the printed patent.

If the residence is not included in the executed oath or declaration filed under 37 CFR 1.63, the Office of **>Patent Application Processing (OPAP)< will normally so indicate on a "Notice of Informal Application," so as to require the submission of the residence information within a set period for reply. If the examiner notes that the residence has not been included in the oath or declaration or in an application data sheet, form paragraphs 6.05 (reproduced in MPEP § 605.01) and 6.05.02 should be used.

MPEP 605.03 Applicant's Mailing or Post Office Address

Each applicant's mailing or post office address is required to be supplied on the oath or declaration, if not stated in an application data sheet. Applicant's mailing address means that address at which he or she customarily receives his or her mail. Either applicant's home or business address is acceptable as the mailing address. The mailing address should include the ZIP Code designation.

605.04(a) Applicant's Signature and Name

There is no requirement that a signature be made in any particular manner. See MPEP § 605.04(d). If applicant signs his or her name using non-English characters, then such a signature will be accepted.

It is improper for an applicant to sign an oath or declaration which is not attached to or does not identify a specification and/or claims.

Attached does not necessarily mean that all the papers must be literally fastened. It is sufficient that the specification, including the claims, and the oath or declaration are physically located together at the time of execution. Physical connection is not required. Copies of declarations are encouraged. See MPEP § 502.01, § 502.02.

* * * * * * * * * * *

37 CFR 1.52(c)(1) states that "[a]ny interlineation, erasure, cancellation or other alteration of the application papers filed must be made before the signing of any accompanying oath or declaration pursuant to § 1.63 referring to those application papers and should be dated and initialed or signed by the applicant on the same sheet of paper. Application papers containing alterations made after the signing of an oath or declaration referring to those application papers must be supported by a supplemental oath or declaration under § 1.67. In either situation, a substitute specification (§ 1.125) is required if the application papers do not comply with paragraphs (a) and (b) of this section." 37 CFR 1.52(c)(2) states that after the signing of the oath or declaration referring to the application papers, amendments may only be made in the manner provided by 37 CFR 1.121.

* * * * * * * * * * *

Any changes made in ink in the application or oath prior to signing should be initialed and dated by the applicants prior to execution of the oath or declaration. The Office　will require a new oath or declaration if the alterations are not initialed and dated.

* * * * * * * * * * *

MPEP 605.04(f) Signature on Joint Applications - Order of Names

The order of names of joint patentees in the heading of the patent is taken from the order in which the typewritten names appear in the original oath or declaration. Care should therefore be exercised in selecting the preferred order of the typewritten names of the joint inventors, before filing, as requests for subsequent shifting of the names would entail changing numerous records in the Office. Since the particular order in which the names appear is of no consequence insofar as the legal rights of the joint applicants are concerned, no changes will be made except when a petition under 37 CFR 1.182 is granted. The petition should be directed to the attention of the Office of Petitions. The petition to change the order of names must be signed by either the attorney or agent of record or all the applicants. Applicants are strongly encouraged to submit an application data sheet showing the new order of inventor names to ensure appropriate printing of the inventor names in any patent to issue. It is suggested that all typewritten and signed names appearing in the application papers should be in the same order as the typewritten names in the oath or declaration.

In those instances where the joint applicants file separate oaths or declarations, the order of names is taken from the order in which the several oaths or declarations appear in the application papers unless a different order is requested at the time of filing.

§2.9　發明人錯誤與專利之有效性

如果一專利申請案之個別發明人之名字錯誤只是因為申請時在宣誓書或申請資料表（application data sheet）上繕打錯誤或音譯錯誤，就不需提出請願或新的宣誓書，只要通知美國專利局更正即可。如果是因為發明人更改名字，則必需提出請願書或宣誓書說明更改名字之原因、過程或官方證明文件（如法院之更名文件），向美國專利局提出更正。

如果專利申請案有些發明人不該列為發明人（非真正的發明人），或有些發明人應該列為發明人而漏列，則需更正此發明人（inventorship）之錯誤，只要發明人本人無欺騙之意圖，不論在申請階段或已核發專利均可向美國專利局提出修正。此發明人之錯誤不會影響專利之有效性。

MPEP 605.04(b)

＊＊＊＊＊＊＊＊＊＊＊

When a typographical or transliteration error in the spelling of an inventor's name is discovered during pendency of an application, a petition is not required, nor is a new oath or declaration under 37 CFR 1.63 needed. The U.S. Patent and Trademark Office should simply be notified of the error and reference to the notification paper will be made on the previously filed oath or declaration by the Office.

＊＊＊＊＊＊＊＊＊＊＊

MPEP 605.04(c) Inventor Changes Name

In cases where an inventor's name has been changed after the application has been filed and the inventor desires to change his or her name on the application, he or she must submit a petition under 37 CFR 1.182. Applicants are also strongly encouraged to submit an application data sheet (37 CFR 1.76) showing the new name. The petition should be directed to the attention of the Office of Petitions. The petition must include an appropriate petition fee and a statement signed by the inventor setting forth both names and the procedure whereby the change of name was effected, or a copy of the court order.

＊＊＊＊＊＊＊＊＊＊＊

35USC § 116 Inventors

＊＊＊＊＊＊＊＊＊＊＊

Whenever through error a person is named in an application for patent as the inventor, or through an error an inventor is not named in an application, and such error arose without any deceptive intention on his part, the Director may permit the application to be amended accordingly, under such terms as he prescribes.

35 U.S.C. 256 Correction of named inventor.

Whenever through error a person is named in an issued patent as the inventor, or through error an inventor is not named in an issued patent and such error arose without any deceptive intention on his part, the Director may, on application of all the parties and assignees, with proof of the facts and such other requirements as may be imposed, issue a certificate correcting such error.

The error of omitting inventors or naming persons who are not inventors shall not invalidate the patent in which such error occurred if it can be corrected as provided in this section. The court before which such matter is called in question may order correction of the patent on notice and hearing of all parties concerned and the Director shall issue a certificate accordingly.

§2.10　非暫時專利申請案之發明人錯誤時之修正

一非暫時專利申請案之發明人如果有錯誤，而此錯誤是多列或漏列了某些發明人，且發明無欺騙意圖所造成的情況下，可檢附以下之文件，提出請願（petition）向美國專利局提出修正：

1. 申請書，註明發明人要如何修改。

2. 聲明書：由要加入的發明人及/或要刪去之發明人聲明此錯誤並無欺騙之意圖。

3. 新的宣誓書：由真正的發明人簽署。

4. 受讓人之同意書。

5. 規費

在專利申請案仍審查中任何時間均可修改發明人，並無一定之期限。且申請書並不必說明此錯誤是在怎樣的情況下發生的。另外，如果無法取得發明人無欺騙之意圖的聲明書時，可依 37 CFR 1.183 提出請願，請求不必呈送聲明書。

---

37 CFR 1.48 Correction of inventorship in a patent application, other than a reissue application, pursuant to 35 U.S.C. 116.

(a) Nonprovisional application after oath/declaration filed. If the inventive entity is set forth in error in an executed §1.63 oath or declaration in a nonprovisional application, and such error arose without any deceptive intention on the part of the person named as an inventor in error or on the part of the person who through error was not named as an inventor, the inventorship of the nonprovisional application may be amended to name only the actual inventor or inventors. Amendment of the inventorship requires:

(1) A request to correct the inventorship that sets forth the desired inventorship change;

(2) A statement from each person being added as an inventor and from each person being deleted as an inventor that the error in inventorship occurred without deceptive intention on his or her part;

(3) An oath or declaration by the actual inventor or inventors as required by § 1.63 or as permitted by §§ 1.42, 1.43 or § 1.47;

(4) The processing fee set forth in § 1.17(i); and

(5) If an assignment has been executed by any of the original named inventors, the written consent of the assignee (see § 3.73(b) of this chapter).

＊＊＊＊＊＊＊＊＊＊＊

§2.11 因請求項之修改刪除部份的發明人

非暫時申請案如未因請求項之修改要刪除部份的發明人時，需提出以下之文件：

1. 申請書：註明要被刪除的發明人，如申請書需由以下之人士簽署：

   (1) 專利代理人；或

   (2) 所有的受讓人；或

   (3) 所有的發明人；或

   (4) 部份的受讓人及剩下的發明人。

2. 一聲明：因為請求項修改，使被刪除之發明人的發明不再存在於專利申請案的請求項。

3. 規費

---

37 CFR§ 1.48

＊＊＊＊＊＊＊＊＊＊＊

(b) Nonprovisional application-fewer inventors due to amendment or cancellation of claims. If the correct inventors are named in a nonprovisional application, and the prosecution of the nonprovisional application results in the amendment or cancellation of claims so that fewer than all of the currently named inventors are the actual inventors of the invention being claimed in the nonprovisional application, an amendment must be filed requesting deletion of the name or names of the person or persons who are not inventors of the invention being claimed. Amendment of the inventorship requires:

(1) A request, signed by a party set forth in §1.33(b), to correct the inventorship that identifies the named inventor or inventors being deleted and acknowledges that the inventor's invention is no longer being claimed in the nonprovisional application; and (2) The processing fee set forth in § 1.17(i).

＊＊＊＊＊＊＊＊＊＊＊

§2.12 因請求項之修改增加發明人

非暫時申請案如果因請求項之修改要增加發明人時需呈送以下文件：

1. 申請書：註明發明人如何修改。

2. 聲明書：由要加入之發明人簽署，聲明①此發明人之增加是由於請求項之改變②此錯誤之發生並無欺騙之意圖。

3. 新的宣誓書真實的發明人全部重簽。

4. 受讓人之同意書。

5. 規費

---

37 CFR 1.48(C)

＊＊＊＊＊＊＊＊＊＊＊

(c) Nonprovisional application-inventors added for claims to previously unclaimed subject matter. If a nonprovisional application discloses unclaimed subject matter by an inventor or inventors not named in the application, the application may be amended to add claims to the subject matter and name the correct inventors for the application. Amendment of the inventorship requires:

(1) A request to correct the inventorship that sets forth the desired inventorship change;

(2) A statement from each person being added as an inventor that the addition is necessitated by amendment of the claims and that the inventorship error occurred without deceptive intention on his or her part;

(3) An oath or declaration by the actual inventors as required by § 1.63 or as permitted by §§ 1.42, 1.43, or §1.47;

(4) The processing fee set forth in § 1.17(i); and

(5) If an assignment has been executed by any of the original named inventors, the written consent of the assignee (see §3.73(b) of this chapter).

＊＊＊＊＊＊＊＊＊＊＊

§2.13　暫時專利申請案增加發明人

一暫時專利申請案之發明人有誤時，通常可不必提出修正，除非在一年內要提出正式申請案主張暫時案之優先權時發明人與暫時案之發明人無一相同時才需要修正。要增加發明人時需呈送以下文件：

1. 申請書：註明要增加的發明人，並由以下人士簽署：

① 專利代理人；或

② 所有的受讓人；或

③ 所有的發明人；或

④ 部份的受讓人及剩下的發明人。並需聲明此錯誤發明人並無欺騙之意圖。

2. 規費

---

37 CFR 1.48(d)

＊＊＊＊＊＊＊＊＊＊＊

(d) Provisional application-adding omitted inventors. If the name or names of an inventor or inventors were omitted in a provisional application through error without any deceptive intention on the part of the omitted inventor or inventors, the provisional application may be amended to add the name or names of the omitted inventor or inventors. Amendment of the inventorship requires:

(1) A request, signed by a party set forth in § 1.33(b), to correct the inventorship that identifies the inventor or inventors being added and states that the inventorship error occurred without deceptive intention on the part of the omitted inventor or inventors; and(2) The processing fee set forth in §1.17(q).

＊＊＊＊＊＊＊＊＊＊＊

§2.14　暫時專利申請案刪除部份的發明人

一暫時專利申請案如果要刪除部份發明人時，需呈送以下文件：

1. 申請書：註明發明人如何修改。

2. 聲明書：由要被刪除之發明人簽署並聲明此錯誤之發生並無欺騙之企圖。

3. 受讓人之同意書。

4. 規費

---

37 CFR 1.48(e)

＊＊＊＊＊＊＊＊＊＊＊

(e) Provisional application-deleting the name or names of the inventor or inventors. If a person or persons were named as an inventor or inventors in a provisional application through error without any deceptive intention on the part of such person or persons, an amendment may be filed in the provisional application deleting the name or names of the person or persons who were erroneously named. Amendment of the inventorship requires:

(1) A request to correct the inventorship that sets forth the desired inventorship change;

(2) A statement by the person or persons whose name or names are being deleted that the inventorship error occurred without deceptive intention on the part of such person or persons;

(3) The processing fee set forth in §1.17(q); and

(4) If an assignment has been executed by any of the original named inventors, the written consent of the assignee (see § 3.73(b) of this chapter).

＊＊＊＊＊＊＊＊＊＊＊

§2.15　修改非暫時專利申請案之發明人的錯誤之其他方法

如果一非暫時專利申請案在提出申請時並未呈送宣誓書而發現申請書所記載之發明人有誤時，在補送宣誓書時由正確的發明人簽署，就可不必提出發明人之修改，因為以補送之宣誓書上之發明人為準。

另外，如果要被刪去的發明人作出無欺騙之意圖之聲明有困難時，可提出一延續案由正確的發明人簽署延續案之宣誓書（要有至少一發明人與母案之發明人相同），並放棄母案即可達到修改發明人之目的。

＊＊＊＊＊＊＊＊＊＊＊

37 CFR 1.48(f)

37 CFR 1.48 Correction of inventorship in a patent application, other than a reissue application, pursuant to 35 U.S.C.

＊＊＊＊＊＊＊＊＊＊＊

(f)(1) Nonprovisional application-filing executed oath/declaration corrects inventorship. If the correct inventor or inventors are not named on filing a nonprovisional application under § 1.53(b) without an executed oath or declaration under § 1.63 by any of the inventors, the first submission of an executed oath or declaration under § 1.63 by any of the inventors during the pendency of the application will act to correct the earlier identification of inventorship. See §§1.41(a) (4) and 1.497(d)and (f) for submission of an executed oath or declaration to enter the national stage under 35 U.S.C. 371 naming an inventive entity different from the inventive entity set forth in the international stage.

＊＊＊＊＊＊＊＊＊＊＊

MPEP 201.03

Correction of Inventorship in an Application [R-5]

＊＊＊＊＊＊＊＊＊＊＊

37 CFR 1.48(f) operates to automatically correct the inventorship upon filing of a first executed oath or declaration under 37 CFR 1.63 by any of the inventors in a nonprovisional application or upon filing of a cover sheet in a provisional application.

Correction of inventorship may also be obtained by the filing of a continuing application under 37 CFR 1.53 without the need for filing a request under 37 CFR 1.48, either in the application containing the inventorship err(using a copy of the executed oath or declaration from the parent application)or (to be abandoned) or in the continuing application. The continuing application must be filed with the correct inventorship named therein. The filing of a continuing application to correct the inventorship is appropriate if at least one of the correct inventors has been named in the prior application (35 U.S.C. 120 and 37 CFR 1.78(a) (1)). That is, at least one of the correct inventors must be named in the executed oath or declaration filed in the prior application, or where no executed oath or declaration has been submitted in the prior application, the name of at least one correct inventor must be set forth in the application papers pursuant to 37 cfr 1.41(a)(1). Where the name of at least one inventor is to be added, correction of inventorship can be accomplished by filing a continuing application under 37 CFR 1.53(b) with a newly executed oath or declaration under 37 CFR 1.63(a). Where the name of an inventor(s) is to be deleted, applicant can file a continuation or divisional application (using a copy of the executed oath or declaration from the parent application)< with a request for deletion of the name of the inventor(s). See 37 CFR 1.63(d)(2). If a continuing application is filed with a new executed oath or declaration properly naming the correct inventors, a request for deletion of the name(s) of the person(s) who are not inventors in the continuing application is not necessary. The continuing application may be filed under 37 CFR 1.53(b) or, if the application is for a design patent, under 37 CFR 1.53(d). Note the requirements of 37 CFR 1.78 (a)(1)(ii).

＊＊＊＊＊＊＊＊＊＊＊

§2.16 已發證之專利之發明人的修正

如果一發證之專利因為錯誤，漏列某些發明人或多列了某些發明人，而這些發明人並無欺騙之意圖時，則可提出請願或依據法院之命令請求美國專利局作修改。美國專利局將發下修正專利證書。此修改需呈送以下文件：

① 請願書，

② 由要加入之發明人簽署之聲明書，聲明其無欺騙之意圖，

③ 由原來之專利申請案所有之發明人簽署之聲明書：聲明其同意發明人之修改，

④ 加入之發明人所讓渡之受讓人簽署之聲明書，聲明同意此發明人之修改及其擁有該專利權，

⑤ 規費。

＊＊＊＊＊＊＊＊＊＊＊

37CFR 1.324 Correction of inventorship in patent, pursuant to 35 U.S.C.256.

(a) Whenever through error a person is named in an issued patent as the inventor, or through error an inventor is not named in an issued patent and such error arose without any deceptive intention on his or her part, the Director, pursuant to 35 U.S.C.256, may, on application of all the parties and assignees, or on order of a court before which such matter is called in question, issue a certificate naming only the actual inventor or inventors. A petition to correct inventorship of a patent involved in an interference must comply with the requirements of this section and must be accompanied by a motion under § 41.121(a)(2) or §41.121(a)(3) of this title.

(b) Any request to correct inventorship of a patent pursuant to paragraph (a) of this section must be accompanied by:

(1) Where one or more persons are being added, a statement from each person who is being added as an inventor that the inventorship error occurred without any deceptive intention on his or her part;

(2) A statement from the current named inventors who have not submitted a statement under paragraph (b)(1) of this section either agreeing to the change of inventorship or stating that they have no disagreement in regard to the requested change;

(3) A statement from all assignees of the parties submitting a statement under paragraphs (b)(1) and (b)(2) of this section agreeing to the change of inventorship in the patent, which statement must comply with the requirements of §3.73(b) of this chapter; and

(4) The fee set forth in　§ 1.20(b).

(c) For correction of inventorship in an application, see §§ 1.48 and 1.497, and in an interference see §1.634.

實例 1

X 公司提出一發明專利申請並以發明人 A 及 B 為發明人，發明人 A 及 B 均已將其專利申請權讓予 X 公司。在專利提出申請後之二個月後，X 公司發現發明人有誤，由於疏忽誤將發明人 C 漏掉，故須將發明人 C 加入。請問須提出哪些文件以修正此發明人之錯誤？

需以下文件
① 申請書；
② 發明人 C 之聲明（statement）聲明此錯誤本人無欺騙之意圖；
③ 37CFR1.63 之宣誓書由發明人 A,B,C 三人簽名；
④ 相關規費；
⑤ X 公司之同意書。

實例 2

A 及 B 均受雇於 X 公司，依照 X 公司之規定，員工如果發明並取得專利並將專利讓予公司，則公司將給予 NTD 10,000 之獎金。B 有一發明，A 知曉後請 B 將其列為共同發明人，以當發明取得專利後可共享 NTD 10,000 之獎金。B 同意並以兩人之名義申請專利，專利獲證，A 與 B 兩人分享獎金。但在取得專利六個月之後，兩人均退休，而 X 公司之專利律師知曉 A 並非真正的發明人後，向美國專利局提出請願並提出 A,B 簽名之聲明及 X 公司之同意書，請求 USPTO 修正發明人。美國專利局是否會修正此專利之發明人？

美國專利局將不修正發明人，因為此專利之發明人有欺騙之意圖。

實例 3

A 與 B 均為 X 公司研發部門之員工，且共同研發了一發明擬申請專利，但 B 在尚未申請時即提出辭呈。研發部門之主管為了節省發給 B 之專利申請獎金乃在 B 離職後以 A 為單獨發明人提出專利申請，A 不知 B 並未被列為發明人，B 亦不知此發明有提出專利申請並已讓渡給 X 公司，請問此情形向美國專利局提出修正發明人是否可接受？需提出哪些文件已修正發明人？

修正發明人可被美國專利局接受，因為有欺騙意圖者並非發明人本人。提出修正可呈送發明人 B 之聲明，發明人 A 及 B 之宣誓書及 X 公司之同意書即可。或者提出延續案由發明人 A 及 B 簽署宣誓書並將母案放棄則是較簡單的方法。

實例 4

A 發明了一用來漂白紙張之化合物的製法（該化合物為已知），B 及 A 又一起發明了使用 A 發明之製法所製出之化合物以清除有毒廢紙之方法。A 與 B 共同列為發明人提出一發明專利申請，其中 claims 1～9 乃為該已知化合物之製法，claims 10～19 為清除有毒廢紙之方法，claim 20 乃為使用由 claim 1 製出之化合物以清除有毒廢紙之方法。之後收到限制性要求之 Office Action，要求選擇 claims 1～9 或 claims 10～20 之發明，結果選擇了 claims 1～9 並刪除 claims 10～20，請問在提出分割案之後是否須修改本案之發明人？需提出哪些文件？

需提出發明人修正，刪除發明人 B 需提出一聲明指出發明人 B 已刪除且發明人 B 之發明未在本案中請求，並繳規費。

第 3 章　讓渡與專利所有權

目　錄

§3.1　所有權（ownership），讓渡（assignment），授權（licensing），個別所有權（individual ownership）及共同所有權（co-ownership）

一專利（或專利申請案）的所有權乃屬於該專利或專利申請案之發明人，但此專利（專利申請案）可藉由書面的文件、讓渡書將所有權讓渡給對此專利（專利申請案）有興趣之受讓人。一專利的所有權人擁有排除他人製造、使用、銷售、要約銷售（offering for sale）、輸入至美國該專利的發明之權利。但並不一定自己擁有製造、使用、銷售、要約銷售、輸入至美國該專利的發明之權利，因為可能有其他法律排除專利所有權人作以上之行為，例如他人擁有支配專利（dominant patent），未取得美國食品藥物管理局（FDA）之核准，法院有禁止令禁止製造，及國家安全等理由。

一專利或專利申請案之讓渡乃指將全部的（entire）所有權轉移另一人。

一專利或專利申請案之授權乃指將少於全部之權利轉移給另一人，例如權利可能被限制在某一時間、地域或使用之領域。授權可視為一合約，當被授權人盡其義務且在授權合約所規定之範圍內操作時，專利權人不會對其製造、銷售、要約銷售、使用、輸入至美國之行為提出告訴。專利所有權人亦可轉移排他的授權（exclusive license）給被授權人，此排他的授權可防止專利所有權人（或其他專利所有權人想授權的人）與排他的被授權人在授權契約所規定的地域、時間及使用領域上競爭。排他的授權並非專利權之讓渡。

所謂個別所有權，指只有一人擁有一專利的權利，例如一專利只有一發明人，而此發明人並未將權利讓渡給任何人，或者是一專利之所有發明人將權利只讓渡給一人（法人）。所謂的共同所有權，指有多人（法人）擁有一專利之權利。例如一專利有多個發明人，且此多個發明人均未將權利讓渡他人，或發明人將權利讓渡給多個部份受讓人（partial assignees）。共同所有權人在美國專利局之所有法律程序必需共同參與。但是除非有合約，各個共同所有權人可分別在美國境內製造、使用、銷售、要約銷售該專利之發明，或將專利之發明輸入至美國或將其所有權讓渡，授權他人而不需其他共同所有權人之同意，且不論其對專利之發明的貢獻占有比率有多少。

MPEP 301

I.OWNERSHIP

Ownership of a patent gives the patent owner the right to exclude others from making, using, offering for sale, selling, or importing into the United States the invention claimed in the patent. 35 U.S.C. 154(a)(1). Ownership of the patent does not furnish the owner with the right to make, use, offer for sale, sell, or import the claimed invention because there may be other legal considerations precluding same (e.g., existence of another patent owner with a dominant patent, failure to obtain FDA approval of the patented invention, an injunction by a court against making the product of the invention, or a national security related issue).

The ownership of the patent (or the application for the patent) initially vests in the named inventors of the invention of the patent. See Beech Aircraft Corp. v. EDO Corp., 990 F.2d 1237, 1248, 26 USPQ2d 1572, 1582 (Fed. Cir. 1993). The patent (or patent application) is then assignable by an instrument in writing, and the assignment of the patent, or patent application, transfers to the assignee(s) an alienable (transferable) ownership interest in the patent or application. 35 U.S.C. 261.

II.ASSIGNMENT

"Assignment," in general, is the act of transferring to another the ownership of one's property, i.e., the interest and rights to the property. In 37 CFR 3.1, assignment of patent rights is defined as "a transfer by a party of all or part of its right, title and interest in a patent or patent application…..." An assignment of a patent, or patent application, is the transfer to another of a party's entire ownership interest or a percentage of that party's ownership interest in the patent or application. In order for an assignment to take place, the transfer to another must include the entirety of the bundle of rights that is associated with the ownership interest, i.e., all of the bundle of rights that are inherent in the right, title and interest in the patent or patent application.

III.LICENSING

As compared to assignment of patent rights, the licensing of a patent transfers a bundle of rights which is less than the entire ownership interest, e.g., rights that may be limited as to time, geographical area, or field of use. A patent license is, in effect, a contractual agreement that the patent owner will not sue the licensee for patent infringement if the licensee makes, uses, offers for sale, sells, or imports the claimed invention, as long as the licensee fulfills its obligations and operates within the bounds delineated by the license agreement.

An exclusive license may be granted by the patent owner to a licensee. The exclusive license prevents the patent owner (or any other party to whom the patent owner might wish to sell a license) from competing with the exclusive licensee, as to the geographic region, the length of time, and/or the field of use, set forth in the license agreement.

A license is not an assignment of the patent. Even if the license is an exclusive license, it is not an assignment of patent rights in the patent or application.

IV. INDIVIDUAL AND JOINT OWNERSHIP

Individual ownership - An individual entity may own the entire right, title and interest of the patent property. This occurs where there is only one inventor, and the inventor has not assigned the patent property. Alternatively, it occurs where all parties having ownership interest (all inventors and assignees) assign the patent property to one party.

Joint ownership - Multiple parties may together own the entire right, title and interest of the patent property. This occurs when any of the following cases exist:

(A) Multiple partial assignees of the patent property;

(B) Multiple inventors who have not assigned their right, title and interest; or

(C) A combination of partial assignee(s), and inventor(s) who have not assigned their right, title and interest.

Each individual inventor may only assign the interest he or she holds; thus, assignment by one joint inventor renders the assignee a partial assignee. A partial assignee likewise may only assign the interest it holds; thus, assignment by a partial assignee renders a subsequent assignee a partial assignee. All parties having any portion of the ownership in the patent property must act together as a composite entity in patent matters before the Office.

35 U.S.C. 262 Joint owners.

In the absence of any agreement to the contrary, each of the joint owners of a patent may make, use, offer to sell, or sell the patented invention within the United States, or import the patented invention into the United States, without the consent of and without accounting to the other owners.

§3.2　美國專利之所有權人

任何美國人、外國人、美國法人、外國法人、美國機構、外國機構、政府機關經被讓渡一美國專利（美國專利申請案）就成為美國專利之所有權人。但是美國專利商標局之官員或受雇人員則有限制。任何美國專利商標局之官員及受雇人員受雇期間及離職之後的一年之內不能申請美國專利亦不能直接或間接取得任何美國專利

或任何美國專利之權益，除非是由於繼承或遺贈。而且亦不能享有任何優先權日早
於在受僱終止之一年之申請案之優先權利益。以下圖說明：

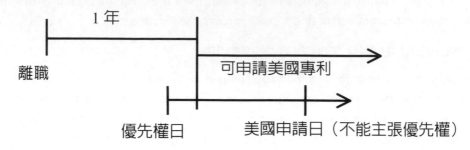

---

35 U.S.C. 4 Restrictions on officers and employees as to interest in patents.

Officers and employees of the Patent and Trademark Office shall be incapable, during the period of their appointments and for one year thereafter, of applying for a patent and of acquiring, directly or indirectly, except by inheritance or bequest, any patent or any right or interest in any patent, issued or to be issued by the Office. In patents applied for thereafter they shall not be entitled to any priority date earlier than one year after the termination of their appointment.

實例 1

A 為 X 公司之研發主管，A 有一發明並在 X 公司同意下以 A 為發明人名義提出專利申請並獲得專利並將其 1/2 之權利讓渡給 X 公司。之後 A 離開 X 公司至 Y 公司任職，Y 公司為 X 公司之競爭廠商，A 將其專利權讓渡給 Y 公司並製造及銷售該專利之產品，請問 X 公司是否可採取任何法律行動？

除非 A 與 X 公司有合約，聲明 A 讓渡其專利需經 X 公司之同意，否則 A 可不經 X 公司之同意將權利讓渡給 Y 公司。在此情況下 X 公司無法採取任何法律行動。

§3.3　專利讓渡文件登錄之效力與利益

專利讓渡文件在美國專利局登錄只是一行政行為，美國專利局對讓渡文件之有效性或讓渡文件對專利之所有權之效力（effect），並不會作決定。但是專利讓渡文

件在美國專利局登錄之後，受讓人在很多情況下在專利申請案之相關法律程序中就有權力進行、主導相關程序（take action），例如，對審定書之答辯可簽名，可提出尾端棄權書，可同意提出再發證案，可同意修正發明人等。此外，未在美國專利局登錄的讓渡對於後續之購買者，承受抵押者（受讓者，mortgagee），其在付了有價值之報酬且未被告知之情況下，為無效，除非讓渡文件於讓渡之後的三個月內在美國專利局登錄，或者比後續購買者（承受抵押者）先在美國專利局登錄。再者，專利讓渡文件之效力及於延續案、分割案，但不及於取代申請案及部份延續案。亦即，如果母案申請時發明人已將權利讓渡給一公司，則其延續案、分割案亦一併讓渡該公司且不必重新簽署讓渡文件，但取代申請案，部份延續案則發明人需重新簽署讓渡文件。

37 CFR 3.54 Effect of recording.

The recording of a document pursuant to Section　3.11 is not a determination by the Office of the validity of the document or the effect that document has on the title to an application, a patent, or a registration. When necessary, the Office will determine what effect a document has, including whether a party has the authority to take an action in a matter pending before the Office.

MPEP 324

The owner or assignee of a patent property can take action in a patent application or patent proceeding in numerous instances. The owner or assignee can sign a reply to an Office action (37 CFR 1.33(b)(3) and (4)), a request for a continued prosecution application under 37 CFR 1.53(d) (MPEP § 201.06(d)), a terminal disclaimer (MPEP § 1490), Fee(s) Transmittal (PTOL-85B) (MPEP § 1306), or a request for status of an application (MPEP § 102). The owner or assignee can file an application under 37 CFR 1.47(b) (MPEP § 409.03(b)), appoint its own registered patent **>practitioner< to prosecute an application (37 CFR 1.32 and MPEP § 402.07), grant a power to inspect an application (MPEP § 104), and acquiesce to express abandonment of an application (MPEP § 711.01). The owner or assignee consents to the filing of a reissue application (MPEP § 1410.01), and to the correction of inventorship (MPEP § 201.03 or § 1481).

35 USC 261

＊＊＊＊＊＊＊＊＊＊＊

An assignment, grant, or conveyance shall be void as against any subsequent purchaser or mortgagee for a valuable consideration, without notice, unless it is recorded in the Patent and

Trademark Office within three months from its date or prior to the date of such subsequent purchase or mortgage

MPEP 306 Assignment of Division, Continuation, Substitute, and Continuation-in-Part in Relation to Parent Application [R-3]

In the case of a division or continuation application, a prior assignment recorded against the original application is applied >(effective)< to the division or continuation application because the assignment recorded against the original application gives the assignee rights to the subject matter common to both applications. >Although the assignment recorded against an original application is applied to the division or continuation application, the Office's assignment records will only reflect an assignment of a division or continuation application (or any other application) if a request for recordation in compliance with 37 CFR 3.28, accompanied by the required fee (37 CFR 3.41), is filed.<

In the case of a substitute or continuation-in-part application, a prior assignment of the original application is not applied >(effective)< to the substitute or continuation-in-part application because the assignment recorded against the original application gives the assignee rights to only the subject matter common to both applications. Substitute or continuation-in-part applications require >the recordation of< a new assignment if they are to be issued to an assignee. >See 37 CFR 3.81.<

實例 1

A 於 1970 年 4 月 1 日將其專利申請權讓渡給 X 公司並於 1970 年 11 月 1 日讓渡書於美國專利局登錄。A 又於 1970 年 10 月 15 日將此之專利申請權讓渡給 Y 公司，Y 公司並於同日於美國專利局登錄此讓渡書。在與 A 協商讓渡事宜之過程中，Y 公司知曉 A 已將其申請權讓渡給 X 公司且亦知道 X 公司在 1970 年 10 月 15 日時並未將其讓渡書在美國專利局登錄。請問 X 公司或 Y 公司擁有該專利申請權？

X 公司，雖然 X 公司未在讓渡之後的三個月內向美國專利局登錄，且在 1970 年 10 月 15 日之時亦尚未向美國專利局登錄，但 Y 公司在與 A 協商專利讓渡事宜時已獲知 A 已將其專利申請權讓渡 X 公司，故專利申請權仍為 X 公司擁有。

實例 2

A 於 1970 年 2 月 1 日將其專利權讓渡 X 公司，X 公司於 1970 年 8 月 1 日於美國專利局登錄。於 1970 年 3 月 1 日 A 又將專利權讓渡給其兒子作為生日禮物。其兒子於 1970 年 4 月 1 日將讓渡書於美國專利局登錄。請問 X 公司或 A 之兒子擁有此專利權？

X 公司，雖然 X 公司未在讓渡之後的三個月內或在後來受讓者之前去登錄，但後來之受讓人並非基於有價值之考量取得專利權，故仍為 X 公司擁有此專利權。

實例 3

A 於 1977 年 6 月 1 日將其專利權讓渡給其兒子作為畢業禮物。其兒子於 1977 年 6 月 20 日將讓渡書於美國專利局登錄。之後，A 又將其專利權於 1977 年 8 月 1 日讓渡給 X 公司。請問誰擁有此專利權？

A 之兒子，有價值之考量只針對後續之專利受讓人，A 將其專利讓渡給他兒子雖非有價值之考量，但其兒子在 X 公司受讓專利前已先去登記，故專利權仍為其兒子擁有。

實例 4

發明人 A 將其發明專利（具有橢圓截面之高爾夫球桿）讓渡給其自己擁有的 X 公司。之後，A 又申請了該申請案之部份延續案（CIP）。該 CIP 案乃是有關於一具有橢圓截面之高爾夫球桿，在其球桿上具有凹陷部以使打出之球可飛得更遠。而 A 最近發現 Y 公司正在製造及銷售其 CIP 案所申請之高爾夫球桿。A 想等此 CIP 案核准後立即以 X 公司之名義向 Y 提出告訴。請問 A 是否需再簽署讓渡書將其 CIP 專利申請案之權利讓與 X 公司？

A 必須重新簽署一讓渡書將其權利讓與 X 公司。

§3.4　專利讓渡文件登錄之注意事項

美國專利局只接受及登錄讓渡書之影本。可送整份之讓渡書或相關部份（extract）之影本（可為紙本或傳真本）。讓渡書影本只能印單面，且大小需為 21.0×29.7cm

（A4 尺寸）或 21.0×27.9cm。紙張之四側需有 2.5 公分之留白，紙張需為白色，堅韌，不反光及耐久的。亦可以電子檔呈送讓渡書其格式為 TIFF （Tagged Image File Format）。讓渡書之內容需為英文，如非英文則需附英文翻譯並由翻譯者簽名。在讓渡書中並需註明要讓渡之專利號碼或專利申請號碼，如果在簽署讓渡時尚不知申請案號，例如申請時同時或申請前簽署讓渡書，則需註明發明人名稱及發明之名稱。

送入美國專利局的讓書需同時附上一首頁（cover sheet）註明以下事項：

(A) 讓渡人之簽名

(B) 受讓人之姓名及地址

(C) 要被登錄之權利之敘述

(D) 要讓渡之專利或專利申請案之專利號碼或申請案號碼

(E) 讓渡登錄之聯絡人姓名及住址

(F) 讓渡文件簽署之日期

(G) 呈送文件之人之簽名

如果有許多件專利或專利申請案要由相同之所有權人讓渡給相同之受讓人，則可包括在同一讓渡書中辦理登錄，但讓渡之規費則仍是依照單一讓渡之規費（USD 40）乘以總件數計算。

讓渡書如果是有條件的讓渡書（conditional assignment）（例如支付多少金額作讓渡）仍可呈送至美國專利局作登錄。美國專利局僅作登錄之動作而不決定所載之條件是否有履行。

此外，當申請人向美國專利局申請專利申請案之認證本（certified copy）作為優先權文件時，讓渡文件不會包括在認證本之中，申請人需另外申請讓渡書之認證本。

MPEP 302.01 Assignment Document Must Be ** Copy for Recording [R-3]

The United States Patent and Trademark Office will accept and record only **>a< copy of an original assignment or other document. See MPEP § 317. >The document submitted for recordation will not be returned to the submitter. If the copy submitted for recordation is illegible, the recorded document will be illegible. Accordingly, applicants and patent owners should ensure that only a legible copy is submitted for recordation.<

MPEP 302.02 Translation of Assignment Document

37 CFR 3.26 English language requirement.

The Office will accept and record non-English language documents only if accompanied by an English translation signed by the individual making the translation.

The assignment document, if not in the English language, will not be recorded unless accompanied by an English translation signed by the translator.

MPEP 302.03 Identifying Patent or Application [R-3]

The patent or patent application to which an assignment relates must be identified by patent number or application number unless the assignment is executed concurrently with or subsequent to the execution of the application but before the application is filed. Then, the application must be identified by ** the name(s) of the inventors, and the title of the invention. If an assignment of a provisional application is executed before the provisional application is filed, it must identify the provisional application by name(s) of the inventors and the title of the invention.

＊＊＊＊＊＊＊＊＊＊＊＊

MPEP 302.05 Address of Assignee

The address of the assignee may be recited in the assignment document and must be given in the required cover sheet. See MPEP § 302.07.

MPEP 302.06 Fee for Recording [R-3]

The recording fee set forth in 37 CFR 1.21(h) is charged for each patent application and patent identified in the required cover sheet except as provided in 37 CFR 3.41(b).

MPEP 302.07 Assignment Document Must Be Accompanied by a Cover Sheet [R-5]

Each assignment document submitted to the Office for recording must be accompanied by a cover sheet as required by 37 CFR 3.28. The cover sheet for patents or patent applications must contain:

(A) The name of the party conveying the interest;

(B) The name and address of the party receiving the interest;

(C) A description of the interest conveyed or transaction to be recorded;

(D) Each patent application number or patent number against which the document is to be recorded, or an indication that the document is filed together with a patent application;

(E) The name and address of the party to whom correspondence concerning the request to record the document should be mailed;

(F) The date the document was executed; and

(G) The signature of the party submitting the document.

* * * * * * * * * * * *

37 CFR 3.56 Conditional assignments.

Assignments which are made conditional on the performance of certain acts or events, such as the payment of money or other condition subsequent, if recorded in the Office, are regarded as absolute assignments for Office purposes until canceled with the written consent of all parties or by the decree of a court of competent jurisdiction. The Office does not determine whether such conditions have been fulfilled.

MPEP 303 Assignment Documents Not Endorsed on Pending Applications

Certified copies of patent applications as filed do not include an indication of assignment documents. Applicants desiring an indication of assignment documents of record should request separately certified copies of assignment documents and submit the fees required by 37 CFR 1.19.

§3.5　專利發證給受讓人

美國專利局通常發證給發明人（申請人），如果要發證給受讓人，則需於繳領證費之前通知美國專利局（通常是在繳領證之 transmittal letter 上註明要發證給受讓人），同時要指出讓渡書已在美國專利局登錄。如果讓渡書尚未登錄，則要在送入美國專利局之通知書上註明已送入讓渡書但尚未登錄。如果在繳領證費之後才通知美國專利局要發證給受讓人，則需同時請求修正證書（certificate of correction）並繳規費，但美國專利局仍不會將受讓人之名字印在專利證書上，但會發下修正之證書於其上註明受讓人之名字。

37 CFR 3.81 Issue of patent to assignee.

(a) With payment of the issue fee: An application may issue in the name of the assignee consistent with the application's assignment where a request for such issuance is submitted with payment of the issue fee, provided the assignment has been previously recorded in the Office. If the assignment has not been previously recorded, the request must state that the document has been filed for recordation as set forth in § 3.11.

(b) After payment of the issue fee: Any request for issuance of an application in the name of the assignee submitted after the date of payment of the issue fee, and any request for a patent to be corrected to state the name of the assignee, must state that the assignment was submitted for recordation as set forth in § 3.11 before issuance of the patent, and must include a request for a certificate of correction under § 1.323 of this chapter (accompanied by the fee set forth in § 1.20(a)) and the processing fee set forth in § 1.17 (i) of this chapter.

(c) Partial assignees. (1) If one or more assignee, together with one or more inventor, holds the entire right, title, and interest in the application, the patent may issue in the names of the assignee and the inventor.(2) If multiple assignees hold the entire right, title, and interest to the exclusion of all the inventors, the patent may issue in the names of the multiple assignees.<

MPEP 307 Issue to Assignee [R-3]

Normally, for a patent to issue to an assignee, a request for issuance of the application in the name* of the assignee* must be filed in the United States Patent and Trademark Office (Office) at a date not later than the day on which the issue fee is paid. **>Such a request must indicate that the assignment has been previously recorded in the Office. If the assignment has not been previously recorded in the Office, the request must state that the document has been filed for recordation as set forth in 37 CFR 3.11. See 37 CFR 3.81(a).

If a request for issuance to an assignee pursuant to 37 CFR 3.81(b) is submitted after the day on which the issue fee is paid, the request under 37 CFR 3.81(b) must include a request for a certificate of correction under 37 CFR 1.323 (accompanied by the fee set forth in 37 CFR 1.20(a)) and the processing fee set forth in 37 CFR 1.17(i). The request under 37 CFR 3.81(b) must state that the assignment was submitted for recordation as set forth in 37 CFR 3.11 before issuance of the patent. The Office will issue a certificate of correction to reflect that the patent issued to the assignee provided the requirements of 37 CFR 3.81(b) and 37 CFR 1.323 are complied with.<

＊＊＊＊＊＊＊＊＊＊＊

§3.6　登錄之讓渡文件之取得，閱覽及影印

有關專利、已公告之專利申請案的相關讓渡紀錄是開放給大眾檢閱的。已發證之專利及已公告之專利申請案的讓渡紀錄可直接自美國專利局之網站上查得，而如果要親自去閱覽或影印讓渡文件則需提出申請。在發證之後的專利權讓渡或申請人若不願將受讓人名稱印在專利證書上時，專利證書不會顯示最新的所有權人。

　　另外需注意的是，如果將數件專利申請案或至商標申請案或註冊之商標併在一讓與文件中作讓渡，只要有一件已取得專利或已公告則此整份文件均會公開供大眾檢閱。

　　放棄的專利申請案如果已公告，則其讓渡文件亦是公開的，分割案、延續案、部份延續案如果已發證則其母案的讓渡文件亦是公開的。再發證的申請案的讓渡文件亦是公開的。

MPEP 301.01 Accessibility of Assignment Records [R-3]

Assignment documents relating to patents, published patent applications, registrations of trademarks, and applications for registration of trademarks are open to public inspection. >Records related to assignments of patents, and patent applications that have been published as patent application publications are available on the USPTO Internet web site. To view the recorded assignment document itself, members of the public must place an order pursuant to 37 CFR 1.12(d).<

The Office will not open only certain parts of an assignment document to public inspection. If such a document contains two or more items, any one of which, if alone, would be open to such inspection, then the entire document will be open. Thus, if a document covers either a trademark or a patent in addition to one or more patent applications, it will be available to the public ab initio; and if it covers a number of patent applications, it will be so available as soon as any one of them is published or patented. Documents relating only to one or more pending applications for patent which have not been published under 35 U.S.C. 122(b) will not be open to public inspection.

Copies of assignment records relating to pending or abandoned patent applications **>which are open to the public pursuant to 37 CFR 1.11 or for which copies or access may be supplied pursuant to 37 CFR 1.14 are available to the public. For pending or abandoned applications which are not open to the public pursuant to 37 CFR 1.11 or for which copies or access may not be supplied pursuant to 37 CFR 1.14,< information related thereto *>is only< obtainable upon a showing of written authority from the applicant or applicant's assignee or from the attorney or agent of either, or upon a showing that the person seeking such information is a bona fide prospective or actual purchaser, mortgagee, or licensee of such application.

If the application on which a patent was granted is a division *>,< continuation>, or continuation-in-part< of an earlier application, the assignment records of that earlier application will be open to public inspection **>because copies or access may be supplied to the earlier application pursuant to 37 CFR 1.14<.

Assignment records relating to reissue applications are open to public inspection >since reissue applications are open to public inspection pursuant to 37 CFR 1.11(b).<

§3.7　其他不影響專利申請人、專利權人之權利之文件的登錄

除了讓渡書及 Executive Order 9424 要求被登錄之文件以外，會影響到一專利或專利申請案之權利之文件亦可被登錄在美國專利商標局之讓渡部門。其他不會影響權利之文件則依局長之選擇（決定），亦可在美國專利商標局登錄。這些文件例如授權協議書（license agreements）及轉讓秘密利益（security interest）之協議書等。這些文件登錄之目的乃是為了公共利益，讓第三者知道一專利或一專利申請案的所有權的相關事由。如果這類文件申請登錄被退回，而請求人認為應該登錄時，可提出請願，請求美國專利商標局予以登錄。

另外，受讓人如為一企業（公司或法人），當公司變更名稱或被合併時，官方所發下的更名（合併）證書都可在美國專利局登錄以顯示權利之轉移。

MPEP 313 Recording of Licenses, Security Interests, and Other Documents Other Than Assignments [R-3]

In addition to assignments and documents required to be recorded by Executive Order 9424, documents affecting title to a patent or application will be recorded in the Assignment Division of the United States Patent and Trademark Office (Office). Other documents not affecting title may be recorded at the discretion of the *>Director<. 37 CFR 3.11(a).

Thus, some documents which relate to patents or applications will be recorded, although they do not constitute a transfer or change of title. Typical of these documents which are accepted for recording are license agreements and agreements which convey a security interest. Such documents are recorded in the public interest in order to give third parties notification of equitable interests or other matters relevant to the ownership of a patent or application.

Any document returned unrecorded, which the sender nevertheless believes represents an unusual case which justifies recordation, may be submitted to the Office of Petitions with a petition under 37 CFR 1.181 requesting recordation of the document.

The recordation of a document is not a determination of the effect of the document on the chain of title. The determination of what, if any, effect a document has on title will be made by the Office at such times as ownership must be established to permit action to be taken by the Office in connection with a patent or an application. See MPEP § 324.

MPEP 314 Certificates of Change of Name or of Merger

Certificates issued by appropriate authorities showing a change of name of a business or a merger of businesses are recordable. Although a mere change of name does not constitute a change in legal entity, it is properly a link in the chain of title. Documents of merger are also proper links in the chain of title. They may represent a change of entity as well as a change of name.

§3.8　專利讓渡文件之登錄程序

專利之讓渡或相關文件及其首頁將被送至美國專利局之讓渡部門作形式之審查以決定所送入之文件是否是可登錄的。如果不可登錄，則會將其退回給申請人並附上不可登錄之理由，如果申請人不同意則可提出請願請求美國專利局作登錄，如果是可登錄的文件，則會再審查是否符合 37CFR 3.21 之註明要件（identification requirement），例如是否讓渡書有註明專利案號，申請號等，如果符合就會登錄並有登錄號碼（reel and frame number），並將讓渡文件退回申請人。登錄之日期乃是讓渡文件符合登錄要件之日期。

37CFR §3.21 Identification of patents and patent applications.

An assignment relating to a patent must identify the patent by the patent number. An assignment relating to a national patent application must identify the national patent application by the application number (consisting of the series code and the serial number, e.g., 07/123,456). An assignment relating to an international patent application which designates the United States of America must identify the international application by the international application number (e.g., PCT/US90/01234). If an assignment of a patent application filed under §1.53(b) is executed concurrently with, or subsequent to, the execution of the patent application, but before the patent application is filed, it must identify the patent application by the name of each inventor and the title of the invention so that there can be no mistake as to the patent application intended. If an assignment of a provisional application under §1.53(c) is executed before the provisional application is filed, it must identify the provisional application by the name of each inventor and the title of the invention so that there can be no mistake as to the provisional application intended.

MPEP 317 Handling of Documents in the Assignment Division [R-3]

All documents and cover sheets submitted for recording are examined for formal requirements in the Assignment Division in order to separate documents which are recordable from those which are not recordable.

Documents and cover sheets that are considered not to be recordable are returned to the sender by the Assignment Division with an explanation. If the sender disagrees or believes that the document represents an unusual case which justifies recordation, the sender may present the question to the *>Director< by way of petition under 37 CFR 1.181, filed with the Office of Petitions.

After an assignment and cover sheet have been recorded, they will be returned to the name and address indicated on the cover sheet to receive correspondence, showing the reel and frame number.

37 CFR 3.51 Recording date.

The date of recording of a document is the date the document meeting the requirements for recording set forth in this part is filed in the Office. A document which does not comply with the identification requirements of § 3.21 will not be recorded. Documents not meeting the other requirements for recording, for example, a document submitted without a completed cover sheet or without the required fee, will be returned for correction to the sender where a correspondence address is available.

§3.9　受讓人建立所有權以進行相關程序

受讓人要在一專利之申請及審查程序中有權力進行相關程序，需先建立所有權（establishment of ownership）。受讓人需呈送一聲明書（statement）註明為受讓人並附上書面證據證明所有權由最初的所有權人經一系列權利轉讓（chain of title）給受讓人。此書面證據可為註明登錄號碼（reel and frame number）之讓渡書影本。此聲明書需受讓人授權的人簽名或由受讓人委託之人簽名。此外，受讓人應是該專利或專利申請案之完全權利受讓人（assignee of the entire right）而不是部份受讓人（partial assignee），如果是由部份受讓人提出此聲明書，則所有部份受讓人均需提出此聲明書，否則美國專利局可能會拒絕此建立所有權之聲明書。

　　並非對所有的申請、審查程序要主動進行程序時，受讓人均需提出建立所有權之聲明書（statement under 37 CFR 3.73(b)）。

　　以下之程序需提出 37 CFR 3.73(b)之建立所有權之聲明書：

(1)　提出 37 CFR 1.53(d)之 CPA 申請案（原未送 37 CFR 373(b)之聲明書）。

(2)　查詢案件之狀態（status），或給予別人權力檢視專利申請案。

(3)　提出主動放棄書。

(4)　指定自己的專利代理人（專利律師）代理案件之申請審查程序。

(5)　簽署尾端棄權書（terminal disclaimer）。

(6)　同意申請再發證（reissue）申請案。

(7)　同意修正發明人。

(8)　找不到發明人或發明人拒絕簽署宣誓書時簽署宣誓書。

(9)　簽署繳費文件。

(10) 簽署對審定書（Office Action）之答辯書。

　　以下程序不必提出 37CFR 3.73(b)之建立所有權之聲明書。

(1)　提出 37 CFR 1.53(d)之 CPA 申請案（已送 37CFR 3.73(b)）之聲明書且所有權未改變。

(2)　簽署小實體聲明。

(3)　簽署兩發明具有共同所有權人之聲明。

(4)　簽署 NASA 或 DOE 財產權聲明。

(5)　發明人找不到時簽署 37CFR 1.131 之宣誓書

(6)　簽署 37CFR 1.8 郵寄或傳送申請書（certificate of mailing or transmission）

(7)　提出再審查（re-examination）申請案。

---

37 CFR 3.73 Establishing right of assignee to take action.

(a) The inventor is presumed to be the owner of a patent application, and any patent that may issue therefrom, unless there is an assignment. The original applicant is presumed to be the owner of a trademark application or registration, unless there is an assignment.

(b) (1) In order to request or take action in a patent or trademark matter, the assignee must establish its ownership of the patent or trademark property of paragraph (a) of this section to the satisfaction of the Director. The establishment of ownership by the assignee may be combined with the paper that requests or takes the action. Ownership is established by submitting to the Office a signed statement identifying the assignee, accompanied by either: **>

(i) Documentary evidence of a chain of title from the original owner to the assignee (e.g., copy of an executed assignment). For trademark matters only, the documents submitted to establish ownership may be required to be recorded pursuant to §3.11 in the assignment records of the Office as a condition to permitting the assignee to take action in a matter pending before the Office. For patent matters only, the submission of the documentary evidence must be accompanied by a statement affirming that the documentary evidence of the chain of title from the original owner to the assignee was or concurrently is being submitted for recordation pursuant to §3.11; or<

(ii) A statement specifying where documentary evidence of a chain of title from the original owner to the assignee is recorded in the assignment records of the Office (e.g., reel and frame number).

(2) The submission establishing ownership must show that the person signing the submission is a person authorized to act on behalf of the assignee by:

(i) Including a statement that the person signing the submission is authorized to act on behalf of the assignee; or

(ii) Being signed by a person having apparent authority to sign on behalf of the assignee, e.g., an officer of the assignee.

(c) For patent matters only:

(1) Establishment of ownership by the assignee must be submitted prior to, or at the same time as, the paper requesting or taking action is submitted.

(2) If the submission under this section is by an assignee of less than the entire right, title and interest, such assignee must indicate the extent (by percentage) of its ownership interest, or the Office may refuse to accept the submission as an establishment of ownership.

＊＊＊＊＊＊＊＊＊＊＊

MPEP 324

VI.WHEN OWNERSHIP MUST BE ESTABLISHED

Examples of situations where ownership must be established under 37 CFR 3.73(b) are when the assignee: signs a request for a continued prosecution application under 37 CFR 1.53(d), unless papers establishing ownership under 37 CFR 3.73(b) were filed in the prior application and ownership has not changed (MPEP § 201.06(d)); signs a request for status of an application or gives a power to inspect an application (MPEP § 102 and § 104); acquiesces to express abandonment of an application (MPEP § 711.01); appoints its own registered attorney or agent to prosecute an application (37 CFR 3.71 and MPEP § 402.07); signs a terminal disclaimer (MPEP § 1490); consents to the filing of a reissue application (MPEP § 1410.01); consents to the correction of inventorship (MPEP § 201.03 or § 1481); files an application under 37 CFR 1.47(b) (MPEP § 409.03(b)) or 37 CFR 1.425; signs a Fee(s) Transmittal (PTOL-85B) (MPEP § 1306); or signs a reply to an Office action.

VII. WHEN OWNERSHIP NEED NOT BE ESTABLISHED

Examples of situations where ownership need not be established under 37 CFR 3.73(b) are when the assignee: signs a request for a continued prosecution application under 37 CFR 1.53(d), where papers establishing ownership under 37 CFR 3.73(b) were filed in the prior application and ownership has not changed (MPEP § 201.06(d)); signs a small entity statement (MPEP § 509.03); signs a statement of common ownership of two inventions (MPEP § 706.02(l)(2)); signs a NASA or DOE property rights statement (MPEP § 151); signs an affidavit under 37 CFR 1.131 where the inventor is unavailable (MPEP § 715.04); signs a certificate under 37 CFR 1.8 (MPEP § 512); or files a request for reexamination of a patent under 37 CFR 1.510 (MPEP § 2210).

實例

發明人 A 及 B 均受雇於 X 公司，C 為 X 公司之專利律師，對 A 及 B 之發明準備專利說明書並提出專利申請。發明人 A 簽署了宣誓書及委任書給 C 並簽署讓渡書將權利讓渡給 X 公司。但是發明人 B 則拒絕簽署宣誓書，委任書及讓渡書。B 與 X 公司之雇傭契約並無條文規定 B 須將其發明讓渡給 X 公司。之後，C 仍提出此一專利申請以 A 及 B 為發明人但 B 未簽署文件。B 之後離職並生產該發明之產品且要求美國專利局告知該專利申請案之審查狀況。請問 X 公司是否可向美國專利局要求排除發明人 B 不讓其知悉該專利申請案之狀態？

由於發明人 B 並未將其權利讓與 X 公司，故 X 公司並非具有完全權利之受讓人，故 X 公司不能要求排除發明人 B 不讓其知悉該專利申請案之狀態。

MPEP 106

The assignee of record of entire interest in an application may intervene in the prosecution of the application, appointing an attorney of his or her own choice. Such intervention, however, does not exclude the applicant from access to the application to see that it is being prosented properly, unless the assignee makes specific request to that effect.

第 4 章　代理人之委任及解任

目　錄

§4.1 美國專利執業人員資格之取得與可代理的範圍

美國專利執業人員（patent practitioner）分為二種：專利律師（patent attorney）及專利代理人（patent agent）。要取得專利代理人（patent agent）之資格需為美國公民或在美國有合法居住權（即具有永久居留權（permanent residence），即綠卡)之外國人，其先向美國專利局提出相關資料及申請，證明符合下列條件：

(i)　具有良好之品德及聲譽，

(ii)　具有法律、科學及技術之背景而能提供申請人有價值之服務，

(iii)　具有能力協助申請人進行美國專利之申請及審查的相關程序。

並通過美國專利局所舉辦之專利代理人考試，且繳交註冊費後才能登錄。但如果擔任美國專利局審查委員四年以上，則可不必通過專利代理人之資格考試。如取得專利代理人資格且同時具有律師資格者即稱為專利律師。專利代理人、專利律師如果為外國人，則只有當其仍繼續合法居住在美國時，其登錄始有效。

如為專利代理人則其可代理申請人進行所有美國專利商標局之專利程序，例如申請、答辯、繳交領證費和年費、讓渡登錄、再審查、抵觸程序、訴願程序等。但不能代理申請人進行專利讓渡書、授權文件之準備，提供專利訴訟相關意見或代理專利訴訟，亦不能代理商標案件之申請及審查相關程序。如為專利律師則可代理申請人進行所有美國專利商標局之專利、商標業務，亦可提供專利商標訴訟相關意見及代理專利商標訴訟等業務。

37CFR§ 10.6 Registration of attorneys and agents.

(a) Attorneys. Any citizen of the United States who is an attorney and who fulfills the requirements of this part may be registered as a patent attorney to practice before the Office. When appropriate, any alien who is an attorney, who lawfully resides in the United States, and who fulfills the requirements of this part may be registered as a patent attorney to practice before the Office, provided: Registration is not inconsistent with the terms upon which the alien was admitted to, and resides in, the United States and further provided: The alien may remain registered only (1) if the alien continues to lawfully reside in the United States and registration does not become inconsistent with the terms upon which the alien continues to lawfully reside in the United States, or (2) if the alien ceases to reside in the United States, the alien is qualified to be registered under paragraph (c) of this section. See also § 10.9(b).

(b) Agents. Any citizen of the United States who is not an attorney and who fulfills the requirements of this part may be registered as a patent agent to practice before the Office. When appropriate, any alien who is not an attorney, who lawfully resides in the United States, and who fulfills the requirements of this part may be registered as a patent agent to practice before the Office, provided: Registration is not inconsistent with the terms upon which the alien was admitted to, and resides in, the United States, and further provided: The alien may remain registered only (1) if the alien continues to lawfully reside in the United States and registration does not become inconsistent with the terms upon which the alien continues to lawfully reside in the United States or (2) if the alien ceases to reside in the United States, the alien is qualified to be registered under paragraph (c) of this section. See also § 10.9(b).

＊＊＊＊＊＊＊＊＊＊＊

37CFR § 10.7 Requirements for registration.

(a) No individual will be registered to practice before the Office unless he or she shall:

(1) Apply to the Commissioner1 in writing on a form supplied by the Director and furnish all requested information and material and

(2) Establish to the satisfaction of the Director that he or she is:

(i) Of good moral character and repute;

(ii) Possessed of the legal, scientific, and technical qualifications necessary to enable him or her to render applicants for patents valuable service; and

(iii) Is otherwise competent to advise and assist applicants for patents in the presentation and prosecution of their applications before the Office.

(b) In order that the Director may determine whether an in-dividual seeking to have his or her name placed upon the register has the qualifications specified in paragraph (a) of this section, satisfactory proof of good moral character and repute and of sufficient basic training in scientific and technical matters must be submitted to the Director. Except as provided in this paragraph, each applicant for registration must take and pass an examination which is held from time to time. Each application for admission to take the examination for registration must be accompanied by the fee set forth in 37 CFR § 1.21(a)(1) of this subchapter. The taking of an examination may be waived in the case of any individual who has actively served for at least four years in the patent examining corps of the Office. The examination will not be administered as a mere academic exercise.

實例 1

　　A 為一電機工程師，其與 X 公司（行動電話製造公司）訂有契約替 X 公司裝設一緊急發電設備。A 在與 X 公司合作期間自行研發出一開關控制機構可用來改善 X 公司所生產之行動電話裝置。A 與 X 公司之契約中並無任何專利條款。X 公司發現 A 之發明對其產品相當重要就延聘一專利代理人（patent agent）B 請其處理 A 之專利讓渡或授權給 X 公司及申請事宜。請問專利代理人 B 可替 X 公司辦理上述事項嗎？

　　B 不能替 X 公司辦理（包括與 A 協商），準備讓渡文件，授權文件及讓渡（授權）條件事宜，只能當 A 欲將其專利申請讓渡給 X 時，代理專利說明書之撰寫，提出專利申請及讓渡書之登錄申請及代理答辯事宜。

§4.2　委任及解任之意義及應簽署之文件

　　申請人要委任一專利律師或專利代理人時需簽署委任書（power of attorney）以讓專利執業者有權力替其進行美國專利局之申請、審查相關程序。委任書需為書面文件且需指名一人或一人以上的專利執業者，而且需賦予所指名之專利執業者權力替其進行相關法律程序。此委任書需由發明人簽名，如有讓渡時需完全之受讓人簽名。被委任之專利執業者必需為登錄之專利代理人或專利律師且其登錄未被吊銷（suspend or disbared），否則委任將無效。當受委託之專利執業者對於受委託之專利案件，本人出席或簽署文件時，則其出席及簽名即代表其被授權替申請人進行相關法律程序。專利執業者在其送入美國專利局之所有文件上均需記載名字、登錄號碼並親自簽名。親自簽名乃保證以下事項：

(1) 所有執業者基於自己知識所作的陳述均是真實的，所有基於資訊所作的陳述，均相信是真實的，而且在作陳述時知道任何人明知或故意作偽證，藉由任何技巧隱藏事實，設計一事實，或製造、使用任何虛假的書面文件，將受到美國法律 18USC 1001 之處罰，而且違反此規定將危及專利申請案或文件之有效性或所得到之專利在執行時的有效性。

(2) 據專利執業者之知識、資訊及了解：

(i)　所呈送之文件並不是為了任何不適當的目的，例如干擾某人或引起不必要的延遲或增加美國專利局審查之成本；

(ii)　所作的請求或法律上的爭論乃是在目前的法律或新的法律上為正當者；

(iii)　所作的主張或事實的爭論乃是基於書面證據為正當者，或在合理的作調查或檢索後是具有證據支持的；

(iv)　基於事實之爭論的反駁乃是基於證據為正當者或者是因為缺乏資訊所致。

如果要解除對一專利執業者之委任，則需向美國專利局提出解任書（revocation of power of attorney）。對專利執業者之解任可在任何時間提出。解任可由申請人（發明人）提出，如果有讓渡則需由完全的受讓人提出，或由完全的所有權人提出。解任在美國專利局收到解任書時生效。美國專利局會通知被解任之專利執業者。解任通常同時會重新委任一新的專利執業者，此二程序可併在一解任/委任書中提出。申請人（發明人）提出之讓渡書本身並不會被美國專利局當作是一解除原申請人所委任之解任書。但受讓人在讓渡書登錄之後卻可解除原來之委任而重新委任新的代理人。

---

37 CFR 1.32 Power of attorney.

＊＊＊＊＊＊＊＊＊＊＊

(b) A power of attorney must:

(1) Be in writing;

(2) Name one or more representatives in compliance with (c) of this section;

(3) Give the representative power to act on behalf of the principal; and

(4) Be signed by the applicant for patent (§ 1.41(b)) or the assignee of the entire interest of the applicant.

＊＊＊＊＊＊＊＊＊＊＊

37 CFR 10.18 Signature and certificate for correspondence filed in the Patent and Trademark Office.

(a) For all documents filed in the Office in patent, trademark, and other non-patent matters, except for correspondence that is required to be signed by the applicant or party, each piece of correspondence filed by a practitioner in the Patent and Trademark Office must bear a signature by such practitioner complying with the provisions of § 1.4(d), §1.4(e), or § 2.193(c)(1) of this chapter.

(b) By presenting to the Office (whether by signing, filing, submitting, or later advocating) any paper, the party presenting such paper, whether a practitioner or non-practitioner, is certifying that-

(1) All statements made therein of the party's own knowledge are true, all statements made therein on information and belief are believed to be true, and all statements made therein are made with the knowledge that whoever, in any matter within the jurisdiction of the Patent and Trademark Office, knowingly and willfully falsifies, conceals, or covers up by any trick, scheme, or device a material fact, or makes any false, fictitious or fraudulent statements or representations, or makes or uses any false writing or document knowing the same to contain any false, fictitious or fraudulent statement or entry, shall be subject to the penalties set forth under 18 U.S.C. 1001, and that violations of this paragraph may jeopardize the validity of the application or document, or the validity or enforceability of any patent, trademark registration, or certificate resulting therefrom; and

(2) To the best of the party's knowledge, information and belief, formed after an inquiry reasonable under the circumstances, that -

(i) The paper is not being presented for any improper purpose, such as to harass someone or to cause unnecessary delay or needless increase in the cost of prosecution before the Office;

(ii) The claims and other legal contentions therein are warranted by existing law or by a nonfrivolous argument for the extension, modification, or reversal of existing law or the establishment of new law;

(iii) The allegations and other factual contentions have evidentiary support or, if specifically so identified, are likely to have evidentiary support after a reasonable opportunity for further investigation or discovery; and

(iv) The denials of factual contentions are warranted on the evidence, or if specifically so identified, are reasonably based on a lack of information or belief.

37CFR§1.36 Revocation of power of attorney; withdrawal of patent attorney or agent.

(a) A power of attorney, pursuant to §1.32(b), may be revoked at any stage in the proceedings of a case by an applicant for patent (§ 1.41 (b)) or an assignee of the entire interest of the applicant, or the owner of the entire interest of a patent. A power of attorney to the patent practitioners associated with a Customer Number will be treated as a request to revoke any powers of attorney previously given. Fewer than all of the applicants (or fewer than all of the assignees of the entire interest of the applicant or, in a reexamination proceeding, fewer than all the owners of the entire interest of a patent) may revoke the power of attorney only upon a showing of sufficient cause, and payment of the petition fee set forth in § 1.17 (f). A patent practitioner will be notified of the revocation of the power of attorney. Where power of attorney is given to the patent practitioners associated with a Customer Number (§ 1.32(c)(2)), the practitioners so appointed will also be notified of the revocation of the power of attorney when the power of attorney to all of the practitioners associated with the Customer Number is revoked. The notice of revocation will be mailed to the correspondence address for the application (§ 1.33) in effect before the revocation. An assignment will not of itself operate as a revocation of a power previously given, but the assignee of the entire interest of the applicant may revoke previous powers of attorney and give another power of attorney of the assignee's own selection as provided in § 1.32(b).

＊＊＊＊＊＊＊＊＊＊＊

§4.3　主動向美國專利局撤回代理

　　專利代理人或專利律師要主動撤回代理時，需附上簡單的理由，例如當客戶不願付款或其作為使得專利執業者要代表此客戶有明顯的困難。美國專利局將決定是否核准此代理之撤回。專利執業者在提出撤回代理之前需先採取合理的步驟防止對於客戶可預見的權利之傷害，且需注意不能洩漏客戶的秘密資訊。

　　代理撤回之生效日是美國專利局核准之日期，而非提出撤回代理之日期，美國專利局通常會要求從代理人撤回代理之核准日至下次回覆審定書（通知書）之截止日（延期後之日期）之間有至少 30 天，以避免造成客戶權利之損失，並讓客戶有時間去轉換代理人。

---

37 CFR　1.36 withdrawal of patent attorney or agent.

＊＊＊＊＊＊＊＊＊＊＊

　　(b) A registered patent attorney or patent agent who has been given a power of attorney pursuant to §1.32(b) may withdraw as attorney or agent of record upon application to and approval by the Director. The applicant or patent owner will be notified of the withdrawal of the registered patent attorney or patent agent. Where power of attorney is given to the patent practitioners associated with a Customer Number, a request to delete all of the patent practitioners associated with the Customer Number may not be granted if an applicant has given power of attorney to the patent practitioners associated with the Customer Number in an application that has an Office action to which a reply is due, but insufficient time remains for the applicant to file a reply. See §41.5 of this title for withdrawal during proceedings before the Board of Patent Appeals and Interferences.

MPEP 402.06 Attorney or Agent Withdraws [R-5]

See 37 CFR 1.36(a) in MPEP § 402.05 for revocation. See 37 CFR 10.40 for information regarding permissive and mandatory withdrawal. When filing a request to withdraw as attorney or agent of record, the patent attorney or agent should briefly state the reason(s) for which he or she is withdrawing so that the Office can determine whether to grant the request. >Note that disciplinary rule, 37 CFR 10.40(a) provides that a "practitioner shall not withdraw from employment until the practitioner has taken reasonable steps to avoid foreseeable prejudice to the rights of the client." Among several scenarios addressed in 37 CFR 10.40(c), subsections (iv) and (vi) permit withdrawal when the client fails to compensate the practitioner, or when "other conduct on the part of the client has rendered the representation unreasonably difficult." When preparing a request for withdrawal for such reasons, the practitioner should also be mindful of 37 CFR 10.57(b)(2), which prohibits the use of a confidence or secret of a client to the disadvantage of a client. Where withdrawal is predicated upon such reasons, the practitioner, rather than divulging confidential or secret information about the client, should identify the reason(s) for requesting to withdraw as being based on "irreconcilable differences." An explanation of and the evidence supporting "irreconcilable differences" should be submitted as proprietary material in accordance with MPEP § 724.02 to ensure that the client's confidences are maintained.<

In the event that a notice of withdrawal is filed by the attorney or agent of record, the file will be forwarded to the **>appropriate official for decision on the request<. The withdrawal is effective when approved rather than when received.

To expedite the handling of requests for permission to withdraw as attorney or agent, under 37 CFR 1.36(b), Form PTO/SB/83 may be used. Because the Office does not recognize law firms, each attorney of record must sign the notice of withdrawal, or the notice of withdrawal must contain a clear indication of one attorney signing on behalf of himself or herself and another. A withdrawal of another attorney or agent of record, without also withdrawing the attorney or agent signing the request is a revocation, not a withdrawal.

The Director of the United States Patent and Trademark Office usually requires that there be at least 30 days between approval of withdrawal and the later of the expiration date of a time period for reply or the expiration date of the period which can be obtained by a petition and fee for extension of time under 37 CFR 1.136(a). This is so that the applicant will have sufficient time to obtain other representation or take other action. If a period has been set for reply and the period may be extended without a showing of cause pursuant to 37 CFR 1.136(a) by filing a petition for extension of time and fee, the practitioner will not be required to seek such extension of time for withdrawal to be approved. In such a situation, however, withdrawal will not be approved unless at least 30 days would remain between the date of approval and the last date on which such a petition for extension of time and fee could properly be filed.

37 CFR 10.40 (a)

A practitioner shall not withdraw from employment in a proceeding before the Office without permission from the Office. In any event, a practitioner shall not withdraw from employment until the practitioner has taken reasonable steps to avoid foreseeable prejudice to the rights of the client, including giving due notice to his or her client, allowing time for employment of another practitioner, delivering to the client all papers and property to which the client is entitled, and complying with applicable laws and rules. A practitioner who withdraws from employment shall refund promptly any part of a fee paid in advance that has not been earned.

實例 1

發明人 A 委託專利律師 X 替其準備專利說明書及提出美國專利申請。X 告知 A 其費用為 USD 2800。此專利於 1990 年 8 月 9 日提出申請，發明人 A 於 1990 年 12 月 31 日支付 X USD2200。X 於 1990 年 1 月 30 日告知 A 尚餘 USD 600 未付。X 於 1991 年 2 月 25 日收到一 Office Action 其日期為 1991 年 2 月 21 日，核駁所有之請求項。請問 X 可否以書面通知 A 告知如果 A 未支付餘款 USD 600 則其將放棄本案不提出答辯？

即使 A 未支付款項，X 仍不能主動放棄申請案。正確的作法是通知 A 已發下 OA 並向美國專利局撤回代理，待美國專利局核准撤回後才可停止代理行為。

§4.4　由部份的發明人或部份的所有權人提出委任及解任

要委任或解任專利代理人時，需通常需所有的申請人（發明人）或所有權人（受讓人）簽名。如果由部份的發明人或部份的所有權人簽名，通常美國專利局不會接受，除非同時提出請願，附上規費並敘述特殊的情況，需要由部份發明人（所有權

人）簽名。一旦美國專利核准了此請願，將會造成有一個以上之代理人或申請人或
所有權人一起進行審查程序。此時美國專利局之文件乃將只寄至其中之一。例如，
由部份發明人簽署委任之結果變成由一代理人及一發明人進行審查程序則美國專利
局之文件會寄至該代理人處。而當結果變成由二個代理人進行審查程序時則美國專
利局之文件會寄至申請書上第 1 個被指名之代理人之處。但是上述二種情況，答辯
書或對美國專利局之信函均需由雙方簽名才可。

MPEP 402.10 Appointment/Revocation by Less Than All Applicants or Owners [R-5]

Papers giving or revoking a power of attorney in an application generally require signature by all the applicants or owners of the application. Papers revoking a power of attorney in an application (or giving a power of attorney) will not be accepted by the Office when signed by less than all of the applicants or owners of the application unless they are accompanied by a petition under 37 CFR 1.36(a) and fee under 37 CFR 1.17(*>f<) with a showing of sufficient cause (if revocation), or a petition under 37 CFR 1.183 and fee under 37 CFR 1.17(f) (if appointment) demonstrating the extraordinary situation where justice requires waiver of the requirement of 37 CFR 1.32(b)(4) that the applicant, or the assignee of the entire interest of the applicant sign the power of attorney. The petition should be directed to the Office of Petitions. The acceptance of such papers by petition under 37 CFR 1.36(a) or 1.183 will result in more than one attorney, agent, applicant, or owner prosecuting the application at the same time. Therefore, each of these parties must sign all subsequent replies submitted to the Office. See In re Goldstein, 16 USPQ2d 1963 (Dep. Assist. Comm'r Pat. 1988). In an application filed under 37 CFR 1.47(a), an assignee of the entire interest of the available inventors (i.e., the applicant) who have signed the declaration may appoint or revoke a power of attorney without a petition under 37 CFR 1.36(a) or 1.183. See MPEP § 402.07. However, in applications accepted under 37 CFR 1.47, such a petition under 37 CFR 1.36(a) or 1.183 submitted by a previously nonsigning inventor who has now joined in the application will not be granted. See MPEP § 409.03(i). Upon accepting papers appointing and/or revoking a power of attorney that are signed by less than all of the applicants or owners, the Office will indicate to applicants who must sign subsequent replies. Dual correspondence will still not be permitted. Accordingly, when the acceptance of such papers results in an attorney or agent and at least one applicant or owner prosecuting the application, correspondence will be mailed to the attorney or agent. When the acceptance of such papers results in more than one attorney or agent prosecuting the application, the correspondence address will continue to be that of the attorney or agent first named in the application, unless all parties agree >to a different correspondence address<. Each attorney or agent signing subsequent papers must indicate whom he or she represents.

The following are examples of who must sign replies when there is more than one person responsible for prosecuting the application:

(A) If coinventor A has given a power of attorney >to a patent practitioner< and coinventor B has not, replies must be signed by the *>patent practitioner< of A and by coinventor B.

(B) If coinventors A and B have each appointed their own *>patent practitioner<, replies must be signed by both *>patent practitioners<.

實例

　　發明人 A，B 及 C 發明了一啤酒蓋打開器，並一起聘請專利代理人 X 替其準備專利說明書並提出申請並簽署委任書給 X 代理此專利案件。不久，發明人 A 在申請答辯上與代理人不合，決定解除代理人 X 之委任，重新委任其他代理人。而發明人 B 及 C 則覺得 X 可繼續代理不願更換代理人。請問在此情形下發明人 A 是否有任何方法可更換代理人？

　　發明人 A 可依據 37 CFR 1.183 之規定敘明充分理由為何要更換代理人向美國專利局提出請願（petition）。在請願核准後即可更換代理人。此時本案將有二個代理人處理後續審查程序。

§4.5　美國專利局的審查委員離職後代理專利案件上的限制

　　美國專利局之審查委員職離後至法律事務所執業時需簽署切結書，承諾 1) 不代理，不以任何形式協助、處理在其服務於美國專利局期間在其審查小組之任何審查中之案件；及 2) 不代理、不以任何方式準備，協助、處理會分配至他在美國專利局服務期間之審查小組來審查之在他離職之後二年之內申請的專利申請案件。

37 CFR 10.10(b)(2)

No individual who has served in the patent examining corps of the office may practice before the office after termination of his or her service, unless he or she assigns a written undertaking,

(1)not to prosecute or aid in any manner in the prosecution of any patent application pending in any patent examining group, during his or her period of the service therein, and

not to prepare or prosecute or to assist in any manner in the preparation or prosecution of any application of another (i)assigned to such group for examination and (ii)filed within two years after the date he or she left such group.

實例

A 與 B 均為某一專利法律事務所之專利律師。A 替其客戶在 1990 年 1 月 11 日申請了一專利並於 1991 年 6 月 8 日收到第 1 次審定書。A 注意到此審定書乃由其同事 B 之前在美國專利局作審查委員之審查小組所發出，而 B 於 1987 年 6 月 30 日離開美國專利局之後立即加入此事務所。由於 A 捲入一專利訴訟之中無法承辦此審定書之答辯。請問 A 可否將此答辯改由 B 承辦，並提出答辯？

B 可代理此案件之承辦，因為 B 承辦此案件時已離開美國專利局二年以上且該案在 B 為審查委員時並非審查中之案件。

§4.6　由非申請人指定的專利代理人或專利律師進行相關申請及審查程序

一美國代理人或專利律師即使不是申請人（受讓人）指定者（patent practitioner not of record），仍可代理申請的相關事宜。美國專利局認為其將文件送入專利局即代表其受申請人委託進行相關程序。例如一原受申請人委任之專利代理人因請假或其他理由不能處理一案件之答辯，則可由同事務所之專利代理人或專利律師簽名後將答辯書送入美國專利局。但如果要由非指定之代理人與審查委員進行面談時，只有當代理人能出示相關文件（卷宗）或有適當之授權（例如申請人（發明人）一起親至專利局）時，才被認為是已受委任進行面談程序。

MPEP 405 *>Patent Practitioner< Not of Record [R-5]

Papers may be filed in patent applications and reexamination proceedings by registered attorneys or agents not of record under 37 CFR 1.34. Filing of such papers is considered to be a representation that the attorney or agent is authorized to act in a representative capacity on behalf of applicant. However, interviews with a registered attorney or agent not of record will ordinarily be conducted based only on the information and files supplied by the attorney or agent in view of 35 U.S.C. 122. Interviews may be conducted with a registered practitioner who does not have a copy of the application file, but has proper authority from the applicant or attorney or agent of record in the form of a paper on file in the application. See also MPEP § 713.05. Such a paper may be an "Authorization to Act in a Representative Capacity." **>Form/PTO/SB/84,< "Authorization to Act in a Representative Capacity" is available from the USPTO Internet web site at http://www.uspto.gov/web/forms/sb0084.pdf.

§4.7　美國專利局之審定書、通知及其他信函之寄送地址

美國專利局規定每一專利申請案件申請時必需指定一個連絡地址（correspondence address）以寄送審定書，通知，證書及其他官方函。此連絡地址必需指明在申請資料表（Application Data Sheet, ADS）之上或與專利申請之文件一起呈送之信函上。如果申請人在不同文件上指定了二個不同的連絡地址，則美國專利局會依下列順序選擇連絡地址(1)ADS 上之連絡地址(2)申請信函（application transmittal）上之連絡地址(3)宣誓書上之連絡地址(4)委任書上之連絡地址。

專利申請案中如果沒有指定連絡地址，但有委任代理人時，則美國專利局會將文件寄至代理人之連絡地址。如果申請人先委任了一代理人，之後又委任另一代理人（未將之前的代理人解任時），文件會寄至之後來被委任之代理人之連絡地址，如果專利申請案沒有指定連絡地址，且未委託代理人，則美國專利局之文件會寄至第 1 個申請人（發明人）之地址。

37 CFR§ 1.33 Correspondence respecting patent applications, reexamination proceedings, and other proceedings.

(a) Correspondence address and daytime telephone number. When filing an application, a correspondence address must be set forth in either an application data sheet (§ 1.76), or elsewhere, in a clearly identifiable manner, in any paper submitted with an application filing. If no correspondence address is specified, the Office may treat the mailing address of the first named inventor (if provided, see §§ 1.76 (b)(1) and 1.63 (c)(2)) as the correspondence address. The Office will direct, or otherwise make available, all notices, official letters, and other communications relating to the application to the person associated with the correspondence address. For correspondence submitted via the Office's electronic filing system, however, an electronic acknowledgment receipt will be sent to the submitter. The Office will generally not engage in double correspondence with an applicant and a patent practitioner, or with more than one patent practitioner except as deemed necessary by the Director. If more than one correspondence address is specified in a single document, the Office will select one of the specified addresses for use as the correspondence address and, if given, will select the address associated with a Customer Number over a typed correspondence address. For the party to whom correspondence is to be addressed, a daytime telephone number should be supplied in a clearly identifiable manner and may be changed by any party who may change the correspondence address.

＊ ＊ ＊ ＊ ＊ ＊ ＊ ＊ ＊ ＊ ＊

MPEP 403.02 Two *>Patent Practitioners< for Same Application [R-5]

If the applicant simultaneously appoints two principal *>patent practitioners<, he or she should indicate with whom correspondence is to be conducted. If one is a local Washington metropolitan area *>patent practitioner< and the applicant fails to indicate either *>patent practitioner<, correspondence will be conducted with the local *>patent practitioner<.

If, after one *>patent practitioner< is appointed, a second *>patent practitioner< is later appointed without revocation of the power of the first *>patent practitioner<, the correspondence address of the second *>patent practitioner< is entered into the application file record (Ex parte Eggan, 1911 C.D. 213, 172 O.G. 1091 (Comm'r Pat. 1911)), so that the Office letters are to be sent to him or her.

第 5 章　說明書之內容與格式

目　錄

§5.1　實用專利申請案之申請書及說明書之內容

一實用專利申請案之申請書需依序包括下列文件（事項）：

(1)　實用專利申請案傳送表

(2)　費用傳送表

(3)　申請資料表

(4)　說明書

(5)　圖式

(6)　簽署之宣誓書

說明書需依序包括以下內容（事項）：

(1)　發明之名稱，可同時包括申請人之姓名、國籍及居住地之簡述（除非已包括在申請資料表中）

(2)　相關申請案之對照參考（除非已包括在申請資料表中）

(3)　關於聯邦資助研究或發展之陳述

(4)　共同研究契約各當事人之名稱

(5)　以光碟（CD）呈送之序列表或電腦程式表格或以光碟呈送之合併參照資料，需註明光碟數目及光碟上之檔案

(6)　發明之背景

(7)　發明之概要

(8)　圖式之簡單說明

(9)　發明之詳細敘述

(10) 請求項

(11) 揭露之摘要

(12) 紙本之序列表

上述說明書中如無(2)、(3)、(4)、(5)、(8)、(12)項目，則可不包括在說明書之中。

37 CFR 1.77 Arrangement of application elements.

(a) The elements of the application, if applicable, should appear in the following order:

(1) Utility application transmittal form.

(2) Fee transmittal form.

(3) Application data sheet (see §1.76).

(4) Specification.

(5) Drawings.

(6) Executed oath or declaration.

(b) The specification should include the following sections in order:

(1) Title of the invention, which may be accompanied by an introductory portion stating the name, citizenship, and residence of the applicant (unless included in the application data sheet).

(2) Cross-reference to related applications (unless included in the application data sheet).

(3) Statement regarding federally sponsored research or development.

(4) The names of the parties to a joint research agreement.

(5) Reference to a "Sequence Listing," a table, or a computer program listing appendix submitted on a compact disc and an incorporation-by-reference of the material on the compact disc (see § 1.52(e)(5)). The total number of compact discs including duplicates and the files on each compact disc shall be specified.

(6) Background of the invention.

(7) Brief summary of the invention.

(8) Brief description of the several views of the drawing.

(9) Detailed description of the invention.

(10) A claim or claims.

(11) Abstract of the disclosure.

(12) "Sequence Listing," if on paper (see §§ 1.821 through 1.825).

(c) The text of the specification sections defined in paragraphs (b)(1) through (b)(12) of this section, if applicable, should be preceded by a section heading in uppercase and without underlining or bold type.

§5.2　專利說明書之格式

專利說明書均需印在相同大小的紙上而且非永久地固結在一齊，所有用紙必需是可撓的，堅韌的，平滑，非亮面且白色的。紙張大小需為 21.0cm×29.7cm（A4）或 21.6cm×27.9cm（$8\frac{1}{2}$×11 英吋），每張紙之上方需留 2.0cm 之空白，左方 2.5cm 之空白，右方 2.0cm 之空白，下方 2.0cm 之空白。每張紙需只有一面，用打字機或印表機清楚地以永久黑色的墨汁印刷，且需有足夠的對比及清晰度以便容易地以影印、掃描或微影片複製。所有送入之說明書將成為美國專利局之永久記錄，故紙張上需無任何孔洞或破裂。如果以電子申請則說明書需作成電子申請之格式，並以符合電子申請系統之方式傳送。

專利說明書需以英語撰寫或需同時附上英文翻譯及修正部份之英譯及一翻譯為正確之聲明。說明書之行距需為 $1\frac{1}{2}$ 或 2 行距，說明書之英文字之字型需為 nonscript type font（如 Arial, Times Roman, or Courier,最好字型大小為 12），而字母大小需至少 0.3175 cm（0.125 英吋）高，但不小於 0.21 cm（0.08 英吋）高（字型大小 6）。說明書每頁只能有一欄（column）。請求項及摘要需從另一頁開始。說明書（包括請求項及摘要）之頁碼需從 1 開始連續編頁碼，且頁碼印在每頁之下面中間或上面中間，最好在下面中間。英文說明書之每個段落，除了請求項及摘要以外，可在專利申請案申請時，就個別且連續地用阿拉伯數字編號，以讓各段落易於辨識。各段落之編號需至少 4 位數並以 0 開始及用方形括弧括起，如[0001]，[0002]-[0021]等。此編號加在各段剛開始之左邊並用粗體字。每一段之第 1 行之最開始與段落編號應留約 4 個 space 之空格。表格，數學式，化學式，化學結構式，序列表被視為段落之一部份，不用另外加上段落編號。段落或章節之標頭（headers），不論置於最左邊或中間，均不視為是一段落故不必加上段落編號。

與英文說明書相關之下列文件可以光碟（CD）之方式呈送美國專利局：

（I）電腦程式表

（Ⅱ）序列表

（Ⅲ）任何獨立的表，其總頁數超過 50 頁或者說明書中表格的總頁數超過 100 頁者。

所呈送的光碟必需是 CD-ROM 或 CD-R。CD-ROM 是唯讀媒體，資料載入後不能改變或抹去，而 CD-R 是單次寫入媒體，一旦資料記錄進去就不能改變或抹去。各 CD 需符合 ISO 9660 標準，而且各 CD 之內容需符合 ASCII 之標準。CD 送入時需裝入盒子以保護且需送二份，註明 copy 1 及 copy 2。如有附上 CD，則說明書需有一段落將 CD 中之資訊作為合併參照（incorporation-by-reference），其需指明各 CD 之檔案名稱，製作日期及檔案大小。CD 需標示下列資訊：

（Ⅰ）發明人之名字

（Ⅱ）發明人之名稱

（Ⅲ）檔號或申請案號

（Ⅳ）CD 之製作日期

（Ⅴ）需註明各 CD 之順序如 1 of X

（Ⅵ）需註明各 copy 1 或 copy 2

如果送入美國專利局之二片 CD 均無法閱讀，則將被視為未呈送 CD。

英文說明書及請求項之中亦可包含化學式或數學式，但不能包含圖式或流程圖。表格如果是以電子型式呈送，則需注意其排列需在顯示之後能夠對齊。表格及化學式或數學式如果不能以直立放置亦可成橫式放置。化學式、數學式或表格中之文字、符號需至少 0.422cm（0.166 英吋）高（最好為 Arial, Times Roman,或 Courier，字型大小 12），但是不能小於 0.21 cm.（0.08 英吋）高（字型 6）。複雜的式子或表格之行距需至少 0.64 cm（1/4 英吋）。

此外，說明書之用語不可侵犯任何種族、宗教、性別或國籍，亦不可包括任何與發明無關之敘述。英文之用字可為英式英語拼法（如 colour），並不一定要用美式英語拼法。如果在說明書中有度量衡之用語，建議使用公制，如果使用英制單位，需於其後標上公制單位。

如果以非英文之說明書提出申請，需後補英文說明書，一翻譯為正確之聲明以及處理費。需注意的是英文翻譯必需是非英文說明書之精準翻譯（literal translation），亦即翻譯版本不能改變非英文說明書之內容，段落之排列前後，且需每句均翻譯出。如果翻譯之英文並不符合美國慣用之英語（idiomatic English）則需附上先前修正，但不能加入任何新事項（new matter）。此外，翻譯者尚需附上一聲明（statement）聲明為正確之翻譯。美國專利局通常不鼓勵以非英文說明書提出申請，因為如果翻譯本並非精準之譯本，則可能對申請人及大眾不利。另外，此種申請案將加重美國專利局之行政作業負擔。

如果說明書或請求項中有圖式，圖解說明或流程圖，則審查委員通常會發下形式核駁（objection），請求修正。請求項只有在無法以文字定義所請求之發明時才能加入圖式（例如合金之金相圖）。說明書本文可加入表格，但如果說明書已有表格，相同之表格，就不能再同時以圖式表示。

再者，說明書中亦不能包括有超連結（hyperlinks）或其他瀏覽器執行碼（browser-executable code）。如果包括此種形式則審查委員會認為是不適當的參照併入（incorporation by reference）而發下形式核駁，要求刪除。

37CFR § 1.52 Language, paper, writing, margins, compact disc specifications.

(a) Papers that are to become a part of the permanent United States Patent and Trademark Oaffice records in the file of a patent application or a reexamination proceeding.

(1) All papers, other than drawings, that are submitted on paper or by facsimile transmission, and are to become a part of the permanent United States Patent and Trademark Office records in the file of a patent application or reexamination proceeding, must be on sheets of paper that are the same size, not permanently bound together, and:

(i) Flexible, strong, smooth, non-shiny, durable, and white;

(ii) Either 21.0 cm by 29.7 cm (DIN size A4) or 21.6 cm by 27.9 cm (8 1/2 by 11 inches), with each sheet including a top margin of at least 2.0 cm (3/4 inch), a left side margin of at least 2.5 cm (1 inch), a right side margin of at least 2.0 cm (3/4 inch), and a bottom margin of at least 2.0 cm (3/4 inch);

(iii) Written on only one side in portrait orientation;

(iv) Plainly and legibly written either by a typewriter or machine printer in permanent dark ink or its equivalent; and

(v) Presented in a form having sufficient clarity and contrast between the paper and the writing thereon to permit the direct reproduction of readily legible copies in any number by use of photographic, electrostatic, photo-offset, and microfilming processes and electronic capture by use of digital imaging and optical character recognition.

(2) All papers that are submitted on paper or by facsimile transmission and are to become a part of the permanent records of the United States Patent and Trademark Office should have no holes in the sheets as submitted.

(3) The provisions of this paragraph and paragraph (b) of this section do not apply to the pre-printed information on paper forms provided by the Office, or to the copy of the patent submitted on paper in double column format as the specification in a reissue application or request for reexamination.

(4) See § 1.58 for chemical and mathematical formulae and tables, and § 1.84 for drawings.

(5) Papers that are submitted electronically to the Office must be formatted and transmitted in compliance with the Office's electronic filing system requirements.

(b) The application (specification, including the claims, drawings, and oath or declaration) or reexamination proceeding and any amendments or corrections to the application or reexamination proceeding.

(1) The application or proceeding and any amendments or corrections to the application (including any translation submitted pursuant to paragraph (d) of this section) or proceeding, except as provided for in § 1.69 and paragraph (d) of this section, must:

(i) Comply with the requirements of paragraph (a) of this section; and

(ii) Be in the English language or be accompanied by a translation of the application and a translation of any corrections or amendments into the English language together with a statement that the translation is accurate.

(2) The specification (including the abstract and claims) for other than reissue applications and reexamination proceedings, and any amendments for applications (including reissue applications) and reexamination proceedings to the specification, except as provided for in §§ 1.821 through 1.825, must have:

(i) Lines that are 1 1/2 or double spaced;

(ii) Text written in a nonscript type font (e.g., Arial, Times Roman, or Courier, preferably a font size of 12) lettering style having capital letters which should be at least 0.3175 cm. (0.125 inch) high, but may be no smaller than 0.21 cm. (0.08 inch) high (e.g., a font size of 6); and

(iii) Only a single column of text.

(3) The claim or claims must commence on a separate physical sheet or electronic page (§ 1.75(h)).

(4) The abstract must commence on a separate physical sheet or electronic page or be submitted as the first page of the patent in a reissue application or reexamination proceeding (§ 1.72(b)).

(5) Other than in a reissue application or reexamination proceeding, the pages of the specification including claims and abstract must be numbered consecutively, starting with 1, the numbers being centrally located above or preferably below, the text.

(6) Other than in a reissue application or reexamination proceeding, the paragraphs of the specification, other than in the claims or abstract, may be numbered at the time the application is filed, and should be individually and consecutively numbered using Arabic numerals, so as to unambiguously identify each paragraph. The number should consist of at least four numerals enclosed in square brackets, including leading zeros (e.g., [0001]). The numbers and enclosing brackets should appear to the right of the left margin as the first item in each paragraph, before the first word of the paragraph, and should be highlighted in bold. A gap, equivalent to approximately four spaces, should follow the number. Nontext elements (e.g., tables, mathematical or chemical formulae, chemical structures, and sequence data) are considered part of the numbered paragraph around or above the elements, and should not be independently numbered. If a nontext element extends to the left margin, it should not be numbered as a separate and independent paragraph. A list is also treated as part of the paragraph around or above the list, and should not be independently numbered. Paragraph or section headers (titles), whether abutting the left margin or centered on the page, are not considered paragraphs and should not be numbered.

＊＊＊＊＊＊＊＊＊＊＊

(e)Electronic documents that are to become part of the permanent United States Patent and Trademark Office records in the file of a patent application or reexamination proceeding.

(1) The following documents may be submitted to the Office on a compact disc in compliance with this paragraph:

(i) A computer program listing (see § 1.96);

(ii) A "Sequence Listing" (submitted under § 1.821(c)); or

(iii) Any individual table (see § 1.58) if the table is more than 50 pages in length, or if the total number of pages of all of the tables in an application exceeds 100 pages in length, where a table page is a page printed on paper in conformance with paragraph (b) of this section and § 1.58(c).

(2) A compact disc as used in this part means a Compact Disc-Read Only Memory (CD-ROM) or a Compact Disc-Recordable (CD-R) in compliance with this paragraph. A CD-ROM is a "read-only" medium on which the data is pressed into the disc so that it cannot

be changed or erased. A CD-R is a "write once" medium on which once the data is recorded, it is permanent and cannot be changed or erased.

(3)(i) Each compact disc must conform to the International Standards Organization (ISO) 9660 standard, and the contents of each compact disc must be in compliance with the American Standard Code for Information Interchange (ASCII). CD-R discs must be finalized so that they are closed to further writing to the CD-R.

(ii) Each compact disc must be enclosed in a hard compact disc case within an unsealed padded and protective mailing envelope and accompanied by a transmittal letter on paper in accordance with paragraph (a) of this section. The transmittal letter must list for each compact disc the machine format (e.g., IBM-PC, Macintosh), the operating system compatibility (e.g., MS-DOS, MS-Windows, Macintosh, Unix), a list of files contained on the compact disc including their names, sizes in bytes, and dates of creation, plus any other special information that is necessary to identify, maintain, and interpret (e.g., tables in landscape orientation should be identified as landscape orientation or be identified when inquired about) the information on the compact disc. Compact discs submitted to the Office will not be returned to the applicant.

(4) Any compact disc must be submitted in duplicate unless it contains only the "Sequence Listing" in computer readable form required by § 1.821(e). The compact disc and duplicate copy must be labeled "Copy 1" and "Copy 2," respectively. The transmittal letter which accompanies the compact disc must include a statement that the two compact discs are identical. In the event that the two compact discs are not identical, the Office will use the compact disc labeled "Copy 1" for further processing. Any amendment to the information on a compact disc must be by way of a replacement compact disc in compliance with this paragraph containing the substitute information, and must be accompanied by a statement that the replacement compact disc contains no new matter. The compact disc and copy must be labeled "COPY 1 REPLACEMENT MM/DD/YYYY" (with the month, day and year of creation indicated), and "COPY 2 REPLACEMENT MM/DD/YYYY," respectively.

(5) The specification must contain an incorporation-by-reference of the material on the compact disc in a separate paragraph (§ 1.77(b)(5)), identifying each compact disc by the names of the files contained on each of the compact discs, their date of creation and their sizes in bytes. The Office may require applicant to amend the specification to include in the paper portion any part of the specification previously submitted on compact disc.

(6) A compact disc must also be labeled with the following information:

(i) The name of each inventor (if known);

(ii) Title of the invention;

(iii) The docket number, or application number if known, used by the person filing the application to identify the application; and

(iv) A creation date of the compact disc.

(v) If multiple compact discs are submitted, the label shall indicate their order (e.g. "1 of X").

(vi) An indication that the disk is "Copy 1" or "Copy 2" of the submission. See paragraph (b)(4) of this section.

＊＊＊＊＊＊＊＊＊＊＊＊

37CFR § 1.58 Chemical and mathematical formulae and tables.

(a) The specification, including the claims, may contain chemical and mathematical formulae, but shall not contain drawings or flow diagrams. The description portion of the specification may contain tables, but the same tables may only be included in both the drawings and description portion of the specification if the application was filed under 35 U.S.C. 371. Claims may contain tables either if necessary to conform to 35 U.S.C. 112 or if otherwise found to be desirable.

(b) Tables that are submitted in electronic form (§§ 1.96(c) and 1.821(c)) must maintain the spatial relationships (e.g., alignment of columns and rows) of the table elements when displayed so as to visually preserve the relational information they convey. Chemical and mathematical formulae must be encoded to maintain the proper positioning of their characters when displayed in order to preserve their intended meaning.

(c) Chemical and mathematical formulae and tables must be presented in compliance with § 1.52(a) and (b), except that chemical and mathematical formulae or tables may be placed in a landscape orientation if they cannot be presented satisfactorily in a portrait orientation. Typewritten characters used in such formulae and tables must be chosen from a block (nonscript) type font or lettering style having capital letters which should be at least 0.422 cm. (0.166 inch) high (e.g., preferably Arial, Times Roman, or Courier with a font size of 12), but may be no smaller than 0.21 cm. (0.08 inch) high (e.g., a font size of 6). A space at least 0.64 cm. (1/4 inch) high should be provided between complex formulae and tables and the text. Tables should have the lines and columns of data closely spaced to conserve space, consistent with a high degree of legibility.

MPEP 608 Disclosure [R-2]

In return for a patent, the inventor gives as consideration a complete revelation or disclosure of the invention for which protection is sought. All amendments or claims must find descriptive basis in the original disclosure, or they involve new matter. Applicant may rely for disclosure upon the specification with original claims and drawings, as filed. See also **>37 CFR 1.121(f)< and MPEP § 608.04.

If during the course of examination of a patent application, an examiner notes the use of language that could be deemed offensive to any race, religion, sex, ethnic group, or nationality, he or she should object to the use of the language as failing to comply with the Rules of Practice. 37 CFR 1.3 proscribes the presentation of papers which are lacking in decorum and courtesy.

＊ ＊ ＊ ＊ ＊ ＊ ＊ ＊ ＊ ＊ ＊ ＊

The specification is a written description of the invention and of the manner and process of making and using the same. The specification must be in such full, clear, concise, and exact terms as to enable any person skilled in the art or science to which the invention pertains to make and use the same. See 35 U.S.C. 112 and 37 CFR 1.71. If a newly filed application obviously fails to disclose an invention with the clarity required by 35 U.S.C. 112, revision of the application should be required. See MPEP § 702.01. The written description must not include information that is not related to applicant's invention, e.g., prospective disclaimers regarding comments made by examiners. If such information is included in the written description, the examiner will object to the specification and require applicant to take appropriate action, e.g., cancel the information. The specification must commence on a separate sheet. Each sheet including part of the specification may not include other parts of the application or other information. The claim(s), abstract and sequence listing (if any) should not be included on a sheet including any other part of the application (37 CFR 1.71(f)). That is, the claim(s), abstract and sequence listings (if any) should each begin on a new page since each of these sections (specification, abstract, claims, sequence listings) of the disclosure are separately indexed in the Image File Wrapper (IFW). There should be no overlap on a single page of more than one section of the disclosure.

The specification does not require a date.

Certain cross references to other related applications may be made. References to foreign applications or to applications identified only by the attorney's docket number should be required to be canceled. U.S. applications identified only by the attorney's docket number may be amended to properly identify the earlier application(s). See 37 CFR 1.78.

As the specification is never returned to applicant under any circumstances, the applicant should retain an accurate copy thereof. In amending the specification, the attorney or the applicant must comply with 37 CFR 1.121 (see MPEP § 714).

Examiners should not object to the specification and/or claims in patent applications merely because applicants are using British English spellings (e.g., colour) rather than

American English spellings. It is not necessary to replace the British English spellings with the equivalent American English spellings in the U.S. patent applications. Note that 37 CFR 1.52(b)(1)(ii) only requires the application to be in the English language. There is no additional requirement that the English must be American English.

MPEP 608.01

IV. USE OF METRIC SYSTEM OF MEASUREMENTS IN PATENT APPLICATIONS

In order to minimize the necessity in the future for converting dimensions given in the English system of measurements to the metric system of measurements when using printed patents as research and prior art search documents, all patent applicants should use the metric (S.I.) units followed by the equivalent English units when describing their inventions in the specifications of patent applications.

* * * * * * * * * * *

37 § 1.52 Language, paper, writing, margins, compact disc specifications.

(d) A nonprovisional or provisional application may be in a language other than English.

(1) Nonprovisional application. If a nonprovisional application is filed in a language other than English, an English language translation of the non-English language application, a statement that the translation is accurate, and the processing fee set forth in § 1.17(i) are required. If these items are not filed with the application, applicant will be notified and given a period of time within which they must be filed in order to avoid abandonment

* * * * * * * * * * *

MPEP 608.01

The translation must be a literal translation and must be accompanied by a statement that the translation is accurate. The translation must also be accompanied by a signed request from the applicant, his or her attorney or agent, asking that the English translation be used as the copy for examination purposes in the Office. If the English translation does not conform to idiomatic English and United States practice, it should be accompanied by a preliminary amendment making the necessary changes without the introduction of new matter prohibited by 35 U.S.C. 132. If such an application is published as a patent application publication, the document that is published is the translation. See 37 CFR 1.215(a) and MPEP § 1121 regarding the content of the application publication. In the event that the English translation and the statement are not timely filed in the nonprovisional application, the nonprovisional application will be regarded as abandoned.

It should be recognized that this practice is intended for emergency situations to prevent loss of valuable rights and should not be routinely used for filing applications. There are at least two reasons why this should not be used on a routine basis. First, there are obvious dangers to applicant and the public if he or she fails to obtain a correct literal translation. Second, the filing of a large number of applications under the procedure will create significant administrative burdens on the Office.

VI. ILLUSTRATIONS IN THE SPECIFICATION

Graphical illustrations, diagrammatic views, flowcharts, and diagrams in the descriptive portion of the specification do not come within the purview of 37 CFR 1.58(a), which permits tables, chemical and mathematical formulas in the specification in lieu of formal drawings. The examiner should object to such descriptive illustrations in the specification and request drawings in accordance with 37 CFR 1.81 when an application contains graphs, drawings, or flow charts in the specification.

The specification, including any claims, may contain chemical formulas and mathematical equations, but >the written description portion of the specification< must not contain drawings or flow diagrams. >A claim may incorporate by reference to a specific figure or table where there is no practical way to define the invention in words. See MPEP § 2173.05(s).< The description portion of the specification may contain tables, but the same tables must not be included in both the drawings as a figure and in the description portion of the specification. Applications filed under 35 U.S.C. 371 are excluded from the prohibition from having the same tables in both the description portion of the specification and drawings. Claims may contain tables either if necessary to conform to 35 U.S.C. 112 or if otherwise found to be desirable. See MPEP § 2173.05(s). When such a patent is printed, however, the table will not be included as part of the claim, and instead the claim will contain a reference to the table number.

See MPEP § 601.01(d) for treatment of applications filed without all pages of the specification.

VII. Hyperlinks and Other Forms of Browser-Executable Code in the Specification

Examiners must review patent applications to make certain that hyperlinks and other forms of browser-executable code, especially commercial site URLs, are not included in a patent application. 37 CFR 1.57(d) states that an incorporation by reference by hyperlink or other form of browser executable code is not permitted.

＊＊＊＊＊＊＊＊＊＊＊

If hyperlinks and/or other forms of browser-executable code are embedded in the text of the patent application, examiners should object to the specification and indicate to applicants that the embedded hyperlinks and/or other forms of browser-executable code are impermissible and require deletion. This requirement does not apply to electronic documents listed on forms PTO-892 and PTO/SB/08 where the electronic document is identified by reference to a URL.

The attempt to incorporate subject matter into the patent application by reference to a hyperlink and/or other forms of browser-executable code is considered to be an improper incorporation by reference.

＊＊＊＊＊＊＊＊＊＊＊

§5.3　發明之名稱

發明之名稱（Title of the Invention）需置於說明書第 1 頁的最上面，除非已在申請資料表（application data sheet）中提供。發明之名稱必需簡短且技術上正確，而且不能多於 500 文字（字母）（characters）。此外，'new','improved', 'improvement of,' 及'improvement in'並不認為是發明名稱的一部份，因此不應加在發明名稱之前面。美國專利局將發明名稱輸入電腦記錄及發證時會將這些用語刪除。同樣的，冠詞'a' 'an'及'the'均不應加入發明名稱之中。

當說明書之發明名稱是非敘述性時，審查委員會要求修改發明名稱。如果修改的發明名稱仍不能令審查委員滿意，審查委員在專利申請案核准時以審查委員修正（examiner's amendment）自行修正發明之名稱。

MPEP 606 Title of Invention [R-5]

The title of the invention should be placed at the top of the first page of the specification unless it is provided in the application data sheet (see 37 CFR 1.76). The title should be brief but technically accurate and descriptive and should contain fewer than 500 characters. Inasmuch as the words >"new,"< "improved," "improvement of," and "improvement in" are not considered as part of the title of an invention, these words should not be included at the beginning of the title of the invention and will be deleted when the Office enters the title into the Office's computer records, and when any patent issues

MPEP 606.01 Examiner May Require Change in Title [R-2]

Where the title is not descriptive of the invention claimed, the examiner should require the substitution of a new title that is clearly indicative of the invention to which the claims are directed.

This may result in slightly longer titles, but the loss in brevity of title will be more than offset by the gain in its informative value in indexing, classifying, searching, etc. If a satisfactory title is not supplied by the applicant, the examiner may, at the time of allowance, change the title by examiner's amendment.

＊＊＊＊＊＊＊＊＊＊＊

§5.4　揭露之摘要

揭露之摘要（abstract of the Disclosure）或摘要（abstract）乃是專利之技術的簡單敘述，所以摘要應敘述發明相關之揭露。如果專利是基礎的發明，整個技術就是新的，摘要就需針對整個技術作摘要敘述。如果專利是裝置、方法、產品之改良，則摘要應對改良部份作敘述。如果專利是化合物，組合物之製法發明，則摘要就應對製法作敘述。摘要不應敘述可疑之優點或不確定之應用，也不應該比較發明與先前技術。此外，發明是有關於機器或裝置，則摘要需包括其構造及操作。如果發明是有關於物品，摘要需包括其製法。如果發明是化合物，摘要需包括其鑑定資料。如果發明是有關混合物，摘要需包括其成份。如果發明是有關方法，摘要需包括其步驟。

摘要需接在請求項之後，從另一頁開始。記載摘要之頁不可包括其他說明書之部份或資訊。

摘要需以敘述方式撰寫且通常限定為一段，且字數在 50~150 字之間，摘要的長度亦不能超過 15 行。摘要如果超過 15 行時，審查委員會再檢查是否超過 150 字。如果超過 150 字則無法以電腦排印，會被退回審查委員準備較短的摘要。一般請求項（claim）之用語，如 means 及 said 應避免。摘要需充份敘述發明之揭露以協助讀者決定是否需閱讀整份說明書。

摘要之用語需清楚且簡潔且不應重複發明名稱之資訊，應避免使用"The disclosure concerns," "The disclosure defined by this invention," "The disclosure describes,"等。

MPEP 608.01(b)

＊＊＊＊＊＊＊＊＊＊＊

B.Content

A patent abstract is a concise statement of the technical disclosure of the patent and should include that which is new in the art to which the invention pertains.

If the patent is of a basic nature, the entire technical disclosure may be new in the art, and the abstract should be directed to the entire disclosure.

If the patent is in the nature of an improvement in old apparatus, process, product, or composition, the abstract should include the technical disclosure of the improvement.

In certain patents, particularly those for compounds and compositions, wherein the process for making and/or the use thereof are not obvious, the abstract should set forth a process for making and/or a use thereof.

If the new technical disclosure involves modifications or alternatives, the abstract should mention by way of example the preferred modification or alternative.

The abstract should not refer to purported merits or speculative applications of the invention and should not compare the invention with the prior art.

Where applicable, the abstract should include the following: (1) if a machine or apparatus, its organization and operation; (2) if an article, its method of making; (3) if a chemical compound, its identity and use; (4) if a mixture, its ingredients; (5) if a process, the steps. Extensive mechanical and design details of apparatus should not be given.

With regard particularly to chemical patents, for compounds or compositions, the general nature of the compound or composition should be given as well as the use thereof, e.g., "The compounds are of the class of alkyl benzene sulfonyl ureas, useful as oral anti-diabetics." Exemplification of a species could be illustrative of members of the class. For processes, the type reaction, reagents and process conditions should be stated, generally illustrated by a single example unless variations are necessary.

C. Language and Format

The abstract must commence on a separate sheet, preferably following the claims, under the heading "Abstract" or "Abstract of the Disclosure." The sheet or sheets presenting the abstract may not include other parts of the application or other material. Form paragraph 6.16.01 (below) may be used if the abstract does not commence on a separate sheet. Note that the abstract for a national stage application filed under 35 U.S.C. 371 may be found on the front page of the Patent Cooperation Treaty publication (i.e., pamphlet). See MPEP § 1893.03(e).

The abstract should be in narrative form and generally limited to a single paragraph within the range of 50 to 150 words. The abstract should not exceed 15 lines of text. Abstracts exceeding 15 lines of text should be checked to see that it does not exceed 150 words in length since the space provided for the abstract on the computer tape by the printer is limited. If the abstract cannot be placed on the computer tape because of its excessive length, the application will be returned to the examiner for preparation of a shorter abstract. The form and legal phraseology often used in patent claims, such as "means" and "said," should be avoided. The abstract should sufficiently describe the disclosure to assist readers in deciding whether there is a need for consulting the full patent text for details.

The language should be clear and concise and should not repeat information given in the title. It should avoid using phrases which can be implied, such as, "This disclosure concerns," "The disclosure defined by this invention," "This disclosure describes," etc.

＊＊＊＊＊＊＊＊＊＊＊

§5.5　發明之背景

發明之背景需包括二部份：

(1) 發明之領域（Field of the Invention）：敘述發明所屬的技術領域，且需針對所請求之發明之標的（subject matter）。此部份說明了美國的專利分類。

(2) 相關技術之敘述（Description of the related art），此部份包括依照 37 CFR 1.97 及 1.98 所揭露之資訊。此部份應敘述先前技術（Prior art）之狀態及其他申請人所知的資訊。先前技術之問題或申請人的發明所解決之其他資訊亦包括在此部份。

另外需注意的是申請人可對先前技術（prior art）作一般之敘述，亦可敘述其發明之優點，但是不能對其他人之發明作誹謗之批評（derogatory remarks）。所謂誹謗之批評乃指對他人之產品或方法作貶抑或不利之評語，或對他人之專利或申請案之有效性作貶抑或不利之評語。單純的比較先前技術則不被認為是貶抑（誹謗）之批評。

MPEP 608.01(c) Background of the Invention

The Background of the Invention ordinarily comprises two parts:

(1) Field of the Invention: A statement of the field of art to which the invention pertains. This statement may include a paraphrasing of the applicable U.S. patent classification definitions. The statement should be directed to the subject matter of the claimed invention.

(2) Description of the related art including information disclosed under 37 CFR 1.97 and 37 CFR 1.98: A paragraph(s) describing to the extent practical the state of the prior art or other information disclosed known to the applicant, including references to specific prior art or other information where appropriate. Where applicable, the problems involved in the prior art or other information disclosed which are solved by the applicant's invention should be indicated. See also MPEP § 608.01(a), § 608.01(p) and § 707.05(b). browse after

＊＊＊＊＊＊＊＊＊＊＊

MPEP 608.01(r) Derogatory Remarks About Prior Art in Specification

The applicant may refer to the general state of the art and the advance thereover made by his or her invention, but he or she is not permitted to make derogatory remarks concerning the inventions of others. Derogatory remarks are statements disparaging the products or processes of any particular person other than the applicant, or statements as to the merits or validity of applications or patents of another person. Mere comparisons with the prior art are not considered to be disparaging, per se.

§5.6　發明之概要

由於發明之概要的目的乃是要告知大眾,特別是對與發明相關或特定技術有興趣的人士有關發明之本質(nature),因此發明之概要需針對所請求之特定發明作敘述。亦即,發明之標的需用一句或二句清楚簡潔之句子敘述。刻版的一般敘述是無用的,而且會被要求刪除。

記載發明的本質(nature),操作及目的之概要說明才能立即了解發明並對將來之檢索上有實質之幫助。發明之概要亦可敘述發明之目的(object)。發明之概要應與請求項(claims)之標的一致。

MPEP 608.01(d) Brief Summary of Invention

37 CFR 1.73 Summary of the invention.

A brief summary of the invention indicating its nature and substance, which may include a statement of the object of the invention, should precede the detailed description. Such summary should, when set forth, be commensurate with the invention as claimed and any object recited should be that of the invention as claimed.

Since the purpose of the brief summary of invention is to apprise the public, and more especially those interested in the particular art to which the invention relates, of the nature of the invention, the summary should be directed to the specific invention being claimed, in contradistinction to mere generalities which would be equally applicable to numerous preceding patents. That is, the subject matter of the invention should be described in one or more clear, concise sentences or paragraphs. Stereotyped general statements that would fit one application as well as another serve no useful purpose and may well be required to be canceled as surplusage, and, in the absence of any illuminating statement, replaced by statements that are directly on point as applicable exclusively to the case at hand.

The brief summary, if properly written to set out the exact nature, operation, and purpose of the invention, will be of material assistance in aiding ready understanding of the patent in future searches. The brief summary should be more than a mere statement of the objects of the invention, which statement is also permissible under 37 CFR 1.73.

The brief summary of invention should be consistent with the subject matter of the claims. Note final review of application and preparation for issue, MPEP § 1302.

§5.7　圖式之簡單說明

美國專利局之專利申請處理辦公室（Office of Patent Application Processing, OPAP）會檢查圖式之簡單說明以決定是否在說明書中所敘述到之圖式皆有附上。如果有缺圖，則依 MPEP 601.01(g)視為圖式不完整之申請案處理。如果申請案完全無圖式，則依 MPEP 601.01(f)視為無圖式之申請案處理。圖式應以 Fig 1,Fig 2……敘述，如果圖式之簡單說明為 Fig 1，而實際上所附之圖式為 Fig 1A, Fig 1B 及 Fig 1C，則審查委員會作形式核駁（objection），並要求申請人將圖式之簡單說明書依 Fig 1A,Fig 1B,Fig 1C 作敘述。另外，如果說明書有彩色之圖式，則在圖式之簡單說明之第 1 段包含下列之文字：

The patent or application file contains at least one drawing executed in color. Copies of this patent or patent application publication with color drawing(s) will be provided by the Office upon request and payment of the necessary fee.

MPEP 608.01(f) Brief Description of Drawings [R-7]

37 CFR 1.74 Reference to drawings.

When there are drawings, there shall be a brief description of the several views of the drawings and the detailed description of the invention shall refer to the different views by specifying the numbers of the figures, and to the different parts by use of reference letters or numerals (preferably the latter).

The Office of **>Patent Application Processing (OPAP)< will review the specification, including the brief description, to determine whether all of the figures of drawings described in the specification are present. If the specification describes a figure which is not present in the drawings, the application will be treated as an application filed without all figures of drawings in accordance with MPEP § 601.01(g), unless the application lacks any drawings, in which case the application will be treated as an application filed without drawings in accordance with MPEP § 601.01(f).

The examiner should see to it that the figures are correctly described in the brief description of the drawing, that all section lines used are referred to, and that all needed section lines are used. **>If the drawings show Figures 1A, 1B, and 1C and the brief description of the drawings refers only to Figure 1, the examiner should object to the brief description, and require applicant to provide a brief description of Figures 1A, 1B, and 1C.<

The specification must contain or be amended to contain proper reference to the existence of drawings executed in color as required by 37 CFR 1.84.

§5.8　發明之詳細說明

申請人可使用自己的用語來敘述所請求之發明，但需此技藝人士可了解所敘述之發明。發明之詳細敘述需滿足下列三要件：

（Ⅰ）敘述要件（Description requirement）：需包含所請求之發明（claimed invention）之書面敘述。

（Ⅱ）據以實施要件（Enablement requirement）：需使熟習該發明之技術人士可製造及使用該發明。

(III) 最佳模式要件（Best mode requirement）：需包含發明人所知之實施該所請求發明之最佳模式。

所謂敘述要件就是需對所請求的發明作書面之敘述，亦即如果請求項包括了不同之發明，例如產品、製法及製造產品之裝置則需三者均作敘述。而此時重要的是說明書中對產品、製法及製造裝置之敘述最好不要混在一起敘述。最佳的方式是具備裝置之構造圖、製法之流程圖及具有製品之組成（成份）並加以分別敘述。分別敘述之好處是較明確，而且如果審查委員認為屬於不同之發明，需分案時每一組請求項均會有明確的支持。另外，如果審查當中要增加請求項時，對不同之發明有分別之敘述則較易獲得明確之支持。

所謂符合據以實施要件就是指說明書本文之敘述，要詳細到能使熟習此技藝人士能夠製造及使用該所請求之發明。此要件有時相當難掌握，因為何謂熟習此技藝人士之認定常是非常不清楚的。基本上，所謂熟習此技藝人士乃是指從事該技術之人士，例如發明是有關於聚合所用之觸媒，則熟習此技術人士就是熟習觸媒化學之化學家。如果發明是有關於汽車安全帶之構造，則熟習此技藝人士就是汽車安全帶之設計師。由於美國專利實務上如果申請時揭露不足致使審查委員認為不符據以實施要件時要再補充資料，基本上均可能造成新事項（new matter），而無法補充，因而應掌握揭露多比揭露少為佳之原則，並於說明書完成草稿時詢問發明人熟習此技藝人士是否可製造及使用所請求之發明。

另外，衡量揭露是否充份到能夠滿足據以實施要件時，只需考慮所請求之標的（發明）。例如所請求者為一種化合物之製法而非其量產之方法，則說明書中只要揭示在實驗室之製造之實例，即可滿足據以實施之要件。再者如果揭露之內容會使熟技藝人士需作不當的實驗（undue experimentation）才能製造及使用該發明，則會被認為是不符合據以實施要件。至於是否會使熟習此技術人士作不當之實施，通常考慮下列因素：

(I) 所需實驗之量；

(II) 所需指示之量；

(III) 是否有實施例；

(IV) 發明之本質（通常化學、生物技術之發明較有可能揭露不足而造成熟習此技術人士需作不當的實驗）；

(V) 先前技術之狀態、水平；

(VI) 此技藝人士可預測性及不可預測性；

(VII) 請求項範圍之寬窄。

最後，未揭露支持發明之科學原理，通常不會被認為不符合據以實施要件，亦即發明的背後科學原理不必揭露。再者，有時在說明書中可敘述假想的實例以滿足據以實施要件。

說明書中要揭露最佳模式要件是美國專利法上特有之規定，而且在說明書故意不揭露最佳模式有可能造成專利之無效。所謂滿足最佳模式要件，就是說明書中需揭露在申請日之前發明人所知道發明之最佳模式，亦即在申請日之後之最佳模式可不必揭露。在說明書中只要將最佳模式揭露即可，不必特別指出哪一個是最佳模式。另外，最佳模式之揭露並不只限定於揭露中最有特徵之元件之最佳模式，還需揭露組合之最佳模式。

此外，一發明如要取得專利需是有用的，因而在說明書中必需揭露發明之用途。通常有關電機、機械之發明均能滿足有用（useful）之要件，但有關化學、生物技術發明之標的是否為有用，則可能造成質疑。然而在實務上此要件極容易滿足。亦即，只要在說明書中揭示一種用途即可滿足此要件。

---

35 U.S.C. 112 Specification.

The specification shall contain a written description of the invention, and of the manner and process of making and using it, in such full, clear, concise, and exact terms as to enable any person skilled in the art to which it pertains, or with which it is most nearly connected, to make and use the same, and shall set forth the best mode contemplated by the inventor of carrying out his invention.

＊＊＊＊＊＊＊＊＊＊＊＊

§5.9　利用參照併入避免說明書之冗長敘述

美國之專利說明書可利用參照併入（incorporation by reference）之方式將一些資訊加入說明書中以避免說明書過於冗長。同時當專利申請案發證之後，可減少大眾要檢索相關資訊之負擔並容易取得作為參照併入之資訊。

　　如果要以參照併入方式加入之資料是重要之內容（essential material）時，則能夠加入之內容只限於美國專利及公告之美國專利申請案，其本身未將此要內容作為參照併入者。所謂重要之內容乃指(1)加入之內容為可提供所請求之發明之書面敘述，且以一完整、清楚、正確之方式敘述而使得此技藝人士可據以實施，並記載了發明人所認為實施此發明之最佳模式之內容，或(2)加入內容為符合 35USC §112 第 2 段之規定，以清楚的用語界定所請求之發明，或(3)加入之內容為可符合 35USC §112 第 6 段之規定，敘述對應於所請求之裝置或方法步驟之結構、材料或動作者。

　　如果要以參照併入加入之內容為不重要之內容（non-essential material）時，則美國專利，公告之美國專利申請案，外國專利，公告之外國專利申請案，及申請人共同擁有之同時審查中的美國專利申請案（未公告者），以及其他非專利之刊物均可以參照併入的方式加入。但是不能以超連接（hyperlink）或瀏覽器執行之編碼的方式將參照併入的內容加入。再者，在過去 20 年內放棄的美國專利申請案亦可以參照併入的方式加入。如果要將上述重要內容或不重要內容以參照併入方式加入說明書中時，需明確地在說明書中註明，亦即，需用"Incorporation by reference"之字眼，且需明確地將作為參照之專利，專利申請案或刊物說明。

　　專利說明書中如有參照併入之資訊時，審查委員可能會要求申請人提供一份參照併入之內容之影本。申請人送入此內容時需同時附上一聲明，聲明所送入之影本與說明書中參照併入之內容為完全相同。之後，申請人要將參照併入之內容插入至說明書本文之中時，需以修正說明書之方式為之，且亦需附上加入之內容與說明書參照併入之內容為完全相同之聲明。再者，如果以參照併入之方式加入內容不符合規定，則需在專利申請案審查結束或放棄之前提出修正，否則參照併入之內容將被視為無效。

　　此外，如果專利申請案有主張國外優先權，且在專利說明書有將國外優先權案內容作為參照併入的話，當專利申請案有內容因疏忽而漏掉時，可主張為參照併入之內容，提出修正而不會被視為是新事項（new matter）。

37CFR§ 1.57 Incorporation by reference.

　　(a) Subject to the conditions and requirements of this paragraph, if all or a portion of the specification or drawing(s) is inadvertently omitted from an application, but the application contains a claim under § 1.55 for priority of a prior-filed foreign application, or a claim under § 1.78 for the benefit of a prior-filed provisional, nonprovisional, or international application, that was present on the filing date of the application, and the inadvertently omitted portion of the specification or drawing(s) is completely contained in the prior-filed application, the claim under § 1.55 or § 1.78 shall also be considered an incorporation by reference of the prior-filed application as to the inadvertently omitted portion of the specification or drawing(s).

(1) The application must be amended to include the inadvertently omitted portion of the specification or drawing(s) within any time period set by the Office, but in no case later than the close of prosecution as defined by §1.114 (b), or abandonment of the application, whichever occurs earlier. The applicant is also required to:

(i) Supply a copy of the prior-filed application, except where the prior-filed application is an application filed under 35 U.S.C. 111;

(ii) Supply an English language translation of any prior-filed application that is in a language other than English; and

(iii) Identify where the inadvertently omitted portion of the specification or drawings can be found in the prior-filed application.

(2) Any amendment to an international application pursuant to this paragraph shall be effective only as to the United States, and shall have no effect on the international filing date of the application. In addition, no request under this section to add the inadvertently omitted portion of the specification or drawings in an international application designating the United States will be acted upon by the Office prior to the entry and commencement of the national stage (§ 1.491) or the filing of an application under 35 U.S.C. 111(a) which claims benefit of the international application. Any omitted portion of the international application which applicant desires to be effective as to all designated States, subject to PCT Rule 20.8(b), must be submitted in accordance with PCT Rule 20.

(3) If an application is not otherwise entitled to a filing date under § 1.53 (b), the amendment must be by way of a petition pursuant to this paragraph accompanied by the fee set forth in § 1.17(f).

(b) Except as provided in paragraph (a) of this section, an incorporation by reference must be set forth in the specification and must:

(1) Express a clear intent to incorporate by reference by using the root words "incorporat(e)" and "reference" (e.g., "incorporate by reference"); and

(2) Clearly identify the referenced patent, application, or publication.

(c) "Essential material" may be incorporated by reference, but only by way of an incorporation by reference to a U.S. patent or U.S. patent application publication, which patent or patent application publication does not itself incorporate such essential material by reference. "Essential material" is material that is necessary to:

(1) Provide a written description of the claimed invention, and of the manner and process of making and using it, in such full, clear, concise, and exact terms as to enable any person skilled in the art to which it pertains, or with which it is most nearly connected, to make and use the same, and set forth the best mode contemplated by the inventor of carrying out the invention as required by the first paragraph of 35 U.S.C. 112;

(2) Describe the claimed invention in terms that particularly point out and distinctly claim the invention as required by the second paragraph of 35 U.S.C. 112; or

(3) Describe the structure, material, or acts that correspond to a claimed means or step for performing a specified function as required by the sixth paragraph of 35 U.S.C. 112.

(d) Other material ("Nonessential material") may be incorporated by reference to U.S. patents, U.S. patent application publications, foreign patents, foreign published applications, prior and concurrently filed commonly owned U.S. applications, or non-patent publications. An incorporation by reference by hyperlink or other form of browser executable code is not permitted.

(e) The examiner may require the applicant to supply a copy of the material incorporated by reference. If the Office requires the applicant to supply a copy of material incorporated by reference, the material must be accompanied by a statement that the copy supplied consists of the same material incorporated by reference in the referencing application.

(f) Any insertion of material incorporated by reference into the specification or drawings of an application must be by way of an amendment to the specification or drawings. Such an amendment must be accompanied by a statement that the material being inserted is the material previously incorporated by reference and that the amendment contains no new matter.

(g) An incorporation of material by reference that does not comply with paragraphs (b), (c), or (d) of this section is not effective to incorporate such material unless corrected within any time period set by the Office, but in no case later than the close of prosecution as defined by § 1.114(b), or abandonment of the application, whichever occurs earlier. In addition:

(1) A correction to comply with paragraph (b)(1) of this section is permitted only if the application as filed clearly conveys an intent to incorporate the material by reference. A mere reference to material does not convey an intent to incorporate the material by reference.

(2) A correction to comply with paragraph (b)(2) of this section is only permitted for material that was sufficiently described to uniquely identify the document.

§5.10　專利說明書中使用商業名稱、商標、著作權標記或光罩著作標記

　　所謂商業上之名稱（names used in trade）乃指在一技藝領域中之業者所熟知且稱呼之非專用的名稱，但此名稱可能並非一般公眾所熟知。商業上之名稱並非是某一特定製造者之產品的專用名稱，而是識別單一物品或產品之名稱。商業上的名稱如果在申請時同時滿足下列二條件的話，是允許被使用在專利說明書之中的：(A)使用商業上之名稱時需同時加上定義，此定義需是正確及清楚而能成為請求項之一部份，(B)在美國，此商業上的名稱是習知的且在文獻上有另人滿意之定義。

　　而所謂商標（trademark），則是製造商或商人所使用之文字、字母、圖樣或記號以用來識別自己的商品或與他人之商品者。商標是一專用的名稱。然而，商標與其所識別之產品之間的關係有時是不明確，不確定，且任意的。一製造商之某一商標之產品的配方，特性可能隨時改變，但是在一專利說明書之中，產品之各元件及各成份需以正確、準確之語言記載，一製造商之商標通常會代表不同之產品，故只有當在說明書中，商標所指的產品之記載是清楚地，可分辨的，才會被准許使用。如果商標具有固定及清楚的意義，則具有足夠之識別性（identification）。如果商標沒有固定及清楚的意義，則需要藉由科學或其他的解釋來確立識別性。當在說明書使用一商標，但該商標不具識別性，而需藉由修正說明書以加入科學或其他的解釋時，只能加入在申請時所習知的，該商標所代表之產品之特性，以避免產生新事項（new matter）。如果在說明書中以商標銷售之產物之適當的識別性未註明，而此識別性又是重要時，審查委員會認為揭露不足而加以核駁。

　　雖然在說明書中使用有明確意義之商標是可被允許的，但應尊重商標的專用特性。如果商標是文字或字母，則商標需大寫，如果是記號或圖樣之商標，則需指明商標之敘述。

　　使用商標時不可敘述成 "the product X commonly known as Y（trademark）"，應敘述成 "the product X sold under the trademark Y"。在發明之名稱中應避免使用商標，亦不可將商標與 "type 一字一齊使用，如 "Band-Aid type bandage"。如果專用的商標是在圖式中之記號或圖樣時，圖式之簡單說明或詳細說明中需特別指出該記號或圖樣是某公司之商標。

美國專利局亦允許在實用專利申請案或新式樣專利申請案之說明書中加入著作權或光罩創作之標記（copyright or mask work notice），但需符合以下規定：

(A) 著作權或光罩創作之標記需置於著作權或光罩創作之物件的旁邊。此標記可加在說明書本文之中亦可在圖式之中。如果是加在圖式之中則需符合 37CFR 1.84(s)之規定（直接放在表示著作權或光罩創作物件之下面，且大小為 0.32cm×0.64cm）

(B) 標記的內容必需只限定於法律所規定之元件，例如："ⓒ1983 John Doe" "(17 U.S.C. 401)及*M* John Doe"（17 U.S.C. 909）。

(C) 需於說明書之最前面包括下列之說明文：

A portion of the disclosure of this patent document contains material which is subject to (copyright or mask work) protection. The (copyright or mask work) owner has no objection to the facsimile reproduction by anyone of the patent document or the patent disclosure, as it appears in the Patent and Trademark Office patent file or records, but otherwise reserves all (copyright or mask work) rights whatsoever.

(D) 如果在核准通知發下後，要加入著作權或光罩創作之標記，需符合 37CFR 1.312 之規定（在繳領證費之前提出，並經審查委員之同意）。

MPEP 608.01(v) Trademarks and Names Used in Trade [R-7]

The expressions "trademarks" and "names used in trade" as used below have the following meanings:

Trademark: a word, letter, symbol, or device adopted by one manufacturer or merchant and used to identify and distinguish his or her product from those of others. It is a proprietary word, letter, symbol, or device pointing distinctly to the product of one producer.

Names Used in Trade: a nonproprietary name by which an article or product is known and called among traders or workers in the art, although it may not be so known by the public, generally. Names used in trade do not point to the product of one producer, but they identify a single article or product irrespective of producer.

Names used in trade are permissible in patent applications if:

(A) Their meanings are established by an accompanying definition which is sufficiently precise and definite to be made a part of a claim, or

(B) In this country, their meanings are well-known and satisfactorily defined in the literature.

Condition (A) or (B) must be met at the time of filing of the complete application.

TRADEMARKS

The relationship between a trademark and the product it identifies is sometimes indefinite, uncertain, and arbitrary. The formula or characteristics of the product may change from time to time and yet it may continue to be sold under the same trademark. In patent specifications, every element or ingredient of the product should be set forth in positive, exact, intelligible language, so that there will be no uncertainty as to what is meant. Arbitrary trademarks which are liable to mean different things at the pleasure of manufacturers do not constitute such language. Ex Parte Kattwinkle, 12 USPQ 11 (Bd. App. 1931).

However, if the product to which the trademark refers is set forth in such language that its identity is clear, the examiners are authorized to permit the use of the trademark if it is distinguished from common descriptive nouns by capitalization. If the trademark has a fixed and definite meaning, it constitutes sufficient identification unless some physical or chemical characteristic of the article or material is involved in the invention. In that event, as also in those cases where the trademark has no fixed and definite meaning, identification by scientific or other explanatory language is necessary. In re Gebauer-Fuelnegg, 121 F.2d 505, 50 USPQ 125 (CCPA 1941).

The matter of sufficiency of disclosure must be decided on an individual case-by-case basis. In re Metcalfe, 410 F.2d 1378, 161 USPQ 789 (CCPA 1969).

Where the identification of a trademark is introduced by amendment, it must be restricted to the characteristics of the product known at the time the application was filed to avoid any question of new matter.

If proper identification of the product sold under a trademark, or a product referred to only by a name used in trade, is omitted from the specification and such identification is deemed necessary under the principles set forth above, the examiner should hold the disclosure insufficient and reject on the ground of insufficient disclosure any claims based on the identification of the product merely by trademark or by the name used in trade. If the product cannot be otherwise defined, an amendment defining the process of its manufacture may be permitted. Such amendments must be supported by satisfactory showings establishing that the specific nature or process of manufacture of the product as set forth in the amendment was known at the time of filing of the application.

Although the use of trademarks having definite meanings is permissible in patent applications, the proprietary nature of the marks should be respected. Trademarks should be identified by capitalizing each letter of the mark (in the case of word or letter marks) or otherwise indicating the description of the mark (in the case of marks in the form of a symbol or device or other nontextual form). Every effort should be made to prevent their use in any manner which might adversely affect their validity as trademarks.

* * * * * * * * * * *

The examiner should not permit the use of language such as "the product X (a descriptive name) commonly known as Y (trademark)" since such language does not bring out the fact that the latter is a trademark. Language such as "the product X (a descriptive name) sold under the trademark Y" is permissible.

The use of a trademark in the title of an application should be avoided as well as the use of a trademark coupled with the word "type", e.g., "Band-Aid type bandage."

In the event that the proprietary trademark is a "symbol or device" depicted in a drawing, either the brief description of the drawing or the detailed description of the drawing should specify that the "symbol or device" is a registered trademark of Company X.

The owner of a trademark may be identified in the specification.

Technology Center Directors should reply to all trademark misuse complaint letters and forward a copy to the editor of this manual. >Where a letter demonstrates a trademark misuse in a patent application publication, the Office should, where the application is still pending, ensure that the trademark is replaced by appropriate generic terminology.

See Appendix I for a partial listing of trademarks and the particular goods to which they apply.

INCLUSION OF COPYRIGHT OR MASK WORK NOTICE IN PATENTS

The U.S. Patent and Trademark Office will permit the inclusion of a copyright or mask work notice in a design or utility patent application, and thereby any patent issuing therefrom, which discloses material on which copyright or mask work protection has previously been established, under the following conditions:

(A) The copyright or mask work notice must be placed adjacent to the copyright or mask work material. Therefore, the notice may appear at any appropriate portion of the patent application disclosure, including the drawing. However, if appearing in the drawing, the notice must comply with 37 CFR 1.84(s). If placed on a drawing in conformance with these provisions, the notice will not be objected to as extraneous matter under 37 CFR 1.84.

(B) The content of the notice must be limited to only those elements required by law. For example, "©1983 John Doe"(17 U.S.C. 401) and "*M* John Doe" (17 U.S.C. 909) would be properly limited, and under current statutes, legally sufficient notices of copyright and mask work respectively.

(C) Inclusion of a copyright or mask work notice will be permitted only if the following authorization in 37 CFR 1.71(e) is included at the beginning (preferably as the first paragraph) of the specification to be printed for the patent:

A portion of the disclosure of this patent document contains material which is subject to (copyright or mask work) protection. The (copyright or mask work) owner has no objection to the facsimile reproduction by anyone of the patent disclosure, as it appears in the Patent and Trademark Office patent files or records, but otherwise reserves all (copyright or mask work) rights whatsoever.

(D) Inclusion of a copyright or mask work notice after a Notice of Allowance has been mailed will be permitted only if the criteria of 37 CFR 1.312 have been satisfied.

****

§5.11　請求項之構成

一請求項（Claim）通常由三部份構成：前言（preamble），連接用語（transitional phrase）及本體（body）。前言乃用來界定發明之類別，例如：裝置、物品、化合物、組合物或方法，且亦可能同時記述了發明之用途，如 A method for increasing the color fastness of……。連接用語乃用來連接前言與本體，例如 comprising, consists of, consisting essentially of。本體則用來記載所請求之發明的構成元件、成份或步驟及其連結關係。

§5.12　請求項前言之效力

請求項前言之效力乃指請求項之前言是否會限制請求項之範圍，亦即前言是否會構成請求項之一限制條件。依照美國專利之實務，前言是否會限制一請求項之範圍會依案例之不同而有所不同（case-by case basis）。故實務上，並沒有一種測試方法可界定前言是否限制了一請求項之範圍。但是,此議題有一指標就是,如果當閱讀一請求項之所有前後文時,前言記載了請求項之限制條件,或者前言帶給請求項生命、意義、或活力時,前言就應被解釋為限制了請求項之範圍。換句話說,如果前言對於指出請求項所定義之發明是重要的時候,前言就限制了請求項之範圍。或者,請求項之本體部份有提到(refer back)前言部份之用語時,前言部份就變成了請求項之限制條件。

另外一個判斷的指標是檢視所有申請案之記錄去了解發明人真正發明的是什麼,真正想要藉由請求項去保護(涵蓋)的範圍是什麼?如果請求項之本體已完全地記載了所請求的發明的元件,而前言僅僅敘述發明之目的或用途,則前言可解釋成對所請求之發明不重要,而不是一限制條件。

MPEP 2111.02 Effect of Preamble [R-3]

The determination of whether a preamble limits a claim is made on a case-by-case basis in light of the facts in each case; there is no litmus test defining when a preamble limits the scope of a claim. Catalina Mktg. Int'l v. Coolsavings.com, Inc., 289 F.3d 801, 808, 62 USPQ2d 1781, 1785 (Fed. Cir. 2002). See id. at 808-10, 62 USPQ2d at 1784-86 for a discussion of guideposts that have emerged from various decisions exploring the preamble's effect on claim scope, as well as a hypothetical example illustrating these principles.

"[A] claim preamble has the import that the claim as a whole suggests for it." Bell Communications Research, Inc. v. Vitalink Communications Corp., 55 F.3d 615, 620, 34 USPQ2d 1816, 1820 (Fed. Cir. 1995). "If the claim preamble, when read in the context of the entire claim, recites limitations of the claim, or, if the claim preamble is 'necessary to give life, meaning, and vitality' to the claim, then the claim preamble should be construed as if in the balance of the claim."

＊＊＊＊＊＊＊＊＊＊＊

I.< PREAMBLE STATEMENTS LIMITING STRUCTURE

Any terminology in the preamble that limits the structure of the claimed invention must be treated as a claim limitation. See, e.g., Corning Glass Works v. Sumitomo Elec. U.S.A., Inc., 868 F.2d 1251, 1257, 9 USPQ2d 1962, 1966 (Fed. Cir. 1989) (The determination of whether preamble recitations are structural limitations can be resolved only on review of the entirety of the application "to gain an understanding of what the inventors actually invented and intended to encompass by the claim.");

＊＊＊＊＊＊＊＊＊＊＊

II.PREAMBLE STATEMENTS RECITING PURPOSE OR INTENDED USE

The claim preamble must be read in the context of the entire claim. The determination of whether preamble recitations are structural limitations or mere statements of purpose or use "can be resolved only on review of the entirety of the [record] to gain an understanding of what the inventors actually invented and intended to encompass by the claim." Corning Glass Works, 868 F.2d at 1257, 9 USPQ2d at 1966. If the body of a claim fully and intrinsically sets forth all of the limitations of the claimed invention, and the preamble merely states, for example, the purpose or intended use of the invention, rather than any distinct definition of any of the claimed invention's limitations, then the preamble is not considered a limitation and is of no significance to claim construction

§5.13　請求項之連接用語及是否限制請求項之範圍

請求項之連接用語通常分為三種：(1)開放式用語(2)封閉式用語及(3)半封閉式用語。

'comprising'，'including'，'containing'，或 'characterized by' 為開放式用語（inclusive or open-ended），其不排除多加的或未記載之元件、成份或步驟。例如一請求項寫成：A device comprising elements A,B and C。而有一產品為一裝置包括了 A,B,C and D 元件，則此產品侵害了請求項之專利，因為該請求項之連接用語為 comprising。同樣地，如果在審查時審查委員找到一先前技術之裝置包括了元件 A,B,C 及 D，則審查委員可用缺乏新穎性核駁該請求項。

'consisting of' 為封閉式用語（close-ended），其排除任何未在請求項中指明之元件，步驟或成份。例如一請求項寫成 A compostion consisting of A, B and C。則當一組合物包括成份 A，B，C 及 D 時，並不侵害該請求項之專利。此外，當在一請求項之本體中之一子句使用 consisting of 時，則其僅限制該子句中記載之元件的範圍。'consisting essentially of' 為半封閉式用語，其會將請求項之範圍限制至所指明之成份，步驟以及不會實質上影響所請求之發明的基本及新穎的特徵之成份（步驟）。例如申請人所請求之發明為一功能性流體 consisting essentially of 成份 A，B 和 C。此功能性流體的基本及新穎的特徵是強化的抗氧化能力及洗淨及分散能力。

引證之先前技術為一水力流體，其包括成份 A，B，C 及一分散劑。但在申請人之說明書中敘述到其功能性流體亦可包括習知之添加劑，例如分散劑。而且並無任何證據顯示分散劑會實質上影響該基本及新穎的特徵（強化之抗氧化能力及洗淨及分散能力）。因此，引證之先前技術將會使所請求之發明喪失新穎性。

　　除了上述三種連接用語外，'having' 及 'composed of' 需依照說明書所揭露之內容來解釋其是開放或封閉之連接用語。

MPEP 2111.03 Transitional Phrases [R-3]

The transitional phrases "comprising", "consisting essentially of" and "consisting of" define the scope of a claim with respect to what unrecited additional components or steps, if any, are excluded from the scope of the claim.

The transitional term "comprising", which is synonymous with "including," "containing," or "characterized by," is inclusive or open-ended and does not exclude additional, unrecited elements or method steps.

The transitional phrase "consisting of" excludes any element, step, or ingredient not specified in the claim. In re Gray, 53 F.2d 520, 11 USPQ 255 (CCPA 1931); Ex parte Davis, 80 USPQ 448, 450 (Bd. App. 1948) ("consisting of" defined as "closing the claim to the inclusion of materials other than those recited except for impurities ordinarily associated therewith.").

＊＊＊＊＊＊＊＊＊＊＊

A claim which depends from a claim which "consists of" the recited elements or steps cannot add an element or step. When the phrase "consists of" appears in a clause of the body of a claim, rather than immediately following the preamble, it limits only the element set forth in that clause; other elements are not excluded from the claim as a whole.

＊＊＊＊＊＊＊＊＊＊＊

The transitional phrase "consisting essentially of" limits the scope of a claim to the specified materials or steps "and those that do not materially affect the basic and novel characteristic(s)" of the claimed invention. In re Herz, 537 F.2d 549, 551-52, 190 USPQ 461, 463 (CCPA 1976) (emphasis in original) (Prior art hydraulic fluid required a dispersant which appellants argued was excluded from claims limited to a functional fluid "consisting essentially of" certain components. In finding the claims did not exclude the prior art dispersant, the court noted that appellants' specification indicated the claimed composition can contain any well-known additive such as a dispersant, and there was no evidence that the presence of a dispersant would materially affect the basic and novel characteristic of the claimed invention. The prior art composition had the same basic and novel characteristic (increased oxidation resistance) as well as additional enhanced detergent and dispersant characteristics.). "A 'consisting essentially of' claim occupies a middle ground between closed claims that are written in a 'consisting of' format and fully open claims that are drafted in a 'comprising' format.

＊＊＊＊＊＊＊＊＊＊＊

OTHER TRANSITIONAL PHRASES

Transitional phrases such as "having" must be interpreted in light of the specification to determine whether open or closed claim language is intended.

＊＊＊＊＊＊＊＊＊＊＊

The transitional phrase "composed of" has been interpreted in the same manner as either "consisting of" or "consisting essentially of," depending on the facts of the particular case.

＊＊＊＊＊＊＊＊＊＊＊

實例

一申請案之請求項如下：

A laminated article comprising layers A, B, C andD.而審查委員找到一專利其揭露了一 laminated article 具有 A, B,C,D 及 E layers。請問如果要克服此引證之核駁，該請求有項之連接用語須如何修改？

應修改成：A laminated article consisting of layers A, B, C, and D.

§5.14　請求項之種類

(1) 依是否依附分：

　　a) 獨立項（independent claim）

　　b) 依附項（dependent claim）包括多項附屬項（multiple-dependent claim）

(2) 依請求之方式分：

　　a) 組合式請求項（combination claim）

b) 吉普森式請求項（Jepson-type claim）

(3) 依請求之內容分：

　　a) 裝置請求項（apparatus or machine claim）

　　b) 物品請求項（article of manufacture claim）包括方法定義產品請求項（product by process claim）

　　c) 方法請求項（method or process claim）

　　d) 組合物請求項（composition of matter or chemical claim）

所謂的獨立項乃指一請求項單獨成立，未依附至任何請求項。本身包括了必需之限制條件，而未包括其他請求項之限制條件。一獨立項通常是範圍最廣的請求項。而所謂依附項乃指一請求項依附至其他的獨立項或依附項，包括了其他請求項之限制條件。其可將所依附之請求項之元件進一步限制或再包括其他之限制條件。

所謂的組合式請求項乃指一請求項之本體中所有元件，不論是習知之元件或特徵之元件均全部列出，而不特別指明哪些是構成發明之特徵之元件者。而吉普森式請求項乃是在前言部份定義出習知之元件，連接用語則使用 wherein the improvement comprises 之請求項。

如果使用吉普森式請求項即表示承認前言部份之標的是他人之先前技術。只有連接用語之後之部份才是發明之特徵部份。但是如果申請人使用吉普森請求項之目的是要避免共同申請中之案件之重複專利之核駁，而且審查委員並未引證出任何與前言部份相同之先前技術時，前言部份敘述之構造可不被視為先前技術。

MPEP 2129

＊＊＊＊＊＊＊＊＊＊＊

Drafting a claim in Jepson format (i.e., the format described in 37 CFR 1.75(e); see MPEP § 608.01(m)) is taken as an implied admission that the subject mater of the preamble is the prior art work of another. In re Fout, 675 F.2d 297, 301, 213 USPQ 532, 534 (CCPA 1982) (holding preamble of Jepson-type claim to be admitted prior art where applicant's specification credited another as the inventor of the subject matter of the preamble). However, this implication may be overcome where applicant gives another credible reason for drafting the claim in Jepson format. In re Ehrreich, 590 F.2d 902, 909-910, 200 USPQ 504, 510 (CCPA 1979) (holding preamble not to be admitted prior art where applicant explained that the Jepson format was used to avoid a double patenting rejection in a co-pending application and the examiner cited no art showing the subject matter of the preamble). Moreover, where the preamble of a Jepson claim describes applicant's own work, such may not be used against the claims. Reading & Bates Construction Co. v. Baker Energy Resources Corp., 748 F.2d 645, 650, 223 USPQ 1168, 1172 (Fed. Cir. 1984); Ehrreich, 590 F.2d at 909-910, 200 USPQ at 510.

§5.15　多項附屬項及其撰寫

多項附屬項（multiple dependent claim）乃是指一請求項其以擇一之方式依附至一個以上之前面的獨立項或依附項者。1978 年 1 月 24 日以後提申之專利申請案可允許使用多項附屬項，只要其是以擇一之方式（in the alternative form）撰寫。例如，A machine according to claims 3 or 4。但是累積請求方式（cumulative claiming），例如，A machine according to claim 3 and 4，則不允許。一多項附屬項只能以擇一方式依附至一組請求項，例如 A device as in claims 1,2,3 or 4, made by a process of claims 5,6,7 or 8，是不允許的。此外，一多項附屬項不能直接或間接作為其他的多項附屬項之基礎。

下述是一些美國專利局可接受及不可接受的多項附屬項之寫法：

A：可接受之多項附屬項：

Claim 5. A gadget according to claims 3 or 4, further comprising--

Claim 5. A gadget as in any one of the preceding claims, in which--

Claim 5. A gadget as in any one of claims 1,2, and 3, in which--

Claim 3. A gadget as in either of claim 1 or claim 2, further comprising --

Claim 4. A gadget as in claim 2 or 3, further comprising--

Claim 16. A gadget as in claims 1,7,12, or 15, further comprising--

Claim 5. A gadget as in any of the preceding claims, in which--

Claim 8. A gadget as in one of claims 4-7, in which--

Claim 5. A gadget as in any preceding claim, in which--

Claim 10. A gadget as in any of claims 1-3 or 7-9, in which--

Claim 11. A gadget as in any one of claims 1,2, or 7-10 inclusive, in which--

B：不能接受之多項附屬項：

1. 並非僅以擇一方式依附

Claim 5. A gadget according to claim 3 and 4, further comprising ---

Claim 9. A gadget according to claims 1-3, in which ---

Claim 9. A gadget as in claims 1 or 2 and 7 or 8, which ---

Claim 6. A gadget as in the preceding claims in which ---

Claim 6. A gadget as in claims 1, 2, 3, 4 and/or 5, in which ---

Claim 10. A gadget as in claims 1-3 or 7-9, in which ---

2. 並非依附至前面之請求項

Claim 3. A gadget as in any of the following claims, in which ---

Claim 5. A gadget as in either claim 6 or claim 8, in which ---

3. 依附至二組不同特徵之請求項

Claim 9. A gadget as in claim 1 or 4 made by the process of claims 5, 6, 7, or 8, in which ---

4. 依附至另一多項附屬項

Claim 8. A gadget as in claim 5 (claim 5 is a multiple dependent claim) or claim 7, in which ---

如果專利申請案包括有多項附屬項，則一多項附屬項中之各實施例均被視為一個單一的依附項。因此審查委員可能會要求對多項附屬項中之各實施例作限制／選擇。此外，某些實施例會被視為撤回，某些實施例會被審查。

一專利申請案中之每一多項附屬項除了需多繳 USD 330（一項）之費用以外，一可接受之多項附屬項之項數乃是以其依附項來計算。如果是一不可接受的多項附屬項，則項數乃以一項計算，以下之例子就可知多項附屬項之一費用計算方式：

Claim no..	Ind.	Dep
1. Independent	1	
2. Dependent on claim 1..		1
3. Dependent on claim 2..		1

4.	Dependent on claim 2 or 3...		2
5.	Dependent on claim 4...		2
6.	Dependent on claim 5...		2
7.	Dependent on claim 4, 5 or 6..		1
8.	Dependent on claim 7...		1
9.	Independent..	1	
10.	Dependent on claim 1 or 9...		2
11.	Dependent on claim 1 and 9...		1
	Total	2	13

MPEP 608.01(n) Dependent Claims [R-7]

I.MULTIPLE DEPENDENT CLAIMS

Generally, a multiple dependent claim is a dependent claim which refers back in the alternative to more than one preceding independent or dependent claim.

The second paragraph of 35 U.S.C. 112 has been revised in view of the multiple dependent claim practice introduced by the Patent Cooperation Treaty. Thus 35 U.S.C. 112 authorizes multiple dependent claims in applications filed on and after January 24, 1978, as long as they are in the alternative form (e.g., "A machine according to claims 3 or 4, further comprising ---"). Cumulative claiming (e.g., "A machine according to claims 3 and 4, further comprising ---") is not permitted. A multiple dependent claim may refer in the alternative to only one set of claims. A claim such as "A device as in claims 1, 2, 3, or 4, made by a process of claims 5, 6, 7, or 8" is improper. 35 U.S.C. 112 allows reference to only a particular claim. Furthermore, a multiple dependent claim may not serve as a basis for any other multiple dependent claim, either directly or indirectly. These limitations help to avoid undue confusion in determining how many prior claims are actually referred to in a multiple dependent claim.

＊＊＊＊＊＊＊＊＊＊＊

C.Restriction Practice

For restriction purposes, each embodiment of a multiple dependent claim is considered in the same manner as a single dependent claim. Therefore, restriction may be required between the embodiments of a multiple dependent claim. Also, some embodiments of a multiple dependent claim may be held withdrawn while other embodiments are considered on their merits.

＊＊＊＊＊＊＊＊＊＊＊

2.Calculation of Fees

(a)Proper Multiple Dependent Claim

35 U.S.C. 41(a), provides that claims in proper multiple dependent form may not be considered as single dependent claims for the purpose of calculating fees. Thus, a multiple dependent claim is considered to be that number of dependent claims to which it refers. Any proper claim depending directly or indirectly from a multiple dependent claim is also considered as the number of dependent claims as referred to in the multiple dependent claim from which it depends.

(b)Improper Multiple Dependent Claim

If none of the multiple dependent claims is proper, the multiple dependent claim fee set forth in 37 CFR 1.16(j) will not be required. However, the multiple dependent claim fee is required if at least one multiple dependent claim is proper.

If any multiple dependent claim is improper, *>OPAP< may indicate that fact by placing an encircled numeral "1" in the "Dep. Claims" column of form PTO/SB/07. The fee for any improper multiple dependent claim, whether it is defective for either not being in the alternative form or for being directly or indirectly dependent on a prior multiple dependent claim, will only be one, since only an objection to the form of such a claim will normally be made. This procedure also greatly simplifies the calculation of fees. Any claim depending from an improper multiple dependent claim will also be considered to be improper and be counted as one dependent claim.

＊＊＊＊＊＊＊＊＊＊＊

實例

一專利申請案中之 claim 乃如下:

1.A device …… (獨立項)

2.The device as claimed in claim 1, ……

3.The device as claimed in claim 1, ……

4.The device as claimed in claim 2, and 3, ……

5.A device …… (獨立項)

6.The device as claimed in claim 1, 2, or 5, ……

7.The device as claimed in claim 1, 2, or 5, ……

請問在計算費用上，此申請案之總項數為幾項？

11 項，claim 4 算 1 項，claim 6 及 claim 7 均算 3 項。

§5.16　方法定義產品請求項

所謂的方法定義產品請求項（product by process claim）乃指一請求項其以製造的方法來界定所請求之產品，例如：A product made by the process comprising step A,B,C, and D。需注意的是，雖然方法定義產品請求項是用方法來限制或定義產品，但是否具有專利性乃由產品本身來決定。亦即如果在方法定義產品請求項中之產品與先前技術之產品相同或是顯而易見的，則此請求項不能給予專利，即使先前之產品是由不同之方法製造。例如，一請求項乃是有關於酚醛顯色劑（novolac color developer）。製造此酚醛顯色劑之方法已核准。本發明之製法與先前技術的製法不同之處乃在於分別加入金屬氧化物（metal oxide）和羧酸（carboxylic acid）以在原地產出羧酸金屬鹽。先前技術的製法是直接加入較昂貴的，預先反應好的羧酸金屬鹽。因此，此方法定義產品之請求項將被核駁，因為雖然製法不同，但所得到之產品與先前技術之產品相同，均含有羧酸金屬鹽。但是當評估方法定義產品請求項之專利性時，亦應考慮方法步驟所隱含之結構特性，特別是當產品只能用方法步驟去定義時，或者製法步驟可被期待賦予最終產品不同的結構特性時。

當申請人以方法義定產品請求項界定一產品時，美國專利局之審查委員在作推斷的顯而易見（prima facie obviousness）核駁時負較少的舉證責任。亦即，審查委員只要推斷所請求之產品與先前技術之產品似乎相同或相似時，雖然是由不同之方法產出，舉證責任就轉移至申請人。申請人需負舉證責任證明其產品與先前技術之產品有所不同。

MPEP2173.05(p) Claim Directed to Product-By- Process or Product and Process [R-5]

＊＊＊＊＊＊＊＊＊＊＊＊

I.PRODUCT-BY-PROCESS

A product-by-process claim, which is a product claim that defines the claimed product in terms of the process by which it is made, is proper. In re Luck, 476 F.2d 650, 177 USPQ 523 (CCPA 1973); In re Pilkington, 411 F.2d 1345, 162 USPQ 145 (CCPA 1969); In re Steppan, 394 F.2d 1013, 156 USPQ 143 (CCPA 1967). A claim to a device, apparatus, manufacture, or composition of matter may contain a reference to the process in which it is intended to be used without being objectionable under 35 U.S.C. 112, second paragraph, so long as it is clear that the claim is directed to the product and not the process.

An applicant may present claims of varying scope even if it is necessary to describe the claimed product in product-by-process terms. Ex parte Pantzer, 176 USPQ 141 (Bd. App. 1972).

＊＊＊＊＊＊＊＊＊＊＊

MPEP 2113 Product-by-Process Claims [R-1]

PRODUCT-BY-PROCESS CLAIMS ARE NOT LIMITED TO THE MANIPULATIONS OF THE RECITED STEPS, ONLY THE STRUCTURE IMPLIED BY THE STEPS

"[E]ven though product-by-process claims are limited by and defined by the process, determination of patentability is based on the product itself. The patentability of a product does not depend on its method of production. If the product in the product-by-process claim is the same as or obvious from a product of the prior art, the claim is unpatentable even though the prior product was made by a different process." In re Thorpe, 777 F.2d 695, 698, 227 USPQ 964, 966 (Fed. Cir. 1985) (citations omitted) (Claim was directed to a novolac color developer. The process of making the developer was allowed. The difference between the inventive process and the prior art was the addition of metal oxide and carboxylic acid as separate ingredients instead of adding the more expensive pre-reacted metal carboxylate. The product-by-process claim was rejected because the end product, in both the prior art and the allowed process, ends up containing metal carboxylate. The fact that the metal carboxylate is not directly added, but is instead produced in-situ does not change the end product.).

>The structure implied by the process steps should be considered when assessing the patentability of product-by-process claims over the prior art, especially where the product can only be defined by the process steps by which the product is made, or where the manufacturing process steps would be expected to impart distinctive structural characteristics to the final product. See, e.g., In re Garnero, 412 F.2d 276, 279, 162 USPQ 221, 223 (CCPA 1979) (holding "interbonded by interfusion" to limit structure of the claimed composite and noting that terms such as "welded," "intermixed," "ground in place," "press fitted," and "etched" are capable of construction as structural limitations.)<

ONCE A PRODUCT APPEARING TO BE SUBSTANTIALLY IDENTICAL IS FOUND AND A 35 U.S.C. 102/ 103 REJECTION MADE, THE BURDEN SHIFTS TO THE APPLICANT TO SHOW AN UNOBVIOUS DIFFERENCE

"The Patent Office bears a lesser burden of proof in making out a case of prima facie obviousness for product-by-process claims because of their peculiar nature" than when a product is claimed in the conventional fashion. In re Fessmann, 489 F.2d 742, 744, 180 USPQ 324, 326 (CCPA 1974). Once the examiner provides a rationale tending to show that the claimed product appears to be the same or similar to that of the prior art, although produced by a different process, the burden shifts to applicant to come forward with evidence establishing an unobvious difference between the claimed product and the prior art product.

＊＊＊＊＊＊＊＊＊＊＊

實例

　　X 之專利申請案請求一多元醇（polyol），說明書中揭露了此多元醇乃用來形成具有結構式為 Z 之硬的聚氨酯發泡體。審查委員引證了一先前技術揭露了 X 之專利申請案之多元醇且亦揭露了多元醇可用來形成具有結構式 Z 之硬聚氨酯發泡體。但是 X 之說明書中亦揭露了該多元醇可由包括步驟 A, B, C 之方法製出，該方法具有新穎性且為非顯而易見。請問將請求項修改如下述是否可克服審查委員之核駁？

A polyol produced by the process comprising the steps A, B, and C, said polyol having the property of forming rigid polyurethane foam having structure formula Z.

　　無法克服審查委員之核駁，因為審查委員不考量 product by process claim 中之方法步驟，只要所製出之產品與先前技術相同，即使加入方法之限制條件亦不能與先前技術區別。

§5.17 請求項之形式

請求項（claims）必需從說明書另一頁開始，而且置於發明之詳細敘述之後。包括有請求項之頁不能再包含任何說明書之其他部份。請求項之標題需為 I claim（We claim），The Invention claimed is，或其他類似之寫法。每一請求項之第 1 字母需大寫且結束時用句點。每一請求項之中間不能有其他句點，除非是縮寫。當一請求項記載了複數個元件或步驟時，請求項之各元件或各步驟之間需留空白（line indentation）分開。

在說明書或圖式中元件之參考數字（標號）亦可用來敘述請求項中相同（對應）之元件，但是需用括弧括起以避免與其他數字混淆，在請求項中使用參考數字標記元件並不影響請求項之範圍。

請求項之安排需將範圍最大的請求項安排為 claim 1。請求之標的相同之依附項需安排在一起。例如 claim 1～10 均為請求產品，claim 11~20 請求方法。請求項如此安排才方便分類及審查。

MPEP 608.01(m) Form of Claims [R-7]

The claim or claims must commence on a separate physical sheet or electronic page and should appear after the detailed description of the invention. Any sheet including a claim or portion of a claim may not contain any other parts of the application or other material. While there is no set statutory form for claims, the present Office practice is to insist that each claim must be the object of a sentence starting with "I (or we) claim," "The invention claimed is" (or the equivalent). If, at the time of allowance, the quoted terminology is not present, it is inserted by the Office of **>Data Management<. Each claim begins with a capital letter and ends with a period. Periods may not be used elsewhere in the claims except for abbreviations. See Fressola v. Manbeck, 36 USPQ2d 1211 (D.D.C. 1995). Where a claim sets forth a plurality of elements or steps, each element or step of the claim should be separated by a line indentation, 37 CFR 1.75(i).

There may be plural indentations to further segregate subcombinations or related steps. In general, the printed patent copies will follow the format used but printing difficulties or expense may prevent the duplication of unduly complex claim formats.

Reference characters corresponding to elements recited in the detailed description and the drawings may be used in conjunction with the recitation of the same element or group of elements in the claims. The reference characters, however, should be enclosed within parentheses so as to avoid confusion with other numbers or characters which may appear in the claims. The use of reference characters is to be considered as having no effect on the scope of the claims.

Many of the difficulties encountered in the prosecution of patent applications after final rejection may be alleviated if each applicant includes, at the time of filing or no later than the first reply, claims varying from the broadest to which he or she believes he or she is entitled to the most detailed that he or she is willing to accept.

Claims should preferably be arranged in order of scope so that the first claim presented is the least restrictive. All dependent claims should be grouped together with the claim or claims to which they refer to the extent practicable. Where separate species are claimed, the claims of like species should be grouped together where possible. Similarly, product and process claims should be separately grouped. Such arrangements are for the purpose of facilitating classification and examination.

§5.18　美國專利圖式之標準

對於實用專利申請案及新式樣申請案有二種圖式是可接受的，一種是墨線圖（black ink drawing），即黑白圖式，使用永久墨水（India ink）在紙上繪製出之圖式。專利之圖式一般使用墨線圖。另一種是彩色圖式，只要申請專利之標的只能用彩色的圖式來表現時才使用彩色圖式。彩色圖式必需足夠清楚且在印成專利文件時可以黑白方式複製才可。在國際申請案（PCT）或以電子申請送件之申請案中則不能使用彩色圖式。如果在實用專利申請案及新式樣申請案中要使用彩色圖式，則需提出請願（petition）說明為何使用彩色圖式是必需的。請願書需包括(i)規費(ii)三份彩色圖式(iii)修正說明書以在圖式之簡單說明之第 1 段插入以下文字：

The patent or application file contains at least one drawing executed in color. Copies of this patent or patent application publication with color drawing(s) will be provided by the Office upon request and payment of the necessary fee.

在每一張圖之上方空白處之前方需作識別標記（identifying indicia），如申請案號。如申請案號尚不知，可註明發明名稱，發明人名稱或檔號。如果在申請之後才送入之圖式則需標示"Replacement Sheet"，"New Sheet"。如果送入修正圖式之註記版本（Marked-up copy）則需標示為"Annotated Sheet"。

另外，化學式，數學式，表格及波形如為了方便表示，亦可當作圖式。各化學式及數學式必需各自歸類為分別的圖式。各波形組必需以單一圖式呈現並使用同一縱軸及時間之橫軸。在說明書中討論之各波形必需以不同之字母（letter）標示在縱軸之旁邊。

繪製圖式所用之紙張必需是可撓的，強度夠，白色，平滑，不發光且堅韌的。所有的紙張需無裂痕，皺摺或折線，且無擦痕（erasures），變質（alterations），重複書寫（over writings），插字於行間（interlineations）。

所有繪圖之紙張必需是相同大小，較短之邊當作上方。紙張可使用二種尺寸：

(1) 21.0 cm. by 29.7 cm. (DIN size A4),

(2) 21.6 cm. by 27.9 cm. (8 1/2 by 11 inches).

在紙張之繪圖區域（insight）不能有框線，各張繪圖紙的上方需留至少 2.5cm 之空白，左方留至少 2.5cm 之空白，右方留至少 1.5cm 之空白，下方留至少 1.0cm 之空白而使得 A4 紙上有 17.0cm×26.2 cm 之繪圖區域，21.6 cm×27.9cm 之紙上有 17.6cm×24.4cm 之繪圖區域。

所附之圖式中需包括足夠的視圖以能完全顯示所要請求專利之發明，且各視圖皆按照在說明書中敘及之順序標以圖號（Fig.1、Fig.2 等）。這些視圖可包括平面圖，側視圖，立體圖，元件細部之視圖或是放大圖、斷面圖等。如果使用爆炸圖（Exploded view），則分開之元件需以括弧括起，以顯示各元件之關係或組合順序。如果機構、機器之組件太大時，亦可以分成數張紙繪製圖式；但當將二張或以上的圖併在一起時，需不失其完整性，且不能有任何元件未顯示。被截取剖面之視圖上需繪以斷線，且斷線之兩端需標以數字，此數字乃對應於剖面圖之圖號，且需標以其剖面方向之箭頭。移動的位置得以以斷線重疊地繪製在適當的圖式上，但如果圖面太擠，則需繪製在另一視圖上。

所有繪製在一張紙上之視圖，需能以直立方式在同一方向閱覽。如果視圖太大，亦可以從左至右橫向安排在同一張紙上之視圖。圖之尺寸大小為當需複製並將之縮小為 2/3 時仍不會造成圖式擁擠之程度。另外，當要顯示一機構之某一部份時，需將其放大以清楚顯示其細節，唯當在一張紙上空間不足以將機構之細節顯示清楚時，可使用二張以上之紙繪製。

　　線必須使用製圖工具或可得到令人滿意之複製效果之方法繪製。所有的線及字母必需為黑色，不可太淡，顏色需均勻且清楚以便能夠複製、影印。

　　陰影線需成分開之斜的平行線，且物品之陰影面最好用較粗之線，而光線應從左上方角落以 45°之角度射入。此外，表面輪廓最好用適當之陰影線顯示。

　　不同之視圖必須以連續之數字標成 Fig.1、Fig.2 等。圖式中所用之參考文字及數字必需清楚、未變形，且不必圈以圓圈；參考之數字及文字之高度需至少 1/8 英吋（3.2mm）並且當縮小時至少高度為 1/24 英吋（1.1mm）。此外，參考之數字及文字不可與圖繪製之線交叉或太接近所繪製之圖式，以致不能清楚辨別。

　　在不同視圖上之相同元件、組件及部份需標以相同之參考文字及數字，且對於不同之組件及部份不可標以相同之數字。在說明書中未敘及之參考記號（如箭頭），則不可標示在圖式中。

　　對於習知之元件、組件可以用符號或圖例表示，但需於說明書中敘及。另外，對於流程圖等，亦可於圖式中加入符號或圖例。如需要，亦可標以箭頭，但其大小不能大於參考數字或文字。

37CFR 1.84 Standards for drawings.

(a) Drawings. There are two acceptable categories for presenting drawings in utility and design patent applications.

(1) Black ink. Black and white drawings are normally required. India ink, or its equivalent that secures solid black lines, must be used for drawings; or

(2) Color. On rare occasions, color drawings may be necessary as the only practical medium by which to disclose the subject matter sought to be patented in a utility or design patent application or the subject matter of a statutory invention registration. The color drawings must be of sufficient quality such that all details in the drawings are reproducible in black and white in the printed patent. Color drawings are not permitted in international applications (see PCT Rule 11.13), or in an application, or copy thereof, submitted under the Office electronic filing system. The Office will accept color drawings in utility or design patent applications and statutory invention registrations only after granting a petition filed under this paragraph explaining why the color drawings are necessary. Any such petition must include the following:

(i) The fee set forth in § 1.17(h);

(ii) Three (3) sets of color drawings;

(iii) An amendment to the specification to insert (unless the specification contains or has been previously amended to contain) the following language as the first paragraph of the brief description of the drawings:

The patent or application file contains at least one drawing executed in color. Copies of this patent or patent application publication with color drawing(s) will be provided by the Office upon request and payment of the necessary fee.

＊＊＊＊＊＊＊＊＊＊＊

(c)Identification of drawings. Identifying indicia should be provided, and if provided, should include the title of the invention, inventor's name, and application number, or docket number (if any) if an application number has not been assigned to the application. If this information is provided, it must be placed on the front of each sheet within the top margin. Each drawing sheet submitted after the filing date of an application must be identified as either "Replacement Sheet" or "New Sheet" pursuant to § 1.121(d). If a marked-up copy of any amended drawing figure including annotations indicating the changes made is filed, such marked-up copy must be clearly labeled as "Annotated Sheet" pursuant to § 1.121(d)(1).

(d) Graphic forms in drawings. Chemical or mathematical formulae, tables, and waveforms may be submitted as drawings and are subject to the same requirements as drawings. Each chemical or mathematical formula must be labeled as a separate figure, using brackets when necessary, to show that information is properly integrated. Each group of waveforms must be presented as a single figure, using a common vertical axis with time extending along the horizontal axis. Each individual waveform discussed in the specification must be identified with a separate letter designation adjacent to the vertical axis.

(e) Type of paper. Drawings submitted to the Office must be made on paper which is flexible, strong, white, smooth, non-shiny, and durable. All sheets must be reasonably free from cracks, creases, and folds. Only one side of the sheet may be used for the drawing. Each sheet must be reasonably free from erasures and must be free from alterations, overwritings, and interlineations. Photographs must be developed on paper meeting the sheet-size requirements of paragraph (f) of this section and the margin requirements of paragraph (g) of this section. See paragraph (b) of this section for other requirements for photographs.

(f) Size of paper. All drawing sheets in an application must be the same size. One of the shorter sides of the sheet is regarded as its top. The size of the sheets on which drawings are made must be:

(1) 21.0 cm. by 29.7 cm. (DIN size A4), or

(2) 21.6 cm. by 27.9 cm. (8 1/2 by 11 inches).

(g) Margins. The sheets must not contain frames around the sight (i.e., the usable surface), but should have scan target points (i.e., cross-hairs) printed on two cater-corner margin corners. Each sheet must include a top margin of at least 2.5 cm. (1 inch), a left side margin of at least 2.5 cm. (1 inch), a right side margin of at least 1.5 cm. (5/8 inch), and a bottom margin of at least 1.0 cm. (3/8 inch), thereby leaving a sight no greater than 17.0 cm. by 26.2 cm. on 21.0 cm. by 29.7 cm. (DIN size A4) drawing sheets, and a sight no greater than 17.6 cm. by 24.4 cm. (6 15/16 by 9 5/8 inches) on 21.6 cm. by 27.9 cm. (8 1/2 by 11 inch) drawing sheets.

(h) Views. The drawing must contain as many views as necessary to show the invention. The views may be plan, elevation, section, or perspective views. Detail views of portions of elements, on a larger scale if necessary, may also be used. All views of the drawing must be grouped together and arranged on the sheet(s) without wasting space, preferably in an upright position, clearly separated from one another, and must not be included in the sheets containing the specifications, claims, or abstract. Views must not be connected by projection lines and must not contain center lines. Waveforms of electrical signals may be connected by dashed lines to show the relative timing of the waveforms.

(1) Exploded views. Exploded views, with the separated parts embraced by a bracket, to show the relationship or order of assembly of various parts are permissible. When an exploded view is shown in a figure which is on the same sheet as another figure, the exploded view should be placed in brackets.

(2) Partial views. When necessary, a view of a large machine or device in its entirety may be broken into partial views on a single sheet, or extended over several sheets if there is no loss in facility of understanding the view. Partial views drawn on separate sheets must always be capable of being linked edge to edge so that no partial view contains parts of another partial view. A smaller scale view should be included showing the whole formed by the partial views and indicating the positions of the parts shown. When a portion of a view is enlarged for magnification purposes, the view and the enlarged view must each be labeled as separate views.

✱ ✱ ✱ ✱ ✱ ✱ ✱ ✱ ✱ ✱ ✱

(k) Scale. The scale to which a drawing is made must be large enough to show the mechanism without crowding when the drawing is reduced in size to two-thirds in reproduction. Indications such as "actual size" or "scale 1/2" on the drawings are not permitted since these lose their meaning with reproduction in a different format.

(l) Character of lines, numbers, and letters. All drawings must be made by a process which will give them satisfactory reproduction characteristics. Every line, number, and letter must be durable, clean, black (except for color drawings), sufficiently dense and dark, and uniformly thick and well-defined. The weight of all lines and letters must be heavy enough to permit adequate reproduction. This requirement applies to all lines however fine, to shading, and to lines representing cut surfaces in sectional views. Lines and strokes of different thicknesses may be used in the same drawing where different thicknesses have a different meaning.

(m) Shading. The use of shading in views is encouraged if it aids in understanding the invention and if it does not reduce legibility. Shading is used to indicate the surface or shape of spherical, cylindrical, and conical elements of an object. Flat parts may also be lightly shaded. Such shading is preferred in the case of parts shown in perspective, but not for cross sections. See paragraph (h)(3) of this section. Spaced lines for shading are preferred. These lines must be thin, as few in number as practicable, and they must contrast with the rest of the drawings. As a substitute for shading, heavy lines on the shade side of objects can be used except where they superimpose on each other or obscure reference characters. Light should come from the upper left corner at an angle of 45°. Surface delineations should preferably be shown by proper shading. Solid black shading areas are not permitted, except when used to represent bar graphs or color.

(n)Symbols. Graphical drawing symbols may be used for conventional elements when appropriate. The elements for which such symbols and labeled representations are used must be adequately identified in the specification. Known devices should be illustrated by symbols which have a universally recognized conventional meaning and are generally accepted in the art. Other symbols which are not universally recognized may be used, subject to approval by the Office, if they are not likely to be confused with existing conventional symbols, and if they are readily identifiable.

(o) Legends. Suitable descriptive legends may be used subject to approval by the Office, or may be required by the examiner where necessary for understanding of the drawing. They should contain as few words as possible.

(p) Numbers, letters, and reference characters.

(1) Reference characters (numerals are preferred), sheet numbers, and view numbers must be plain and legible, and must not be used in association with brackets or inverted commas, or enclosed within outlines, e.g., encircled. They must be oriented in the same direction as the view so as to avoid having to rotate the sheet. Reference characters should be arranged to follow the profile of the object depicted.

(2) The English alphabet must be used for letters, except where another alphabet is customarily used, such as the Greek alphabet to indicate angles, wavelengths, and mathematical formulas.

(3) Numbers, letters, and reference characters must measure at least.32 cm. (1/8 inch) in height. They should not be placed in the drawing so as to interfere with its comprehension.

Therefore, they should not cross or mingle with the lines. They should not be placed upon hatched or shaded surfaces. When necessary, such as indicating a surface or cross section, a reference character may be underlined and a blank space may be left in the hatching or shading where the character occurs so that it appears distinct.

(4) The same part of an invention appearing in more than one view of the drawing must always be designated by the same reference character, and the same reference character must never be used to designate different parts.

(5) Reference characters not mentioned in the description shall not appear in the drawings. Reference characters mentioned in the description must appear in the drawings.

(q) Lead lines. Lead lines are those lines between the reference characters and the details referred to. Such lines may be straight or curved and should be as short as possible. They must originate in the immediate proximity of the reference character and extend to the feature indicated. Lead lines must not cross each other. Lead lines are required for each reference character except for those which indicate the surface or cross section on which they are placed. Such a reference character must be underlined to make it clear that a lead line has not been left out by mistake. Lead lines must be executed in the same way as lines in the drawing. See paragraph (l) of this section.

(r) Arrows. Arrows may be used at the ends of lines, provided that their meaning is clear, as follows:

(1) On a lead line, a freestanding arrow to indicate the entire section towards which it points;

(2) On a lead line, an arrow touching a line to indicate the surface shown by the line looking along the direction of the arrow; or

(3) To show the direction of movement.

(s) Copyright or Mask Work Notice. A copyright or mask work notice may appear in the drawing, but must be placed within the sight of the drawing immediately below the figure representing the copyright or mask work material and be limited to letters having a print size of 32 cm. to 64 cm. (1/8 to 1/4 inches) high. The content of the notice must be limited to only those elements provided for by law. For example, "©1983 John Doe" (17 U.S.C. 401) and "*M* John Doe" (17 U.S.C. 909) would be properly limited and, under current statutes, legally sufficient notices of copyright and mask work, respectively. Inclusion of a copyright or mask work notice will be permitted only if the authorization language set forth in § 1.71(e) is included at the beginning (preferably as the first paragraph) of the specification.

(t) Numbering of sheets of drawings. The sheets of drawings should be numbered in consecutive Arabic numerals, starting with 1, within the sight as defined in paragraph (g) of this section. These numbers, if present, must be placed in the middle of the top of the sheet, but not in the margin. The numbers can be placed on the right-hand side if the drawing extends too close to the middle of the top edge of the usable surface. The drawing sheet numbering must be clear and larger than the numbers used as reference characters to avoid confusion. The number of each sheet should be shown by two Arabic numerals placed on either side of an oblique line, with the first being the sheet number and the second being the total number of sheets of drawings, with no other marking.

(u) Numbering of views.

(1) The different views must be numbered in consecutive Arabic numerals, starting with 1, independent of the numbering of the sheets and, if possible, in the order in which they appear on the drawing sheet(s). Partial views intended to form one complete view, on one or several sheets, must be identified by the same number followed by a capital letter. View numbers must be preceded by the abbreviation "FIG." Where only a single view is used in an application to illustrate the claimed invention, it must not be numbered and the abbreviation "FIG." must not appear.

(2) Numbers and letters identifying the views must be simple and clear and must not be used in association with brackets, circles, or inverted commas. The view numbers must be larger than the numbers used for reference characters.

(v) Security markings. Authorized security markings may be placed on the drawings provided they are outside the sight, preferably centered in the top margin.

(w) Corrections. Any corrections on drawings submitted to the Office must be durable and permanent.

(x) Holes. No holes should be made by applicant in the drawing sheets.

(y) Types of drawings. See § 1.152 for design drawings, § 1.165 for plant drawings, and § 1.173(a)(2) for reissue drawings.

§5.19　使用黑白或彩色相片作為圖式

當用墨線圖（繪製之圖）無法正確或充份地圖示發明時，可用相片或顯微相片（photomicrographs）取代。例如電泳凝膠圖，墨點圖（blots），如免疫西方、南方、北方墨點圖，放射線圖，細胞培養圖（染色或未染色），組織截面圖（染色或未染色），動物、植物、活體影像圖，薄膜色層分析圖，結晶構造圖，合金微結構圖，紡織纖維圖，晶粒結構圖及顯示裝飾效果之新式樣圖式等。相片必需較用繪製圖更能清楚地顯示發明時才能使用。

　　如果用黑白相片取代繪製之圖式，並不需要提出請願，且只要呈送一份黑白相片即可。黑白相片需製作在表面光滑，白色的紙之上。2001 年 10 月 1 日以後申請之案件，相片可不必再粘於厚紙板（Bristol Board）上。如果數張黑白相片構成一圖式，則此數張黑白相片需製作在單張紙上。

　　在實用專利申請案及新式樣專利申請案中使用彩色相片(圖式)是有限制的。除非提出請願且請願核准，才可使用彩色相片，不然審查委員就會要求申請人將彩色相片刪除或補送取代之黑白圖式。

　　申請人請願時需繳規費及三份彩色相片。是否准予使用彩色相片將由主管級審查委員（supervisory patent examiner）決定。

　　如果在延續案（包括 CIP 案）中使用彩色相片，申請人需再提一次請願，在請願未核准前，審查委員會對使用彩色相片作形式核駁（objection）。

　　由於要在專利證書上印刷彩色相片（圖式）較麻煩，在證書上美國專利局會以黑白印刷，但在發給申請人之專利證書正本上會附上一份彩色相片。此外，任何人如果提出特殊要求及繳交規費，美國專利局可提供一份附有彩色相片之專利證書。

　　再者，在提出請願時，申請人需於圖式之簡單說明書第 1 段以修正之方式加以下述之說明：

The patent or application file contains at least one drawing executed in color. Copies of this patent or patent application publication with color drawing(s) will be provided by the U.S. Patent and Trademark Office upon request and payment of the necessary fee.

　　美國專利局只有當認為彩色相片是唯一可行的方式來顯示專利申請案所要請求之標的時，才會核准此請願。另外，需強調的是核准使用彩色相片不應被認為彩色相片（圖式）是符合法律要件所必需的。使用彩色相片在可避免，例如，缺乏據以實施要件，或新事項等法律上之缺失時，才是適宜的。

37CFR 1.84 VII. BLACK AND WHITE PHOTOGRAPHS

　　Photographs or photomicrographs (not photolithographs or other reproductions of photographs made by using screens) printed on sensitized paper are acceptable as final drawings, in lieu of India ink drawings, to illustrate inventions which are incapable of being accurately or adequately depicted by India ink drawings, e.g., electrophoresis gels, blots, (e.g., immunological, western, Southern, and northern), autoradiographs, cell cultures (stained and unstained), histological tissue cross sections (stained and unstained), animals, plants, in vivo imaging, thin layer chromatography plates, crystalline structures, metallurgical microstructures, textile fabrics, grain structures and ornamental effects. The photographs or photomicrographs must show the invention more clearly than they can be done by India ink drawings and otherwise comply with the rules concerning such drawings.

Black and white photographs submitted in lieu of ink drawings must comply with 37 CFR 1.84(b). There is no requirement for a petition or petition fee, and only one set of photographs is required. See 1213 O.G. 108 (Aug. 4, 1998) and 1211 O.G. 34 (June 9, 1998) and 37 CFR 1.84(b)(1).

Such photographs to be acceptable must be made on photographic paper having the following characteristics which are generally recognized in the photographic trade: double weight paper with a surface described as smooth with a white tint. Note that photographs filed on or after October 1, 2001 may no longer be mounted on Bristol Board. See 37 CFR 1.84(e) and 1246 O.G. 106 (May 22, 2001). If several photographs are used to make one sheet of drawings, the photographs must be contained (i.e., developed) on a single sheet.

＊ ＊ ＊ ＊ ＊ ＊ ＊ ＊ ＊ ＊ ＊ ＊

VIII.COLOR DRAWINGS OR COLOR PHOTOGRAPHS

Limited use of color drawings in utility patent applications is provided for in 37 CFR 1.84(a)(2) and (b)(2). Unless a petition is filed and granted, color drawings or color photographs will not be accepted in a utility or design patent application. The examiner must object to the color drawings or color photographs as being improper and require applicant either to cancel the drawings or to provide substitute black and white drawings.

Under 37 CFR 1.84(a)(2) and (b)(2), the applicant must file a petition with fee requesting acceptance of the color drawings or color photographs. Three sets of color drawings or color photographs must also be submitted(37 CFR1.84(a)(2)(ii)). The petition is decided by a Supervisory Patent Examiner. See MPEP § 1002.02(d).

If the application is an IFW application, the color photographs are maintained in an artifact folder.

Where color drawings or color photographs are filed in a continuing application, applicant must renew the petition under 37 CFR 1.84(a)(2) and (b)(2) even though a similar petition was filed in the prior application. Until the renewed petition is granted, the examiner must object to the color drawings or color photographs as being improper.

In light of the substantial administrative and economic burden associated with printing a utility patent with color drawings or color photographs, the patent copies which are printed at issuance of the patent will depict the drawings in black and white only. However, a set of color drawings or color photographs will be attached to the Letters Patent. Moreover, copies of the patent with color drawings or color photographs attached thereto will be provided by the U.S. Patent and Trademark Office upon special request and payment of the fee necessary to recover the actual costs associated therewith.

Accordingly, the petition must also be accompanied by a proposed amendment to insert the following language as the first paragraph in the portion of the specification containing a brief description of the drawings:

The patent or application file contains at least one drawing executed in color. Copies of this patent or patent application publication with color drawing(s) will be provided by the U.S. Patent and Trademark Office upon request and payment of the necessary fee.

＊＊＊＊＊＊＊＊＊＊＊

It is emphasized that a decision to grant the petition should not be regarded as an indication that color drawings or color photographs are necessary to comply with a statutory requirement. In this latter respect, clearly it is desirable to file any desired color drawings or color photographs as part of the original application papers in order to avoid issues concerning statutory defects (e.g., lack of enablement under 35 U.S.C. 112 or new matter under 35 U.S.C. 132).

§5.20　圖式與所請求之發明的關係

專利申請所附之圖式必需要顯示請求項之每一特徵，亦即請求項所請求之發明之每一元件均需在圖式中顯示。然而，如果在說明書本文及請求項中敘及之習知的特徵（conventional features），其詳細之圖示對於了解所請求之發明並不重要時，可用一方框來表示。此外，在說明書中之表格及序列表，除非是 PCT 進入美國階段專利申請案，不允許被包括在圖式之中。再者，如果發明是一舊的機器之改良，則圖式可在一視圖中只顯示該改良部份，而在另外的視圖中只顯示舊的構造及其連接關係即可。

如果圖式不符合上述之規定時，審查委員會發下通知，給申請人不少於二個月之時間補充修正圖式。

37 CFR 1.83 Content of drawing.

(a) **>The drawing in a nonprovisional application must show every feature of the invention specified in the claims. However, conventional features disclosed in the description and claims, where their detailed illustration is not essential for a proper understanding of the invention, should be illustrated in the drawing in the form of a graphical drawing symbol or a labeled representation (e.g., a labeled rectangular box). In addition, tables and sequence listings that are included in the specification are, except for applications filed under 35 U.S.C. 371, not permitted to be included in the drawings.<

(b) When the invention consists of an improvement on an old machine the drawing must when possible exhibit, in one or more views, the improved portion itself, disconnected from the old structure, and also in another view, so much only of the old structure as will suffice to show the connection of the invention therewith.

(c) Where the drawings in a nonprovisional application do not comply with the requirements of paragraphs (a) and (b) of this section, the examiner shall require such additional illustration within a time period of not less than two months from the date of the sending of a notice thereof. Such corrections are subject to the requirements of § 1.81(d)

第6章　主張國外優先權

目　錄

§6.1　主張國外優先權之意義及利益

　　一美國專利申請案之申請人如果對相同之發明在美國專利申請案之申請日之前已在外國提出專利申請時，主張國外之優先權可以取得國外申請案之申請日之利益。所謂取得國外申請案之申請日之利益乃指一旦國外優先權之主張完備之後，申請人就可依賴國外申請日作為申請人之發明的發明日之證據。例如，當審查委員引證一 35 USC §102 之先前技術，而該先前技術之日期在申請人之美國專利申請案之申請日之前，但在申請人國外優先權日之後時，申請可能藉由國外優先權之主張以克服該引證資料而建立其專利性。

MPEP 201.13

　　Under certain conditions and on fulfilling certain requirements, an application for patent filed in the United States may be entitled to the benefit of the filing date of a prior application filed in a foreign country, to overcome an intervening reference or for similar purposes.

＊＊＊＊＊＊＊＊＊＊＊＊

實例

　　發明人 X 為德國公民於 1997 年 7 月 25 日發明人了一訂書機，並於 1998 年 1 月 22 日提出一德國發明專利申請。X 於 1999 年 1 月 22 日提出一美國專利申請，並主張德國之優先權。美國專利局於 1999 年 4 月 16 日發下一審定書並以一雜誌作為引證以 35 USC§ 102(a)核駁本案所有的請求項。該雜誌於 1998 年 2 月於美國發行，且揭示了與相同之訂書機之製造及使用。請問 X 應如何克服此 35 USC§ 102(a)之核駁？

　　X 可於 1999 年 7 月 16 日前提出該德國專利申請案之優先權文件以及其認證翻譯本主張該引證並非本案之先前技術。

§6.2　藉由主張國外優先權取得國外申請日之利益之要件

一美國專利申請案要藉由主張國外優先權取得國外申請日之利益，需滿足以下之要件：

(A) 外國申請案必需在經承認之外國（對於在美國之專利申請案或對美國公民提供相同之優先權利益之外國，或 WTO 之會員國），提出專利申請者；

(B) 此外國申請案必需與美國專利案有相同之申請人（發明人），或由其法律上之代表人所提出申請者；

(C) 此美國專利申請案或其較早申請之母案，必需是在經承認外國之外國申請之最早申請日之一年內提出申請者（如為新式樣申請案則需於 6 個月內申請）；

(D) 此外國申請案必需與美國申請案之發明相同（美國專利申請案請求之發明必需在外國申請案中有揭露）；

(E) 2000 年 11 月 29 日當天或以後申請之專利申請案（不包括新式樣專利申請案）需於專利審查期間提出優先權之主張，且此優先權主張需於美國專利申請案之實際申請日起 4 個月或外國申請案之申請日起 16 個月之較晚日期之前提出（不可延期）。

(F) 對於 2000 年 11 月 29 日當天或之後申請之 PCT 專利申請案，進入美國國家階段之申請案，在符合 35USC§371 之規定之後需於申請案審查期間提出優先權之主張，且此主張需於 PCT 法律及細則所規定之期限內提出；

(G) 對於所主張外國申請案為發明人證書之申請案者，亦需符合 37CFR 1.55(b) 之要件。

---

35 U.S.C. 119 Benefit of earlier filing date; right of priority.

(a) An application for patent for an invention filed in this country by any person who has, or whose legal representatives or assigns have, previously regularly filed an application for a patent for the same invention in a foreign country which affords similar privileges in the case of applications filed in the United States or to citizens of the United States, or in a WTO member country, shall have the same effect as the same application would have if filed in this country on the date on which the application for patent for the same invention was first filed in such foreign country, if the application in this country is filed within twelve months from the earliest date on which such foreign application was filed; but no patent shall be granted on any application for patent for an invention which had been patented or described in a printed publication in any country more than one year before the date of the actual filing of the application in this country, or which had been in public use or on sale in this country more than one year prior to such filing.

(b)(1) No application for patent shall be entitled to this right of priority unless a claim is filed in the Patent and Trademark Office, identifying the foreign application by specifying the application number on that foreign application, the intellectual property authority or country in or for which the application was filed, and the date of filing the application, at such time during the pendency of the application as required by the Director.

(2) The Director may consider the failure of the applicant to file a timely claim for priority as a waiver of any such claim. The Director may establish procedures, including the payment of a surcharge, to accept an unintentionally delayed claim under this section.

(3) The Director may require a certified copy of the original foreign application, specification, and drawings upon which it is based, a translation if not in the English language, and such other information as the Director considers necessary. Any such certification shall be made by the foreign intellectual property authority in which the foreign application was filed and show the date of the application and of the filing of the specification and other papers.

(c) In like manner and subject to the same conditions and requirements, the right provided in this section may be based upon a subsequent regularly filed application in the same foreign country instead of the first filed foreign application, provided that any foreign application filed prior to such subsequent application has been withdrawn, abandoned, or otherwise disposed of, without having been laid open to public inspection and without leaving any rights outstanding, and has not served, nor thereafter shall serve, as a basis for claiming a right of priority.

(d) Applications for inventors' certificates filed in a foreign country in which applicants have a right to apply, at their discretion, either for a patent or for an inventor's certificate shall be treated in this country in the same manner and have the same effect for purpose of the right of priority under this section as applications for patents, subject to the same conditions and requirements of this section as apply to applications for patents, provided such applicants are entitled to the benefits of the Stockholm Revision of the Paris Convention at the time of such filing.

＊ ＊ ＊ ＊ ＊ ＊ ＊ ＊ ＊ ＊ ＊

MPEP 210.13

＊＊＊＊＊＊＊＊＊＊＊

The period of 12 months specified in this section is 6 months in the case of designs, 35 U.S.C. 172. See MPEP § 1504.10.

The conditions, for benefit of the filing date of a prior application filed in a foreign country, may be listed as follows:

(A) The foreign application must be one filed in "a foreign country which affords similar privileges in the case of applications filed in the United States or to citizens of the United States or in a WTO member country."

(B) The foreign application must have been filed by the same applicant (inventor) as the applicant in the United States, or by his or her legal representatives or assigns.

(C) The application, or its earliest parent United States application under 35 U.S.C. 120, must have been filed within 12 months from the date of the earliest foreign filing in a "recognized" country as explained below.

(D) The foreign application must be for the same invention as the application in the United States.

(E) For an original application filed under 35 U.S.C. 111(a) (other than a design application) on or after November 29, 2000, the claim for priority must be presented during the pendency of the application, and within the later of four months from the actual filing date of the application or sixteen months from the filing date of the prior foreign application. This time period is not extendable.

(F) For applications that entered the national stage from an international application filed on or after November 29, 2000, after compliance with 35 U.S.C. 371, the claim for priority must be made during the pendency of the application and within the time limit set forth in the PCT Article and Regulations.

(G) In the case where the basis of the claim is an application for an inventor's certificate, the requirements of 37 CFR 1.55(b) must also be met.

＊＊＊＊＊＊＊＊＊＊＊

§6.3　經承認之國家（recognized country）

　　各國之間專利申請可主張優先權乃是起源於 1883 年由世界智慧財產權組織（World Intellectual Property Organization（WIPO）所主導之巴黎協約（Paris Convention）。美國乃於 1887 年加入。巴黎協約最近一次之修訂乃在 1967 年在斯德哥爾摩（Stockholm）。依照斯德哥爾摩修法，會員國需給予其他會員國主張優先權之權利。

　　美國與其他各國之間之優先權主張除了基於上述多邊協定（multilateral treaty），亦可基於國與國之間的雙邊協定或互惠協定。例如美國於 1996 年 4 月 10 日與台灣訂有專利、商標互相主張優先權之協定。美國與一些拉丁美洲之國家亦訂有互相主張優先權之協定。

　　以下是與美國有互相主張優先權協定之經承認之國家（recognized country）之一覽表。其其中國家之後之字母 "I" 表示是基於巴黎協約，"P"表示基於 Inter-American Convention，"L" 表示為互惠協定（reciprocal legislation），"W" 表示基於 WTO（世界貿易組織）之協定，"W°" 表示該國在 1996 年 1 月 1 日後成為 WTO 之會員會。

Albania (I, W°),
Algeria (I),
Angola (W°),
>Andorra (I),<
Antigua and Barbuda (I, W),
Argentina (I, W),
Armenia (I>, W°<),
Australia (I, W),
Austria (I, W),
Azerbaijan (I),
Bahamas (I),
Bahrain (I, W),
Bangladesh (I, W),
Barbados (I, W),
Belarus (I),
Belgium (I, W),

Burkina Faso (I, W),
Burundi (I, W),
Cambodia (I>, W°<),
Cameroon (I, W),
Canada (I, W),
Central African Republic (I, W),
Chad (I, W°),
Chile (I, W),
China (I, W°)
Colombia (I, W),
>Comoros (I),<
Congo (I, W°),
Costa Rica (I, P, W),
Cote d'Ivoire (I, W),
Croatia (I, W°),
Cuba (I, P, W),

Belize (I, W),

Benin (I, W°),

Bhutan (I),

Bolivia (I, P, W),

Bosnia and Herzegovina (I),

Botswana (I, W),

Brazil (I, P, W),

Brunei Darussalam (W),

Bulgaria (I, W°),

Egypt (I, W),

El Salvador (I, W),

Equatorial Guinea (I),

Estonia (I, W°),

European Community (W),

Fiji (W°),

Finland (I, W),

France (I, W),

Gabon (I, W),

Gambia (I, W°),

Georgia (I, W°),

Germany (I, W),

Ghana (I, W),

Greece (I, W),

Grenada (I, W°),

Guatemala (I, P, W),

Guinea (I, W),

Guinea-Bissau (I , W),

Guyana (I, W),

Haiti (I, P, W°),

Holy See (I),

Honduras (I, P, W),

Hong Kong Special

Administrative Region of China (I, W)

Hungary (I, W),

Madagascar (I, W),

Malawi (I, W),

Malaysia (I, W),

Maldives (W),

Mali (I, W),

Malta (I, W),

Mauritania (I, W),

Cyprus (I, W),

Czech Republic (I, W),

Democratic People's Republic of Korea (I),

Democratic Republic of the Congo (I, W°),

Denmark (I, W),

Djibouti (I, W),

Dominica (I, W),

Dominican Republic (I, P, W),

Ecuador (I, P, W°),

India (I, W),

Indonesia (I, W),

Iran (Islamic Republic of) (I),

Iraq (I),

Ireland (I, W),

Israel (I, W),

Italy (I, W),

Jamaica (I, W),

Japan (I, W),

Jordan (I, W°),

Kazakstan (I),

Kenya (I, W),

Kuwait (W),

Kyrgyzstan (I, W°),

Lao People's Democratic Republic (I),

Latvia (I, W°),

Lebanon (I),

Lesotho (I, W),

Liberia (I),

Libya (I),

Libyan Arab Jamahiriya (I),

Liechtenstein (I, W),

Lithuania (I, W°),

Luxembourg (I, W),

Papua New Guinea (I, W°),

Paraguay (I, P, W),

Peru (I, W),

Philippines (I, W),

Poland (I, W),

Portugal (I, W),

Qatar (I, W°),

Mauritius (I, W),

Mexico (I, W),

Monaco (I),

Mongolia (I, W°),

Morocco (I, W),

Mozambique (I, W),

Myanmar (W),

Namibia (I, W),

Nepal (I>, W°<),

Netherlands (I, W,),

New Zealand (I, W),

Nicaragua (I, P, W),

Niger (I, W°),

Nigeria (I, W),

Norway (I, W),

Oman (I, PW°),

Pakistan (>I,< W),

Panama (I, W°),

Solomon Islands (W°),

South Africa (I, W),

Spain (I, W),

Sri Lanka (I, W),

Sudan (I),

Suriname (I, W),

Swaziland (I, W),

Sweden (I, W),

Switzerland (I, W),

Syrian Arab Republic (I),

Taiwan>, Province of China (Chinese Taipei)< (L, W°),

Tajikistan (I),

Tanzania, United Republic of (I, W),

Thailand (L, W),

The former Yugoslav Republic of Macedonia (I, W°),

Togo (I, W),

Republic of Korea (I, W),

Republic of Moldova (I, W°),

Romania (I, W),

Russian Federation (I),

Rwanda (I, W°),

Saint Kitts and Nevis (I, W°),

Saint Lucia (I, W),

Saint Vincent and the Grenadines (I, W),

San Marino (I),

Sao Tome and Principe (I),

Saudi Arabia (I)

Senegal (I, W),

Serbia and Montenegro (I)

Seychelles (I)

Sierra Leone (I, W),

Singapore (I, W),

Slovakia (I, W),

Slovenia (I, W),

Tonga (I),

Trinidad and Tobago (I, W),

Tunisia (I, W),

Turkey (I, W),

Turkmenistan (I),

Uganda (I, W),

Ukraine (I),

United Arab Emirates (I, W°),

United Kingdom (I, W),

Uruguay (I, P, W),

Uzbekistan (I),

Venezuela (I, W),

Viet Nam (I),

Zambia (I, W),

Zimbabwe (I, W).

實例 1

申請人 A 於 1980 年 1 月 15 日申請一英國專利，其關於一新的感光薄膜。之後在 1981 年 7 月 5 日其又申請一加拿大專利，有關於該新的感光薄膜及一新穎的鹵化銀結晶其可用於感光薄膜之光敏層。A 於 1981 年 10 月提出一美國專利申請案請求具有該鹵化銀結晶之感光薄膜。請問 A 之美國專利申請案可主張哪一國外申請案之優先權？

可主張加拿大申請案之優先權，因為美國專利申請案只請求加拿大申請案所加入的新穎的鹵化銀結晶部份，且美國申請案之申請日仍在加拿大申請案申請日之後的一年之內。

實例 2

A 為法國之公民，其於 1990 年 1 月 3 日於法國提出一專利申請案，且此法國專利申請案於 1991 年 2 月 4 日取得專利。A 同時於 1991 年 1 月 2 日提出英國專利主張該法國之優先權，此英國專利於 1991 年 8 月 12 日獲准專利。A 之後於 1992 年 1 月 2 日提出美國專利申請案。請問此美國專利申請案是否可主張國外優先權？

不能主張法國或英國專利申請案之優先權。因為發明之內容相同。故需於最早申請日（法國專利申請案）1990 年 1 月 3 日起一年之內提出美國專利申請案才可主張優先權。

實例 3

北韓發明人 A 於 2000 年 1 月 10 日於北韓提出一發明專利申請，之後又於 2000 年 10 月 10 日於南韓提出一發明專利申請，後於 2001 年 5 月 1 日於美國提出一發明專利申請。請問發明人 A 的美國專利申請案可主張哪一國外優先權？

可主張南韓申請案之優先權，但不可主張北韓申請案之優先權。因為北韓並非 35 USC §119 規定之承認之國家，故 12 個月之主張期限並非從北韓專利申請日起算而是從南韓專利申請日起算。

§6.4　主張已撤回或放棄之國外申請案之優先權

如果美國專利申請案要主張已撤回或已放棄之國外申請案之優先權的話，只要該先前之國外申請案未被公開，則在該國外申請案之申請日一年之內，此國外優先權之主張仍會被接受。35 USC§119 之規定並未規定國外申請案與美國專利申請案需為共同審查中（co pending）之情形。

---

35 U.S.C. 119 Benefit of earlier filing date; right of priority.

In like manner and subject to the same conditions and requirements, the right provided in this section may be based upon a subsequent regularly filed application in the same foreign country instead of the first filed foreign application, provided that any foreign application filed prior to such subsequent application has been withdrawn, abandoned, or otherwise disposed of, without having been laid open to public inspection and without leaving any rights outstanding, and has not served, nor thereafter shall serve, as a basis for claiming a right of priority.

實例 1

發明人 A 於 2002 年 1 月 10 日提出一台灣發明專利申請，後於 2002 年 6 月 10 日撤回此申請案且並未在其他國家提出專利申請。發明人 A 又於 2003 年 2 月 10 日再度提出台灣發明專利申請。請問發明人 A 如於 2004 年 2 月 10 日前於美國提出發明專利申請可否主張 2003 年 2 月 10 日提申之台灣發明專利申請案之優先權？

可以。因為 2002 年 1 月 10 日提申之台灣申請案已撤回，且並未公開。

實例 2

發明人 A 於 2002 年 1 月 10 日提出一專利申請，並於 2002 年 6 月 10 日撤回此專利申請案，請問如果發明人 A 於 2003 年 1 月 10 日前提出一美國發明專利申請案是否可主張該已撤回之台灣專利申請案之優先權？

可以。依據 35 USC§119 之優先權並未要求國外申請案須與美國專利申請案為共同審查中。

§6.5 主張複數的國外優先權

一非暫時專利申請案之申請人只要符合 35§119(e)、(d)及(f)以及 35USC §172 及 365(a)及(b)之規定可以主張複數的國外優先權。

---

35 U.S.C. 172 Right of priority.

The right of priority provided for by subsections (a) through (d) of section 119 of this title and the time specified in section 102(d) shall be six months in the case of designs. The right of priority provided for by section 119(e) of this title shall not apply to designs.

35 U.S.C. 365 Right of priority; benefit of the filing date of a prior application.

(a) In accordance with the conditions and requirements of subsections (a) through (d) of section 119 of this title, a national application shall be entitled to the right of priority based on a prior filed international application which designated at least one country other than the United States.

(b) In accordance with the conditions and requirements of section 119(a) of this title and the treaty and the Regulations, an international application designating the United States shall be entitled to the right of priority based on a prior foreign application, or a prior international application designating at least one country other than the United States.

＊＊＊＊＊＊＊＊＊＊＊

37 CFR§1.55 Claim for foreign priority.

(a) An applicant in a nonprovisional application may claim the benefit of the filing date of one or more prior foreign applications under the conditions specified in 35 U.S.C. 119(a) through (d) and (f), 172, and 365(a) and (b).

＊＊＊＊＊＊＊＊＊＊＊

實例 1

　　發明人 A 於 2007 年 1 月 14 日申請了一台灣發明專利申請案 TW1，並於 2007 年 10 月 1 日申請了一美國發明專利申請案（US1）主張該台灣專利申請案之優先權。之後，發明人 A 又於 2008 年 1 月 10 日提申一台灣發明專利申請案（TW2）主張該 2007 年 1 月 14 日申請之台灣專利申請案之國內優先權。請問發明人如果要就 2008 年 1 月 10 日提申之台灣發明專利申請案提出美國專利申請案（US2），應如何主張優先權？

　　如果該美國專利申請案（US2）在 2008 年 1 月 14 日前提申則可同時主張 TW1 及 TW2 之優先權。如果在 2008 年 1 月 14 日以後提出則建議主張 US1 之優先權及 TW2 之優先權。

§6.6　主張國外優先權之程序要件（formal requirement）

　　依照 35USC119(b)，申請人要主張國外優先權時需在一特定的時間滿足一些程序之要求。例如優先權文件國外申請案之認證本（certified copy）必需在專利發證之前送入美國專利局。但是美國專利局亦可在專利申請案還在審查期間設定一較早的時間要求申請人送入優先權文件，例如審查委員發現在美國專利申請案之申請日與國外申請案之申請日之期間有一先前技術（intervening reference）時。另外，如果優先權文件為非英文時，美國專利局會要求呈送優先權文件之英文譯本。

　　依照 37CFR§1.55，主張優先權時申請人需註明國外申請案之申請案號，國外申請案提出申請之專利局（智慧財產局）名稱及國家以及國外申請案之申請日，這些資訊必需在美國申請案之申請日之 4 個月或國外申請日之 16 個月內（視哪一期限為晚）之前呈送美國專利局。上述期限是不能延期的，如果未在上述期限內提供上述資訊，則將被視為放棄優先權日，如果為未於上述期限內呈送國外優先權之資訊，而此未呈送優先權資訊不是故意的（unintentionally）時以可提出請願（petition）之方式請求恢復優先權之主張。

　　上述優先權主張及優先權資訊可包括在申請資料表（Application Data Sheet）中或者包括在宣誓書中均可。

MPEP 201.14 Right of Priority, Formal Requirements [R-5]

Under the statute (35 U.S.C. 119(b)), an applicant who wishes to secure the right of priority must comply with certain formal requirements within a time specified. If these requirements are not complied with the right of priority is lost and cannot thereafter be asserted.

For nonprovisional applications filed prior to November 29, 2000, the requirements of the statute are (a) that the applicant must file a claim for the right and (b) he or she must also file a certified copy of the original foreign application; these papers must be filed within a certain time limit. The maximum time limit specified in the statute is that the claim for priority and the priority papers must be filed before the patent is granted, but the statute gives the Director authority to set this time limit at an earlier time during the pendency of the application.

* * * * * * * * * * *

It should be particularly noted that these papers must be filed in all cases even though they may not be necessary during the pendency of the application to overcome the date of any reference. The statute also gives the Director authority to require a translation of the foreign documents if not in the English language and such other information as the Director may deem necessary.

For original applications filed under 35 U.S.C.111(a) (other than a design application) on or after November 29, 2000, the requirements of the statute are that the applicant must (a) file a claim for the right of priority and (b) identify the original foreign application by specifying the application number of the foreign application, the intellectual property authority or country in which the application was filed and the date of filing of the application. These papers must be filed within a certain time limit. The time limit specified in 35 U.S.C.119(b)(1) is that the claim for priority and the required identification information must be filed at such time during the pendency of the application as set by the Director. The Director has by rule set this time limit as the later of four months from the actual filing date of the application or sixteen months from the filing date of the prior foreign application. See 37 CFR 1.55(a)(1)(i). This time period is not extendable.

* * * * * * * * * * *

Unless provided in an application data sheet, 37 CFR 1.63 requires that the oath or declaration must identify the foreign application for patent or inventor's certificate for which priority is claimed under 37 CFR 1.55, and any foreign applications having a filing date before that of the application on which priority is claimed, by specifying the application number, country, day, month, and year of its filing.

§6.7 主張國外優先權與參照併入

依照 37CFR§1.57(a)，如果一美國專利申請案因疏忽（inadvertently）而將說明書之全部或一部份省略掉，但此專利申請案在申請時有主張國外之優先權，而且比省略的部份完全地包括在國外申請案之中時，則主張國外優先權之利益應被認為包括了參照併入國外申請案所包含之被疏忽而省略部份。所以，申請人可再藉由修正將省略的部份加入說明書中而不被認為是新事項。然而，申請人要依 37CFR§1.57(a)將省略的部份加入說明書之中，需符合以下要件：

(A) 美國專利申請案必需是在 2004 年 9 月 21 日當天或之後申請者；

(B) 所有或部份之說明書或圖式之必需是因為疏忽而省略者；

(C) 依照 37CFR§1.55 之優先權主張必需是在申請時就主張者；

(D) 因疏忽而省略之部份必需完全包括在國外申請案之中；

(E) 申請人必需在美國專利局設定之期限內提出修正，將被省略之部份加入，提出修正之時間必需在審查結束之前或放棄之前（視哪一先發生）；

(F) 如果申請人無法依 37CFR§1.53(d)取得申請日（OIPE 發出申請案不完全之通知指出申請案缺說明或圖式），申請人需依 37 CFR§1.57(a)提出請願並附上請願之費用；

(G) 申請人需呈送優先權文件；

(H) 如果優先權文件非英文，需提供英文翻譯本；

(I) 申請人需註明在國外申請案中哪裡可找到因疏忽被省略之部份。

MPEP 201.17 Incorporation by Reference Under 37 CFR 1.57(a) [R-3]

＊＊＊＊＊＊＊＊＊＊＊＊

I.IN GENERAL

37 CFR 1.57(a) provides that, if all or a portion of the specification or drawing(s) is inadvertently omitted from an application, but the application contains a claim under 37 CFR 1.55 for priority of a prior-filed foreign application, or a claim under 37 CFR 1.78 for the

benefit of a prior-filed provisional, nonprovisional, or international application, that was present on the filing date of the application, and the inadvertently omitted portion of the specification or drawing(s) is completely contained in the prior-filed application, the claim for priority or benefit shall be considered an incorporation by reference of the prior-filed application as to the inadvertently omitted portion of the specification or drawings.

＊＊＊＊＊＊＊＊＊＊＊＊

II.CONDITIONS AND REQUIREMENTS OF 37 CFR 1.57(a)

The following conditions and requirements need to be met for an applicant to add omitted material to an application pursuant to 37 CFR 1.57(a):

(A) the application must have been filed on or after September 21, 2004;

(B) all or a portion of the specification or drawing(s) must have been inadvertently omitted from the application;

(C) a claim under 37 CFR 1.55 for priority of a prior-filed foreign application, or a claim under 37 CFR 1.78 for the benefit of a prior-filed provisional, nonprovisional, or international application, must have been present on the filing date of the application;

(D) the inadvertently omitted portion of the specification or drawing(s) must be completely contained in the prior-filed application;

(E) applicant must file an amendment to include the inadvertently omitted portion of the specification or drawing(s) within any time period set by the Office, but in no case later than the close of prosecution as defined by 37 CFR 1.114(b), or abandonment of the application, whichever occurs earlier;

(F) if the application is not otherwise entitled to a filing date, applicant must also file a petition under 37 CFR 1.57(a) accompanied by the petition fee set forth in 37 CFR 1.17(f);

(G) applicant must supply a copy of the prior-filed application, except where the prior-filed application is an application filed under 35 U.S.C. 111;

(H) applicant must supply an English language translation of any prior-filed application that is in a language other than English; and

(I) applicant must identify where the inadvertently omitted portion of the specification or drawing(s) can be found in the prior-filed application.

第 7 章　暫時申請案

目　錄

§7.1 暫時申請案之由來及特點

美國為配合關稅暨貿易總協定（General Agreement on Tariffs and Trade; GATT），於 1995 年 6 月 8 日起導入“暫時申請案（provisional application）”制度，可稱之為國內優先權（domestic priority）制度，其主要係修改 35U.S.C§119 及 35U.S.C§111，而創造出所謂的“暫時申請案”。一般提出美國專利申請所依據的規定是 35U.S.C§111(a)，而暫時申請案的提出則是依據 35U.S.C§111(b)。暫時申請案不能直接獲准專利，但可在申請日之 12 個月之內變更為正式申請案，或者是在暫時申請案申請日之 12 個月內申請正式申請案主張暫時申請案之優先權。由於暫時申請案可不必包括有請求項，且專利說明書可以任何形式提出（包括以非英文之資料），因而可迅速取得美國專利之申請日以排除相關之先前技術，故對於技術競爭激烈且時效上急迫之發明，不失為一有利的申請管道。

申請暫時申請案需注意以下事項：

(1) 一暫時申請案需由發明人書面提出申請，申請書需包括(A)符合 35U.S.C§112 第 1 段之說明書，(B)符合 35U.S.C§113 之圖式；

(2) 暫時申請案並不要求有請求項；

(3) 申請費可在提出說明書及圖式之後才呈送，如未在指定之期限內呈送申請費，暫時申請案視同放棄；

(4) 暫時申請案之申請日為說明書及圖式被美國專利局收到之日期；

(5) 如果申請人未依 35U.S.C§119(e)(3)主張暫時申請案之優先權，則暫時申請案自其申請日起 12 個月被視為放棄，且不能復活；

(6) 依照 35U.S.C§111(a)申請之專利申請案亦可被視為一暫時申請案；

(7) 一暫時申請案不能依照 35U.S.C§119 或 365(a)主張任何其他申請案之優先權或取得依照 35U.S.C§120，121 或 365(c)較早申請日之利益；

(8) 35U.S.C 對於申請案之相關法律均適用於暫時申請案，但是，暫時申請案不適用 35U.S.C§115（宣誓書），§131（審查），§135（抵觸程序）及§157（法定發明註冊）之相關法律。

35 U.S.C§111

(b) PROVISIONAL APPLICATION.-

(1) AUTHORIZATION.-A provisional application for patent shall be made or authorized to be made by the inventor, except as otherwise provided in this title, in writing to the Director. Such application shall include-

(A) a specification as prescribed by the first paragraph of section 112 of this title; and

(B) a drawing as prescribed by section 113 of this title.

(2) CLAIM.-A claim, as required by the second through fifth paragraphs of section 112, shall not be required in a provisional application.

(3) FEE.-

(A) The application must be accompanied by the fee required by law.

(B) The fee may be submitted after the specification and any required drawing are submitted, within such period and under such conditions, including the payment of a surcharge, as may be prescribed by the Director.

(C) Upon failure to submit the fee within such prescribed period, the application shall be regarded as abandoned, unless it is shown to the satisfaction of the Director that the delay in submitting the fee was unavoidable or unintentional.

(4) FILING DATE.-The filing date of a provisional application shall be the date on which the specification and any required drawing are received in the Patent and Trademark Office.

(5) ABANDONMENT.-Notwithstanding the absence of a claim, upon timely request and as prescribed by the Director, a provisional application may be treated as an application filed under subsection (a). Subject to section 119(e)(3) of this title, if no such request is made, the provisional application shall be regarded as abandoned 12 months after the filing date of such application and shall not be subject to revival after such 12-month period.

(6) OTHER BASIS FOR PROVISIONAL APPLICATION.-Subject to all the conditions in this subsection and section 119(e) of this title, and as prescribed by the Director, an application for patent filed under subsection (a) may be treated as a provisional application for patent.

(7) NO RIGHT OF PRIORITY OR BENEFIT OF EARLIEST FILING DATE.-A provisional application shall not be entitled to the right of priority of any other application

under section 119 or 365(a) of this title or to the benefit of an earlier filing date in the United States under section 120, 121, or 365(c) of this title.

(8) APPLICABLE PROVISIONS.-The provisions of this title relating to applications for patent shall apply to provisional applications for patent, except as otherwise provided, and except that provisional applications for patent shall not be subject to sections 115, 131, 135, and 157 of this title.

$$* * * * * * * * * * * *$$

§7.2　取得一暫時申請案之有效的申請日所需之文件

只要有符合 35 USC §112 第 1 段之說明書及必要的圖式即可取得一暫時案之申請日。但是，如要使此申請日為有效之申請日須後補下列文件：①申請費②一首頁（cover sheet）註明為暫時申請案及註明發明人之姓名、住址及發明名稱。暫時申請案之說明書不必包含請求項，亦不必發明人之宣誓書。

---

37 CFR 1.51 (c)

A complete provisional application filed under § 1.53(c) comprises:

(1) A cover sheet identifying:

(i) The application as a provisional application,

(ii) The name or names of the inventor or inventors,

(iii) The residence of each named inventor,

(iv) The title of the invention

(2) A specification as prescribed by the first paragraph of 35 U.S.C. 112,

(3) Drawings, when necessary, and

(4) The prescribed filing fee and application size fee.

§7.3　以非英文之發明揭露書取得暫時申請案之申請日

可用非英文之發明揭露資料取得一暫時案之申請日，但在一年內提出正式申請案主張暫時案之優先權日時，美國專利局將要求提出暫時案說明書之認證翻譯本（certified translation），如未提出則該正式申請案將視同放棄

---

37 CFR 1.78(iv)

If the prior-filed provisional application was filed in a language other than English and an English-language translation of the prior-filed provisional application and a statement that the translation is accurate were not previously filed in the prior-filed provisional application or the later-filed nonprovisional application, applicant will be notified and given a period of time within which to file an English-language translation of the non-English-language prior-filed provisional application and a statement that the translation is accurate. In a pending nonprovisional application, failure to timely reply to such a notice will result in abandonment of the application.

§7.4　正式申請案主張暫時申請案之優先權需具備之條件

一正式申請案要主張先前申請之暫時申請案之優先權而得到暫時申請案之申請日的利益的話，需符合以下條件：

(A)　暫時申請案必需符合 35USC§112 第 1 段之規定揭露了正式申請案所請求之發明；

(B)　正式申請案需在暫時申請案之申請日之 12 個月內提出申請；

(C)　正式申請案必需在說明書之第 1 行或在申請資料表（Application Data Sheet）中註明主張暫時申請案之優先權；

(D)　正式申請案必需由暫時申請案之一個發明人或數個發明人提出。

MPEP 201.11 Claiming the Benefit of an Earlier Filing Date Under 35 U.S.C. 120 and 119(e) [R-5]

There are several conditions for a later-filed application to receive the benefit of the filing date of a prior-filed application under 35 U.S.C. 120, 121, or 365(c), or, provided the later-filed application is not a design application (see 35 U.S.C. 172), under 35 U.S.C. 119(e). The conditions are briefly summarized as follows:

(A) The prior-filed application must disclose the claimed invention of the later-filed application in the manner provided by the first paragraph of 35 U.S.C. 112 for a benefit claim under 35 U.S.C. 120, 121, or 365(c), and also for a benefit claim under 35 U.S.C. 119(e).

(B) The later-filed application must be copending with the prior-filed nonprovisional application for a benefit claim under 35 U.S.C. 120, 121, or 365(c). For a benefit claim under 35 U.S.C. 119(e), the later-filed application must be filed not later than 12 months after the filing date of the prior provisional application.

(C) The later-filed application must contain a reference to the prior-filed application in the first sentence(s) of the specification or in an application data sheet, for a benefit claim under 35 U.S.C. 120, 121, or 365(c), and also for a benefit claim under 35 U.S.C. 119(e).

(D) The later-filed application must be filed by an inventor or inventors named in the prior-filed application for a benefit claim under 35 U.S.C. 120, 121, or 365(c), and also for a benefit claim under 35 U.S.C. 119(e).

＊＊＊＊＊＊＊＊＊＊＊

實例 1

2004 年 1 月 14 日發明人 A 在一科技雜誌上發表了一論文，揭示了其發明之新的醫療用化合物。於 2004 年 2 月 10 日發明人以該論文提出一暫時專利申請案，於 2005 年 2 月 10 日發明人 A 提出一正式之專利申請案主張該暫時申請案之優先權。請問發明人 A 之正式申請案是否會被其自己發表之論文以 35 USC § 102(b)核駁？

不會，因為正式申請案乃在暫時申請案之申請日一年內提申，故其有效申請日為 2004 年 2 月 10 日，而論文發表為 2004 年 1 月 14 日並非在其有效申請日之一年之前。

實例 2

發明人 A 於 2005 年 11 月 5 日提出一暫時申請案,接著於 2006 年 11 月 4 日提出一正式專利申請案主張該暫時申請案之優先權,該正式申請案於 2008 年 2 月 1 日收到核准通知,並於 2008 年 7 月 21 日發證。請問此專利之專利權限從哪一天至哪一天?

自 2008 年 7 月 21 日起至 2026 年 11 月 3 日止。美國專利之專利權期限終止日為正式案之申請日起 20 年而非暫時案之申請日。

MPEP 2701

Domestic priority under 35 U.S.C. 119 (e) to one or more U.S. provisional applications is not considered in the calculation of the twenty-year term.

實例 3

申請人 A 於 2005 年 7 月 1 日於美國申請了一暫時申請案,該暫時申請案之內容於 2006 年 6 月 16 日發表,申請人 A 於 2006 年 6 月 30 日提出台灣發明專利申請並主張該美國暫時申請案之優先權。請問此台灣申請案是否會被其發表之內容核駁?

不會,因為美國的暫時申請案可作為台灣專利申請案主張優先權之基礎。

實例 4

發明人 A 於 2005 年 4 月 1 日發明了一筆記型電腦適合於寒冷的天氣下操作且防水,故於 2005 年 6 月 1 日提出一暫時申請案,之後於 2006 年 1 月時發現此電腦之防水性並不理想,於是再經研究於 2006 年 3 月 27 日發明了改良此筆記型電腦防水之裝置,請問發明人 A 應如何保護其發明?

發明人 A 應於 2006 年 6 月 1 日之前提出一正式專利申請案主張該暫時申請案之優先權,且於正式申請案中同時請求暫時申請案中之發明及其在 2006 年 3 月 27 日所完成之改良的防水裝置

實例 5

　　發明人 A 於 2004 年 6 月 30 日於台灣申請了一發明專利申請案，後於 2005 年 1 月 10 日於美國提出一暫時申請案主張該台灣申請案之優先權，並於 2006 年 1 月 9 日提出美國正式申請案主張該暫時申請案之優先權。請問如果審查委員引證一日期為 2004 年 12 月 1 日之先前技術可否核駁本正式申請案？

　　可以，因為暫時申請案不能主張國外優先權，而且正式申請案之申請日從台灣申請案之申請日起已超過一年亦不能主張台灣之優先權。

§7.5　將正式申請案轉換成暫時申請案

　　依照 35CFR§1.53(b)申請之專利申請案（非暫時申請案及其延續案，分割案,CIP 申請案）可轉換成暫時申請案。轉換之目的是可使得非暫時申請案在一年之內不被審查。請求轉換及費用必需在①非暫時申請案放棄②非暫時申請案領證③非暫時申請案申請日之 12 個月內④請求法定發明註冊（SIR）四者較早發生之日期之前為之。請求轉換核准之後，原申請之非暫時申請案之規費不能退費，而且暫時申請案之申請日即為非暫時申請案之申請日。

MPEP 201.04(b) Provisional Application

＊＊＊＊＊＊＊＊＊＊＊

　　An application filed under 37 CFR 1.53(b) may be converted to a provisional application provided a request for conversion is submitted along with the fee as set forth in 37 CFR 1.17(q). The request and fee must be submitted prior to the earlier of the abandonment of the nonprovisional application, the payment of the issue fee, the expiration of 12 months after the filing date of the nonprovisional application, or the filing of a request for statutory invention registration. The grant of any such request will not entitle applicant to a refund of the fees which were properly paid in the application filed under 37 CFR 1.53(b). See MPEP § 601.01(c)

§7.6　將暫時申請案轉換成正式申請案

　　1995 年 6 月 8 日當天或之後申請的暫時申請案可轉換成非暫時（正式）申請案。此轉換必需繳費且需提出修正以加入至少一請求項。此轉換必需在暫時申請案放棄之前或暫時申請案之申請日起 12 個月（視哪一日期為早）之前請求。轉換時需繳交非暫時申請案之申請費，簽署之宣誓書。轉換核准後，申請人繳交暫時申請案之規費不能退費，而且專利權期限之起算乃自原暫時申請案之申請日起算 20 年結束。

MPEP 201.04(b) Provisional Application

＊＊＊＊＊＊＊＊＊＊＊＊

　　Public Law 106-113 amended 35 U.S.C. 111(b)(5) to permit a provisional application filed under 37 CFR 1.53(c) be converted to a nonprovisional application filed under 37 CFR 1.53(b). 35 U.S.C. 111(b)(5) as amended by Public Law 106-113 is effective as of November 29, 1999 and applies to any provisional applications filed on or after June 8, 1995. A request to convert a provisional application to a nonprovisional application must be accompanied by the fee set forth in 37 CFR 1.17(i) and an amendment including at least one claim as prescribed by 35 U.S.C. 112, unless the provisional application otherwise contains at least one such claim. The request must be filed prior to the earliest of the abandonment of the provisional application or the expiration of twelve months after the filing date of the provisional application. The filing fee for a nonprovisional application, an executed oath or declaration under 37 CFR 1.63, and the surcharge under 37 CFR 1.16(f), if appropriate, are also required. The grant of any such request will not entitle applicant to a refund of the fees which were properly paid in the application filed under 37 CFR 1.53(c). Conversion of a provisional application to a nonprovisional application will result in the term of any patent issuing from the application being measured from at least the filing date of the provisional application. This adverse patent term impact can be avoided by filing a nonprovisional application claiming the benefit of the provisional application under 35U.S.C. 119(e), rather than requesting conversion of the provisional application to a nonprovisional application. See 37 CFR 1.53(c)(3).

＊＊＊＊＊＊＊＊＊＊＊＊

第 8 章　核准前公告與法定發明註冊（SIR）

目　錄

§8.1 核准前公告公之專利申請案

實用專利（utility application）、植物申請案及延續申請案，其申請日為 2000 年 11 月 29 日當天及之後者，國際申請案之進入美國國家階段之申請案，其國際申請日為 2000 年 11 月 29 日當天及之後者，以及上述之申請案申請日在 2000 年 11 月 29 日之前但主動請求核准前公告者，均會在其最早申請日（優先權日）起 18 個月作核准前公告。已放棄之申請案、受秘密命令之保護之申請案、暫時申請案、新式樣申請案、再發證申請案以及請求不作核准前公告之專利申請案均不會作核准前公告。

---

35 USC §122 (b)

(A)Subject to paragraph (2), each application for a patent shall be published, in accordance with procedures determined by the Director, promptly after the expiration of a period of 18 months from the earliest filing date for which a benefit is sought under this title. At the request of the applicant, an application may be published earlier than the end of such 18-month period.

(B)No information concerning published patent applications shall be made available to the public except as the Director determines.

(C)Notwithstanding any other provision of law, a determination by the Director to release or not to release information concerning a published patent application shall be final and nonreviewable.

EXCEPTIONS.—

(A)An application shall not be published if that application is—

(i)no longer pending;

(ii)subject to a secrecy order under section 181 of this title;

(iii)a provisional application filed under section 111(b) of this title; or

(iv)an application for a design patent filed under chapter 16 of this title.

§8.2 請求專利申請案不核准前公告

申請人可以請求美國專利局對其專利申請案不作核准前公告，但需於申請時同時提出，明確地指出不作核准前公告，且須聲明無任何對應專利申請案已在任何有核准前公告制度之國家提出申請或將在其他國家申請。申請人可隨時撤銷此請求，而讓其專利申請案作核准前公告。

---

37 CFR 1.213

If the invention disclosed in an application has not been and will not be the subject of an application filed in another country, or under a multilateral international agreement, that requires publication of applications eighteen months after filing, the application will not be published under 35 U.S.C. 122(b) and § 1.211 provided:

(1) A request (nonpublication request) is submitted with the application upon filing;

(2) The request states in a conspicuous manner that the application is not to be published under 35 U.S.C. 122(b);

(3) The request contains a certification that the invention disclosed in the application has not been and will not be the subject of an application filed in another country, or under a multilateral international agreement, that requires publication at eighteen months after filing; and

(4) The request is signed in compliance with § 1.33(b).

§8.3 違反不核准前公告的規定之後果

申請人如果請求不核准前公告之後又在其他國家提出專利申請，則須於其他有核准前公告制度之國家申請後之 45 天之內告知美國專利局。如果未及時告知美國專利局且此未及時告知是故意的，則該專利申請案將被視為放棄。如果未及時通知美國專利局是非故意的（unintentional），則可請求復活專利申請案。美國專利局收到申請人之通知後會立即安排該專利申請案之核准前公告事宜。

37 CFR 1.213 (c)

If an applicant who has submitted a nonpublication request under paragraph (a) of this section subsequently files an application directed to the invention disclosed in the application in which the nonpublication request was submitted in another country, or under a multilateral international agreement, that requires publication of applications eighteen months after filing, the applicant must notify the Office of such filing within forty-five days after the date of the filing of such foreign or international application. The failure to timely notify the Office of the filing of such foreign or international application shall result in abandonment of the application in which the nonpublication request was submitted.

實例 1

發明人 A 於 2005 年 6 月 30 日於美國提出一發明專利申請案同時聲明本案將不在任何有核准前公告制度之國家申請，請求美國專利局對本案不作核准前公告。之後發明人對該美國專利申請案之內容作了一些改良並於台灣提出專利申請（該改良並未揭示於原美國專利申請案之中）並於 90 天之後通知美國專利局提出一台灣專利申請案。請問該美國專利申請案是否會被放棄，及如何復活此美國專利申請案？

雖然在台灣申請之專利申請案包括了改良部份，但仍違反 37 CFR 1.213 (c)之規定。如果發明人 A 未及時通知在台灣提出專利申請並非故意，則可依據 37 CFR 1.137(b)之規定提出復活之程序。

37 CFR 1.137(b)

(b) Unintentional. If the delay in reply by applicant or patent owner was unintentional, a petition may be filed pursuant to this paragraph to revive an abandoned application, a reexamination prosecution terminated under §§ 1.550(d) or 1.957(b) or limited under § 1.957(c), or a lapsed patent. A grantable petition pursuant to this paragraph must be accompanied by:

(1) The reply required to the outstanding Office action or notice, unless previously filed;

(2) The petition fee as set forth in § 1.17(m);

(3) A statement that the entire delay in filing the required reply from the due date for the reply until the filing of a grantable petition pursuant to this paragraph was unintentional. The

Director may require additional information where there is a question whether the delay was unintentional; and

(4) Any terminal disclaimer required pursuant to paragraph (d) of this section.

實例 2

　　發明人 A 於 2006 年 3 月 20 日提出一美國專利申請案，於 2007 年 7 月 10 日發明人基於某些理由不願讓其發明之內容於申請日起 18 個月公告。請問有何程序可使該美國專利申請案於 2007 年 9 月 20 日左右不會被公告？

　　發明人 A 可提出一延續案（CA）同時請求美國專利局不作核准前公告，並主動放棄原申請案，但須注意提出放棄專利申請案通常須於預定公告日之前四週提出，因為一專利申請案之放棄日期乃是美國專利局官員確實收到及承認之時點才是放棄之日期。另一方式可將此專利申請案轉換成暫時申請案就可避免其被公告。

---

37 CFR 1.138 (a),(c)

　　(a) An application may be expressly abandoned by filing a written declaration of abandonment identifying the application in the United States Patent and Trademark Office. Express abandonment of the application may not be recognized by the Office before the date of issue or publication unless it is actually received by appropriate officials in time to act.

　　(c) An applicant seeking to abandon an application to avoid publication of the application must submit a declaration of express abandonment by way of a petition under this paragraph including the fee set forth in § 1.17(h) in sufficient time to permit the appropriate officials to recognize the abandonment and remove the application from the publication process. Applicants should expect that the petition will not be granted and the application will be published in regular course unless such declaration of express abandonment and petition are received by the appropriate officials more than four weeks prior to the projected date of publication.

§8.4　修訂之公告（redacted publication）

　　申請人提出之美國專利申請案如果揭露之內容比對應之外國申請案為多時，可於最早優先權日起算之 16 個月內提出修訂之公告，請求多揭露之部份從公告本刪除，請求作修訂之公告需提出以下文件：

① 電子形式之修訂公告版本；

② 各對應國外申請案之認證本；

③ 各國外對應申請案之英譯本；

④ 將美國專利申請案多揭露部份用括弧括起之修訂版；

⑤ 此修訂版只刪除了未包含在對應國外申請案之揭露之聲明

---

37 CFR 1.217

(a) If an applicant has filed applications in one or more foreign countries, directly or through a multilateral international agreement, and such foreign-filed applications or the description of the invention in such foreign-filed applications is less extensive than the application or description of the invention in the application filed in the Office, the applicant may submit a redacted copy of the application filed in the Office for publication, eliminating any part or description of the invention that is not also contained in any of the corresponding applications filed in a foreign country. The Office will publish the application as provided in § 1.215(a) unless the applicant files a redacted copy of the application in compliance with this section within sixteen months after the earliest filing date for which a benefit is sought under title 35, United States Code.

(b) The redacted copy of the application must be submitted in compliance with the Office electronic filing system requirements. The title of the invention in the redacted copy of the application must correspond to the title of the application at the time the redacted copy of the application is submitted to the Office. If the redacted copy of the application does not comply with the Office electronic filing system requirements, the Office will publish the application as provided in § 1.215(a).

(c) The applicant must also concurrently submit in paper (§ 1.52(a)) to be filed in the application:

(1) A certified copy of each foreign-filed application that corresponds to the application for which a redacted copy is submitted;

(2) A translation of each such foreign-filed application that is in a language other than English, and a statement that the translation is accurate;

(3) A marked-up copy of the application showing the redactions in brackets; and

(4) A certification that the redacted copy of the application eliminates only the part or description of the invention that is not contained in any application filed in a foreign country, directly or through a multilateral international agreement, that corresponds to the application filed in the Office.

§8.5 專利申請案資料之保密，檢閱及影印

未公告之專利申請案之所有資訊都是保密的，但不包括以下案件:

① SIR（Statutory Invention Registration）申請案；

② 再發證申請案；

③ 再審查程序之案件；

④ 抵觸程序（interference）之案件；及

⑤ 開放給公眾檢視之專利申請案。

下列專利申請案，任何人提出書面請求及付費均可檢閱及影印:

① 已發證之專利申請案；

② SIR 申請案；

③ 已公告但放棄之專利申請案；及

④ 未公告但放棄而在另一已發證之專利申請案，SIR 申請案或已公告之專利申請案中被指明（identified）或所主張優先權之專利申請案。

下列專利申請案，任何人均可以書面請求及付費取得影本:

① 已公告之專利申請案及其全部檔案資料（限定於修訂版）；

② 未公告且申請中，而所主張優先權之專利請案為已發證，SIR 申請案或已公告之申請案者；及

③ 未公告且申請中（包括暫時申請案），而在一已發證之專利申請案，SIR 申請案或已公告之專利申請案中被指明或被參照併入（incorporated by reference）者。

下列專利申請案任何人均可詢問狀態（status，乃指是否申請中、是否已公告、申請案號及是否有其他專利申請案主張此申請案之優先權）:

① 所有可檢閱及影印之專利申請案均可詢問狀態；及

② 已公告之專利申請案，已發證之專利申請案，已發證之國際專利申請案（指定美國）及開放公共檢視之申請案，只要告知申請案號均可詢問狀態。

§8.6　檢閱及影印未公告之專利申請案之資料

　　只有該專利申請案之發明人、代理人及向美國專利局登記之讓渡書上之受讓人之授權的代表人，可檢閱及影印未公告之專利申請案之資料。另外，除非有特殊情況（例如，收到專利申請人警告其有專利申請案而欲了解其專利申請案之內容）時可提出請願（petition）及繳規費，經核准後，檢閱未公告之專利申請案。

---

37 CFR 1.14(c), 1.14(i)

　　(c) Power to inspect a pending or abandoned application. Access to an application may be provided to any person if the application file is available, and the application contains written authority (e.g., a power to inspect) granting access to such person. The written authority must be signed by:

　　(1) An applicant;

　　(2) An attorney or agent of record;

　　(3) An authorized official of an assignee of record (made of record pursuant to § 3.71 of this chapter); or

　　(4) A registered attorney or agent named in the papers accompanying the application papers filed under § 1.53 or the national stage documents filed under § 1.495, if an executed oath or declaration pursuant to § 1.63 or § 1.497 has not been filed.

　　(i) Access or copies in other circumstances.

　　The Office, either sua sponte or on petition, may also provide access or copies of all or part of an application if necessary to carry out an Act of Congress or if warranted by other special circumstances. Any petition by a member of the public seeking access to, or copies of, all or part of any pending or abandoned application preserved in confidence pursuant to paragraph (a) of this section, or any related papers, must include:

The fee set forth in §1.17(h); and

　　A showing that access to the application is necessary to carry out an Act of Congress or that special circumstances exist which warrant petitioner being granted access to all or part of the application.

§8.7　提早公告、重新公告及主動公告專利申請案

申請人如果提供申請案之電子檔及支付公告費可請求提早於18個月之前公告專利申請案。但是美國專利局並不會依照申請人所希望或指定之日期提早公告，只會早於18個月公告而已。如果美國專利局公告之版本包含實質上由美國專利局造成之大的錯誤，則申請人可於公告之後之二個月內請求美國專利局重新公告。另外，如果是在2000年11月29日之前所提申之專利申請案，基於需要，亦可請求主動公告，但須附上專利申請書之電子檔、公告費及處理費。

---

37 CFR 1.219 , 1.221

§ 1.219 Early publication.

Applications that will be published under § 1.211 may be published earlier than as set forth in § 1.211(a) at the request of the applicant. Any request for early publication must be accompanied by the publication fee set forth in § 1.18(d). If the applicant does not submit a copy of the application in compliance with the Office electronic filing system requirements pursuant to § 1.215(c), the Office will publish the application as provided in § 1.215(a). No consideration will be given to requests for publication on a certain date, and such requests will be treated as a request for publication as soon as possible.

§ 1.221 Voluntary publication or republication of patent application publication.

(a) Any request for publication of an application

filed before, but pending on, November 29, 2000, and any request for republication of an application previously published under § 1.211, must include a copy of the application in compliance with the Office electronic filing system requirements and be accompanied by the publication fee set forth in § 1.18(d) and the processing fee set forth in § 1.17(i). If the request does not comply with the requirements of this paragraph or the copy of the application does not comply with the Office electronic filing system requirements, the Office will not publish the application and will refund the publication fee.

(b) The Office will grant a request for a corrected

or revised patent application publication other than as provided in paragraph (a) of this section only when the Office makes a material mistake which is apparent from Office records. Any request for a corrected

or revised patent application publication other than as provided in paragraph (a) of this section must be filed within two months from the date of the patent application publication. This period is not extendable.

§8.8　核准前公告之後專利申請人之權利

如果在專利核准前公告至專利發證之期間內有疑似侵權者，專利獲證之後可向該侵權者求償在該期間內之合理的權利金，但須符合下列二要件：

(a)　侵權者確實被告知該公告之專利申請案；及

(b)　獲證之專利中至少有一請求項實質上與該公告之專利申請案的一請求項相同。

---

35 USC 154(d)

(1) IN GENERAL.– In addition to other rights provided by this section, a patent shall include the right to obtain a reasonable royalty from any person who, during the period beginning on the date of publication of the application for such patent under section 122(b), or in the case of an international application filed under the treaty defined in section 351(a) designating the United States under Article 21(2)(a) of such treaty, the date of publication of the application, and ending on the date the patent is issued-

(A) (i)makes, uses, offers for sale, or sells in the United States the invention as claimed in the published patent application or imports such an invention into the United States; or

(ii)if the invention as claimed in the published patent application is a process, uses, offers for sale, or sells in the United States or imports into the United States products made by that process as claimed in the published patent application; and

(B) had actual notice of the published patent application and, in a case in which the right arising under this paragraph is based upon an international application designating the United States that is published in a language other than English, had a translation of the international application into the English language.

§8.9 第三者對公告之專利申請案提出先前技術

第三者在發現核准前公告之專利申請案之請求項不具專利性時，並不能依照 35USC§122(C)提出發證前異議（protest）。但是可在公告之二個月之內，將所知道 的先前技術（限定為書面之資料，最多 10 篇，不能作專利性之討論）送交美國專利 局供審查委員作參考。呈送之先前技術之第三者不會被通知此審查委員對此先前技 術之意見，亦不能參與該公告之專利申請案之審查。此外，第三者可以匿名方式（通 常由專利代理人之名義提出）對公告之專利申請案提出先前技術。

---

35 U.S.C. 122 Confidential status of applications; publication of patent applications.

* * * * * * * * * * *

(c) PROTEST AND PRE-ISSUANCE OPPOSITION.- The Director shall establish appropriate procedures to ensure that no protest or other form of pre-issuance opposition to the grant of a patent on an application may be initiated after publication of the application without the express written consent of the applicant.

* * * * * * * * * * *

§8.10 法定發明註冊（SIR）

依照 35USC157，如果一非暫時申請案之申請人在該專利申請案提出申請時或是 之後於審查期間（pending）向美國專利局提出請求，並符合下列條件：
(1) 說明書之揭露符合 35USC112 之要件；
(2) 說明書符合印刷之要件；
(3) 放棄對該發明取得專利之權利；及
(4) 繳交申請費、公告費及其他處理費。

則美國專利局長會公告包含此專利申請案之說明書及圖式之法定發明註冊（Statutory Invention Registration, SIR）且不對此申請案作實質之審查。

此種法定發明註冊之公告其實是一種防衛性之公告。事實上自從美國專利局採取核准前公告（pre-grant publication）之後，非暫時申請案之核准前公告亦為一種防衛性之公告而且其可不必放棄專利權。

---

35 U.S.C. 157 Statutory invention registration.

(a) Notwithstanding any other provision of this title, the Director is authorized to publish a statutory invention registration containing the specification and drawings of a regularly filed application for a patent without examination if the applicant -

(1) meets the requirements of section 112 of this title;

(2) has complied with the requirements for printing, as set forth in regulations of the Director;

(3) waives the right to receive a patent on the invention within such period as may be prescribed by the Director; and

(4) pays application, publication, and other processing fees established by the Director. If an interference is declared with respect to such an application, a statutory invention registration may not be published unless the issue of priority of invention is finally determined in favor of the applicant.

* * * * * * * * * * *

MPEP 1101 Request for Statutory Invention Registration (SIR) [R-2]

* * * * * * * * * * *

A request for a statutory invention registration (SIR) may be filed at the time of filing a nonprovisional application for patent, or may be filed later during pendency of a nonprovisional application. The fee required (37 CFR 1.17(n) or (o)) depends on when the request is filed. The application to be published as a SIR must be complete as set forth in 37 CFR 1.51(b) including a specification with a claim or claims, an oath or declaration, and drawings when necessary.

* * * * * * * * * * *

An applicant may find publication of an application to be a desirable alternative to requesting a SIR since publication of the application is achieved without any waiver of patent rights.

* * * * * * * * * * *

§8.11　SIR 之審查

對於 SIR 申請案，美國專利局將先審查請求 SIR 之申請案是否仍在審查中（pending），如果在請求時已放棄，或已核准，繳領證費，則將退回 SIR 之請求。如果仍在審查中，審查委員會審查是否有發下審定書（Office Action），其中是否包含有 35USC112 之核駁。如果有 35USC112 之核駁，且申請人尚未回答，則審查委員會再通知申請人儘速回答。在審查委員確認請求 SIR 之專利申請案符合 35USC112 之要件之後，會再檢查是否符合 37CFR 1.293 之程式要件，例如是否簽名有缺陷，是否繳交規費等，如不符合要件則會給予一個月之時間補正。

在申請案符合 37CFR 1.293，37CFR 1.294 及 35USC112 之要件後，審查委員會再決定是否該申請案有涉入抵觸程序，如果有涉入抵觸程序，審查委員就會一直等到抵觸程序終止後才對 SIR 之請求作決定。

另外，受秘密命令之申請案，如果提出 SIR 請求，則不會公告一直到秘密命令解除為止。申請人如果對審查委員最後拒絕公告 SIR 之決定不服而拒絕公告 SIR 之理由如不是不符合 35USC112 之要件時，可提出請願（petition）請求再審理。如果拒絕公告之理由是不符合 35USC112 之要件，則需提出訴願（appeal）請求再審理。

MPEP 1103 Examination of a SIR [R-2]

＊＊＊＊＊＊＊＊＊＊＊

An examiner in Art Unit 3641 or 3662, where appropriate, will determine whether the application in which a request for a statutory invention registration has been filed is a pending nonprovisional application. If the application was abandoned at the time the request was filed, has been patented, or has been allowed and the issue fee paid, the examiner should return the SIR request to the requester accompanied by a Return of Statutory Invention Registration Request to Requester notice (form *>SIR-C<).

If the application is pending, the examiner should ascertain whether an Office action with a rejection under 35 U.S.C. 112 has been issued and not replied to. If so, and if there remains any time to reply to the rejection, the examiner should send the applicant a courtesy notice requiring a timely reply. If no time for reply remains, the application is abandoned and the examiner should inform the applicant of this fact.

After the examiner handling the SIR has ascertained that all outstanding rejections under 35 U.S.C. 112 have been replied to, the examiner should verify that the request for a SIR meets the requirements of 37 CFR 1.293. First, applicant should be notified of any defects in the signature on the SIR request or of any inadequacy of the SIR fee. A 1-month time period should be set for applicant to correct the signature or fee before any further consideration of the SIR request is given. **>A Notice of Improper Request for a Statutory Invention Registration (form SIR-E)< may be used for this purpose. Next, applicant should be given 1 month to correct any other informalities in the SIR request under 37 CFR 1.293 and any informalities in the application under 37 CFR 1.294 using a Notice of Informal Statutory Invention Registration (SIR) Request, form *>SIR-F<.

＊＊＊＊＊＊＊＊＊＊＊

After the application complies with 37 CFR 1.293, 37 CFR 1.294, and 35 U.S.C. 112, the examiner should determine whether the application is involved in a pending interference. If so, applicant should be notified, using form *>SIR-J<, that no decision will be made on the SIR request until the interference proceedings are concluded.

＊＊＊＊＊＊＊＊＊＊

An application under secrecy order will be withheld from publication during such period as the national interest requires, and the applicant should be informed of this fact by using a Notice of Statutory Invention Registration * Acceptance (Form D-11), form *>SIR-N (Form D-11)<.

1105 Review of Final Refusal to Publish SIR [R-5]

＊＊＊＊＊＊＊＊＊＊

An applicant who is dissatisfied with a final refusal to publish a SIR for reasons other than compliance with 35 U.S.C. 112 may obtain review by filing a petition as set forth in 37 CFR 1.295(a). The petition should be directed to the TC Director responsible for the art unit handling the SIR.

An applicant who is dissatisfied with a decision finally rejecting claims under 35 U.S.C. 112 may obtain review by filing an appeal with the Board of Patent Appeals and Interferences as set forth in 37 CFR 1.295(b).

§8.12 SIR 公告之效果

依照 35USC 157(c)公告之 SIR 被作為防衛用時之效果有如美國專利，亦即和美國專利一樣作為先前技術時其先前技術日期為其申請日。亦即公告之 SIR 是 35USC102(g)之推斷的付諸實施，而且可作為 35USC102 之各項之先前技術，包括 35USC102(e)。SIR 之資料會被分類入檢索之檔案，且會分發給為外國之專利局並存在美國專利局之電腦檔案中作為商業資料庫，並公告在公報之上。

SIR 申請案之放棄取得專利之權利乃自公告日生效，且將影響相關案件之請求項之專利性。例如一專利申請案之請求項請求一上位概念之發明，其請求了 SIR 並公告之後，其他被此上位概念之請求項涵蓋之相關申請案之相同發明的請求項將均不能取得專利。

另外，如果一專利申請案及一 SIR 包括有相同之發明標的，且此申請案與 SIR 有共同之發明人時，則申請案之該請求項將被以重複專利核駁，且此重複專利之核駁不能以尾端棄權書克服。

MPEP1111 SIR Publication and Effect

＊＊＊＊＊＊＊＊＊＊＊

In accordance with 35 U.S.C. 157(c), a published SIR will be treated the same as a U.S. patent for all defensive purposes, usable as a reference as of its filing date in the same manner as a patent. A SIR is a "constructive reduction to practice" under 35 U.S.C. 102(g) and "prior art" under all applicable sections of 35 U.S.C. 102 including section 102(e). SIRs are classified, cross-referenced, and placed in the search files, disseminated to foreign patent offices, stored in U.S. Patent and Trademark Office computer tapes, made available in commercial data bases, and announced in the Official Gazette.

The waiver of patent rights to the subject matter claimed in a statutory invention registration takes effect on publication (37 CFR 1.293(c)) and may affect the patentability of claims in related applications without SIR requests, such as divisional or other continuing applications, since the waiver of patent rights is effective for all inventions claimed in the SIR and would effectively waive the right of the inventor to obtain a patent on the invention claimed in the same application or on the same invention claimed in any other application not issued before the publication date of the SIR. If an application containing generic claims is published as a SIR, the waiver in that application applies to any other related applications to the extent that the same invention claimed in the SIR is claimed in the other application. Examiners should apply standards similar to those applied in making "same invention" double patenting determinations to determine whether a waiver by an inventor to claims in a SIR precludes patenting by the same inventor to subject matter in any related application. If the same subject matter is claimed in an application and in a published statutory invention registration naming a common inventor, the claims in the application should be rejected as being precluded by the waiver in the statutory invention registration. See 37 CFR 1.104(c)(5). A rejection as being precluded by a waiver in a SIR cannot be overcome by a terminal disclaimer.

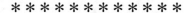

＊＊＊＊＊＊＊＊＊＊＊

第 9 章　專利申請案之審查

目　錄

§9.1　美國專利申請案之審查流程

　　美國專利申請案之審查通常分為形式審查（formalities）及實質審查（substantive）二部份。形式審查是為一初步的審查，由初步審查辦公室（OIPE, Office of Initial Patent Examination）針對專利申請案文件，例如說明書是否為英文，是否包括圖式，請求項（claim），是否包括宣誓書，是否繳費等形式要件作審查。如果初步審查發現缺少文件則會發下補文通知（notice to file missing parts），並給予申請人二個月之時間補正（需繳規費），如果初步審查通過即發交各技術中心（Technology Center）由審查長（director）分配給審查委員作實質審查。審查委員會先作檢索，再依檢索之資料針對專利申請案之專利性（單一性，新穎性，實用性，非顯而易見性，揭露性等）作審查。如果審查委員認為請求項（claims）包括二個以上之獨立發明或專利性上不同之發明，則會先發下限制／選擇之審定書（Office Action），請申請人作限制／選擇之後再進行實質審查並發下第一次審定書，並給申請人三個月的時間答辯及修正。申請人如果在三個月內（或延期至六個月內）提出答辯及修正，則審查委員會針對答辯理由作再次之審查，如果仍有請求項（claim）不可核准則通常會發下最終之審定書。申請人可於三個月（或延期至六個月）內對最終審查作答辯或修正或提延續案或申請繼續審查（Request for Continued Examination,RCE）或提出訴願聲明。提出訴願聲明後之二個月內要提出訴願理由，如經訴願仍不能核准則可向聯邦法庭上訴。如果對最終審定之答辯、修正或訴願後可核准，則會發下核准通知（Notice of Allowance），申請人需於三個月內繳領證費及公告費後發下專利證書取得專利。如取得者為實用專利（Utility）則需於發證日起 3.5 年，7.5 年及 11.5 年屆滿之六個月前繳維護費。而新式樣專利及植物專利則不必繳任何維護費。

　　上述大致的審查流程乃如圖所示：

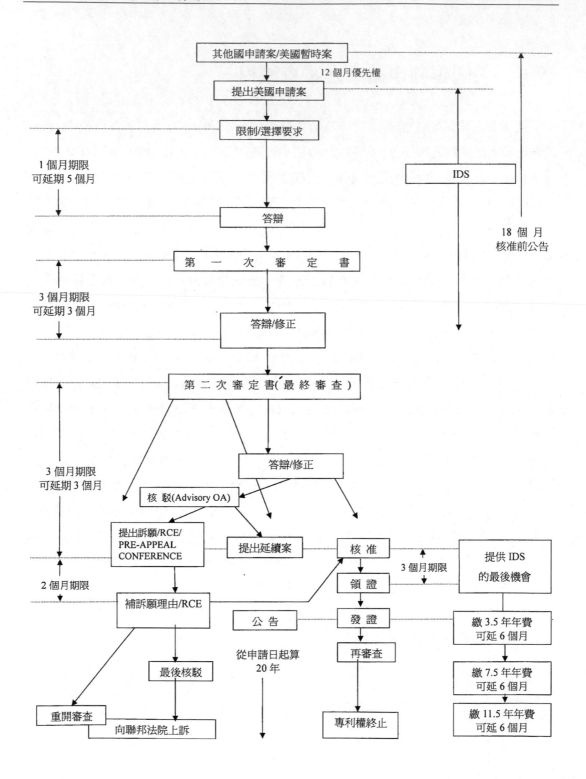

214

§9.2　申請案不符合形式要件，揭露不全或請求項界定不清楚時之審查

初步審查辦公室（OIPE）通常只針對表面之形式要件作審查，但仍有許多專利申請案之缺失，需由技術中心的審查委員實際閱讀專利說明書及申請文件後才會發現。例如，由說明書之英文很難了解其發明內容，發明之揭露不足，請求項未清楚界定要請求之發明，漏掉圖式等。如有上述情況，審查委員採作如下之處理：

(a) 即使說明書很難懂，仍需儘可能去檢索先前技術（prior art）並審查請求項，但若確實無法依照說明書去作檢索時可發下審定書，指出未作前案檢索。

(b) 如果有缺少文件（圖式）或文件不符規定，審查委員會以審定書之附件之方式告知申請人。

(c) 如果說明書之英文無法了解，審查委員會通知申請人補送修正之英文說明書或取代之英文說明書（substitute specification）。

(d) 如果請求項不符合 35 USC §112 清楚界定所請求之發明之規定，可直接將所有申請項核駁。

MPEP 702.01 Obviously Informal Cases

When an application is reached for its first Office action and it is then discovered to be impractical to give a complete action on the merits because of an informal or insufficient disclosure, the following procedure may be followed:

(A)A reasonable search should be made of the invention so far as it can be understood from the disclosure, objects of invention and claims and any apparently pertinent art cited. In the rare case in which the disclosure is so incomprehensible as to preclude a reasonable search, the Office action should clearly inform applicant that no search was made;

(B)Informalities noted by the Office of Initial Patent Examination (OIPE) and deficiencies in the drawing should be pointed out by means of attachments to the Office action (see MPEP § 707.07(a));

(C)A requirement should be made that the specification be revised to conform to idiomatic English and United States practice;

(D)The claims should be rejected as failing to define the invention in the manner required by 35 U.S.C. 112 if they are informal. A blanket rejection is usually sufficient.

§9.3 要求申請人提供相關資訊以便檢索

審查委員在檢索時如果無法找到足夠的先前技術以判斷專利性時，可要求申請人提供資訊，其可另外發函要求提供資訊，亦可在審定書中作要求。但是審查委員不能只依賴申請人提供資訊而不去自己作檢索。申請人可能被要求之資訊包括：①商業之資料庫，②申請人檢索之資料，③相關資訊，④用來撰寫說明書之相關資訊，⑤在發明過程中所用之資訊，⑥發明所改良之部份，⑦發明之用途。

申請人收到資訊要求函（Requirements for information）後如果可提供所要求之資訊，則需善意地（不隱瞞地）提供。而如果無法取得所要求之資訊則可回覆說無法獲知該些資訊或無法容易取得該些資訊即可。

---

37 CFR 1.105 Requirements for information.

(a)(1) In the course of examining or treating a matter in a pending or abandoned application filed under 35 U.S.C. 111 or 371 (including a reissue application), in a patent, or in a reexamination proceeding, the examiner or other Office employee may require the submission, from individuals identified under § 1.56(c), or any assignee, of such information as may be reasonably necessary to properly examine or treat the matter, for example:

(i) Commercial databases: The existence of any particularly relevant commercial database known to any of the inventors that could be searched for a particular aspect of the invention.

(ii) Search: Whether a search of the prior art was made, and if so, what was searched.

(iii) Related information: A copy of any non-patent literature, published application, or patent(U.S. or foreign), by any of the inventors, that relates to the claimed invention.

(iv) Information used to draft application: A copy of any non-patent literature, published application, or patent (U.S. or foreign) that was used to draft the application.

(v) Information used in invention process: A copy of any non-patent literature, published application, or patent (U.S. or foreign) that was used in the invention process, such as by designing around or providing a solution to accomplish an invention result.

(vi) Improvements: Where the claimed invention is an improvement, identification of what is being improved.

(vii) In Use: Identification of any use of the claimed invention known to any of the inventors at the time the application was filed notwithstanding the date of the use.

(2) Where an assignee has asserted its right to prosecute pursuant to § 3.71(a) of this chapter, matters such as paragraphs (a)(1)(i), (iii), and (vii) of this section may also be applied to such assignee.

(3) Any reply to a requirement for information pursuant to this section that states either that the information required to be submitted is unknown to or is not readily available to the party or parties from which it was requested may be accepted as a complete reply.

(b) The requirement for information of paragraph (a)(1) of this section may be included in an Office action, or sent separately.

MPEP 704.12(b) What Constitutes a Complete Reply

A complete reply to a 37 CFR 1.105 requirement is a reply to each enumerated requirement for information giving either the information required or a statement that the information required to be submitted is unknown and/or is not readily available to the party or parties from which it was requested. There is no requirement for the applicant to show that the required information was not, in fact, readily attainable, but applicant is required to make a good faith attempt to obtain the information and to make a reasonable inquiry once the information is requested.

A reply stating that the information required to be submitted is unknown and/or is not readily available to the party or parties from which it was requested will generally be sufficient unless, for example, it is clear the applicant did not understand the requirement, or the reply was ambiguous and a more specific answer is possible.

§9.4　專利申請案的審查順序

非暫時專利申請案在完成形式審查之後就會送到依專利分類之審查技術中心（examining Technology Center（TC））交給審查委員審查。審查委員審查案件之順序乃依照專利申請案之有效申請日（effective filing date），先申請的先審查。但依照

37 CFR 1.102 請求優先審查之案件（advancement of examination）則例外。對於部份延續案（CIP），則依其部份延續案之申請日作為審查順序之依據，但審查委員如果認為必要時亦可依其有效申請日（母案之申請日）作為審查順序之依據。

　　另外，對於再審查案件（re-examination）及再發證案件（reissue）需優先審查，涉及訴訟程序之再審查案件及再發證案件則列為最優先。

　　此外，對於最終審定書（Final Action）有提出修正之案件需於二個月內審查。

MPEP 708 Order of Examination

Each examiner will give priority to that application in his or her docket, whether amended or new, which has the oldest effective U.S. filing date. Except as rare circumstances may justify Technology Center Directors in granting individual exceptions, this basic policy applies to all applications.

The actual filing date of a continuation-in-part application is used for docketing purposes. However, the examiner may act on a continuation-in-part application by using the effective filing date, if desired.

If at any time an examiner determines that the "effective filing date" status of any application differs from what the records show, the technical support staff should be informed, who should promptly amend the records to show the correct status, with the date of correction.

The order of examination for each examiner is to give priority to reissue applications and to reexamination proceedings, with top priority to reissue applications in which litigation has been stayed (MPEP § 1442.03) and to reexamination proceedings involved in litigation (MPEP § 2261), then to those special cases having a fixed 30-day due date, such as examiner's answers and decisions on motions. Most other cases in the "special" category (for example, interference cases, cases made special by petition, cases ready for final conclusion, etc.) will continue in this category, with the first effective U.S. filing date among them normally controlling priority.

All amendments before final rejection should be responded to within two months of receipt.

§9.5　專利申請案之加速審查

美國專利局的審查委員一般依照專利申請案之（有效）申請日之先後順序逐件審查，亦即先申請的會先審查，而如果有下列四種情況，審查委員會對個案加速審查。

(a) 由美國專利局長命令要加速審查之案件。

(b) 美國政府部門首長認為對於公共服務部門特別重要之專利申請案件。

(c) 申請人因為年齡較大（65 歲以上）或健康不佳，或者是其專利申請案之內容會實質上增進環境品質或對節約能源有貢獻者，申請人可提出請願（petition，不必規費）請求加速審查。

(d) 其他申請人認為需加速審查之案件則可提出請願，並繳規費請求加速審查，一般之情形有下列數種：

(1) 申請人計劃要製造請求項之內容之產品，而且已有足夠之資金及設備，打算專利一核准就生產，如果不核准則不生產，且已對先前技術已作了檢索並附上檢索報告。

(2) 申請人申請專利之內容已被他人侵權（市場上已有侵權之產品或方法）的話，申請人可將侵權產品與專利申請內容作比較，並且對先前技術作檢索後，提出請願並繳規費請求加速審查。

(3) 申請人之專利申請內容如果是有關於重組 DNA (recombinant DNA)研究之安全性者，亦可提出請願並繳規費請求加速審查。

(4) 申請人之專利申請內容如果是有關於 HIV/AIDS 及癌症治療者，亦可提出請願及繳規費請求加速審查。

(5) 申請人之專利申請是有關於生物技術且申請人(或受讓人)是小實體時，則可提出請願，說明其專利申請之技術如果不能很快得到專利將對此技術之發展不利，並繳規費請求加速審查。

(6) 申請人之專利申請內容如有關於打擊恐怖主義之方法或設備，亦可提出請願並繳交規費請求加速審查。

(7) 申請人之發明是有關於超導體技術，亦可提出請願及繳交規費請求加速審查。

(8) 即使沒有上述特殊之情況，申請人也可提出請願，繳交規費請求加速審查，但需符合下列要件:①請求項必須只涵蓋單一發明，或如果非單一發明同意作限制/選擇，②要自行作檢索並提出先前技術之檢索報告，③對先前技術之分析及為何發明具有專利性之討論。

---

37 CFR 1.102 Advancement of examination.

(a) Applications will not be advanced out of turn for examination or for further action except as provided by this part, or upon order of the Director to expedite the business of the Office, or upon filing of a request under paragraph (b) of this section or upon filing a petition under paragraphs (c) or (d) of this section with a showing which, in the opinion of the Director, will justify so advancing it.

(b) Applications wherein the inventions are deemed of peculiar importance to some branch of the public service and the head of some department of the Government requests immediate action for that reason, may be advanced for examination.

(c) A petition to make an application special may be filed without a fee if the basis for the petition is the applicant's age or health or that the invention will materially enhance the quality of the environment or materially contribute to the development or conservation of energy resources.

(d) A petition to make an application special on grounds other than those referred to in paragraph (c) of this section must be accompanied by the fee set forth in § 1.17(h).

MPEP 708.02

New applications ordinarily are taken up for examination in the order of their effective United States filing dates. Certain exceptions are made by way of petitions to make special, which may be granted under the conditions set forth below.

I. Prospective Manufacture

II. Actual Infringement

III. Applicant's health

IV. Applicant's age is over 65

V. Environment quality

VI. Energy

VII. Inventions relating to recombinant DNA

VIIII. Special Examining procedure for certain new applications-accelerated Examination

IX. Special status for patent applications relating to super conductivity

X. Inventions relating to HIV/AIDS and Cancer

XI. Inventions for countering Terrorism

XII. Special status for applications relating to biotechnology filed by applicants who are small entities

§9.6　利用專利審查高速公路計劃請求加速審查

美國專利商標局與許多國家之專利局訂有專利審查高速公路計劃（Patent Prosecution Highway（PPH）pilot program）以用來加速專利案件之審查。依照此計劃，一專利申請人在收到第一申請國之專利局（Office of First Filing,OFF）之審定書中當有至少一請求項（claim）是可核准的時，就可請求第二申請國之專利局（Office of Second Filing,OSF）加速審查該對應國之申請案。目前美國與澳洲，加拿大，丹麥，芬蘭，德國，匈牙利，日本，韓國，新加坡，英國之專利局以及歐洲專利局已共同簽定了此 PPH 計劃。例如，當一申請人之對應歐洲專利申請案，歐洲專利局發下審定書指出有些請求項可核准時，此申請人可提出下列文件：①歐洲專利局之審定書影本，②可核准之請求項之影本，③審查委員所引證之先前技術影本，④對應於歐洲專利申請案之可核准請求項之對應請求項表列，向美國專利局提出請願（petition），請求依照此 PPH 計劃加速審查。

依照美國專利局之官方資料，向美國專利局提出依照 PPH 計劃請求加速審查之美國專利申請案，從請願核准到美國專利局發下第 1 次審定書（First Office Action）之期間為 2～3 個月，而且 95%之案件均核准。但注意的是，請求項需完全對應於 OFF 之申請案之可核准的請求項。

§9.7　請求延緩審查及暫停審查

在 2000 年 11 月 29 日當天或之後申請之發明專利申請案或植物專利申請案在未發下審定書或核准通知之前，可以繳交公告費及處理費，請求延緩進行審查，但最多只能延緩三年（自優先權日或申請日起算）。但是請求延緩審查之專利申請案必須未曾請求不作核准前公告且其申請案是在可適合核准前公告之狀態。

　　已進行審查中之發明專利申請案及植物專利申請案，如果申請人具有充份之理由可繳交費用提出請願，請求審查委員暫停審查最多 6 個月，但是請求暫停審查時必須沒有尚未答辯之審定書方可。另外，欲提出請求繼續審查（RCE）或 CPA 延續案（CPA）請求之專利案件可繳交處理費在提出請求繼續審查或 CPA 延續案之同時提出暫停審查之要求，最多三個月（例如要補充非顯而易見之答辯證據等情況）。

37 CFR 1.103 Suspension of action by the Office.

(a) Suspension for cause. On request of the applicant, the Office may grant a suspension of action by the Office under this paragraph for good and sufficient cause. The Office will not suspend action if a reply by applicant to an Office action is outstanding. Any petition for suspension of action under this paragraph must specify a period of suspension not exceeding six months. Any petition for suspension of action under this paragraph must also include:

(1) A showing of good and sufficient cause for suspension of action; and

(2) The fee set forth in § 1.17(g), unless such cause is the fault of the Office.

(b) Limited suspension of action in a continued prosecution application (CPA) filed under § 1.53(d). On request of the applicant, the Office may grant a suspension of action by the Office under this paragraph in a continued prosecution application filed under § 1.53(d) for a period not exceeding three months. Any request for suspension of action under this paragraph must be filed with the request for an application filed under § 1.53(d), specify the period of suspension, and include the processing fee set forth in § 1.17(i).

(c) Limited suspension of action after a request for continued application (RCE) under § 1.114. On request of the applicant, the Office may grant a suspension of action by the Office under this paragraph after the filing of a request for continued examination in compliance with § 1.114 for a period not exceeding three months. Any request for suspension of action under this paragraph must be filed with the request for continued examination under § 1.114, specify the period of suspension, and include the processing fee set forth in § 1.17(i).

(d) Deferral of examination. On request of the applicant, the Office may grant a deferral of examination under the conditions specified in this paragraph for a period not extending beyond three years from the earliest filing date for which a benefit is claimed under title 35, United States Code. A request for deferral of examination under this paragraph must include the publication fee set forth in § 1.18(d) and the processing fee set forth in § 1.17(i). A request for deferral of examination under this paragraph will not be granted unless:

(1) The application is an original utility or plant application filed under § 1.53(b) or resulting from entry of an international application into the national stage after compliance with § 1.495;

(2) The applicant has not filed a nonpublication request under § 1.213(a), or has filed a request under § 1.213(b) to rescind a previously filed nonpublication request;

(3) The application is in condition for publication as provided in § 1.211(c); and

(4) The Office has not issued either an Office action under 35 U.S.C. 132 or a notice of allowance under 35 U.S.C. 151.

§9.8　初審審定書及最終審定書

審查委員對於一專利申請案之專利要件作實質審查（是否符合 35 USC§101、102、103、112）而發下之第一次審定書則為最初審定書（first Office Action），而審查委員發下的第二次或以後之審定書，一般審查終結，則為最終審定書（final Office Action），除非審查委員在第二次或以後之審定書所述之新的核駁理由不是因為申請人對請求項（claims）之修改所造成的。例如，審查委員新的核駁理由是由於在最初審查時未檢索到的先前技術；或者除非審查委員在第二次或以後之審定書所述之新的核駁理由不是根據申請人在 37 CFR 1.97(c)之期間（最初審定書至最終審定書之期間）送進 IDS（Information Disclosure Statement）資料者。

MPEP 706.07(a)

Under present practice, second or any subsequent actions on the merits shall be final, except where the examiner introduces a new ground of rejection that is neither necessitated by applicant's amendment of the claims, nor based on information submitted in an information disclosure statement filed during the period set forth in 37 CFR 1.97(c) with the fee set forth in 37 CFR 1.17(p). Where information is submitted in an information disclosure statement during the period set forth in 37 CFR 1.97(c) with a fee, the examiner may use the information submitted, e.g., a printed publication or evidence of public use, and make the next Office action final whether or not the claims have been amended, provided that no other new ground of rejection which was not necessitated by amendment to the claims is introduced by the examiner.

§9.9 初審審定書及最終審定書發下時之修正限制

申請人對於最初審定書答辯時對於說明書、圖式及請求項之修正只要是不會造成新事項（new matter）者均可進入審查（entered）。例如，說明書之文法、用字之修改，增加請求項以請求說明書原揭露未請求之發明，刪除請求項，修正圖式之不清楚等。然而，一旦發下最終審定書，則申請人就不能無限制地對請求項作修正，但是不意味審查委員不考慮任何對請求項之修正。任何可使申請案成可核准狀態之修正（例如刪除被核駁的請求項及對請求項作形式，如打字錯誤、文法錯誤之修正）或適合訴願之修正均有可能進入審查。其他的修正，除非申請人可提出說明為何未在初審時就提出的充份理由，否則通常不能進入審查。

MPEP 714.12

Once a final rejection that is not premature has been entered in an application, applicant or patent owner no longer has any right to unrestricted further prosecution. This does not mean that no further amendment or argument will be considered. Any amendment that will place the application either in condition for allowance or in better form for appeal may be entered. Also, amendments complying with objections or requirements as to form are to be permitted after final action in accordance with 37 CFR 1.116(b). Ordinarily, amendments filed after the final action are not entered unless approved by the examiner.

實例 1

審查委員在一最初審定書中引證一先前技術 A 以顯而易見核駁一專利申請案之二獨立項 1 及 2。於答辯時申請人僅對獨立項 1 作修改並未修改獨立項 2。審查委員於第二次審定書中除了引用原引證 A 外另引證一先前技術 B 再以顯而易見核駁獨立項 1 及獨立項 2，請問審查委員之第二次審定書是否可作為最終審定書（final Action）？

審查委員之第二次審定書不能作為最終審定書，因為獨立項 2 並未修改，故審查委員另外引證之先前技術 B 並非因為申請人對獨立項 2 作修正所必需。

實例 2

申請人之專利申請案包括下列二請求項：

claim 1:A composition comprising wheat flour and corn flour.

claim 2:The composition of claim 1, further comprising molasses.

審查委員引證了一先前技術 A 以顯而易見核駁 Claim 1 及 Claim 2。申請人於答辯時刪除 claim 2，並修正 claim 1 如下：

Claim 1: A composition comprising wheat flour; corn flour and molasses.

審查委員發下最終審定書，於最終審定書中審查委員除了先前技術 A 外又引證了一先前技術 B，再以顯而易見核駁 claim 1，請問審查委員是否可以此第二次審定書作為最終審定書？

審查委員不能以此第二次審定書作為最終審定書。申請人雖將 claim 1 修正，但此修正只是將 claim 2 之限制條件併入 claim 1 之中，審查委員在作最初審定書時就應檢索完全，不應在第二次審定書時才檢索出先前技術 B，故此先前技術 B 之引證並非因為申請人修正 claim 所必須者。

§9.10　初審審定書為最終審定書

符合在下列二種條件下，審查委員可將初審審定書作為最終審定書：

(A) 該新專利申請案乃為一延續案（Continuing Application）或 RCE 案或為一先前申請案之取代案（substitute for an earlier application）；及

(B) 該新專利申請案之請求項請求與一先前申請案之請求項相同之發明，而且該請求項如果在先前申請案就已進入審查（entered），且應被適當地最終核駁的話。

MPEP 706.07 (b)

The claims of a new application may be finally rejected in the first Office action in those situations where (A) the new application is a continuing application of, or a substitute for, an earlier application, and (B) all claims of the new application (1) are drawn to the same invention claimed in the earlier application, and (2) would have been properly finally rejected on the grounds and art of record in the next Office action if they had been entered in the earlier application.

實例 1

審查委員引證先前技術 A 及 B 以顯而易見核駁一專利申請案之 Claims 1-5。申請人申請一延續案，其 Claims 1-5 與原母案完全相同，審查委員發下一最初審定書作為最終核駁，審查委員此次除了引證原有之先前技術 A 及 B 以顯而易見核駁外，又引證了先前技術 C 以 35USC§102 核駁 Claims 1-5。請問申請人以請願要求審查委員撤回此最終之審定書是否會成功？

申請人提出請願要求撤回最終審定書會成功。雖然申請人之延續案的 Claims 1-5 與原母案相同，但審查委員又引證了先前技術 C 並作為最終核駁，致使申請人對先前技術 C 之核駁無法充分答辯。

實例 2

申請人之專利申請案所有之請求項均被審查委員以 35 USC§103(a)以顯而易見作最終核駁，申請人將請求項修改提出答辯以克服核駁，審查委員不接受所提之修正發下建議性審定書（Advisory Action），原因是此修正產生了新議題（new issue），使得審查委員需重新檢索及考慮。申請人決定不訴願而是提出延續案。申請人在延續案中提出先前修正將請求項修改成母案在最終審定時提出修正之版本。審查委員在審查此延續案後未再引證其他先前技術，但以相同之顯而易見核駁理由核駁，並將第一次審定書作為最終核駁。請問申請人如果提出請願請求撤回最終核駁，是否會成功？

申請人提出請願會成功。因為延續案的請求項乃是在母案中未進入審查且產生新議題的版本。

§9.11　關於國家安全之專利申請案之審查

　　如果一美國專利申請案（不論由美國人或外國人申請）所揭露之發明的技術美國政府機關具有財產上之所有權，且美國政府機關認為其公告或獲准專利將有害於國家安全時，就會通知美國專利局長發下保密命令（secrecy order）將該發明保密並不准（withhold）公告或獲准專利一年，如需要可續行延展每次一年。如果美國專利申請案所揭露的發明技術，美國政府機關不具有財產上之所有權，但美國專利局長認為公告或獲准將有害國家安全時，會交原子能委員會，國防委員會或其他政府機關審查，如這些機關認為有害於國家安全就會通知專利局局長發下保密秘令將該發明保密並不准公告或獲准專利一年，如需要會續行延展每次一年。所有保密命令之專利申請案仍會作專利性之審查，但是即使具有專利性，亦不會發證書亦不會與其他案件進入抵觸程序。如果發下最終審定書則必需提出訴願，以防止案件放棄，但是訴願期間不會進行口頭聽證程序。如果保密命令之案件可核准則會發下核准通知，但不會發證，一直到保密命令廢止為止。

　　如果是國際申請案則程序只進行到檢索報告發下，但不會將申請案寄至國際局。

　　申請人亦可提出請願，請求廢止或修正此保密命令。例如申請人相信在某些存在之事實或情況下此保密命令是無效的或者將申請案中敏感之部份刪除即可，不必列為保密命令之案件，此請願如經專利局局長核准即可廢止保密命令。

　　如果在保密命令發下之前或之後，專利申請案之保密內容被洩漏給在美國境內之美國公民時，申請人需立即告知該美國公民有保密命令及不當洩密將受處罰。如果保密內容被洩露給在外國之任何人或在美國之外國人時，申請人不需通知此人，只要通知美國專利局即可。

---

35 U.S.C. 181 Secrecy of certain inventions and withholding of patent.

　　Whenever publication or disclosure by the publication of an application or by the grant of a patent on an invention in which the Government has a property interest might, in the opinion of the head of the interested Government agency, be detrimental to the national security, the Commissioner of Patents upon being so notified shall order that the invention be kept secret and shall withhold the publication of an application or the grant of a patent therefor under the conditions set forth hereinafter.

Whenever the publication or disclosure of an invention by the publication of an application or by the granting of a patent, in which the Government does not have a property interest, might, in the opinion of the Commissioner of Patents, be detrimental to the national security, he shall make the application for patent in which such invention is disclosed available for inspection to the Atomic Energy Commission, the Secretary of Defense, and the chief officer of any other department or agency of the Government designated by the President as a defense agency of the United States.

Each individual to whom the application is disclosed shall sign a dated acknowledgment thereof, which acknowledgment shall be entered in the file of the application. If, in the opinion of the Atomic Energy Commission, the Secretary of a Defense Department, or the chief officer of another department or agency so designated, the publication or disclosure of the invention by the publication of an application or by the granting of a patent therefor would be detrimental to the national security, the Atomic Energy Commission, the Secretary of a Defense Department, or such other chief officer shall notify the Commissioner of Patents and the Commissioner of Patents shall order that the invention be kept secret and shall withhold the publication of the application or the grant of a patent for such period as the national interest requires, and notify the applicant thereof. Upon proper showing by the head of the department or agency who caused the secrecy order to be issued that the examination of the application might jeopardize the national interest, the Commissioner of Patents shall thereupon maintain the application in a sealed condition and notify the applicant thereof. The owner of an application which has been placed under a secrecy order shall have a right to appeal from the order to the Secretary of Commerce under rules prescribed by him.

An invention shall not be ordered kept secret and the publication of an application or the grant of a patent withheld for a period of more than one year. The Commissioner of Patents shall renew the order at the end thereof, or at the end of any renewal period, for additional periods of one year upon notification by the head of the department or the chief officer of the agency who caused the order to be issued that an affirmative determination has been made that the national interest continues to so require.

＊ ＊ ＊ ＊ ＊ ＊ ＊ ＊ ＊ ＊ ＊

MPEP 120

V. CHANGES IN SECRECY ORDERS

Applicants may petition for rescission or modification of the Secrecy Order. For example, if the applicant believes that certain existing facts or circumstances would render the Secrecy Order ineffectual, he or she may informally contact the sponsoring agency to discuss these facts or formally petition the Commissioner for Patents to rescind the Order. Rescission of a Secrecy Order may also be effected in some circumstances by expunging the sensitive subject matter from the disclosure, provided the sensitive subject matter is not necessary for an enabling disclosure under 35 U.S.C. 112, first paragraph. See MPEP § 724.05. The defense agency identified with the Secrecy Order as sponsoring the Order should be contacted directly for assistance in determining what subject matter in the application is sensitive, and whether the agency would agree to rescind the Order upon expunging this subject matter. The applicant may also petition the Commissioner for Patents for a permit to disclose the invention to another or to modify the Secrecy Order stating fully the reason or purpose for disclosure or modification. An example of such a situation would be a request to file the application in a foreign country. The requirements for petitions are described in 37 CFR 5.4 and 5.5. The law also provides that if an appeal is necessary, it may be taken to the Secretary of Commerce. Any petition or appeal should be addressed to the Mail Stop L&R, Commissioner for Patents, P.O. Box 1450, Alexandria, Virginia, 22313-1450.

VI. IMPROPER OR INADVERTENT DISCLOSURE

If, prior to or after the issuance of the Secrecy Order, any significant part of the subject matter or material information relevant to the application has been or is revealed to any U.S. citizen in the United States, the principals must promptly inform such person of the Secrecy Order and the penalties for improper disclosure. If such part of the subject matter was or is disclosed to any person in a foreign country or foreign national in the U.S., the principals must not inform such person of the Secrecy Order, but instead must promptly furnish to Mail Stop L&R, Commissioner for Patents, P.O. Box 1450, Alexandria, Virginia, 22313-1450 the following information to the extent not already furnished: date of disclosure; name and address of the disclosee; identification of such subject matter; and any authorization by a U.S. government agency to export such subject matter. If the subject matter is included in any foreign patent application or patent, this should be identified.

VII. EXPIRATION

Under the provision of 35 U.S.C. 181, a Secrecy Order remains in effect for a period of 1 year from its date of issuance. A Secrecy Order may be renewed for additional periods of not more than 1 year upon notice by a government agency that the national interest so requires. The applicant is notified of any such renewal.

The expiration of or failure to renew a Secrecy Order does not lessen in any way the responsibility of the principals for the security of the subject matter if it is subject to the provisions of Exec. Order No. 12958 or the Atomic Energy Act of 1954, as amended, 42 U.S.C. 141 et. seq. and 42 U.S.C. 2181 et. seq. or other applicable law unless the principals have been expressly notified that the subject patent application has been declassified by the proper authorities and the security markings have been authorized to be canceled or removed.

MPEP 130 Examination of Secrecy Order Cases [R-5]

All applications in which a Secrecy Order has been imposed are examined in Technology Center (TC) Working *>Groups< 3640 and **>3660<. If the Order is imposed subsequent to the docketing of an application in another TC, the application will be transferred to TC Working Group 3640 >or 3660<.

Secrecy Order cases are examined for patentability as in other cases, but may not be passed to issue; nor will an interference be declared where one or more of the conflicting cases is classified or under Secrecy Order. See >37 CFR 5.3 and< MPEP § *>2306<. **

In case of a final rejection, while such action must be properly replied to, and an appeal, if filed, must be completed by the applicant to prevent abandonment, such appeal will not be set for hearing by the Board of Patent Appeals and Interferences until the Secrecy Order is removed, unless specifically ordered by the Commissioner for Patents.

When a Secrecy Order case is in condition for allowance, a notice of allowability (Form D-10) is issued, thus closing the prosecution. Any amendments received thereafter are not entered or responded to until such time as the Secrecy Order is rescinded. At such time, amendments which are free from objection will be entered; otherwise they are denied entry.

MPEP 120

IV.PCT APPLICATIONS

If the Secrecy Order is applied to an international application, the application will not be forwarded to the International Bureau as long as the Secrecy Order remains in effect. If the Secrecy Order remains in effect at the end of the time limit under PCT Rule 22.3, the international application will be considered withdrawn (abandoned) because the Record Copy of the international application was not received in time by the International Bureau. 37 CFR 5.3(d), PCT Article 12(3), and PCT Rule 22.3. If the United States of America has been designated, however, it is possible to save the U.S. filing date, by fulfilling the requirements of 35 U.S.C. 371(C) prior to the withdrawal.

第 10 章　限制／選擇實務

目　錄

§10.1　專利申請案之限制要求

當申請人之專利申請案之請求項（claims）包含二個以上獨立之發明（independent invention）或專利性上不同之發明（patentably distinct invention），而且審查員在審查此專利申請案時會有嚴重之負擔時，就會通知申請人對申請案作限制。而所謂獨立之發明，例如電腦與藥品是為獨立之發明，當一申請案有二獨立之發明時，審查委員就可視為審查上有嚴重之負擔，因而在要求申請人作限制時可不必另作說明。而當申請案包括二個以上相關之發明而審查委員作審查時認為此二個以上之相關發明乃屬於(a)不同之分類或(b)雖可分類在一齊但在技藝上屬不同之狀態，或(c)需作不同領域之前案檢索時，審查委員就可主張為專利性上不同之發明，審查上有嚴重之負擔故需作限制。

MPEP 803 Restriction – When Proper

Under the statute an application may properly be required to be restricted to one of two or more claimed inventions only if they are able to support separate patents and they are either independent.

If the search and examination of an entire application can be made without serious burden, the examiner must examine on the merits, even though it includes claims to independent or distinct inventions.

MPEP 808.02 Related Inventions

Where, as disclosed in the application, the several inventions claimed are related, and such related inventions are not patentably distinct as claimed, restriction under 35 U.C.S. 121 is never proper (MPEP § 806.05).　If applicant optionally restricts, double patenting may be held.

Where the related inventions as claimed are shown to be distinct under the criteria of MPEP § 806.05(c)- § 806.05(i), the examiner, in order to establish reasons for insisting upon restriction, must show by appropriate explanation one of the following:

(A)Separate classification thereof: This shows that each distinct subject has attained recognition in the art as a separate subject for inventive effort, and also a separate field of search. Patents need not be cited to show separate classification.

(B)A separate status in the art when they are classifiable together: Even though they are classified together, each subject can be shown to have formed a separate subject for inventive effort when an explanation indicated a recognition of separate inventive effort by inventors.

Separate status in the art may be shown by citing patents which are evidence of such separate status, and also of a separate field of search.

(C)A different field of search: Where it is necessary to search for one of the distinct subjects in places where no pertinent art to the other subject exists, a different field of search is shown, even though the two are classified together. The indicated different field of search must in fact be pertinent to the type of subject matter covered by the claims. Patents need not be cited to show different fields of search.

Where, however, the classification is the same and the field of search is the same and there is no clear indication of separate future classification and field of search, no reasons exist for dividing among related inventions.

§10.2 不同意限制要求時之因應之道

不論審查委員是發下書面之限制要求或者是以電話通知作限制要求，申請人即使不同意，仍須先選一（組）發明作為審查之對象，同時告知審查委員哪些請求項涵蓋所選之發明。申請人雖可反駁（traverse）審查委員之限制要求，但需注意的是不能只告知審查委員要反駁，必須提出確實的反駁理由，例如二發明並非專利上不同之發明或審查委員審查此二發明並無審查上嚴重之負擔等。一旦有提出反駁的理由，如果審查委員不接受，只針對所選之發明作審查時，申請人可保留提出請願（petition）之權利（可在訴願前或核准前提出）。申請人在針對限制要求答辯時可同時修改請求項，但審查委員只會針對可涵蓋所選之發明的新增請求項作審查。另外，要注意的是，一旦作了請求項之限制，而審查委員也針對所限制之發明作了審查，申請人就不能在後續階段將審查對象改變成未選之發明，只能在所選擇之發明之申請案未放棄或發證之前對於未限制之發明提出分割案。另外，美國專利局對於未選擇之發明並不會將之刪除，只有當選擇之發明要準備發證時才會依職權將未選擇之發明（claims）刪除。

MPEP 818.03(b) Must Elect, Even When Requirement Is Traversed

As noted in the second sentence of 37 CFR 1.143, a provisional election must be made even though the requirement is traversed.

All requirements for restriction should include form paragraph 8.22.

¶ 8.22 Requirement for Election and Means for Traversal

Applicant is advised that the reply to this requirement to be complete must include (i) an election of a species or invention to be examined even though the requirement be traversed (37 CFR 1.143)

MPEP 818.03(c)Must Traverse To Preserve Right of Petition

37 CFR 1.144. Petition from requirement for restriction.

After a final requirement for restriction, the applicant, in addition to making any reply due on the remainder of the action, may petition the Director to review the requirement. Petition may be deferred until after final action on or allowance of claims to the invention elected, but must be filed not later than appeal. A petition will not be considered if reconsideration of the requirement was not requested (see § 1.181).

If applicant does not distinctly and specifically point out supposed errors in the restriction requirement, the election should be treated as an election without traverse and be so indicated to the applicant by use of form paragraph 8.25.02.

¶ 8.25.02 Election Without Traverse Based on Incomplete Reply

Applicant's election of [1] in the reply filed on [2] is acknowledged. Because applicant did not distinctly and specifically point out the supposed errors in the restriction requirement, the election has been treated as an election without traverse (MPEP § 818.03(a)).

MPEP 819 Office Generally Does Not Permit Shift

The general policy of the Office is not to permit the applicant to shift to claiming another invention after an election is once made and action given on the elected subject matter. Note that the applicant cannot, as a matter of right, file a request for continued examination (RCE) to obtain continued examination on the basis of claims that are independent and distinct from the claims previously claimed and examined (i.e., applicant cannot switch inventions by way of an RCE as a matter of right).

實例 1

審查委員對一專利申請案發下限制要求，指出該專利申請案包括產品請求項及方法請求項，申請人選擇了產品請求項而且並未反駁（traverse），之後於 2007 年 10 月 20 日審查委員發下審定書（Office Action）核駁了所有的產品請求項，申請人經檢討覺得要答辯成功相當困難，故於 2008 年 1 月 19 日提出一答辯刪除原有之產品請求項而只保留方法請求項，請問此答辯是否符合美國專利局之規定？

　　此答辯將被視為答辯不完全（non-responsive），申請人如果要將未選之請求項被審查，則需提出分割案。

MPEP 821.03

An amendment canceling all claims drawn to the elected invention and presenting only claims drawn to the nonelected invention should not be entered. Such an amendment is nonresponsive.

實例 2

　　申請人之美國專利申請案之申請專利範圍包括二組 claims，方法與產品，其獨立項分別如下所述：

1. A preparation method for biochips comprising：

(a) providing a substrate;

(b) applying a micro-injecting process to XXXXXXX

　　XXXXXXXXXXXXXXXXXXXXXXXXXXXXXXXXXXXXXXXXXXXXXXXXXXXX XXXXXXXXXXXXXXXXXXXXXXXX, XXXXXXXXXXXXXXXXXXXXXXXXXXXXXX XXXXXXXXXXXXXXXXXXXXXXXXXXXXXX; and immobilizing a probe XXXXXXXXXX XXXXXXXXXXXXXXXXXXXXXXXXXXXXXXX.

　　18. A biochip comprising a substrate, a plurality of hydrophobic regions XXXXXXXXXX XXXXXXXXXXXXXXXXXXXXXXXXXXXXXXXXXXXXXXXXXXXXXXXXXXXX XXXXXXXXXXXXXXXXXXX; a plurality of hydrophilic partitions XXXXXXXXXXXX XXXXXXXXXXXX XXXXXXXXXXXXXXXXXXXXXXXXXXXXXXXXXX ; and a probe immobilized onXXXXXXXXXXXXXXXXXXXXXXXXXXXXXX XXXXXXXXXXXXXXXXXX.

　　審查委員發下限制之審定書，要求申請人選一發明作審查，理由是 claim 18 項之生物晶片亦可用微影技術製出，且此二發明屬不同之分類，審查委員在審查上有嚴重之負擔。請問此限制 OA 是否恰當？應如何答辯？

　　依照 MPEP 之規定，當一專利申請案包括有製法與產品二組發明時，審查委員只要能指出下列二者之一，即可做出限制要求：

(1)　所請求之方法可用來製造實質上不同之產品。

(2)　所請求之產品可用實質上不同之方法製出。

由於審查委員只指出該生物晶片實質上可用微影製程技術（photo lithographic techniques）製出，因而申請人如果不同意此申請案包括二個發明仍需先選產品或製法作審查，並提出反駁（traverse），反駁的理由需證明本案之晶片實際上無法用微影製程技術製造方可。至於要如何選，則視申請人想保護的發明是哪一項或哪一項發明具有特徵或哪一項發明在核准專利後較易防止他人侵權而定，但是在產品與製法之限制選擇時，如果申請人選擇產品，而之後經審查產品可核准時，申請人可將包括所有核准之產品之限制條件之製法再加入（rejoin）申請案中。

MPEP 806.05(f)

A process of making and a product made by the process can be shown to be distinct inventions if either or both of the following can be shown: (A) that the process as claimed is not an obvious process of making the product and the process as claimed can be used to make another materially different product; or (B) that the product as claimed can be made by another materially different process.

Allegations of different processes or products need not be documented.

A product defined by the process by which it can be made is still a product claim (In re Bridgeford, 357 F.2d 679, 149 USPQ 55 (CCPA 1966)) and can be restricted from the process if the examiner can demonstrate that the product as claimed can be made by another materially different process; defining the product in terms of a process by which it is made is nothing more than a permissible technique that applicant may use to define the invention.

If applicant convincingly traverses the requirement, the burden shifts to the examiner to document a viable alternative process or product, or withdraw the requirement.

MPEP 821.04 Rejoinder

Where product and process claims drawn to independent and distinct inventions are presented in the same application, applicant may be called upon under 35 U.S.C. 121 to elect claims to either the product or process. See MPEP §806.05(f) and §806.05(h). The claims to the nonelected invention will be withdrawn from further consideration under 37 CFR 1.142. See MPEP §809.02(c) and §821 through §821.03. However, if applicant elects claims directed to the product, and a product claim is subsequently found allowable, withdrawn process claims which depend from or otherwise include all the limitations of the allowable product claim will be rejoined.

實例 3

　　一美國申請案包括了一組合及次組合二個發明，其獨立項分別如下：

1.A direct backlight module, comprising:

　　a reflector having an opening; at least one light source XXXXXXXXXXXXXXXX XXXXXXXXXXXXXXXX, XXXXXXXXXXXXXXXXXXXXXXXXXXXXXXXXXXXXX XXXXXXXXX a plurality of transparent supports XXXXXXXXXXXXXXXXXXXXXXXX XXXX the light source.

6.A liquid crystal display, comprising:

　　a liquid crystal panel; and

　　a direct backlight module, comprising:

　　a reflector having an opening XXXXXXXXX;

　　at least one light source XXX, XXXXXXXXXXXXXXXXXXXXXXXXX; and

　　a plurality of transparent supports XXXXXXXXXXXXXXXXXXXXXXXXXXXXXXXX XXXXXXXXXXXXXXXXXXXXXXXXXXXXXXXXXXXXXXXXX.

　　審查委員發下限制要求之審定書，指出 claim 6 之液晶顯示器（組合）本身具有之功能不同於直接背光模組（次組合），而且直接背光模組（次組合），亦可用在投影機系統之中，因而申請人需作限制。請問此限制 OA 是否恰當？及如何答辯？

　　依照 MPEP 806.05(c)如果申請案包括一組合及次組合，而如果審查委員可證明(1)組合之專利性不需藉次組合之特徵獲得，而且(2)次組合具有其自己的用途時，就可發下限制性審定書。因而申請人必須選一發明先作審查，而如果要反駁（traverse）需證明"組合"之專利性在於"次組合"或"次組合並無自己的用途"方可。

MPEP 806.05(c)

　　In order to establish that combination and subcombination inventions are distinct, two-way distinctness must be demonstrated.

　　To support a requirement for restriction, both two-way distinctness and reasons for insisting on restriction are necessary, i.e., separate classification, status, or field of search,

　　The inventions are distinct if it can be shown that a combination as claimed:

(A)does not require the particulars of the subcombination as claimed for patentability (to show novelty and unobviousness), and

(B)the subcombination can be shown to have utility either by itself or in another and different relations.

When these factors cannot be shown, such inventions are not distinct.

§10.3　專利申請案之群組選擇

如果申請人之專利申請案之請求項（claims）包括許多群組（species）而在說明書中並未揭示群組間之關係（一起使用，或為顯而易見之變形例），則審查委員將視這些群組為獨立之發明，而不管是否有上位請求項（generic claim）（或馬庫西請求項（markush claim））涵蓋各群組。對於群組選擇之要求，即使申請人不同意是獨立之發明，仍需作群組選擇，同時需告知審查委員哪些請求項涵蓋所選之群組。則審查委員就會針對涵蓋所選之群組之請求項作檢索及審查。而如果涵蓋所選之群組之請求項中亦有涵蓋未選之群組之上位請求項，而且此上位請求項是可核准的話，則未選之群組之後可再加入（rejoin）。如果無上位請求項，則未選之群組就不會被審查。

MPEP 808.01(a)

Where there is no disclosure of a relationship between species (see MPEP § 806.04(b)), they are independent inventions and election of one invention following requirement for restriction is mandatory even though applicant disagree with the examiner.　There must be a patentable difference between the species as claimed.　See MPEP § 806.04(h).　Since the claims are directed to independent inventions, restriction is proper pursuant to 35 U.S.C. 121, and it is not cecessary to shown a separate status in the art or separate classification.

A single disclosed species must be elected as a prerequisite to applying the provisions of 37 CFR 1.141 to additional species if a generic claim is allowed.Even if the examiner rejects the generic claims, and even if the applicant cancels the same and admits that the genus is unpatentable, where there is a relationship disclosed between species, such disclosed relation must be discussed and reasons advanced leading to the conclusion that the disclosed relation does not prevent restriction, in order to establish the propriety of restriction.

Election of species should not be required between claimed species that are considered clearly unpatentable (obvious) over each other. In making a requirement for restriction in an application claiming plural species, the examiner should group together species considered clearly unpatentable over each other, with the statement that restriction as between those species is not required.

Election of species may be required prior to a search on the merits (A) in applications containing claims to a plurality of species with no generic claims, and (B) in applications containing both species claims and generic or Markush claims.

In all applications in which no species claims are present and a generic claim recites such a multiplicity of species that an unduly extensive and burdensome search is required, a requirement for an election of species should be made prior to a search of the generic claim.

In all applications where a generic claim is found allowable, the application should be treated as indicated in MPEP §809.02(b), § 809.02(c) or § 809.02(e). If an election is made pursuant to a telephone requirement, the next action should include a full and complete action one the elected species as well as on any generic claim that may be present.

§10.4　馬庫西請求項之群組選擇

當請求項包括有馬庫西表示法（markush expression），而馬庫西表示法中之各元件（member）(1)並未有共同之用途或(2)並未有作為共同用途之實質上結構的共同特徵時，審查委員可將之視為不同之群組而要求申請人選一群組作審查。申請人選了一元件後，審查委員就會就所選之元件及涵蓋此元件之請求項作審查，如果審查可核准則會自行選下一個元件作審查（如果元件不是很多），而如果該元件不准，則其他未選之元件就視為撤回。

MPEP 803.02

In applications containing claims of that nature, the examiner may require a provisional election of a single species prior to examination on the merits. The provisional election will be given effect in the event that the Markush-type claim should be found not allowable. Following election, the Markush-type claim will be examined fully with respect to the elected species and further to the extent necessary to determine patentability. If the Markush-type claim is not allowable over the prior art , examination will be limited to the Markush-type claim and claims to the elected species, with claims drawn to species patentably distinct from the elected species held withdrawn from further consideration.

On the other hand, should no prior art be found that anticipates or renders obvious the elected species, the search of the Markush-type claim will be extend.　If prior art is then found that anticipates or renders obvious the Markush-type claim with respect to a nonelected species, the Markush-type claim shall be rejected and claims to the nonelected species held withdrawn from further consideration. The prior art search, however, will not be extended unnecessarily to cover all nonelected species. Should applicant, in response to this rejection of the Markush-type claim, overcome the rejection, as by amending the Markush-type claim to exclude the species anticipated or rendered obvious by the prior art, the amended Markush-type claim will be reexamined. The prior art search will be extended to the extent necessary to determine patentability of the Markush-type claim. In the event prior art is found during the reexamination that anticipates or renders obvious the amended Markush-type claim, the claim will be rejected and the action made final.　Amendments submitted after the final rejection further restricting the scope of the claim may be denied entry.

實例 1

申請人之美國申請案請求了下式之化合物：

1. A compound having following formula (I):

$$(I),$$

Wherein

one of R_1, R_2, R_3, R_4, R_5, R_6, R_7, and R_8 is a moiety of formula (II)

(II),

審查委員發下選組之審定書，要求申請於上述結構所代表之化合物中選擇單一化合物，請問申請人若只選擇單一化合物，則審查委員將如何審查本案？

申請人只要選擇單一化合物並告知該單一化合物中各取代基對應於該一般式中的哪一個取代基即可。審查委員將會依據所選的化合物確認涵蓋所選之化合物之發明之範圍（亦即包括與所選的化合物結構類似之化合物）並進行審查，審查委員會在下次審定書告知申請人所審查之化合物之範圍。

§10.5　連接請求項之限制要求

連結請求項（linking claims）通常有下列四種型式：

(a)　屬請求項（genus claim）連接群組請求項（Species claim）

(b)　製造一產品之必要步驟之方法請求項連接適當的產品請求項及方法請求項，

(c)　實施一方法之裝置加功能請求項（means claim）連接適當之裝置請求項及方法請求項，

(d)　產品請求項連接產品之製造請求項及產品之使用請求項。

如果一申請案中有連接請求項，則審查委員在要求作限制時，該連接請求項與所選擇之請求項會一齊被審查。如果連接請求項可核准，則申請人撤回之未選的請求項，審查委員會一齊審查。如果未選的請求項已刪除，審查委員亦會通知申請人修正加入並再審查未選之請求項之專利性。

MPEP 809 Claims Linking Distinct Inventions

Where, upon examination of an application containing claims to distinct inventions, linking claims are found, restriction can nevertheless be required. See MPEP § 809.03 for definition of linking claims.

A letter including only a restriction requirement or a telephoned requirement to restrict (the latter being encouraged) will be effected, specifying which claims are considered linking. See MPEP § 812.01 for telephone practice in restriction requirements.

No art will be indicated for this type of linking claim and no rejection of these claims made.

A 1-month (not less than 30 days) shortened statutory period will be set for reply to a written requirement. Such action will not be an "action on the merits" for the purpose of the second action final program.

To be complete, a reply a requirement made according to this section need only include a proper election.

The linking claims must be examined with the invention elected, and should any linking claim be allowed, the restriction requirement must be withdrawn. Any claim(s) directed to the nonelected invention(s), previously withdrawn from consideration, which depends from or includes all the limitation of the allowable linking claim must be rejoined and will be fully examined for patentability. Where such withdrawn claims have been canceled by applicant pursuant to the restriction requirement, upon the allowance of the linking claims(s), the examiner must notify applicant that any canceled, nonelected claims(s) which depends from or includes all the limitations of the allowable linking claim may be reinstated by submitting the claim(s) in an amendment. Upon entry of the amendment, the amended claim(s) will be fully examined for patentability.

MPEP 809.03

The most common types of linking claims which, if allowed, act to prevent restriction between inventions that can otherwise be shown to be divisible, are

(A)genus claims linking species claims;

(B)a claim to the necessary process of making a product linking proper process and product claims;

(C)a claim to "means" for practicing a process linking proper apparatus and process claims; and

(D)a claim to the product linking a process of making and a use (process of using).

實例

專利申請案包括三項請求項，claim 1 乃有關一種熱塑性塑膠產品之射出成型方法，claim 2 乃是有關於一種熱塑性塑膠之熱交換器之射出成型方法，claim 3 乃有關於一種熱塑性塑膠之熱交換器。審查委員發下限制審定書要求自產品及製法作選擇，申請人應如何因應？

申請人可選擇 claim 1 或 claim 3，同時指出 claim 2 乃為連接請求項，請審查委員一齊審查。如果經審查連接請求項亦可核准則申請人可將未選之請求項加入請審查委員再審查其專利性。

第 11 章　揭露性與請求項之界定

目　錄

§11.1 35USC § 112 第 1 段及第 2 段關於揭露性與請求項界定之規定

美國專利法 35USC§112 之第 1 段規定專利說明書需包含以完整、簡約及正確之用語,對發明及製造和使用發明的方式及方法之書面敘述,以使得與此發明有關之此技藝人士可製造及使用此發明,且需記載發明人認為實施此發明之最佳模式。亦即,依照 35USC§112 之第 1 段說明書需符合下列之要件:

(1) 敘述要件(description requirement):需包含所請求之發明之書面敘述。

(2) 據以實施要件(enablement requirement):需使此技藝人士可製造及使用所請求之發明。

(3) 最佳模式要件(best mode requirement):需包括發明人所認為實施所請求之發明的最佳模式。

說明書要符合上述三要件之理由是專利之核准會強化新構想之揭露與發展,也會促進科學知識之進步。而專利一旦核准,包含在專利中之資訊就是公眾可取得來作為更進一步研究及發展之參考,故相對於專利核准後申請人具有排他之權利,35USC§112 第 1 段規定了包括在專利說明書中資訊的質和量的最小要求。所以專利權人必需揭露充份的資訊以使大眾可擁有所請求之發明的內容,使此技藝人士可製造及使用所請求的發明,而且專利權人不可隱藏在申請專利時所知道的實施所請求之發明的最佳方式。

美國專利法 35USC§112 第 2 段,規定說明書需包含至少一請求項(claim),特別地指出並清楚地請求申請人認為是他的發明的標的(subject matter)。亦即,依照 35USC§112 第 2 段,請求項需符合下列二要件:

(1) 請求項需記載申請人認為是其發明的標的。

(2) 請求項需特別指出,且清楚地界定要以專利保護之標的之範圍與界限。

請求項需符合上述二要件的原因是:審查時除了要決定所請求之發明是否新穎,是否非顯而易見外,還需決定請求項之界定是否準確、清楚、正確且不模糊。

35 U.S.C. 112 Specification.

The specification shall contain a written description of the invention, and of the manner and process of making and using it, in such full, clear, concise, and exact terms as to enable any person skilled in the art to which it pertains, or with which it is most nearly connected, to make and use the same, and shall set forth the best mode contemplated by the inventor of carrying out his invention.

The specification shall conclude with one or more claims particularly pointing out and distinctly claiming the subject matter which the applicant regards as his invention.

＊＊＊＊＊＊＊＊＊＊＊

MPEP 2162 Policy Underlying 35 U.S.C. 112, First Paragraph

To obtain a valid patent, a patent application must be filed that contains a full and clear disclosure of the invention in the manner prescribed by 35 U.S.C. 112, first paragraph. The requirement for an adequate disclosure ensures that the public receives something in return for the exclusionary rights that are granted to the inventor by a patent. The grant of a patent helps to foster and enhance the development and disclosure of new ideas and the advancement of scientific knowledge. Upon the grant of a patent in the U.S., information contained in the patent becomes a part of the information available to the public for further research and development, subject only to the patentee's right to exclude others during the life of the patent.

In exchange for the patent rights granted, 35 U.S.C. 112, first paragraph, sets forth the minimum requirements for the quality and quantity of information that must be contained in the patent to justify the grant. As discussed in more detail below, the patentee must disclose in the patent sufficient information to put the public in possession of the invention and to enable those skilled in the art to make and use the invention. The applicant must not conceal from the public the best way of practicing the invention that was known to the patentee at the time of filing the patent application. Failure to fully comply with the disclosure requirements could result in the denial of a patent, or in a holding of invalidity of an issued patent.

MPEP 2171 Two Separate Requirements for Claims Under 35 U.S.C. 112, Second Paragraph

The second paragraph of 35 U.S.C. 112 is directed to requirements for the claims:

The specification shall conclude with one or more claims particularly pointing out and distinctly claiming the subject matter which the applicant regards as his invention.

There are two separate requirements set forth in this paragraph:

(A) the claims must set forth the subject matter that applicants regard as their invention; and

(B) the claims must particularly point out and distinctly define the metes and bounds of the subject matter that will be protected by the patent grant.

The first requirement is a subjective one because it is dependent on what the applicants for a patent regard as their invention. The second requirement is an objective one because it is not dependent on the views of applicant or any particular individual, but is evaluated in the context of whether the claim is definite - i.e., whether the scope of the claim is clear to a hypothetical person possessing the ordinary level of skill in the pertinent art.

Although an essential purpose of the examination process is to determine whether or not the claims define an invention that is both novel and nonobvious over the prior art, another essential purpose of patent examination is to determine whether or not the claims are precise, clear, correct, and unambiguous. The uncertainties of claim scope should be removed, as much as possible, during the examination process.

The inquiry during examination is patentability of the invention as applicant regards it. If the claims do not particularly point out and distinctly claim that which applicants regard as their invention, the appropriate action by the examiner is to reject the claims under 35 U.S.C. 112, second paragraph. In re Zletz, 893 F.2d 319, 13 USPQ2d 1320 (Fed. Cir. 1989). If a rejection is based on 35 U.S.C. 112, second paragraph, the examiner should further explain whether the rejection is based on indefiniteness or on the failure to claim what applicants regard as their invention. Ex parte Ionescu, 222 USPQ 537, 539 (Bd. App. 1984).

§11.2　滿足敘述之要件

說明書之揭露要滿足敘述要件之目的是要清楚地傳達申請人（發明人）已經完成了所請求之發明的標的之資訊，同時讓大眾擁有（in the possession of）申請人所請求之發明。因而，要滿足敘述要件，一說明書必需相當詳細敘述所請求的發明以使得此技藝人士可合理地了解發明人已擁有（完成）了所請求之發明。因而，一說明書之揭露是否滿足敘述要件之判斷之重點乃在於所請求之發明，亦即請求項。所以，只要將請求項的請求之標的逐一地包括在說明書本文中就可很簡單地滿足敘述要件。即使，申請時之說明書並未包括所有請求項之敘述，亦可很容易地加以補充修正而不會違反"新事項"（new matter）之規定，因為提申時之請求項亦為原始揭露之一部份。所以，審查委員如果要以說明書不符合敘述要件核駁一請求項，則需負最初的舉證責任證明此技藝人士會認為發明之敘述不能提供請求項充份之支持。因此，以不符合敘述要件核駁一最初申請之請求項之例子很少見。

一說明書之揭露不能滿足敘述要件之情形大都是在審查過程中對請求項作修正或加入新的請求項時發生，例如所加入之新元件或新的限制條件或新的請求項在原本申請時之說明書中沒有明示、暗示，或固有的揭露來支持。依照 MPEP 714.02，2163.06 之規定，申請人有新的請求項或修正請求項時，需主動指出在原始之揭露中哪些部份可支持新的請求項或修正之請求項。

美國專利局之審查委員在決定說明書之揭露是否滿足敘述要件時，會先分析各請求項，決定請求項整體涵蓋之範圍，包括前言部份及連接用語，以評估請求項是否有充份的結構、動作、功能之敘述以使請求項之範圍及意義很清楚。接下來，實審查委員會檢視整份說明書，包括實施例、圖式、序列表等以了解申請人對所請求之發明之各特徵如何提供支持。最後再決定說明書是否有足夠的敘述可讓此技藝人士知道在專利申請時申請人已擁有所請求之發明。擁有所請求之發明可由各種方式顯示，例如裝置，構造發明之實施例（embodiment）化合物發明之結構式，化合物發明之化合物的物理、化學、光譜性質。

MPEP 2163

I. GENERAL PRINCIPLES GOVERNING COMPLIANCE WITH THE "WRITTEN DESCRIPTION" REQUIREMENT FOR APPLICATIONS

* * * * * * * * * * *

The written description requirement has several policy objectives. "[T]he 'essential goal' of the description of the invention requirement is to clearly convey the information that an applicant has invented the subject matter which is claimed." In re Barker, 559 F.2d 588, 592 n.4, 194 USPQ 470, 473 n.4 (CCPA 1977). Another objective is to put the public in possession of what the applicant claims as the invention. See Regents of the University of California v. Eli Lilly, 119 F.3d 1559, 1566, 43 USPQ2d 1398, 1404 (Fed. Cir. 1997), cert. denied, 523 U.S. 1089 (1998). *>"The 'written description' requirement implements the principle that a patent must describe the technology that is sought to be patented; the requirement serves both to satisfy the inventor's obligation to disclose the technologic knowledge upon which the patent is based, and to demonstrate that the patentee was in possession of the invention that is claimed." Capon v. Eshhar, 418 F.3d 1349, 1357, 76 USPQ2d 1078, 1084 (Fed. Cir. 2005). Further, the< written description requirement ** promotes the progress of the useful arts by ensuring that patentees adequately describe their inventions in their patent specifications in exchange for the right to exclude others from practicing the invention for the duration of the patent's term.

To satisfy the written description requirement, a patent specification must describe the claimed invention in sufficient detail that one skilled in the art can reasonably conclude that the inventor had possession of the claimed invention.

A.Original Claims

There is a strong presumption that an adequate written description of the claimed invention is present when the application is filed. In re Wertheim, 541 F.2d 257, 263, 191 USPQ 90, 97 (CCPA 1976) ("we are of the opinion that the PTO has the initial burden of presenting evidence or reasons why persons skilled in the art would not recognize in the disclosure a description of the invention defined by the claims"). However, as discussed in paragraph I., supra, the issue of a lack of adequate written description may arise even for an original claim when an aspect of the claimed invention has not been described with sufficient particularity such that one skilled in the art would recognize that the applicant had possession of the claimed invention. The claimed invention as a whole may not be adequately described if the claims require an essential or critical feature which is not adequately described in the specification and which is not conventional in the art or known to one of ordinary skill in the art. For example, consider the claim

B.New or Amended Claims

The proscription against the introduction of new matter in a patent application (35 U.S.C. 132 and 251) serves to prevent an applicant from adding information that goes beyond the subject matter originally filed. See In re Rasmussen, 650 F.2d 1212, 1214, 211 USPQ 323, 326 (CCPA 1981). See MPEP § 2163.06 through § 2163.07 for a more detailed discussion of the written description requirement and its relationship to new matter. The claims as filed in the original specification are part of the disclosure and, therefore, if an application as originally filed contains a claim disclosing material not found in the remainder of the specification, the applicant may amend the specification to include the claimed subject matter. In re Benno, 768 F.2d 1340, 226 USPQ 683 (Fed. Cir. 1985). Thus, the written description requirement prevents an applicant from claiming subject matter that was not adequately described in the specification as filed. New or amended claims which introduce elements or limitations which are not supported by the as-filed disclosure violate the written description requirement. See, e.g., In re Lukach, 442 F.2d 967, 169 USPQ 795 (CCPA 1971) (subgenus range was not supported by generic disclosure and specific example within the subgenus range); In re Smith, 458 F.2d 1389, 1395, 173 USPQ 679, 683 (CCPA 1972) (a subgenus is not necessarily described by a genus encompassing it and a species upon which it reads).

While there is no in haec verba requirement, newly added claim limitations must be supported in the specification through express, implicit, or inherent disclosure

* * * * * * * * * * *

II. METHODOLOGY FOR DETERMINING ADEQUACY OF WRITTEN DESCRIPTION

* * * * * * * * * * *

1.For Each Claim, Determine What the Claim as a Whole Covers

* * * * * * * * * * *

2.Review the Entire Application to Understand How Applicant Provides Support for the Claimed Invention Including Each Element and/or Step

* * * * * * * * * * *

3.Determine Whether There is Sufficient Written Description to Inform a Skilled Artisan That Applicant was in Possession of the Claimed Invention as a Whole at the Time the Application Was Filed

* * * * * * * * * * *

MPEP 714.02 Must Be Fully Responsive [R-3]

* * * * * * * * * * *

The prompt development of a clear issue requires that the replies of the applicant meet the objections to and rejections of the claims. Applicant should also specifically point out the support for any amendments made to the disclosure. See MPEP § 2163.06.

* * * * * * * * * * *

MPEP 2163.06 Relationship of Written Description Requirement to New Matter

* * * * * * * * * * *

When an amendment is filed in reply to an objection or rejection based on 35 U.S.C. 112, first paragraph, a study of the entire application is often necessary to determine whether or not "new matter" is involved. Applicant should therefore specifically point out the support for any amendments made to the disclosure.

＊＊＊＊＊＊＊＊＊＊＊＊

§11.3　請求項範圍之改變對滿足敘述要件之影響

通常說明書之揭露不能滿足敘述要件是在申請之後對請求項作修改時發生。對請求項之修改包括擴大請求項之範圍，縮小請求項之範圍，改變數值範圍限定及改變請求項之用語等。要符合 35USC§112 第 1 段之揭露要件或是有權利依 35USC§119,120,365(c)主張前案之優先權，每一請求項之限制條件需在說明書中有明示地、暗示地或固有的支持。

Ⅰ. 修改請求項後範圍變大

將請求項修改使範圍變大的一種方式就是將請求項中某些之限制條件或元件刪除（省略），如果刪除之限制條件是不重要的限制條件或元件則不違反敘述要件。例如在 In re peters 之判例中，所請求之發明為一顯示器裝置，於申請再發證時，申請人將傾斜元件之特定的楔形形狀之限制條件刪除並未違反敘述要件，因為在說明書中並未敘述到此楔形形狀對發明之操作或專利性是重要的（critical）。相反地，如果將一重要的限制條件（元件）刪除則不滿足敘述要件。例如在 Gentry Glallery Inc v. Berkline Corp 之判例中，請求項乃請求一組合式沙發，其包括一座架及一控制裝置。申請人將控制裝置之位置的限制條件刪除。但在說明書中清楚地敘述到"控制裝置之位置是唯一可能的位置而且改變位置則非本發明之目的。因此，此擴大請求項之範圍是違反敘述要件的。另一種擴大請求項之範圍的方式是加入一上位的請求項（generic claim）。加入上位的請求項有時可符合敘述要件，有時則不符合敘述要件。例如在 In re Curtis 之判例中，說明書中揭露的是用微結晶蠟塗覆之牙線，但加入之請求項是具有強化摩擦力塗覆的 PTFE 牙線。但是說明書並未有任何其他塗覆適合作為 PTFE 牙線的塗覆之揭露，故不符合敘述要件。但在 In re Scythe 之判例中，說明書之揭露是空氣或對此流體為鈍性的之其他氣體，加入之請求項則為"鈍性流體介質"。

但是說明書所敘述之空氣及其他氣體之性質及功能對此技藝人士來說可判斷申請人之發明涵蓋鈍性流體介質，故此擴大範圍的請求項是符合敘述要件的。

II. 修正請求項後範圍變小

當將一元件或限制條件加入請求項中以縮小請求項之範圍或者加入一較小範圍之請求項時，是否違反敘述要件端視加入元件或限制條件或較小範圍之請求項是否在申請時之說明書中有揭露而定。例如在 Ex Part Ohshiro 之判例中，請求項乃請求一內燃機，其加入此限制條件"至少該活塞及該汽缸之一具有凹陷之通道"，但是說明書中揭露的是具有凹陷的汽缸及不具有凹陷通道的活塞，故此修正違反敘述要件。相反地在 Ex parte Sorenson 之判例中，將請求項限制為脂肪族羧酸及芳香族羧酸則未違反敘述要件，因為申請時之說明書中之揭露為羧酸。

III. 改變數值範圍之限定

改變數值範圍之限定是否違反敘述要件則需視對此技藝人士來說此數值範圍之改變是否是原來（inherently）就由申請時之說明書之揭露是可支持的而定。例如在 In re Wertheim 之判例中，在申請時之說明書中所露的是 25%- 60%，並有二實施例分別揭示 36%及 50%，如果將請求項限制為至少 35%則不符合敘述要件，但若將範圍修改為 35%-60%則符合敘述要件，因為對此技藝人士 35%至 60%之範圍是可由原來說明書中支持的。

MPEP 2163.05 Changes to the Scope of Claims [R-2]

The failure to meet the written description requirement of 35 U.S.C. 112, first paragraph, commonly arises when the claims are changed after filing to either broaden or narrow the breadth of the claim limitations, or to alter a numerical range limitation or to use claim language which is not synonymous with the terminology used in the original disclosure. To comply with the written description requirement of 35 U.S.C. 112, para. 1, or to be entitled to an earlier priority date or filing date under 35 U.S.C. 119, 120, or 365(c), each claim limitation must be expressly, implicitly, or inherently supported in the originally filed disclosure. See MPEP § 2163 for examination guidelines pertaining to the written description requirement.

I. BROADENING CLAIM

Omission of a Limitation

Under certain circumstances, omission of a limitation can raise an issue regarding whether the inventor had possession of a broader, more generic invention. See, e.g., Gentry Gallery, Inc. v. Berkline Corp., 134 F.3d 1473, 45 USPQ2d 1498 (Fed. Cir. 1998) (claims to a sectional sofa comprising, inter alia, a console and a control means were held invalid for failing to satisfy the written description requirement where the claims were broadened by removing the location of the control means.); Johnson Worldwide Associates v. Zebco Corp., 175 F.3d 985, 993, 50 USPQ2d 1607, 1613 (Fed. Cir. 1999) (In Gentry Gallery, the "court's determination that the patent disclosure did not support a broad meaning for the disputed claim terms was premised on clear statements in the written description that described the location of a claim element--the 'control means'--as 'the only possible location' and that variations were 'outside the stated purpose of the invention.

＊＊＊＊＊＊＊＊＊＊＊

Compare In re Peters, 723 F.2d 891, 221 USPQ 952 (Fed. Cir. 1983) (In a reissue application, a claim to a display device was broadened by removing the limitations directed to the specific tapered shape of the tips without violating the written description requirement. The shape limitation was considered to be unnecessary since the specification, as filed, did not describe the tapered shape as essential or critical to the operation or patentability of the claim.). A claim which omits matter disclosed to be essential to the invention as described in the specification or in other statements of record may also be subject to rejection under 35 U.S.C. 112, para. 1, as not enabling, or under 35 U.S.C. 112, para. 2. See In re Mayhew, 527 F.2d 1229, 188 USPQ 356 (CCPA 1976); In re Venezia, 530 F.2d 956, 189 USPQ 149 (CCPA 1976); and In re Collier, 397 F.2d 1003, 158 USPQ 266 (CCPA 1968). See also MPEP § 2172.01.

Addition of Generic Claim

354 F.3d 1347, 1358, 69 USPQ2d 1274, 1282 (Fed. Cir. 2004) (Claims directed to PTFE dental floss with a friction-enhancing coating were not supported by a disclosure of a microcrystalline wax coating where there was no evidence in the disclosure or anywhere else in the record showing applicant conveyed that any other coating was suitable for a PTFE dental floss.

＊＊＊＊＊＊＊＊＊＊＊

In re Smythe, 480 F.2d 1376, 1383, 178 USPQ 279, 285 (CCPA 1973) (the pHrase "air or other gas which is inert to the liquid" was sufficient to support a claim to "inert fluid media"

II.NARROWING OR SUBGENERIC CLAIM

The introduction of claim changes which involve narrowing the claims by introducing elements or limitations which are not supported by the as-filed disclosure is a violation of the written description requirement of 35 U.S.C. 112, first paragraph.

＊＊＊＊＊＊＊＊＊＊＊

In Ex parte Ohshiro, 14 USPQ2d 1750 (Bd. Pat. App. & Inter. 1989), the Board affirmed the rejection under 35 U.S.C. 112, first paragraph, of claims to an internal combustion engine which recited "at least one of said piston and said cylinder (head) having a recessed channel." The Board held that the application which disclosed a cylinder head with a recessed channel and a piston without a recessed channel did not specifically disclose the "species" of a channeled piston.

＊＊＊＊＊＊＊＊＊＊＊

On the other hand, in Ex parte Sorenson, 3 USPQ2d 1462 (Bd. Pat. App. & Inter. 1987), the subgeneric language of "aliphatic carboxylic acid" and "aryl carboxylic acid" did not violate the written description requirement because species falling within each subgenus were disclosed as well as the generic carboxylic acid.

＊＊＊＊＊＊＊＊＊＊＊

See also In re Smith, 458 F.2d 1389, 1395, 173 USPQ 679, 683 (CCPA 1972) ("Whatever may be the viability of an inductive-deductive approach to arriving at a claimed subgenus, it cannot be said that such a subgenus is necessarily described by a genus encompassing it and a species upon which it reads." (emphasis added)). Each case must be decided on its own facts in terms of what is reasonably communicated to those skilled in the art. In re Wilder, 736 F.2d 1516, 1520, 222 USPQ 369, 372 (Fed. Cir. 1984).

III. RANGE LIMITATIONS

With respect to changing numerical range limitations, the analysis must take into account which ranges one skilled in the art would consider inherently supported by the discussion in the original disclosure. In the decision in In re Wertheim, 541 F.2d 257, 191 USPQ 90 (CCPA 1976), the ranges described in the original specification included a range of "25%- 60%" and specific examples of "36%" and "50%." A corresponding new claim limitation to "at least 35%" did not meet the description requirement because the phrase "at least" had no upper limit and caused the claim to read literally on embodiments outside the "25% to 60%" range, however a limitation to "between 35% and 60%" did meet the description requirement.

實例 1

一專利申請案中說明書中記載之壓力範圍為 10 psi 至 50 psi，而實施例中之壓力記載為 42 psi。原 claim 中壓力記載為從 10 psi 至 50 psi，今擬將其壓力範圍修正為在 35 psi 和 45 ps i 之間，請問此修正是否可由原說明書中支持？

可由說明書中支持，因為 35 psi-45 psi 之範圍在原範圍之內，且有 42 psi 之實例，故對熟習此技藝人士是有固有的支持的。

實例 2

發明人 X 發明了一汽車用之千斤頂，發明人 X 之千斤頂主要是利用剪刀型升降機構，故其準備了一些機構之圖式及說明委託代理人 Y 提出專利申請。代理人 Y 於專利說明書詳細敘述了此千斤頂之構造、原理，並將此剪刀型升降機構作成圖式。但是因為發明人認為剪刀型之機構尚不十分理想效果，故請求項只針對此升降機構以 lifting means 界定，並未針對剪刀型機構作請求。在專利提出申請之後，發明人基於同樣的升降原理設計出螺旋型式之千斤頂，發現更穩定，更小型化。之後，該專利申請案發下初步審定書，代理人 Y 就將原來的請求項刪除，新加入一請求項，請求此螺旋型式的升降裝置。請問此請求項之修正是否符合敘述要件？

此請求項之修正不符合為 35USC§112 第 1 段規定之敘述要件，因為在原始之揭露中並未包括螺紋型式升降機構之敘述，故是為新事項（new matter）。

§11.4　滿足據以實施要件

一說明書之揭露要能滿足據以實施之要件（enablement），需此技藝人士在閱讀了說明書之揭露之後，能夠製造及使用所請求之發明。規定此要件的目的是要以有意義的方式將所請求的發明傳送給有興趣之大眾。要注意的是，說明書之揭露只要使此技藝人士可製造及使用所請求之發明即可，並非要求至此技藝人士可製造及使用一發明的完美、商業上可應用的實施例之程度。而測試是否滿足據以實施之要件之標準是此技藝人士可製造及使用所請求之發明而不會有不恰當的實驗（undue experimentation）。所謂不恰當的實驗會因發明的本質或內容而有所不同。例如發明是有關於化學或生物技術，如果其技術細節未充份揭露在說明書中，則有可能此技藝人士需作相當多的試驗，不恰當的實驗結果仍不能製造及使用該發明。又例如，一發明乃利用電腦程式來控制一系統，但說明書之揭露並未揭露此電腦程式，則此技藝人士可能花數年之功夫仍無法製造及使用此發明。說明書揭露包括發明之背景之資訊，雖然發明之背景並非敘述所請求之發明，但可能此部份之資訊有助於此技藝人士製造及使用所請求之發明。通常在判斷是否揭露之不足會造成此技藝人士的不恰當的實驗，需考慮至少下列因素：

(A) 請求項之範圍

(B) 發明之本質

(C) 先前技術之狀態

(D) 此技藝人士之水平

(E) 此技藝之可預測性之水平

(F) 發明人提供之指示之量

(G) 是否有實施例；及

(H) 基於揭露之內容要製造及使用發明所需之實驗之量

此外，只要說明書揭露了至少一種製造及使用所請求之發明的方法而此方法與請求項之整個範圍有合理之關聯，即已滿足據以實施之要件。再者，發明之本質影響是否會使此技藝人士有不恰當之實驗甚鉅。例如，發明是利用一特殊之菌株來生產一種藥物，如果此菌株並不是很容易取得的，則將造成此技藝人士之不恰當實驗。又例如發明是有關於一化合物之中間產物（intermediate），則不應要求說明書需揭露如何製造及使用穩定及可分離形式之該中間產物。因為中間產物本身就是不穩定，暫時狀態之化合物。例如，發明如果是有關藥物，則不必要求說明書需揭露使用之劑量（dosage），如果此技藝人士在不必作不恰當的實驗就可得到此資訊的話。

MPEP 2164 The Enablement Requirement [R-2]

The enablement requirement refers to the requirement of 35 U.S.C. 112, first paragraph that the specification describe how to make and how to use the invention. The invention that one skilled in the art must be enabled to make and use is that defined by the claim(s) of the particular application or patent.

The purpose of the requirement that the specification describe the invention in such terms that one skilled in the art can make and use the claimed invention is to ensure that the invention is communicated to the interested public in a meaningful way. The information contained in the disclosure of an application must be sufficient to inform those skilled in the relevant art how to both make and use the claimed invention. >However, to comply with 35 U.S.C. 112, first paragraph, it is not necessary to "enable one of ordinary skill in the art to make and use a perfected, commercially viable embodiment absent a claim limitation to that effect.

＊＊＊＊＊＊＊＊＊＊＊

MPEP 2164.01 Test of Enablement [R-5]

Accordingly, even though the statute does not use the term "undue experimentation," it has been interpreted to require that the claimed invention be enabled so that any person skilled in the art can make and use the invention without undue experimentation.

＊＊＊＊＊＊＊＊＊＊＊

Any part of the specification can support an enabling disclosure, even a background section that discusses, or even disparages, the subject matter disclosed therein.

＊＊＊＊＊＊＊＊＊＊＊

MPEP 2164.01(a) Undue Experimentation Factors

There are many factors to be considered when determining whether there is sufficient evidence to support a determination that a disclosure does not satisfy the enablement requirement and whether any necessary experimentation is "undue." These factors include, but are not limited to:

(A) The breadth of the claims;

(B) The nature of the invention;

(C) The state of the prior art;

(D) The level of one of ordinary skill;

(E) The level of predictability in the art;

(F) The amount of direction provided by the inventor;

(G) The existence of working examples; and

(H) The quantity of experimentation needed to make or use the invention based on the content of the disclosure.

＊＊＊＊＊＊＊＊＊＊＊

MPEP 2164.01(b) How to Make the Claimed Invention

As long as the specification discloses at least one method for making and using the claimed invention that bears a reasonable correlation to the entire scope of the claim, then the enablement requirement of 35 U.S.C. 112 is satisfied. In re Fisher, 427 F.2d 833, 839, 166 USPQ 18, 24 (CCPA 1970). Failure to disclose other methods by which the claimed invention may be made does not render a claim invalid under 35 U.S.C. 112. Spectra-Physics, Inc. v. Coherent, Inc., 827 F.2d 1524, 1533, 3 USPQ2d 1737, 1743 (Fed. Cir.), cert. denied, 484 U.S. 954 (1987).

Naturally, for unstable and transitory chemical intermediates, the "how to make" requirement does not require that the applicant teach how to make the claimed product in stable, permanent or isolatable form. In re Breslow, 616 F.2d 516, 521, 205 USPQ 221, 226 (CCPA 1980).

A key issue that can arise when determining whether the specification is enabling is whether the starting materials or apparatus necessary to make the invention are available. In the biotechnical area, this is often true when the product or process requires a particular strain of microorganism and when the microorganism is available only after extensive screening.

＊＊＊＊＊＊＊＊＊＊＊

MPEP 2164.01(c) How to Use the Claimed Invention

If a statement of utility in the specification contains within it a connotation of how to use, and/or the art recognizes that standard modes of administration are known and contemplated, 35 U.S.C. 112 is satisfied. In re Johnson, 282 F.2d 370, 373, 127 USPQ 216, 219 (CCPA 1960); In re Hitchings, 342 F.2d 80, 87, 144 USPQ 637, 643 (CCPA 1965). See also In re Brana, 51 F.2d 1560, 1566, 34 USPQ2d 1437, 1441 (Fed. Cir. 1993).

§11.5　實際的實施例

據以實施要件並不要求說明書之揭露需包括實際的實施例（working example）。所謂實際的實施例乃指實際上實施發明所得到之例子。例如，如為一化合物之發明，則其實際的實施例（working example）為記載由原料製造出該化合物之實際過程之實例。通常乃以過去式敘述此實施例以表示為實際製造及使用該發明。但是，申請人在提出專利申請之前並不需要真正的將發明付諸實施，亦即並不一定要有實際的實施例。只要說明書之內容之揭露足以使此技藝人士可以實施該發明，而不必作不恰當之實驗即可。因此說明書之揭露亦可用預測結果的假想實施例（prophetic or paper example）來滿足據以實施要件。但是，在某些不可預測性高或尚未完全發展之技術領域之發明，說明書之揭露沒有實際的實施例卻是需考慮的因素。所以，在判斷是否滿足據以實施要件時，如果其他因素均顯示是可據以實施的，惟獨並無實際的實施例時，並不能以缺乏實際的實施例作為判斷不能據以實施之理由。此外，如果只有一個或很少的實施例，雖已滿足據以實施要件，但有可能會被認為據以實施只限定在一特定的範圍，而需縮減請求項之範圍。但是，如果請求一群組（genus），但只有少數物種（species）之實際實施例時，附上一聲明主張由此技藝人士之水平及說明書所揭露的資訊看來此實際的實施例可應用至整個群組上，通常就能滿足據以實施要件。除非審查委員有足夠的理由證明此技藝人士無法使用此群組中所有的物種時，才需再補充其他物種之實際的實施例。

再者，在治療方法之發明中，活體外（in vitro）或活體內（in vivo）之動物模式試驗（animal model assay）實際上構成了實際之實施例（working example），如果此試驗和所請求之方法有關聯的話（correlation）。所謂有關聯乃指在此技藝中，此特殊之動物模式試驗是被承認的。如果審查委員認為無關聯，則審查委員需負最初的舉證責任。

MPEP 2164.02 Working Example

Compliance with the enablement requirement of 35 U.S.C. 112, first paragraph, does not turn on whether an example is disclosed. An example may be "working" or "prophetic." A working example is based on work actually performed. A prophetic example describes an embodiment of the invention based on predicted results rather than work actually conducted or results actually achieved.

An applicant need not have actually reduced the invention to practice prior to filing. In Gould v. Quigg, 822 F.2d 1074, 1078, 3 USPQ 2d 1302, 1304 (Fed. Cir. 1987), as of Gould's filing date, no person had built a light amplifier or measured a population inversion in a gas discharge. The Court held that "The mere fact that something has not previously been done clearly is not, in itself, a sufficient basis for rejecting all applications purporting to disclose how to do it." 822 F.2d at 1078, 3 USPQ2d at 1304 (quoting In re Chilowsky, 229 F.2d 457, 461, 108 USPQ 321, 325 (CCPA 1956)).

The specification need not contain an example if the invention is otherwise disclosed in such manner that one skilled in the art will be able to practice it without an undue amount of experimentation. In re Borkowski, 422 F.2d 904, 908, 164 USPQ 642, 645 (CCPA 1970).

Lack of a working example, however, is a factor to be considered, especially in a case involving an unpredictable and undeveloped art. But because only an enabling disclosure is required, applicant need not describe all actual embodiments.

NONE OR ONE WORKING EXAMPLE

When considering the factors relating to a determination of non-enablement, if all the other factors point toward enablement, then the absence of working examples will not by itself render the invention non-enabled. In other words, lack of working examples or lack of evidence that the claimed invention works as described should never be the sole reason for rejecting the claimed invention on the grounds of lack of enablement. A single working example in the specification for a claimed invention is enough to preclude a rejection which states that nothing is enabled since at least that embodiment would be enabled. However, a rejection stating that enablement is limited to a particular scope may be appropriate.

The presence of only one working example should never be the sole reason for rejecting claims as being broader than the enabling disclosure, even though it is a factor to be considered along with all the other factors. To make a valid rejection, one must evaluate all the facts and evidence and state why one would not expect to be able to extrapolate that one example across the entire scope of the claims.

CORRELATION: IN VITRO/IN VIVO

The issue of "correlation" is related to the issue of the presence or absence of working examples. "Correlation" as used herein refers to the relationship between in vitro or in vivo animal model assays and a disclosed or a claimed method of use. An in vitro or in vivo animal model example in the specification, in effect, constitutes a "working example" if that example "correlates" with a disclosed or claimed method invention. If there is no correlation, then the examples do not constitute "working examples." In this regard, the issue of "correlation" is also dependent on the state of the prior art. In other words, if the art is such that a particular model is recognized as correlating to a specific condition, then it should be accepted as correlating unless the examiner has evidence that the model does not correlate. Even with such evidence, the examiner must weigh the evidence for and against correlation and decide whether one skilled in the art would accept the model as reasonably correlating to the condition. In re Brana, 51 F.3d 1560, 1566, 34 USPQ2d 1436, 1441 (Fed. Cir. 1995) (reversing the PTO decision based on finding that in vitro data did not support in vivo applications).

Since the initial burden is on the examiner to give reasons for the lack of enablement, the examiner must also give reasons for a conclusion of lack of correlation for an in vitro or in vivo animal model example. A rigorous or an invariable exact correlation is not required, as stated in Cross v. Iizuka, 753 F.2d 1040, 1050, 224 USPQ 739, 747 (Fed. Cir. 1985):

[B]ased upon the relevant evidence as a whole, there is a reasonable correlation between the disclosed in vitro utility and an in vivo activity, and therefore a rigorous correlation is not necessary where the disclosure of pharmacological activity is reasonable based upon the probative evidence. (Citations omitted.)

WORKING EXAMPLES AND A CLAIMED GENUS

For a claimed genus, representative examples together with a statement applicable to the genus as a whole will ordinarily be sufficient if one skilled in the art (in view of level of skill, state of the art and the information in the specification) would expect the claimed genus could be used in that manner without undue experimentation. Proof of enablement will be required for other members of the claimed genus only where adequate reasons are advanced by the examiner to establish that a person skilled in the art could not use the genus as a whole without undue experimentation.

§11.6 不同領域的發明如何滿足據以實施要件

I. 電機，機械裝置或電機，機械方法之發明

這類發明不能滿足據以實施之情形，為說明書未記載某些重要的元件，或未記載元件之間的連接關係，互動關係。例如請求一電路裝置但只在圖式中用方塊圖表示其動作而且在說明書中未具體說明元件要如何構建，連接，控制以達到申請人所希望的特定操作。另外，例如發明是有關於決定在土壤中水平鑽孔之位置的方法，但是在說明書中並未記載執行此方法之電腦程式之細節，以致此技藝人士無法知道如何比較及再量測資料，故被認為是不符合據以實施要件。

II. 微生物之發明

如果發明是有關使用活的生物產物（living biological products），例如微生物，來製造所請求之發明，則據以實施之問題大都在於此微生物之取得（availability）。例如一發明是使用一特定的微生物以發酵的方法生產抗生素。但是要充份地敘述如何從大自然得到該特定之微生物卻不容易。因此，要滿足據以實施要件的方法就是要在該專利申請案發證前將微生物託存（deposit）。

III. 化學醫藥相關發明

　　這類發明不能滿足據以實施之要件主要在於說明書中關於可據以實施之揭露不足以支持整個請求項之範圍。亦即，請求項之範圍太大，只有部份能滿足據以實施要件。例如在 Enzo Biochem, Inc V.Celgene Inc.之判例中，說明書只揭露利用 antisense 技術調節大腸桿菌（E.Coli）中的三個基因，但於說明書中聲稱此技術可應用至任何包含可表現的基因物質之有機體，如細菌、酵母菌等。而請求項則涵蓋很大範圍之有機體。但最終法院認為(1)說明書中所提供之實際的實施例和請求項之範圍比較起來非常的狹窄(2)antisense 基因技術為一不可預測性很高的技術(3)需要將從大腸桿菌產生 antisense DNA 之技術適用至其他情形之細胞所需作的實驗的量非常大。故判定說明書中可據以實施之揭露並非與請求項之範圍同等大小。

　　在 In re Goodman 之判例中，在 1985 年申請之申請案中請求了一在植物細胞中藉由表現外來基因而產出蛋白質的方法。在說明書中揭露了在雙子葉植物細胞中表現基因以產生蛋白質之方法。但審查委員提供了證據證明在 1987 年時使用該請求項之方法在單子葉植物的細胞中是不能據以實施的。故請求項之範圍不能完全由說明書之揭露據以實施。

　　另外在 In re Colianni 之判例中，說明書只揭露了藉由施加足夠的超音波能量至骨格以修復破碎的骨格之方法，但並未揭露足夠的劑量為多少，且未告知此技藝人士超音波能量之適當強度、頻率及施加之時間，故判定說明書之揭露是無法據以實施的。

IV. 電腦程式相關發明

　　電腦程式相關的發明之系統元件通常用方塊圖（block diagram）表示，而這類發明通常可分為二種：（I）包括電腦及其他軟、硬體元件之系統(II)方塊元件均包括在電腦中之系統。第(I)類之發明是否滿足據以實施之要件需檢視每一方塊元件（組件）的各種功能及說明書中關於如何執行各元件功能之揭露。如果對於此技藝人士要執行元件之功能需要較一般例行實驗更多的實驗，則不符合據以實施之要件。此外構成方塊元件的軟、硬體如果本身就是一複雜的元件組合，具有許多功能及特性時，還需檢視這些軟、硬體是否能與其他元件準確地互動。例如，如果所請求之系統包括了一微處理器及其他由此微處理器控制之系統元件，但是在說明書中只揭露了微處理器是商業上可取得之微處理器，而並未揭露此微處理器所執行之操作，亦未揭露此微處理器如何程式化以執行計算或與其他系統元件互動以執行所請求之發明的功能，則不能滿足據以實施要件。如果在此系統中有揭露一特定的電腦程式，則仍需檢視此電腦程式是否和請求項中記載之程式達成之功能是一致的。第二類的電腦程式相關發明通常是純粹的資料處理的發明，其方塊元件之組合完全在電腦之中，且除了輸入／輸出裝置外並無與外在裝置連接之介面。此類的發明要滿足據以實施之要件，除了揭露所使用之電腦系統外，尚需詳細揭露電腦系統操作之電腦程式。例如揭露每一程式步驟及電腦之結構元件之關聯，流程圖與電腦程式列表（program listing）以及各硬體元件與軟體元件之關聯。如果請求項是方法，則說明書之揭露要教示如何執行此方法，如果執行此方法要用一特殊裝置，則要對此裝置作充份之揭露。

MPEP 2164.06(a) Examples of *>Enablement< Issues-Missing Information [R-1]

It is common that doubt arises about enablement because information is missing about one or more essential parts or relationships between parts which one skilled in the art could not develop without undue experimentation. In such a case, the examiner should specifically identify what information is missing and why the missing information is needed to provide enablement.

I.ELECTRICAL AND MECHANICAL DEVICES OR PROCESSES

For example, a disclosure of an electrical circuit apparatus, depicted in the drawings by block diagrams with functional labels, was held to be nonenabling in In re Gunn, 537 F.2d 1123, 1129, 190 USPQ 402, 406 (CCPA 1976). There was no indication in the specification as to whether the parts represented by boxes were "off the shelf" or must be specifically constructed or modified for applicant's system. Also there were no details in the specification of how the parts should be interconnected, timed and controlled so as to obtain the specific operations desired by the applicant. In In re Donohue, 550 F.2d 1269, 193 USPQ 136 (CCPA 1977), the lack of enablement was caused by lack of information in the specification about a single block labelled "LOGIC" in the drawings. See also Union Pacific Resources Co. v. Chesapeake Energy Corp., 236 F.3d 684, 57 USPQ2d 1293 (Fed. Cir. 2001) (Claims directed to a method of determining the location of a horizontal borehole in the earth failed to comply with enablement requirement of 35 U.S.C. 112 because certain computer programming details used to perform claimed method were not disclosed in the specification, and the record showed that a person of skill in art would not understand how to "compare" or "rescale" data as recited in the claims in order to perform the claimed method.).

* * * * * * * * * * *

II.MICROORGANISMS

Patent applications involving living biological products, such as microorganisms, as critical elements in the process of making the invention, present a unique question with regard to availability. The issue was raised in a case involving claims drawn to a fermentative method of producing two novel antibiotics using a specific microorganism and claims to the novel antibiotics so produced. In re Argoudelis, 434 F.2d 1390, 168 USPQ 99 (CCPA 1970). As stated by the court, "a unique aspect of using microorganisms as starting materials is that a sufficient description of how to obtain the microorganism from nature cannot be given." 434 F.2d at 1392, 168 USPQ at 102. It was determined by the court that availability of the biological product via a public depository provided an acceptable means of meeting the written description and the enablement requirements of 35 U.S.C. 112, first paragraph.

To satisfy the enablement requirement a deposit must be made "prior to issue" but need not be made prior to filing the application. In re Lundak, 773 F.2d 1216, 1223, 227 USPQ 90, 95 (Fed. Cir. 1985).

* * * * * * * * * * *

MPEP 2164.06(b) Examples of Enablement Issues - Chemical Cases

The following summaries should not be relied on to support a case of lack of enablement without carefully reading the case.

SEVERAL DECISIONS RULING THAT THE DISCLOSURE WAS NONENABLING

(A) In Enzo Biochem, Inc. v. Calgene, Inc., 188 F.3d 1362, 52 USPQ2d 1129 (Fed. Cir. 1999), the court held that claims in two patents directed to genetic antisense technology (which aims to control gene expression in a particular organism), were invalid because the breadth of enablement was not commensurate in scope with the claims. Both specifications disclosed applying antisense technology in regulating three genes in E. coli. Despite the limited disclosures, the specifications asserted that the "[t]he practices of this invention are generally applicable with respect to any organism containing genetic material which is capable of being expressed . such as bacteria, yeast, and other cellular organisms." The claims of the patents encompassed application of antisense methodology in a broad range of organisms. Ultimately, the court relied on the fact that (1) the amount of direction presented and the number of working examples provided in the specification were very narrow compared to the wide breadth of the claims at issue, (2) antisense gene technology was highly unpredictable, and (3) the amount of experimentation required to adapt the practice of creating antisense DNA from E. coli to other types of cells was quite high, especially in light of the record, which included notable examples of the inventor's own failures to control the expression of other genes in E. coli and other types of cells. Thus, the teachings set forth in the specification provided no more than a "plan" or "invitation" for those of skill in the art to experiment using the technology in other types of cells.

(B) In In re Wright, 999 F.2d 1557, 27 USPQ2d 1510 (Fed. Cir. 1993), the 1983 application disclosed a vaccine against the RNA tumor virus known as Prague Avian Sarcoma Virus, a member of the Rous Associated Virus family. Using functional language, Wright claimed a vaccine "comprising an immunologically effective amount" of a viral expression product. Id., at 1559, 27 USPQ2d at 1511. Rejected claims covered all RNA viruses as well as avian RNA viruses. The examiner provided a teaching that in 1988, a vaccine for another retrovirus (i.e., AIDS) remained an intractable problem. This evidence, along with evidence that the RNA viruses were a diverse and complicated genus, convinced the Federal Circuit that the invention was not enabled for either all retroviruses or even for avian retroviruses.

＊＊＊＊＊＊＊＊＊＊＊

(C) In In re Goodman, 11 F.3d 1046, 29 USPQ2d 2010 (Fed. Cir. 1993), a 1985 application functionally claimed a method of producing protein in plant cells by expressing a foreign gene. The court stated: "[n]aturally, the specification must teach those of skill in the art 'how to make and use the invention as broadly as it is claimed.'" Id. at 1050, 29 USPQ2d at 2013. Although protein expression in dicotyledonous plant cells was enabled, the claims covered any plant cell. The examiner provided evidence that even as late as 1987, use of the claimed method in monocot plant cells was not enabled. Id. at 1051, 29 USPQ2d at 2014.

＊＊＊＊＊＊＊＊＊＊＊

(D) In In re Vaeck, 947 F.2d 488, 495, 20 USPQ2d 1438, 1444 (Fed. Cir. 1991), the court found that several claims were not supported by an enabling disclosure "[t]aking into account the relatively incomplete understanding of the biology of cyanobacteria as of appellants' filing date, as well as the limited disclosure by appellants of the particular cyanobacterial genera operative in the claimed invention......" The claims at issue were not limited to any particular genus or species of cyanobacteria and the specification mentioned nine genera and the working examples employed one species of cyanobacteria.

＊＊＊＊＊＊＊＊＊＊＊

(E) In In re Colianni, 561 F.2d 220, 222-23, 195 USPQ 150, 152 (CCPA 1977), the court affirmed a rejection under 35 U.S.C. 112, first paragraph, because the specification, which was directed to a method of mending a fractured bone by applying "sufficient" ultrasonic energy to the bone, did not define a "sufficient" dosage or teach one of ordinary skill how to select the appropriate intensity, frequency, or duration of the ultrasonic energy.

MPEP 2164.06(c) Examples of Enablement Issues - Computer Programming Cases [R-5]

* * * * * * * * * * *

In a typical computer application, system components are often represented in a "block diagram" format, i.e., a group of hollow rectangles representing the elements of the system, functionally labeled, and interconnected by lines. Such block diagram computer cases may be categorized into (A) systems which include but are more comprehensive than a computer and (B) systems wherein the block elements are totally within the confines of a computer.

I.BLOCK ELEMENTS MORE COMPREHENSIVE THAN A COMPUTER

The first category of such block diagram cases involves systems which include a computer as well as other system hardware and/or software components. In order to meet his or her burden of establishing a reasonable basis for questioning the adequacy of such disclosure, the examiner should initiate a factual analysis of the system by focusing on each of the individual block element components. More specifically, such an inquiry should focus on the diverse functions attributed to each block element as well as the teachings in the specification as to how such a component could be implemented. If based on such an analysis, the examiner can reasonably contend that more than routine experimentation would be required by one of ordinary skill in the art to implement such a component or components, that component or components should specifically be challenged by the examiner as part of a 35 U.S.C. 112, first paragraph rejection. Additionally, the examiner should determine whether certain of the hardware or software components depicted as block elements are themselves complex assemblages which have widely differing characteristics and which must be precisely coordinated with other complex assemblages. Under such circumstances, a reasonable basis may exist for challenging such a functional block diagram form of disclosure.

* * * * * * * * * * *

For example, in a block diagram disclosure of a complex claimed system which includes a microprocessor and other system components controlled by the microprocessor, a mere reference to a prior art, commercially available microprocessor, without any description of the precise operations to be performed by the microprocessor, fails to disclose how such a microprocessor would be properly programmed to either perform any required calculations or to coordinate the other system components in the proper timed sequence to perform the functions disclosed and claimed.

II.BLOCK ELEMENTS WITHIN A COMPUTER

The second category of block diagram cases occurs most frequently in pure data processing applications where the combination of block elements is totally within the confines of a computer, there being no interfacing with external apparatus other than normal input/output devices. In some instances, it has been found that particular kinds of block diagram disclosures were sufficient to meet the enabling requirement of 35 U.S.C. 112, first paragraph. See In re Knowlton, 481 F.2d 1357, 178 USPQ 486 (CCPA 1973), In re Comstock, 481 F.2d 905, 178 USPQ 616 (CCPA 1973). Most significantly, however, in both the Comstock and Knowlton cases, the decisions turned on the appellants' disclosure of (A) a reference to and reliance on an identified prior art computer system and (B) an operative computer program for the referenced prior art computer system. Moreover, in Knowlton the disclosure was presented in such a detailed fashion that the individual program"s steps were specifically interrelated with the operative structural elements in the referenced prior art computer system. The court in Knowlton indicated that the disclosure did not merely consist of a sketchy explanation of flow diagrams or a bare group of program listings together with a reference to a proprietary computer in which they might be run. The disclosure was characterized as going into considerable detail in explaining the interrelationships between the disclosed hardware and software elements. Under such circumstances, the Court considered the disclosure to be concise as well as full, clear, and exact to a sufficient degree to satisfy the literal language of 35 U.S.C. 112, first paragraph.

＊＊＊＊＊＊＊＊＊＊＊

It must be emphasized that because of the significance of the program listing and the reference to and reliance on an identified prior art computer system, absent either of these items, a block element disclosure within the confines of a computer should be scrutinized in precisely the same manner as the first category of block diagram cases discussed above.

Regardless of whether a disclosure involves block elements more comprehensive than a computer or block elements totally within the confines of a computer, USPTO personnel, when analyzing method claims, must recognize that the specification must be adequate to teach how to practice the claimed method. If such practice requires a particular apparatus, then the application must provide a sufficient disclosure of that apparatus if such is not already available.

＊＊＊＊＊＊＊＊＊＊＊

§11.7 單一裝置請求項

所謂單一裝置請求項（Single Means Claim）乃指請求項之本體只包括單一元件且以裝置加功能（means plus function）之方式撰寫，此種請求項將被視為不能符合據以實施之要件，亦即範圍太廣。因為此單一裝置請求項涵蓋了可以達到所欲之目的（功能）之所有可想到的裝置，但是說明書之揭露最多只能涵蓋發明人所知道的裝置。

MPEP 2164.08(a) Single Means Claim

A single means claim, i.e., where a means recitation does not appear in combination with another recited element of means, is subject to an undue breadth rejection under 35 U.S.C. 112, first paragraph. In re Hyatt, 708 F.2d 712, 714-715, 218 USPQ 195, 197 (Fed. Cir. 1983) (A single means claim which covered every conceivable means for achieving the stated purpose was held nonenabling for the scope of the claim because the specification disclosed at most only those means known to the inventor.). When claims depend on a recited property, a fact situation comparable to Hyatt is possible, where the claim covers every conceivable structure (means) for achieving the stated property (result) while the specification discloses at most only those known to the inventor.

§11.8 請求項中包含不能操作之標的

請求項之範圍中如果包含有不可操作之實例（inoperative embodiments）不一定會使該請求項不能據以實施。例如說明書之揭露中包括了許多請求項所請求之標的的實施例，其中大部份都是可操作的，僅有少部份是不可操作的，此技藝人士在決定這些可操作的實施例是否可製造及使用（據以實施）時並不會有不恰當的實驗。但是，如果說明書中包括了相當數量的不能操作之實施例，而申請人又未明確指出哪些實施例是可操作的時候，此技藝人士在要決定請求項是否可據以實施就可能會有不恰當之實驗，就可能不能滿足據以實施要件。

MPEP2164.08(b) Inoperative Subject Matter

The presence of inoperative embodiments within the scope of a claim does not necessarily render a claim nonenabled. The standard is whether a skilled person could determine which embodiments that were conceived, but not yet made, would be inoperative or operative with expenditure of no more effort than is normally required in the art. Atlas Powder Co. v. E.I. du Pont de Nemours & Co., 750 F.2d 1569, 1577, 224 USPQ 409, 414 (Fed. Cir. 1984) (prophetic examples do not make the disclosure nonenabling).

Although, typically, inoperative embodiments are excluded by language in a claim (e.g., preamble), the scope of the claim may still not be enabled where undue experimentation is involved in determining those embodiments that are operable. A disclosure of a large number of operable embodiments and the identification of a single inoperative embodiment did not render a claim broader than the enabled scope because undue experimentation was not involved in determining those embodiments that were operable. In re Angstadt, 537 F.2d 498, 502-503, 190 USPQ 214, 218 (CCPA 1976). However, claims reading on significant numbers of inoperative embodiments would render claims nonenabled when the specification does not clearly identify the operative embodiments and undue experimentation is involved in determining those that are operative. Atlas Powder Co. v. E.I. duPont de Nemours & Co., 750 F.2d 1569, 1577, 224 USPQ 409, 414 (Fed. Cir. 1984); In re Cook, 439 F.2d 730, 735, 169 USPQ 298, 302 (CCPA 1971).

§11.9　請求項未記載發明之重要特徵

如果在說明書之揭露中明確指出發明之某一特徵是重要的（critical），但是在請求項中卻未記載此特徵，則會被認為不能滿足據以實施要件。在決定請求項記載之發明特徵是否為重要時，需檢視整份說明書之揭露，僅被註明是較佳的（preferred）特徵，並不一定是重要的特徵。只有當說明書中明確地指出該特徵在達到所請求之發明的目的（功能）是不可或缺時，該特徵才該被認為是重要的特徵。

MPEP 2164.08(c) Critical Feature Not Claimed

A feature which is taught as critical in a specification and is not recited in the claims should result in a rejection of such claim under the enablement provision section of 35 U.S.C.

112. See In re Mayhew, 527 F.2d 1229, 1233, 188 USPQ 356, 358 (CCPA 1976). In determining whether an unclaimed feature is critical, the entire disclosure must be considered. Features which are merely preferred are not to be considered critical. In re Goffe, 542 F.2d 564, 567, 191 USPQ 429, 431 (CCPA 1976).

Limiting an applicant to the preferred materials in the absence of limiting prior art would not serve the constitutional purpose of promoting the progress in the useful arts. Therefore, an enablement rejection based on the grounds that a disclosed critical limitation is missing from a claim should be made only when the language of the specification makes it clear that the limitation is critical for the invention to function as intended. Broad language in the disclosure, including the abstract, omitting an allegedly critical feature, tends to rebut the argument of criticality.

§11.10 滿足最佳模式要件

　　說明書之揭露要能夠滿足最佳模式之要件乃是因為專利權人取得專利之後在某一段期間內可排除他人實施其所請求之發明。因此，當然相對地，大眾應該要能夠知道實施該發明之最佳模式。通常說明書之揭露是否滿足最佳模式要件可分兩方面來判斷，第1是在專利申請時發明人是否知道實施發明之最佳模式。第2是如果發明人有最佳模式，說明書是否揭露了實施該發明之最佳模式。如果在提出專利申請時，申請人並不知道實施發明之最佳模式，或者並不知道哪個實施就是最佳模式，因而未將之揭露，則未揭露最佳模式並不會使之後取得之專利無效。但是，並不是說申請人要刻意地隱藏最佳模式（active concealment）或有不正當行為（grossly inequitable conduct）才足以造成專利無效。例如發明人知道有一特定材料可使發明之效果更好，但是將此特定材料以較廣範圍之用語敘述之即有可能違反最佳模式要件。又例如，說明書中不但不揭露一發明之重要成份，反而在實施例中揭露一虛假的，不能操作的成份，則不但會使該專利無效不能執行，亦可能使其他處理此技術的相關專利無效或不能執行。

　　以下幾點是滿足最佳模式要件需考慮的：

　　Ⅰ：只要所請求之發明（亦即請求項所定義的發明）能滿足最佳模式要件即可，未請求之發明元件（成份）並不需要滿足最佳模式要件。

　　Ⅱ：說明書並不需要揭露特定的實施例才能滿足最佳模式要件。例如，不必去揭露實際生產之實施例。例如揭露一生產條件之較佳範圍或一群可使用之反應物就能滿足最佳模式要件。

Ⅲ：說明書並不需指出哪一實施例是最佳模式，只要包括了申請人認為是最佳模式之揭露即可。

Ⅳ：在專利申請案提出申請後，申請人不必，也不能再提供最新的最佳模式（updating best mode is not required），從優先權日至美國專利申請日之間發現的最佳模式亦不必揭露。但是在部份延續案（CIP）中則需將最新的最佳模式加入。

Ⅴ：如果不能滿足最佳模式要件亦不能以加入新事項（new matter）來修正此缺失。

審查委員在審查一專利案件時是先假定最佳模式已揭露在說明書之中，除非有證據顯示並非如此。因而在申請之程序中（exparte prosecution）中審查委員會發下不符最佳模式要件之情形非常稀少。通常會有不能滿足滿最佳模式之要件情形，是在抵觸程序（interference），訴訟程序或其他 inter partes 程序中發生。

MPEP 2165 The Best Mode Requirement

"The best mode requirement creates a statutory bargained-for-exchange by which a patentee obtains the right to exclude others from practicing the claimed invention for a certain time period, and the public receives knowledge of the preferred embodiments for practicing the claimed invention." Eli Lilly & Co. v. Barr Laboratories Inc., 251 F.3d 955, 963, 58 USPQ2d 1865, 1874 (Fed. Cir. 2001).

The best mode requirement is a safeguard against the desire on the part of some people to obtain patent protection without making a full disclosure as required by the statute. The requirement does not permit inventors to disclose only what they know to be their second-best embodiment, while retaining the best for themselves. In re Nelson, 280 F.2d 172, 126 USPQ 242 (CCPA 1960).

Determining compliance with the best mode requirement requires a two-prong inquiry. First, it must be determined whether, at the time the application was filed, the inventor possessed a best mode for practicing the invention. This is a subjective inquiry which focuses on the inventor's state of mind at the time of filing. Second, if the inventor did possess a best mode, it must be determined whether the written description disclosed the best mode such that a person skilled in the art could practice it. This is an objective inquiry, focusing on the scope of the claimed invention and the level of skill in the art. Eli Lilly & Co. v. Barr Laboratories Inc., 251 F.3d 955, 963, 58 USPQ2d 1865, 1874 (Fed. Cir. 2001).

The failure to disclose a better method will not invalidate a patent if the inventor, at the time of filing the application, did not know of the better method OR did not appreciate that it was the best method. All applicants are required to disclose for the claimed subject matter the best mode contemplated by the inventor even though applicant may not have been the discoverer of that mode. Benger Labs. Ltd. v. R.K. Laros Co., 209 F. Supp. 639, 135 USPQ 11 (E.D. Pa. 1962).

ACTIVE CONCEALMENT OR GROSSLY INEQUITABLE CONDUCT IS NOT REQUIRED TO ESTABLISH FAILURE TO DISCLOSE THE BEST MODE

Failure to disclose the best mode need not rise to the level of active concealment or grossly inequitable conduct in order to support a rejection or invalidate a patent. Where an inventor knows of a specific material that will make possible the successful reproduction of the effects claimed by the patent, but does not disclose it, speaking instead in terms of broad categories, the best mode requirement has not been satisfied. Union Carbide Corp. v. Borg-Warner, 550 F.2d 555, 193 USPQ 1 (6th Cir. 1977).

If the failure to set forth the best mode in a patent disclosure is the result of inequitable conduct (e.g., where the patent specification omitted crucial ingredients and disclosed a fictitious and inoperable slurry as Example 1), not only is that patent in danger of being held unenforceable, but other patents dealing with the same technology that are sought to be enforced in the same cause of action are subject to being held unenforceable. Consolidated Aluminum Corp. v. Foseco Inc., 910 F.2d 804, 15 USPQ2d 1481 (Fed. Cir. 1990).

MPEP 2165.01 Considerations Relevant to Best Mode [R-2]

I.DETERMINE WHAT IS THE INVENTION

Determine what the invention is - the invention is defined in the claims. The specification need not set forth details not relating to the essence of the invention. In re Bosy, 360 F.2d 972, 149 USPQ 789 (CCPA 1966). See also Northern Telecom Ltd. v. Samsung Electronics Co., 215 F.3d 1281, 55 USPQ2d 1065 (Fed. Cir. 2000) (Unclaimed matter that is unrelated to the operation of the claimed invention does not trigger the best mode requirement);

＊＊＊＊＊＊＊＊＊＊＊

II.SPECIFIC EXAMPLE IS NOT REQUIRED

There is no statutory requirement for the disclosure of a specific example - a patent specification is not intended nor required to be a production specification. In re Gay, 309 F.2d 768, 135 USPQ 311 (CCPA 1962).

The absence of a specific working example is not necessarily evidence that the best mode has not been disclosed, nor is the presence of one evidence that it has. Best mode may be represented by a preferred range of conditions or group of reactants. In re Honn, 364 F.2d 454, 150 USPQ 652 (CCPA 1966).

III.DESIGNATION AS BEST MODE IS NOT REQUIRED

There is no requirement in the statute that applicants point out which of their embodiments they consider to be their best; that the disclosure includes the best mode contemplated by applicants is enough to satisfy the statute. Ernsthausen v. Nakayama, 1 USPQ2d 1539 (Bd. Pat. App. & Inter. 1985).

IV.UPDATING BEST MODE IS NOT REQUIRED

There is no requirement to update in the context of a foreign priority application under 35 U.S.C. 119, Standard Oil Co. v. Montedison, S.p.A., 494 F.Supp. 370, 206 USPQ 676 (D.Del. 1980) (better catalyst developed between Italian priority and U.S. filing dates), and continuing applications claiming the benefit of an earlier filing date under 35 U.S.C. 120, Transco Products, Inc. v. Performance Contracting Inc., 38 F.3d 551, 32 USPQ2d 1077 (Fed. Cir. 1994) (continuation under >former< 37 CFR 1.60); Sylgab Steel and Wire Corp. v. Imoco-Gateway Corp., 357 F.Supp. 657, 178 USPQ 22 (N.D. Ill. 1973) (continuation); Johns-Manville Corp. v. Guardian Industries Corp., 586 F.Supp. 1034, 221 USPQ 319 (E.D. Mich. 1983) (continuation and CIP). In the last cited case, the court stated that applicant would have been obliged to disclose an updated refinement if it were essential to the successful practice of the invention and it related to amendments to the CIP that were not present in the parent application. In Carter-Wallace, Inc. v. Riverton Labs., Inc., 433 F.2d 1034, 167 USPQ 656 (2d Cir. 1970), the court assumed, but did not decide, that an applicant must update the best mode when filing a CIP application.

V.DEFECT IN BEST MODE CANNOT BE CURED BY NEW MATTER

If the best mode contemplated by the inventor at the time of filing the application is not disclosed, such a defect cannot be cured by submitting an amendment seeking to put into the specification something required to be there when the patent application was originally filed. In re Hay, 534 F.2d 917, 189 USPQ 790 (CCPA 1976).

＊＊＊＊＊＊＊＊＊＊＊

MPEP 2165.03 Requirements for Rejection for Lack of Best Mode [R-1]

ASSUME BEST MODE IS DISCLOSED UNLESS THERE IS EVIDENCE TO THE CONTRARY

The examiner should assume that the best mode is disclosed in the application, unless evidence is presented that is inconsistent with that assumption. It is extremely rare that a best mode rejection properly would be made in ex parte prosecution. The information that is necessary to form the basis for a rejection based on the failure to set forth the best mode is rarely accessible to the examiner, but is generally uncovered during discovery procedures in interference, litigation, or other inter partes proceedings.

＊＊＊＊＊＊＊＊＊＊＊

§11.11 請求項界定的發明被認為不是申請人發明的標的

通常只有在說明書之中申請人敘述之發明與請求項所界定之發明有所不同時才會有請求項之發明被認為不是發明人之標的之情事，亦即除非有其他相反的證據顯示，一般請求項所界定的發明就被認為是申請人之發明的標的。

而申請人之說明書之內容如果其範圍與所請求之發明不一致，並不構成請求項界定之發明不是申請人的發明之標的之證據。說明書所揭露之發明的內容的範圍與請求項所界定之範圍不一致（不一樣大），只構成 35USC§112 第 1 段之不符據以實施之要件之問題。

此外，35USC§112 第 2 段之規定並未禁止申請人在專利申請案之審查程序中將發明之標的改變。申請人在審查過程中可以將原來揭露在說明書中的發明（此發明在原先之請求項中未請求者）以延續案之方式請求，而可主張原母案之申請日之優先權之利益。

MPEP 2172 Subject Matter Which Applicants Regard as Their Invention

I.FOCUS FOR EXAMINATION

A rejection based on the failure to satisfy this requirement is appropriate only where applicant has stated, somewhere other than in the application as filed, that the invention is something different from what is defined by the claims. In other words, the invention set forth in the claims must be presumed, in the absence of evidence to the contrary, to be that which applicants regard as their invention. In re Moore, 439 F.2d 1232, 169 USPQ 236 (CCPA 1971).

II.EVIDENCE TO THE CONTRARY

Evidence that shows that a claim does not correspond in scope with that which applicant regards as applicant's invention may be found, for example, in contentions or admissions contained in briefs or remarks filed by applicant, Solomon v. Kimberly-Clark Corp., 216 F.3d 1372, 55 USPQ2d 1279 (Fed. Cir. 2000); In re Prater, 415 F.2d 1393, 162 USPQ 541 (CCPA 1969), or in affidavits filed under 37 CFR 1.132, In re Cormany, 476 F.2d 998, 177 USPQ 450 (CCPA 1973). The content of applicant's specification is not used as evidence that the scope of the claims is inconsistent with the subject matter which applicants regard as their invention. As noted in In re Ehrreich, 590 F.2d 902, 200 USPQ 504 (CCPA 1979), agreement, or lack thereof, between the claims and the specification is properly considered only with respect to 35 U.S.C. 112, first paragraph; it is irrelevant to compliance with the second paragraph of that section.

III.SHIFT IN CLAIMS PERMITTED

The second paragraph of 35 U.S.C. 112 does not prohibit applicants from changing what they regard as their invention during the pendency of the application. In re Saunders, 444 F.2d 599, 170 USPQ 213 (CCPA 1971) (Applicant was permitted to claim and submit comparative evidence with respect to claimed subject matter which originally was only the preferred embodiment within much broader claims (directed to a method).). The fact that claims in a continuation application were directed to originally disclosed subject matter which applicants had not regarded as part of their invention when the parent application was filed was held not to prevent the continuation application from receiving benefits of the filing date of the parent application under 35 U.S.C. 120. In re Brower, 433 F.2d 813, 167 USPQ 684 (CCPA 1970).

§11.12 特別指出及清楚界定所請求之發明

35USC§112 第 2 段之目的一方面是要確保請求項的範圍很清楚而使得大眾知道構成侵權之界限,另一方面是提供一完整的衡量所請求之標的是否符合專利要件及說明書是否符合 35USC§112 第 1 段之要件的方法。

申請人(代理人)可以用自己的用語來界定一所請求之發明,例如用功能性語言,用選擇式語法或用負面限定之語法去界定所請求之發明的範圍與界限。但是重點是,請求項之界定需很清楚(definite)。所謂很清楚乃是指整體來看請求項是否可使此技藝人士知道所請求之發明的界限及範圍,而提供其他人清楚的警告怎樣構成侵害此專利。如果請求項之用語不能使此技藝人士解讀此請求項之範圍及界限而去了解如何避免侵權，則請求項之界定不清楚。審查委員如果判定請求項之界定不清楚，則其在審定書中需載明為何不清楚之理由及分析。

此外，如果請求項之用語與說明書之揭露或先前技術之教示不一致時，亦可能使得請求項之界定不清楚。例如在 In Cohn 之判例中其請求項乃是有關於用一腐蝕溶液處理一表面直至金屬外觀變為不透明的方法。但是法院發現在說明書中之敘述、定義及實施例顯示之經處理過的表面之外觀並非為不透明，故判定此請求項為不清楚，而違反 35USC§112 第 2 段之規定。

再者，請求項之範圍過廣（undue bredth）是違反 35USC§112 第 1 段，未由說明書中支持（不符合敘述要件）及此技藝人士不能製造及使用（違反據以實施要件），而不一定是違反 35USC§112 第 2 段之界定不清楚之規定。

此外，如果請求項缺少說明書中所敘述之必要元件的話，除了違反 35USC§112、第 1 段之據以實施要件，亦違反 35USC§112、第 2 段之未清楚界定發明之要件。

MPEP 2173 Claims Must Particularly Point Out and Distinctly Claim the Invention

The primary purpose of this requirement of definiteness of claim language is to ensure that the scope of the claims is clear so the public is informed of the boundaries of what constitutes infringement of the patent. A secondary purpose is to provide a clear measure of what applicants regard as the invention so that it can be determined whether the claimed invention meets all the criteria for patentability and whether the specification meets the criteria of 35 U.S.C. 112, first paragraph with respect to the claimed invention.

MPEP 2173.01 Claim Terminology [R-2]

A fundamental principle contained in 35 U.S.C. 112, second paragraph is that applicants are their own lexicographers. They can define in the claims what they regard as their invention essentially in whatever terms they choose so long as **>any special meaning assigned to a term is clearly set forth in the specification. See MPEP § 2111.01.< Applicant may use functional language, alternative expressions, negative limitations, or any style of expression or format of claim which makes clear the boundaries of the subject matter for which protection is sought. As noted by the court in In re Swinehart, 439 F.2d 210, 160 USPQ 226 (CCPA 1971), a claim may not be rejected solely because of the type of language used to define the subject matter for which patent protection is sought.

MPEP 2173.02 Clarity and Precision [R-3]

＊＊＊＊＊＊＊＊＊＊＊

In reviewing a claim for compliance with 35 U.S.C. 112, second paragraph, the examiner must consider the claim as a whole to determine whether the claim apprises one of ordinary skill in the art of its scope and, therefore, serves the notice function required by 35 U.S.C. 112, second paragraph, by providing clear warning to others as to what constitutes infringement of the patent.

＊＊＊＊＊＊＊＊＊＊＊

If the language of the claim is such that a person of ordinary skill in the art could not interpret the metes and bounds of the claim so as to understand how to avoid infringement, a rejection of the claim under 35 U.S.C. 112, second paragraph, would be appropriate.

＊＊＊＊＊＊＊＊＊＊＊

If upon review of a claim in its entirety, the examiner concludes that a rejection under 35 U.S.C. 112, second paragraph, is appropriate, such a rejection should be made and an analysis as to why the phrase(s) used in the claim is "vague and indefinite" should be included in the Office

＊＊＊＊＊＊＊＊＊＊＊

MPEP 2173.03 Inconsistency Between Claim *>and< Specification Disclosure or Prior Art [R-1]

Although the terms of a claim may appear to be definite, inconsistency with the specification disclosure or prior art teachings may make an otherwise definite claim take on an unreasonable degree of uncertainty. In re Cohn, 438 F.2d 989, 169 USPQ 95 (CCPA 1971); In re Hammack, 427 F.2d 1378, 166 USPQ 204 (CCPA 1970). In Cohn, the claim was directed to a process of treating a surface with a corroding solution until the metallic appearance is supplanted by an "opaque" appearance. Noting that no claim may be read apart from and independent of the supporting disclosure on which it is based, the court found that the description, definitions and examples set forth in the specification relating to the appearance of the surface after treatment were inherently inconsistent and rendered the claim indefinite.

MPEP 2173.04 Breadth Is Not Indefiniteness

Breadth of a claim is not to be equated with indefiniteness. In re Miller, 441 F.2d 689, 169 USPQ 597 (CCPA 1971). If the scope of the subject matter embraced by the claims is clear, and if applicants have not otherwise indicated that they intend the invention to be of a scope different from that defined in the claims, then the claims comply with 35 U.S.C. 112, second paragraph.

Undue breadth of the claim may be addressed under different statutory provisions, depending on the reasons for concluding that the claim is too broad. If the claim is too broad because it does not set forth that which applicants regard as their invention as evidenced by statements outside of the application as filed, a rejection under 35 U.S.C. 112, second paragraph, would be appropriate. If the claim is too broad because it is not supported by the original description or by an enabling disclosure, a rejection under 35 U.S.C. 112, first paragraph, would be appropriate. If the claim is too broad because it reads on the prior art, a rejection under either 35 U.S.C. 102 or 103 would be appropriate.

MPEP 2172.01 Unclaimed Essential Matter [R-1]

A claim which omits matter disclosed to be essential to the invention as described in the specification or in other statements of record may be rejected under 35 U.S.C. 112, first paragraph, as not enabling. In re Mayhew, 527 F.2d 1229, 188 USPQ 356 (CCPA 1976). See also MPEP § 2164.08(c). Such essential matter may include missing elements, steps or necessary structural cooperative relationships of elements described by the applicant(s) as necessary to practice the invention.

In addition, a claim which fails to interrelate essential elements of the invention as defined by applicant(s) in the specification may be rejected under 35 U.S.C. 112, second paragraph, for failure to point out and distinctly claim the invention.

＊＊＊＊＊＊＊＊＊＊＊＊

§11.13 請求項如何定義及使用新的用語

在請求項中所使用之每一用語（term）應該是從先前技術或從說明書及請求項，在專利申請案提申時，其意義（定義）是很明顯，清楚的才行。申請人（代理人）在請求項中界定所請求之發明時並不必被局限於先前技術之用語，但用來界定發明之用語需很明顯，很清楚以能確定發明之界限與範圍。需注意的是在專利審查中，審查委員是以能與說明書之揭露一致之最廣的方式去詮釋請求項之用語。亦即以最廣方式詮釋請求項之用語為基礎去作先前技術之檢索及審查。

在請求項中使用新的用語（指在先前技術中未曾出現之用語）界定發明有時會造成不易對先前技術及所請求之發明作比較。但是在請求項中使用新的用語是可被允許的，且是時常被使用的，因為新的用語對於剛發展出之新技術之發明可更準確地敘述且界定新的發明。其前提是，新的用語如果在閱讀說明書之相關揭露之後，可使此技藝人士合理地知道所請求之發明目的，使用及範圍，則此新用語的使用並不違反 35USC§112、第 2 段之清楚界定發明之規定。

另外，請求項中之用語雖非新的用語，但是申請人所用之用語與一般的意義有所不同時，亦即，用語事實上有二個以上之定義時，則需在說明書中清楚地再定義此用語以免造成 35USC§112、第 2 段界定不清楚之核駁。

MPEP 2173.05(a) New Terminology [R-3]

I. THE MEANING OF EVERY TERM SHOULD BE APPARENT

The meaning of every term used in a claim should be apparent from the prior art or from the specification and drawings at the time the application is filed. Applicants need not confine themselves to the terminology used in the prior art, but are required to make clear and precise the terms that are used to define the invention whereby the metes and bounds of the claimed invention can be ascertained. During patent examination, the pending claims must be given the broadest reasonable interpretation consistent with the specification.

* * * * * * * * * * *

II. THE REQUIREMENT FOR CLARITY AND PRECISION MUST BE BALANCED WITH THE LIMITATIONS OF THE LANGUAGE

Courts have recognized that it is not only permissible, but often desirable, to use new terms that are frequently more precise in describing and defining the new invention. In re Fisher, 427 F.2d 833, 166 USPQ 18 (CCPA 1970). Although it is difficult to compare the claimed invention with the prior art when new terms are used that do not appear in the prior art, this does not make the new terms indefinite.

* * * * * * * * * * *

III. TERMS USED CONTRARY TO THEIR ORDINARY MEANING MUST BE CLEARLY REDEFINED IN THE WRITTEN DESCRIPTION

Consistent with the well-established axiom in patent law that a patentee or applicant is free to be his or her own lexicographer, a patentee or applicant may use terms in a manner contrary to or inconsistent with one or more of their ordinary meanings if the written description clearly redefines the terms.

* * * * * * * * * * *

§11.14 請求項中使用相對用語

在請求項中使用表示程度（degree）之相對用語（relative terminology）雖然並不精確，但是不一定會造成請求項界定不清楚。是否會造成請求項界定不清楚端視此技藝人士在閱讀了說明書之揭露之後，是否能了解所請求之發明是什麼，發明的範圍有多大而定。但是如果該用語用來表示一可改變之物體（標的）時，則可能造成界定不清楚。例如在 Ex Parte Brummer 之判例中，有關一腳踏車之請求項中敘述該前輪及後輪分開使輪子基座之距離為騎乘者之身高的 58%至 75%之間（so spaced as to give a wheelbase that is between 58% and 75% of the height of the rider）則為界定不清楚。因為騎乘者之身高並不知道，相反地在 Orthokinetics, Inc v. Safety Travel Chair, Inc.之判例中，請求項敘述為"小孩輪椅之某一元件的尺寸為可插入汽車之門框與一座椅之空間者"，則被認為清楚地界定了其發明。

以下討論常使用之相對用語：

(A) "about"，此用語有時會被認為是不清楚（indefinite），有時被認為是清楚的（definite），視用在說明書中之前後文，先前技術之狀態及審查過程中之歷史（prosecution history）而定。

(B) "Essentially"，此用語通常會被認為是清楚的，例如在 In re Marosi 之判例中，請求項敘述為"a silicon dioxide source that is essentially free of alkali metal"。因為從說明書及實施例中可很容易知道其意義。

(C) "Similar"，此用語用在請求項之前言（實用專利申請案或新式樣專利申請案）之中均被視為不清楚。例如"A nozzle for high-pressure cleaning units or similar apparatus" "was held to be indefinite since it was not clear what "The ornamental design for a feed bunk or similar structure as shown and described.

(D) "Substantially"，此用語通常會被認為是清楚的。例如在 In re Nehrenberg 之判例中，請求項敘述為"to substantially increase the efficiency of the compound as a copper extractant"是清楚的。因為從說明書之敘述可判定其意義。

(E) " Type"，此用語通常被認為是"不清楚"的。例如：Friedel-Crafts-type catalyst, "ZSM-5-type aluminosilicate zeolites"。加上 type 來擴大其範圍被認為是不清楚的。

(F) 其他用語"relatively shallow," "of the order of," "the order of about 5 mm," "substantial portion",”or like material", “comparable” ,”superior”,” aesthetically pleasing"通常會被認為是不清楚的。

MPEP 2173.05(b) Relative Terminology [R-6]

The fact that claim language, including terms of degree, may not be precise, does not automatically render the claim indefinite under 35 U.S.C. 112, second paragraph. Seattle Box Co., v. Industrial Crating & Packing, Inc., 731 F.2d 818, 221 USPQ 568 (Fed. Cir. 1984). Acceptability of the claim language depends on whether one of ordinary skill in the art would understand what is claimed, in light of the specification.

＊＊＊＊＊＊＊＊＊＊＊

REFERENCE TO AN OBJECT THAT IS VARIABLE MAY RENDER A CLAIM INDEFINITE

A claim may be rendered indefinite by reference to an object that is variable. For example, the Board has held that a limitation in a claim to a bicycle that recited "said front and rear wheels so spaced as to give a wheelbase that is between 58 percent and 75 percent of the height of the rider that the bicycle was designed for" was indefinite because the relationship of parts was not based on any known standard for sizing a bicycle to a rider, but on a rider of unspecified build. Ex parte Brummer, 12 USPQ2d 1653 (Bd. Pat. App. & Inter. 1989). On the other hand, a claim limitation specifying that a certain part of a pediatric wheelchair be "so dimensioned as to be insertable through the space between the doorframe of an automobile and one of the seats" was held to be definite. Orthokinetics, Inc. v. Safety Travel Chairs, Inc., 806 F.2d 1565, 1 USPQ2d 1081 (Fed. Cir. 1986).

A. "About"

**>In determining the range encompassed by the term "about", one must consider the context of the term as it is used in the specification and claims of the application......

B. Essentially"

The phrase "a silicon dioxide source that is essentially free of alkali metal" was held to be definite because the specification contained guidelines and examples that were considered sufficient to enable a person of ordinary skill in the art to draw a line between unavoidable impurities in starting materials and essential ingredients. In re Marosi, 710 F.2d 799, 218

USPQ 289 (CCPA 1983). The court further observed that it would be impractical to require applicants to specify a particular number as a cutoff between their invention and the prior art.

C. "Similar"

The term "similar" in the preamble of a claim that was directed to a nozzle "for high-pressure cleaning units or similar apparatus" was held to be indefinite since it was not clear what applicant intended to cover by the recitation "similar" apparatus. Ex parte Kristensen, 10 USPQ2d 1701 (Bd. Pat. App. & Inter. 1989).

A claim in a design patent application which read: "The ornamental design for a feed bunk or similar structure as shown and described." was held to be indefinite because it was unclear from the specification what applicant intended to cover by the recitation of "similar structure." Ex parte Pappas, 23 USPQ2d 1636 (Bd. Pat. App. & Inter. 1992).

D. "Substantially"

The term "substantially" is often used in conjunction with another term to describe a particular characteristic of the claimed invention. It is a broad term. In re Nehrenberg, 280 F.2d 161, 126 USPQ 383 (CCPA 1960). The court held that the limitation "to substantially increase the efficiency of the compound as a copper extractant" was definite in view of the general guidelines contained in the specification.

＊＊＊＊＊＊＊＊＊＊＊

E. "Type"

The addition of the word "type" to an otherwise definite expression (e.g., Friedel-Crafts catalyst) extends the scope of the expression so as to render it indefinite.

＊＊＊＊＊＊＊＊＊＊

F. Other Terms

The phrases "relatively shallow," "of the order of," "the order of about 5mm," and "substantial portion" were held to be indefinite because the specification lacked some standard for measuring the degree intended and, therefore, properly rejected as indefinite under 35 U.S.C. 112, second paragraph. Ex parte Oetiker, 23 USPQ2d 1641 (Bd. Pat. App. & Inter. 1992).

The term "or like material" in the context of the limitation "coke, brick, or like material" was held to render the claim indefinite since it was not clear how the materials other than coke or brick had to resemble the two specified materials to satisfy the limitations of the claim. Ex parte Caldwell, 1906 C.D. 58 (Comm'r Pat. 1906).

The terms "comparable" and "superior" were held to be indefinite in the context of a limitation relating the characteristics of the claimed material to other materials - "properties that are superior to those obtained with comparable" prior art materials. Ex parte Anderson, 21 USPQ2d 1241 (Bd. Pat. App. & Inter. 1991). It was not clear from the specification which properties had to be compared and how comparable the properties would have to be to determine infringement issues. Further, there was no guidance as to the meaning of the term "superior."

The phrase "aesthetically pleasing" was held indefinite because the meaning of a term cannot depend on the unrestrained, subjective opinion of the person practicing the invention. Datamize LLC v. Plumtree Software, Inc., 417 F.3d 1342, 1347-48, 75 USPQ2d 1801, 1807 (Fed. Cir. 2005).

§11.15 請求項中使用數值範圍之限定

通常說來，在一請求項中記載一特定的數值範圍並不會造成請求項界定不清楚之問題。以下討論可能造成請求項界定不清楚之情況：

Ⅰ. 在同一請求項中有寬及窄的數值範圍

如果在單一請求項中同時記載了寬及窄的數值範圍之限定，例如"a temperature between 45℃ and 78℃, preferably between 50℃ and 60℃，則可能會造成請求項之界定不清楚。因為不能確定申請人之數值範圍到底是要限定在哪一個。通常一較窄或較佳之數值範圍應記載在說明書中或者以依附項或獨立項之方式另外記載。

Ⅱ. 開放端之數值範圍

當在請求項中使用開放端之數值範圍作限定時需注意一致性以免造成界定不清楚。例如獨立項為一組合物包括至少 20%之鈉。但在依附項中記載了非鈉之成份可占最多至 100%，則顯然不一致，造成請求項之界定不清楚。

Ⅲ. 有效量（effective amount）之用語

在請求項中使用"有效量"是否會造成請求項之界定不清楚，要視此技藝人士基於說明書之揭露是否可決定特定量為多少而定。當由說明書之揭露此技藝人士可知道其量，而且並無先前技術可使此範圍變得不確定時，此用語通常被認為是清楚的。

MPEP 2173.05(c) Numerical Ranges and Amounts Limitations

Generally, the recitation of specific numerical ranges in a claim does not raise an issue of whether a claim is definite.

I.NARROW AND BROADER RANGES IN THE SAME CLAIM

Use of a narrow numerical range that falls within a broader range in the same claim may render the claim indefinite when the boundaries of the claim are not discernible. Description of examples and preferences is properly set forth in the specification rather than in a single claim. A narrower range or preferred embodiment may also be set forth in another independent claim or in a dependent claim. If stated in a single claim, examples and preferences lead to confusion over the intended scope of the claim. In those instances where it is not clear whether the claimed narrower range is a limitation, a rejection under 35 U.S.C. 112, second paragraph should be made.

* * * * * * * * * * *

II.OPEN-ENDED NUMERICAL RANGES

Open-ended numerical ranges should be carefully analyzed for definiteness. For example, when an independent claim recites a composition comprising "at least 20% sodium" and a dependent claim sets forth specific amounts of nonsodium ingredients which add up to 100%, apparently to the exclusion of sodium, an ambiguity is created with regard to the "at least" limitation (unless the percentages of the nonsodium ingredients are based on the weight of the nonsodium ingredients).

* * * * * * * * * * *

III."EFFECTIVE AMOUNT"

The common phrase "an effective amount" may or may not be indefinite. The proper test is whether or not one skilled in the art could determine specific values for the amount based on the disclosure

＊ ＊ ＊ ＊ ＊ ＊ ＊ ＊ ＊ ＊ ＊

The more recent cases have tended to accept a limitation such as "an effective amount" as being definite when read in light of the supporting disclosure and in the absence of any prior art which would give rise to uncertainty about the scope of the claim.

＊ ＊ ＊ ＊ ＊ ＊ ＊ ＊ ＊ ＊ ＊

實例

一專利申請案之說明書中揭露了一組合物包括成份 A，B 及 C，另包括了成份 D。D 成份有二功用，一是用來降低組合物之殘餘水份，另一功用是可調整組合物之 PH 值。其在說明書分別揭露了 D 成份作為二種不同之功用之時所需之有效量。此 D 成份作為此二種不同功用時，其有效量並不相同。如果請求項敘述為：一種組合物包括成份 A，B，C 及有效量之成份 D，是否會被 35USC§112、第 2 段以未清楚界定而核駁？

會被以未清楚界定請求項核駁，因為 D 成份作為不同之功用時，有效量不相同。請求項應記載為一種組合物包括成份 A，B，C 及有效量的 D 以降低水份。或一種組合物包括成份 A，B，C 及有效量的 D 以調整 PH 值。

§11.16 請求項中使用例示的用語

於請求項中使用例示的用語（exemplary language），例如" for example" ,"such as" 等會導致申請人欲保護之範圍混淆，因而會造成請求項之界定不清楚。此種模範或較佳之例子應記載於說明書本文之中或以依附項之方式知以限定。以下之例子均造成請求項界定不清楚：

(A) "R is halogen, for example, chlorine";

(B) "material such as rock wool or asbestos";

287

(C) "lighter hydrocarbons, such, for example, as the vapors or gas produced"

(D) "normal operating conditions such as while in the container of a proportioner".

MPEP 2173.05(d) Exemplary Claim Language ("for example," "such as") [R-1]

Description of examples or preferences is properly set forth in the specification rather than the claims. If stated in the claims, examples and preferences >may< lead to confusion over the intended scope of a claim. In those instances where it is not clear whether the claimed narrower range is a limitation, a rejection under 35 U.S.C. 112, second paragraph should be made.

＊＊＊＊＊＊＊＊＊＊＊

§11.17 請求項之用語缺乏前置基礎

所謂請求項之用語缺乏前置基礎（Antecedent Basis）乃指一請求項中使用了 said, the 來指定一元件（限制條件），但在請求項中此 said, the 之前並無該被指定的元件（限制條件）出現。例如，在一請求項之中使用了"said lever," 但在此請求項之"said lever"之前的敘述並未有出現任何 lever 之用語。又例如，在之前雖有"lever"一詞出現，但有二個不同之"lever"。另一例子：請求項中使用"said aluminum lever"，但在之前出現"a lever"。這些情形均會造成請求項之界定不清楚。

另外，如果在請求項之用語在說明書本文中無前置基礎並不一定會造成請求項之界定不清楚。美國專利局並未要求請求項之用語需與說明書之中用語一致。申請人可自由使用，創造請求項之用語，只要不會造成所請求之標的不清楚即可。

MPEP 2173.05(e) Lack of Antecedent Basis [R-5]

A claim is indefinite when it contains words or pHrases whose meaning is unclear. The lack of clarity could arise where a claim refers to "said lever" or "the lever," where the claim contains no earlier recitation or limitation of a lever and where it would be unclear as to what element the limitation was making refercnce. Similarly, if two different levers are recited earlier in the claim, the recitation of "said lever" in the same or subsequent claim would be unclear where it is uncertain which of the two levers was intended. A claim which refers to "said aluminum lever," but recites only "a lever" earlier in the claim, is indefinite because it is uncertain as to the lever to which reference is made.

＊＊＊＊＊＊＊＊＊＊＊

A CLAIM TERM WHICH HAS NO ANTECEDENT BASIS IN THE DISCLOSURE IS NOT NECESSARILY INDEFINITE

The mere fact that a term or phrase used in the claim has no antecedent basis in the specification disclosure does not mean, necessarily, that the term or phrase is indefinite. There is no requirement that the words in the claim must match those used in the specification disclosure. Applicants are given a great deal of latitude in how they choose to define their invention so long as the terms and phrases used define the invention with a reasonable degree of clarity and precision.

＊＊＊＊＊＊＊＊＊＊＊

§11.18 加入其他請求項作為限制條件

例如"The product produced by the method of claim 1." "A method of producing ethanol comprising contacting amylose with the culture of claim 1 under the following conditions......"為加入其他請求項作為限制條件之例子。此類請求項是可接受的，並不會有 35USC§112 第 2 段之界定不清楚之問題。

MPEP 2173.05(f) Reference to Limitations in Another Claim

A claim which makes reference to a preceding claim to define a limitation is an acceptable claim construction which should not necessarily be rejected as improper or confusing under 35 U.S.C. 112, second paragraph. For example, claims which read: "The product produced by the method of claim 1." or "A method of producing ethanol comprising contacting amylose with the culture of claim 1 under the following conditions......" are not indefinite under 35 U.S.C. 112, second paragraph, merely because of the reference to another claim.

＊＊＊＊＊＊＊＊＊＊＊

§11.19 請求項中使用功能性限制條件

請求項中使用功能性限制條件是想藉由元件所執行之功能來界定發明之某些部份，而不直接指出其特定結構或成份。藉由功能性限制條件界定一發明之某些部份並不一定會造成界定不清楚。是否會造成界定不清楚應由此技藝人士是否可知道發明之界限及範圍而定。例如""operatively connected" "adapted to be positioned" "being resiliently dilatable"," whereby said housing may be slidably positioned"通常不會造成界定不清楚。要注意的是使用功能性限制條件時，需同時敘述元件、成份或方法之步驟以定義該元件成份，步驟所達成之特定能力及目的，而且功能性用語所界定之功能要確實能從所敘述之元件、成份及步驟達成。

MPEP 2173.05(g) Functional Limitations [R-3]

A functional limitation is an attempt to define something by what it does, rather than by what it is (e.g., as evidenced by its specific structure or specific ingredients). There is nothing inherently wrong with defining some part of an invention in functional terms. Functional language does not, in and of itself, render a claim improper. In re Swinehart, 439 F.2d 210, 169 USPQ 226 (CCPA 1971).

A functional limitation must be evaluated and considered, just like any other limitation of the claim, for what it fairly conveys to a person of ordinary skill in the pertinent art in the context in which it is used. A functional limitation is often used in association with an element, ingredient, or step of a process to define a particular capability or purpose that is served by the recited element, ingredient or step.

＊＊＊＊＊＊＊＊＊＊＊

§11.20 請求項中使用選擇式限制條件

在請求項中使用選擇式限制條件如果不會對於請求項之範圍或是否清楚（clarity）造成問題，則是可被允許的。以下說明：

Ⅰ. 馬庫西群組（MARKUSH GROUPS）

所謂馬庫西群組（MARKUSH GROUPS）之寫法有下列二種：

① 　wherein R is a material selected from the group consisting of A, B, C and D；

② 　wherein R is A, B, C or D;

被包括在馬庫西群組中之材料通常必需屬於具有同一物理、化學性質之族，或特定技藝之族者，或者說明書本文揭露這些材料具有至少一種相同之性質、特性者。如果馬庫西群組中之元件如果均能由說明書支持且數量並不是很多時，則使用馬庫西群組並不會造成請求項界定不清楚。但如果此群組中之元件數目很龐大，則有可能造成界定不清楚。

此外，馬庫西群組中之一元件如果亦涵蓋其他元件之範圍，例如："selected from the group consisting of amino, halogen, nitro, chloro and alkyl"時並不一定會造成界定不清楚，雖然 halogen 包括了 chloro。

Ⅱ. "OR" 用語

此用語 "OR" 是可接受的，不會造成請求項之界定不清楚，例如 "wherein R is A, B, C, or D" "made entirely or in part of" 等。

Ⅲ. "OPTIONALLY" 用語

使用 "optionally" 是否會造成界定不清楚，視情況而定。當可選擇（optional）之元件（成份）會變化而造成含糊時，則可能會造界定不清楚。一般寫成 "containing A, B, and optionally C" 都是可接受的。

MPEP 2173.05(h) Alternative Limitations

I.MARKUSH GROUPS

Alternative expressions are permitted if they present no uncertainty or ambiguity with respect to the question of scope or clarity of the claims. One acceptable form of alternative expression, which is commonly referred to as a Markush group, recites members as being "selected from the group consisting of A, B and C." See Ex parte Markush, 1925 C.D. 126 (Comm'r Pat. 1925).

* * * * * * * * * * *

The use of Markush claims of diminishing scope should not, in itself, be considered a sufficient basis for objection to or rejection of claims. However, if such a practice renders the claims indefinite or if it results in undue multiplicity, an appropriate rejection should be made.

Similarly, the double inclusion of an element by members of a Markush group is not, in itself, sufficient basis for objection to or rejection of claims. Rather, the facts in each case must be evaluated to determine whether or not the multiple inclusion of one or more elements in a claim renders that claim indefinite. The mere fact that a compound may be embraced by more than one member of a Markush group recited in the claim does not necessarily render the scope of the claim unclear. For example, the Markush group, "selected from the group consisting of amino, halogen, nitro, chloro and alkyl" should be acceptable even though "halogen" is generic to "chloro."

The materials set forth in the Markush group ordinarily must belong to a recognized physical or chemical class or to an art-recognized class. However, when the Markush group occurs in a claim reciting a process or a combination (not a single compound), it is sufficient if the members of the group are disclosed in the specification to possess at least one property in common which is mainly responsible for their function in the claimed relationship, and it is clear from their very nature or from the prior art that all of them possess this property.

* * * * * * * * * * *

When materials recited in a claim are so related as to constitute a proper Markush group, they may be recited in the conventional manner, or alternatively. For example, if "wherein R is a material selected from the group consisting of A, B, C and D" is a proper limitation, then "wherein R is A, B, C or D" shall also be considered proper.

Subgenus Claim

Genus, subgenus, and Markush-type claims, if properly supported by the disclosure, are all acceptable ways for applicants to claim their inventions.

＊＊＊＊＊＊＊＊＊＊＊

II. "OR" TERMINOLOGY

Alternative expressions using "or" are acceptable, such as "wherein R is A, B, C, or D." The following phrases were each held to be acceptable and not in violation of 35 U.S.C. 112, second paragraph in In re Gaubert, 524 F.2d 1222, 187 USPQ 664 (CCPA 1975): "made entirely or in part of"; "at least one piece"; and "iron, steel or any other magnetic material."

III."OPTIONALLY"

An alternative format which requires some analysis before concluding whether or not the language is indefinite involves the use of the term "optionally." In Ex parte Cordova, 10 USPQ2d 1949 (Bd. Pat. App. & Inter. 1989) the language "containing A, B, and optionally C" was considered acceptable alternative language because there was no ambiguity as to which alternatives are covered by the claim. A similar holding was reached with regard to the term "optionally" in Ex parte Wu, 10 USPQ2d 2031 (Bd. Pat. App. & Inter. 1989). In the instance where the list of potential alternatives can vary and ambiguity arises, then it is proper to make a rejection under 35 U.S.C. 112, second paragraph, and explain why there is confusion.

§11.21 請求項中使用負面限制條件

只要請求項尋求保護之發明的界限及範圍是清楚的，在請求中使用負面限制（negative limitation）並不會造成請求項之界定不清楚。例如"said homopolymer being free from the proteins, soaps, resins, and sugars present in natural Hevea rubber" "incapable of forming a dye with said oxidized developing agent"均不會造成請求項界定不清楚。通常使用負面限制的目的是要與審查委員所引證之先前技術區別。

MPEP 2173.05(i) Negative Limitations

The current view of the courts is that there is nothing inherently ambiguous or uncertain about a negative limitation. So long as the boundaries of the patent protection sought are set forth definitely, albeit negatively, the claim complies with the requirements of 35 U.S.C. 112, second paragraph.

****

A claim which recited the limitation "said homopolymer being free from the proteins, soaps, resins, and sugars present in natural Hevea rubber" in order to exclude the characteristics of the prior art product, was considered definite because each recited limitation was definite. In re Wakefield, 422 F.2d 897, 899, 904, 164 USPQ 636, 638, 641 (CCPA 1970). In addition, the court found that the negative limitation "incapable of forming a dye with said oxidized developing agent" was definite because the boundaries of the patent protection sought were clear. In re Barr, 444 F.2d 588, 170 USPQ 330 (CCPA 1971).

實例

一專利申請案之說明書揭露了一種層板，其包括了一基板，一光敏層沈積在基板上，以及一透明的保護層與光敏層接觸並藉壓力形成在光敏層之上。其請求項之敘述為：A laminate comprising a substrate, a light-sensitive layer deposited on said substrate, and a transparent protective layer formed on said light-sensitive layer。審查委員引證了一先前技術，其揭示了一基板，一光敏層沈積在基板上，及一透明保護層藉由一粘合層粘合在光敏層之上，並以 35USC§102 核駁此請求項。如果將請求項改為：A laminate comprising a substrate, a light-sensitive layer deposed on said substrate, and a transparent, protective layer formed on said light-sensitive layer, but not including an adhesive-layer 是否可克服此核駁？

可以克服 35USC§102 之核駁，且此負面限制條件不會造成請求項之界定不清楚。

§11.22 請求項中之元件重複敘述

在請求項中重複（多次）敘述（double（multiple）inclusion）元件是否會造成界定不清楚，必需視實際狀況而定。例如在馬庫西族群中敘述為 selected from the group consisting of halogen, hitro, chloro and alkyl。雖然 halogen 包括了 chloro，但在此處並不會造成界定不清楚。。

MPEP 2173.05(o) Double Inclusion

＊＊＊＊＊＊＊＊＊＊＊＊

The facts in each case must be evaluated to determine whether or not the multiple inclusion of one or more elements in a claim gives rise to indefiniteness in that claim. The mere fact that a compound may be embraced by more than one member of a Markush group recited in the claim does not lead to any uncertainty as to the scope of that claim for either examination or infringement purposes. On the other hand, where a claim directed to a device can be read to include the same element twice, the claim may be indefinite. Ex parte Kristensen, 10 USPQ2d 1701 (Bd. Pat. App. & Inter. 1989).

§11.23 用途請求項

所謂用途（use）請求項乃指不用步驟（steps）去定義一方法（process），但只用其用途來界定一請求項。例如"A process for using monoclonal antibodies of claim 4 to isolate and purify human fibroblast interferon." 此類的用途請求項會被認為是界定不清楚，因為其未記載任何主動、正面之步驟以敘述此用途實際上如何運作。此種用途請求項另外亦會被 35USC§101 以非適當的方法請求項（process claim）核駁，因為 35USC§102 定義之可予專利標的中包含 process，但不包括此 use。

但是，在 Ex parte Porter 的判例中判定"使用之步驟"（step of utilizing）並不會界定不清楚。該判例之請求項為 "A method for unloading nonpacked, nonbridging and packed, bridging flowable particle catalyst and bead material from the opened end of a reactor tube ,which comprises utilizing the nozzle of claim 7.").

MPEP 2173.05(q) "Use" Claims

Attempts to claim a process without setting forth any steps involved in the process generally raises an issue of indefiniteness under 35 U.S.C. 112, second paragraph. For example, a claim which read: "A process for using monoclonal antibodies of claim 4 to isolate and purify human fibroblast interferon." was held to be indefinite because it merely recites a use without any active, positive steps delimiting how this use is actually practiced. Ex parte Erlich, 3 USPQ2d 1011 (Bd. Pat. App. & Inter. 1986).

Other decisions suggest that a more appropriate basis for this type of rejection is 35 U.S.C. 101. In Ex parte Dunki, 153 USPQ 678 (Bd. App. 1967), the Board held the following claim to be an improper definition of a process: "The use of a high carbon austenitic iron alloy having a proportion of free carbon as a vehicle brake part subject to stress by sliding friction."

＊＊＊＊＊＊＊＊＊＊＊

BOARD HELD STEP OF "UTILIZING" WAS NOT INDEFINITE

It is often difficult to draw a fine line between what is permissible, and what is objectionable from the perspective of whether a claim is definite. In the case of Ex parte Porter, 25 USPQ2d 1144 (Bd. Pat. App. & Inter. 1992), the Board held that a claim which clearly recited the step of "utilizing" was not indefinite under 35 U.S.C. 112, second paragraph. (Claim was to "A method for unloading nonpacked, nonbridging and packed, bridging flowable particle catalyst and bead material from the opened end of a reactor tube which comprises utilizing the nozzle of claim 7.").

§11.24 請求項以圖式或實施例界定

以圖式或實施例界定請求項即所謂的綜合請求項（Omnibus Claim）。例如 A device substantially as shown（in figure）and described（in example），會被以 35USC§112(2)以界定不清楚核駁。因為其未指出請求項所包括或排除哪些元件（圖式通常會包括非本發明之部份）。只有在很特殊之情況下（例如請求一合金，但某成份，比率很難用文字敘述時可用其金相圖）才能在請求項中參照併入說明書中之圖式及表格作為界定。另外，請求項中之元件可用說明書中對應參考文字（數字）標註以易於閱讀，但是此參考文字（數字）需用括弧括起。

MPEP 2173.05(r) Omnibus Claim

Some applications are filed with an omnibus claim which reads as follows: A device substantially as shown and described. This claim should be rejected under 35 U.S.C. 112, second paragraph, because it is indefinite in that it fails to point out what is included or excluded by the claim language.

MPEP 2173.05(s) Reference to Figures or Tables

Where possible, claims are to be complete in themselves. Incorporation by reference to a specific figure or table "is permitted only in exceptional circumstances where there is no practical way to define the invention in words and where it is more concise to incorporate by reference than duplicating a drawing or table into the claim. Incorporation by reference is a necessity doctrine, not for applicant's convenience." Ex parte Fressola, 27 USPQ2d 1608, 1609 (Bd. Pat. App. & Inter. 1993) (citations omitted).

Reference characters corresponding to elements recited in the detailed description and the drawings may be used in conjunction with the recitation of the same element or group of elements in the claims. See MPEP § 608.01(m).

§11.25 請求項中使用商標或商品名

商標及商品名乃是用來表彰商品之來源，而不是用來代表產品本身，而且一商標及商品名亦可能代表許多不同產品（其可能有不同之構造或者不同之組成、成份）。因而，如果在請求項中使用一商標作為限制條件以識別（identify）一特定材料、產品，則將使請求項之界定不清楚而違反 35USC§112(2)之規定。反過來說，如果在請求項中使用商標、商品名之目的並不是要作為請求項之限制條件，那又為什麼要在請求項中敘述產品、材料之商標或商品名。

MPEP 2173.05(u) Trademarks or Trade Names in a Claim

The presence of a trademark or trade name in a claim is not, per se, improper under 35 U.S.C. 112, second paragraph, but the claim should be carefully analyzed to determine how the mark or name is used in the claim. It is important to recognize that a trademark or trade name is used to identify a source of goods, and not the goods themselves. Thus a trademark or trade name does not identify or describe the goods associated with the trademark or trade name. See definitions of trademark and trade name in MPEP § 608.01(v). A list of some trademarks is found in Appendix I.

If the trademark or trade name is used in a claim as a limitation to identify or describe a particular material or product, the claim does not comply with the requirements of the 35 U.S.C. 112, second paragraph. Ex parte Simpson, 218 USPQ 1020 (Bd. App. 1982). The claim scope is uncertain since the trademark or trade name cannot be used properly to identify any particular material or product. In fact, the value of a trademark would be lost to the extent that it became descriptive of a product, rather than used as an identification of a source or origin of a product. Thus, the use of a trademark or trade name in a claim to identify or describe a material or product would not only render a claim indefinite, but would also constitute an improper use of the trademark or trade name.

If a trademark or trade name appears in a claim and is not intended as a limitation in the claim, the question of why it is in the claim should be addressed. Does its presence in the claim cause confusion as to the scope of the claim? If so, the claim should be rejected under 35 U.S.C. 112, second paragraph.

實例

申請人之專利申請案乃有關於一種用於建築之預成形水密結構，此結構使用了一複合材料嵌板，其在一不銹鋼板上黏附一層彈性的氯磺化聚乙烯（elastomeric chlorosulphonated polyethylene）。此彈性的氯磺聚乙烯乃以商標"HYPALON"在市面上販售，且此行業人士皆知如何使用此材料作為建築上之應用。請問下述之請求項哪一項會造成請求項之界定不清楚？

(A) A prefabricated panel for a building system having a surface comprising HYPALON continuously bonded to a surface of a thin sheet steel member by an adhesive which is resistant to corrosive fluids.

(B) A prefabricated water-tight structure comprising a prefabricated panel, said panel comprising a thin sheet steel member bonded by an adhesive to an elastomeric chlorosulphonated polythene, said elastomeric chlorosulpHonated polythene being sold under the trademarkHYPALON, and said adhesive being resistant to corrosive fluids.

(C) A prefabricated water-tight structure comprising anelastomeric chlorosulphonated polythene, and adhesive, and a thin sheet steel member, said elastomeric chlorosulphonated polythene being sold under the trademark HYPALON, and characterized by good wear resistance, and said elastomeric chloroshulphonated polythene being continuously bonded by said adhesive to a surface of a thin sheet steel member, said member, said adhesive being resistant to corrosive fluids.

(A).因為在(A)中直接以商標"HYPALON"作為請求項之限制條件。

§11.26 請求項中同時請求裝置及使用裝置之方法

一請求項如果同時包括了結構及步驟作為元件（element）則為一混合請求項（hybrid claim），將被視為界定不清楚，違反 35USC§112 第 2 段而核駁。但是如果一裝置（產品）請求項中包括了方法之限制（條件）（所謂限制乃指界定或修飾一元件之敘述）或者一方法請求項中包括了裝置之限制（條件）則並非混合請求不會被 35USC§112 第 2 段核駁。

例如：①A product comprising A, B, and C wherein C is made by step …

②A process comprising A, B, C and step D which is performed in an (apparatus)。

MPEP 2173.05(p) Claim Directed to Product-By- Process or Product and Process [R-5]

＊＊＊＊＊＊＊＊＊＊＊

II. PRODUCT AND PROCESS IN THE SAME CLAIM

A single claim which claims both an apparatus and the method steps of using the apparatus is indefinite under 35 U.S.C. 112, second paragraph. *>IPXL Holdings v. Amazon.com, Inc., 430 F.2d 1377, 1384, 77 USPQ2d 1140, 1145 (Fed. Cir. 2005);< Ex parte Lyell, 17 USPQ2d 1548 (Bd. Pat. App. & Inter. 1990) *>(< claim directed to an automatic transmission workstand and the method * of using it * held ** ambiguous and properly rejected under 35 U.S.C. 112, second paragraph>)<.

＊＊＊＊＊＊＊＊＊＊＊

MPEP 2106

＊＊＊＊＊＊＊＊＊＊＊

Note that an apparatus claim with process steps is not classified as a "hybrid" claim; instead, it is simply an apparatus claim including functional limitations. See, e.g., R.A.C.C. Indus. v. Stun-Tech, Inc., 178 F.3d 1309 (Fed. Cir. 1998) (unpublished).

＊＊＊＊＊＊＊＊＊＊＊

§11.27 請求項之元件無連接（互動）關係，敘述過於冗長或請求項數目過多且重複

　　一請求項如果各元件無連接關係，亦即僅為一集合（aggregation）時，則會被以35USC§112(2)以界定不清楚核駁。一請求項如果包括了非常冗長的敘述或不重要之細節（prolix）而使得所請求之標的之界限與範圍無法確定時，則此請求項之界定不清楚。另外，如果請求項之數目非常龐大（過多），且請求項之內容互相重複時，審查委員亦會以35USC§112(2)認為界定不清楚而核駁。此時，審查委員會以電話與申請人（代理人）連絡，請申請人選擇將特定的請求項作審查。如果申請人同意作選擇，審查委員會在下次之審定書中以請求項過多（undue multiplicity）核駁，同時針對所選擇之請求項作審查。如果申請人拒絕作選擇，則審查委員會發下書面之審定書以請求項過多為由核駁，並要求申請人選擇少於審查委員所特定數目之請求項。

　　如果對申請人之回復審查委員不贊成，審查委員仍堅持此請求項過多時，審查委員會只針對選擇之請求項作審查，並讓申請人有保留提出訴願之權利。

MPEP 2173.05(k) Aggregation [R-1]

　　**>A claim should not be rejected on the ground of "aggregation." In re Gustafson, 331 F.2d 905, 141 USPQ 585 (CCPA 1964) (an applicant is entitled to know whether the claims are being rejected under 35 U.S.C. 101, 102, 103, or 112); In re Collier, 397 F.2d 1003, 1006, 158 USPQ 266, 268 (CCPA 1968) ("[A] rejection for 'aggregation' is non-statutory.").

　　If a claim omits essential matter or fails to interrelate essential elements of the invention as defined by applicant(s) in the specification, see MPEP § 2172.01.<

MPEP 2173.05(m) Prolix

　　Examiners should reject claims as prolix only when they contain such long recitations or unimportant details that the scope of the claimed invention is rendered indefinite thereby. Claims are rejected as prolix when they contain long recitations that the metes and bounds of the claimed subject matter cannot be determined.

MPEP 2173.05(n) Multiplicity [R-2]

＊＊＊＊＊＊＊＊＊＊＊

Where, in view of the nature and scope of applicant's invention, applicant presents an unreasonable number of claims which ** are repetitious and multiplied, the net result of which is to confuse rather than to clarify, a rejection on undue multiplicity based on 35 U.S.C. 112, second paragrapH, may be appropriate.

＊＊＊＊＊＊＊＊＊＊＊

If an undue multiplicity rejection under 35 U.S.C. 112, second paragrapH, is appropriate, the examiner should contact applicant by telepHone explaining that the claims are unduly multiplied and will be rejected under 35 U.S.C. 112, second paragrapH. Note MPEP § 408. The examiner should also request that applicant select a specified number of claims for purpose of examination. If applicant is willing to select, by telepHone, the claims for examination, an undue multiplicity rejection on all the claims based on 35 U.S.C. 112, second paragrapH, should be made in the next Office action along with an action on the merits on the selected claims. If applicant refuses to comply with the telepHone request, an undue multiplicity rejection of all the claims based on 35 U.S.C. 112, second paragrapH, should be made in the next Office action. Applicant's reply must include a selection of claims for purpose of examination, the number of which may not be greater than the number specified by the examiner. In response to applicant's reply, if the examiner adheres to the undue multiplicity rejection, it should be repeated and the selected claims will be examined on the merits. This procedure preserves applicant's right to have the rejection on undue multiplicity reviewed by the Board of Patent Appeals and Interferences.

＊＊＊＊＊＊＊＊＊＊＊

MPEP 2172.01 Unclaimed Essential Matter [R-1]

＊＊＊＊＊＊＊＊＊＊＊

In addition, a claim which fails to interrelate essential elements of the invention as defined by applicant(s) in the specification may be rejected under 35 U.S.C. 112, second paragrapH, for failure to point out and distinctly claim the invention.

§11.28 請求項之元件在圖式中未顯示

依據 37CFR 1.83 一請求項之元件需均顯示在圖式之中，若未完整顯示在圖式之中則會被 35USC§112(2)以未清楚界定所請求之發明而核駁。

37CFR §1.83 Content of drawing.

(a) The drawing in a nonprovisional application must show every feature of the invention specified in the claims. However, conventional features disclosed in the description and claims, where their detailed illustration is not essential for a proper understanding of the invention, should be illustrated in the drawing in the form of a grapHical drawing symbol or a labeled representation (e.g., a labeled rectangular box).

＊＊＊＊＊＊＊＊＊＊＊＊

§11.29 裝置（步驟）加功能用語，其使用及範圍

所謂裝置（步驟）加功能用語（means（step）-plus-function clauses）乃是在請求項中以執行一特定功能之裝置或步驟來定義一元件（步驟）而不敘述其構造或動作之用語。例子如下：

(1)　A device comprising:

element A, element B, and means for combining element A and element B.

(2)　The method of corrugating polyethylene terephthalate film which comprises:

shaping said film at a temperature in the range of about 100 to 175 degrees C by pressing said film between two coacting rotating surfaces, and reducing the coefficient of friction of the resulting film to below about 0.40 as determined by the Bell test.

在上述(1)之例子中，其不特別指出將元件 A 及元件 B 結合之構造（元件）為何，而只用一種結合 A 及 B 之裝置（means）表之。而在上述(2)之例子中，其不敘明用什麼動作（act）來降低薄膜之摩擦係數，只用降低摩擦係數之步驟表之。

當一請求項中使用裝置（步驟）加功能用語來定義一元件或動作時，在以往，美國專利局之審查委員判斷專利性時會將此類用語作最大範圍之解讀，亦即只要執行同樣的功能（動作）之任何先前技術均會被用來作引證，而不管說明書本文中記載之對應之結構，材料或動作為何。但是自 Donaldison 之判例後，美國專利局的審查委員在解讀此類用語時會考慮功能及其對應之結構（材料，動作）以檢索先前技術。（give claims their broadest reasonable interpretation, in light of and consistent with the written description）。而裝置加功能用語要能被視為符合以 35USC§112，第 6 段之規定，需符合下列三項準則：

(A)必需寫成 "means for" 或 "step for;"

(B) "means for" 及 "step for" 之後必需接著功能用語；

(C) "means for" 及 "step for" 不能再用達成該功能之結構，材料及動作法修飾，

上述(A)，(B)項之規定並不是非常嚴格的，例如 "ink delivery means" 可被視為是 35USC§112 第 6 段所指之用語，因為其相當於 "means for ink delivery"。但是第(C)項，如果在 means for 之後再使用本身為結構性之用語去修飾，則可能不被認為是 35USC§112 第 6 段之裝置加功能用語。例如 "perforation means for tearing" said layers and removing said diaper from the wearer.因為 perforation（穿孔）之功能就是可撕開（tearing）。

依照 35USC§112 第 6 段，如果一請求項之限制條件以裝置（步驟）加功能用語表示，則其被解讀為涵蓋在說明書中所敘述到之對應之結構，材料或動作及其對等物。因而，當請求項中使用裝置步驟加功能用語時，在說明書中需有充份之對應的揭露。否則將被視為未在說明書中有支持及界定不清楚，而被 35USC§112 第 1 段及第 2 段核駁。此說明書中的對應揭露可以是明顯地（explicitly），亦可以是暗示的（implicit），或固有的（inherent），只要此技藝人士可清楚地知道其對應的結構（材料，動作）是什麼即可。但是，如果揭露不是明顯的，審查委員會要求申請人修正說明書以將其對應結構（材料，動作）明顯地揭露出來。

此外，如果一請求項只有一元件，且此元件用裝置（步驟）加功能用語撰寫，所謂的 single means claims，將被以 3535USC§112 第 1 段以不能滿足據以實施要件核駁。

　　當一請求項使用裝置（步驟）加功能用語定義一元件（步驟）時，審查委員要找到什麼樣的先前技術才能推斷該裝置加功能用語定義之元件與先前技術等同（equivalent），而據以核駁呢？審查委員找到的先前技術必需符合下列三條件：

(A) 其執行裝置加功能用語相同之功能；

(B) 並未被排除在說明書中對於相等物（equivalent）之明確定義之外，以及

(C) 其是裝置功能用語之對等物（equivalent）。

　　當然，審查委員必需告知其認為先前技術與用裝置加功能用語定義之元件為對等之理由與觀點。通常，審查委員可舉出下列四種因素之一，即可支持其理由及觀點。

(A) 先前技術元件以實質上相同之方式執行相同之功能，且得到實質上相同之結果。

(B) 此技藝人士承認先前技術之元件與說明書揭露之對應元件是可互換的。

(C) 先前技術元件與說明書中揭露之對應元件並無實質上之差異。

(D) 先前技術之元件與揭露在說明書之對應元件在結構上是相等的。

MPEP 2181 Identifying a 35 U.S.C.112, Sixth ParagrapH Limitation [R-6]

＊＊＊＊＊＊＊＊＊＊＊

I.LANGUAGE FALLING WITHIN 35 U.S.C.112, SIXTH PARAGRAPH

The USPTO must apply 35 U.S.C. 112, sixth paragrapH in appropriate cases, and give claims their broadest reasonable interpretation, in light of and consistent with the written description of the invention in the application.

＊＊＊＊＊＊＊＊＊＊＊

A claim limitation will be presumed to invoke 35 U.S.C. 112, sixth paragrapH, if it meets the following 3-prong analysis:

(A) the claim limitations must use the pHrase "means for" or "step for;"

(B) the "means for" or "step for" must be modified by functional language; and

(C) the pHrase "means for" or "step for" must not be modified by sufficient structure, material, or acts for achieving the specified function.

With respect to the first prong of this analysis, a claim element that does not include the pHrase "means for" or "step for" will not be considered to invoke 35 U.S.C.112, sixth paragrapH. If an applicant wishes to have the claim limitation treated under 35 U.S.C.112, sixth paragrapH, applicant must either: (A) amend the claim to include the pHrase "means for" or "step for" in accordance with these guidelines; or (B) show that even though the pHrase "means for" or "step for" is not used, the claim limitation is written as a function to be performed and does not recite sufficient structure, material, or acts which would preclude application.

＊＊＊＊＊＊＊＊＊＊＊

With respect to the second prong of this analysis, it must be clear that the element in the claims is set forth, at least in part, by the function it performs as opposed to the specific structure, material, or acts that perform the function.

＊＊＊＊＊＊＊＊＊＊＊

With respect to the third prong of this analysis, see Seal-Flex, 172 F.3d at 849, 50 USPQ2d at 1234 (Radar, J., concurring) ("Even when a claim element uses language that generally falls under the step-plus-function format, however, 112 ¶ 6 still does not apply when the claim limitation itself recites sufficient acts for performing the specified function.")

＊＊＊＊＊＊＊＊＊＊＊

II. *DESCRIPTION NECESSARY TO SUPPORT A CLAIM LIMITATION WHICH INVOKES 35 U.S.C. 112, SIXTH PARAGRAPH

35 U.S.C.112, sixth paragrapH states that a claim limitation expressed in means-plus-function language "shall be construed to cover the corresponding structure.described in the specification and equivalents thereof." "If one employs means plus function language in a claim, one must set forth in the specification an adequate disclosure showing what is meant by that language. If an applicant fails to set forth an adequate disclosure, the applicant has in effect failed to particularly point out and distinctly claim the invention as required by the second paragrapH of section 112." In re Donaldson Co., 16 F.3d 1189, 1195, 29 USPQ2d 1845, 1850 (Fed. Cir. 1994) (in banc).

The proper test for meeting the definiteness requirement is that the corresponding structure (or material or acts) of a means (or step)-plus-function limitation

＊＊＊＊＊＊＊＊＊＊＊

The disclosure of the structure (or material or acts) may be implicit or inherent in the specification if it would have been clear to those skilled in the art what structure (or material or acts) corresponds to the means (or step)-plus-function claim limitation.

＊＊＊＊＊＊＊＊＊＊＊

III. DETERMINING 35 U.S.C. 112 SECOND PARAGRAPH COMPLIANCE WHEN 35 U.S.C. 112 SIXTH PARAGRAPH IS INVOKED

The following guidance is provided to determine whether applicant has complied with the requirements of 35 U.S.C. 112, second paragrapH, when 35 U.S.C. 112, sixth paragrapH, is invoked:

(A) If the corresponding structure, material or acts are described in the specification in specific terms (e.g., an emitter-coupled voltage comparator) and one skilled in the art could identify the structure, material or acts from that description, then the requirements of 35 U.S.C. 112, second and sixth paragrapHs and are satisfied. See Atmel, 198 F.3d at 1382, 53 USPQ2d 1231.

(B) If the corresponding structure, material or acts are described in the specification in broad generic terms and the specific details of which are incorporated by reference to another document (e.g., attachment means disclosed in U.S. Patent No. X, which is hereby incorporated by reference, or a comparator as disclosed in the IBM article, which is hereby incorporated by reference), Office personnel must review the description in the specification, without relying on any material from the incorporated document, and apply the "one skilled in the art" analysis to determine whether one skilled in the art could identify the corresponding structure (or material or acts) for performing the recited function to satisfy the definiteness requirement of 35 U.S.C. 112, second paragrapH.

＊＊＊＊＊＊＊＊＊＊＊

IV. DETERMINING WHETHER 35 U.S.C. 112, FIRST PARAGRAPH SUPPORT EXISTS

The claims must still be analyzed to determine whether there exists corresponding adequate support for such claim under 35 U.S.C. 112, first paragrapH. In considering whether there is 35 U.S.C. 112, first paragrapH support for the claim limitation, the examiner must consider not only the original disclosure contained in the summary and detailed description of the invention portions of the specification, but also the original claims, abstract, and drawings.

＊＊＊＊＊＊＊＊＊＊＊

Even if the disclosure implicitly sets forth the structure, materials, or acts corresponding to a means- (or step-) plus-function claim element in compliance with 35 U.S.C. 112, first and second paragrapHs, the USPTO may still require the applicant to amend the specification pursuant to 37 CFR 1.75(d) and MPEP § 608.01(o) to explicitly state, with reference to the terms and pHrases of the claim element, what structure, materials, or acts perform the function recited in the claim element.

＊＊＊＊＊＊＊＊＊＊＊

V.SINGLE MEANS CLAIMS

Donaldson does not affect the holding of In re Hyatt, 708 F.2d 712, 218 USPQ 195 (Fed. Cir. 1983) to the effect that a single means claim does not comply with the enablement requirement of 35 U.S.C.112, first paragrapH.

* * * * * * * * * * * *

MPEP 2183 Making a Prima Facie Case of Equivalence

If the examiner finds that a prior art element

(A) performs the function specified in the claim,

(B) is not excluded by any explicit definition provided in the specification for an equivalent, and

(C) is an equivalent of the means- (or step-) plus-function limitation,

the examiner should provide an explanation and rationale in the Office action as to why the prior art element is an equivalent. Factors that will support a conclusion that the prior art element is an equivalent are:

(A) the prior art element performs the identical function specified in the claim in substantially the same way, and produces substantially the same results as the corresponding element disclosed in the specification.

* * * * * * * * * * * *

(B) a person of ordinary skill in the art would have recognized the interchangeability of the element shown in the prior art for the corresponding element disclosed in the specification.

* * * * * * * * * * * *

(C) there are insubstantial differences between the prior art element and the corresponding element disclosed in the specification.

* * * * * * * * * * * *

(D) the prior art element is a structural equivalent of the corresponding element disclosed in the specification.

* * * * * * * * * * * *

實例 1

假設下述之 claims 均在說明書中有支持且並無缺乏前置基礎之問題，請問下列哪一 claim 不符合 35 USC§112 第 6 款之 means-plus-function 之規定？

1. A boring device for deep boring an object rotating about an axis, comprising: a force generating means adapted to provide a force acting in the cutting head to cause radial displacement of said cutting head……

2. In an aircraft having a bladed rotor adapted under at least one translational flight conduction to provide both left and propulsive thrust, a jet driving device so constructed and located on the rotor as to drive the rotor……

3. In a pressure responsive instrument having a pressure responsive chamber, including a wall portion movable in reply to change in fluid pressure thereon,the improvements comprising a plate means and a leaf spring wing means on said plate means……

Claim 3 不符合 35 USC§112.6 之規定，因為 plate means, wing means 並未指出其執行何種功能（function）。

實例 2

一專利申請案乃有關於機械式連接件（fastener），用來將一橡膠鞋根連接至一鞋子之底部。此連接件之特殊構造使鞋跟可牢固地連接至鞋底，同時當鞋底磨擦時可提供平衡效果。說明書中包括了此機械連結件之詳細構造之圖式，而說明書本文亦詳細敘述了其構造、連接方式及如何達到緩衝效果。說明書又提到可同時使用粘合劑以使此機械連結件之效果加強。在說明書中並未詳述可使用何種粘合劑，因為粘合劑是習知的。說明書中並敘述本發明之範圍並不限定於所述之較佳實施例，因為可加以改變而不脫離構發明之精神中。其 Claim 1 如下：

1.A system for securely attaching a rubber heel to the bottom of a shoe and providing a cushioning effect when worn, said system comprising cushionin means for mechanically fastening said heel to said shoe.

請問此請求項是否會被 35USC§112 第 1 段核駁？

此請求項為 Single means claim，將被以不滿足據以實施要件核駁。因為此請求項涵蓋任何可達到所述之結果之裝置（means）。

第 12 章　新穎性與權利喪失

目　錄

§12.1　喪失取得專利的權利及喪失新穎性之事件

美國專利法 35 USC §102 法條乃有關於新穎性及權利喪失。

(一)具有下列四種情形之一者，美國專利申請人將喪失取得專利之權利：

(1)　專利申請人在其美國專利申請案之申請日的一年之前，該發明已在美國或其他地區取得專利或敘述在一刊物之上，在美國已公開使用或銷售（35 USC §102(b)）。

(2)　發明人放棄了該發明（35 USC §102(c)）。

(3)　發明人或其受讓人在美國提出專利申請之前已在外國取得專利而且該外國專利是在美國專利申請日一年之前提出申請者（35 USC §102(d)）。

(4)　發明人並非自己發明了要申請專利之標的 USC §102(f)。

(二)具有下列三者情形中任何一種者，專利申請案喪失新穎性：

(1)　專利申請人在其發明日之前該發明在美國已為他人所知或使用，或該發明在美國及其他國家已取得專利或已見於刊物（35 USC §102(a)）。

(2)　專利申請人在其發明日之前，該發明已敘述於他人在美國申請且依本法第 122(b)條公告之申請案中，或專利申請人在其發明日之前，該發明已敘述在他人較早申請且已取得美國專利之申請案之中;但依照本法第 351(c)條之國際申請案只有當該國際申請案指定美國且依本法第 21(2)(a)條以英文公告者，始具有與依本節之美國申請案之相同之效果。(35 USC §102(e))

(3)　在符合第 351 條或第 291 條進行之抵觸程序期間，在專利申請人之發明日之前，其他的發明人以依本法第 104 條確立該發明已被其他發明人發明，且其未放棄、不發表或隱藏者，或在專利申請人之發明日之前，該項發明已在本國由他人完成且其未在本國放棄、不發表或隱藏者。在決定發明之優先權時，不僅需分別考慮該發明之構想及實施日期，並且需考慮先構想出但較晚付諸實施者，在該他人構想前之合理的努力度。35USC§102(g)

35 U.S.C. 102

A person shall be entitled to a patent unless —

(a) the invention was known or used by others in this country, or patented or described in a printed publication in this or a foreign country, before the invention thereof by the applicant for patent, or

(b) the invention was patented or described in a printed publication in this or a foreign country or in public use or on sale in this country, more than one year prior to the date of the application for patent in the United States, or

(c) he has abandoned the invention, or

(d) the invention was first patented or caused to be patented, or was the subject of an inventor's certificate, by the applicant or his legal representatives or assigns in a foreign country prior to the date of the application for patent in this country on an application for patent or inventor's certificate filed more than twelve months before the filing of the application in the United States, or

(e) the invention was described in — (1) an application for patent, published under section 122(b), by another filed in the United States before the invention by the applicant for patent or (2) a patent granted on an application for patent by another filed in the United States before the invention by the applicant for patent, except that an international application filed under the treaty defined in section 351(a) shall have the effects for the purposes of this subsection of an application filed in the United States only if the international application designated the United States and was published under Article 21(2) of such treaty in the English language; or

(f) he did not himself invent the subject matter sought to be patented, or

(g)(1)during the course of an interference conducted under section 135 or section 291, another inventor involved therein establishes, to the extent permitted in section 104, that before such person's invention thereof the invention was made by such other inventor and not abandoned, suppressed, or concealed, or (2) before such person's invention thereof, the invention was made in this country by another inventor who had not abandoned, suppressed, or concealed it. In determining priority of invention under this subsection, there shall be considered not only the respective dates of conception and reduction to practice of the invention, but also the reasonable diligence of one who was first to conceive and last to reduce to practice, from a time prior to conception by the other.

§12.2　申請日，發明日，有效申請日，他人之定義

申請日（filing date）乃指專利申請案在美國專利局提出之日期。有效申請日（effective filing date）乃指其可主張之最早申請日，例如一分割案、延續案之有效申請日乃為其母案之申請日，一申請案主張暫時申請案之優先權時，其有效申請日乃為暫時申請案之申請日。發明日（date of invention）乃為一發明人將其發明實際付諸實施之日（actual reduction to practice），而如果從構想日（conception）至實際付諸實施之日之期間有合理勤勉度時，則構想日即為其發明日。美國專利之申請日則為推斷的發明日（constructive reduction to practice），美國專利局在應用 35 USC §102(a)、(e)及(g)時乃推斷發明日為美國專利申請案之申請日以進行審查。

他人（others, another）乃指發明人之實體（entity）有一者不相同，例如一申請案 X 之發明人為 A,B,C 三人，另一申請案 Y 之發明人為 A,B,D 三人，則 Y 申請案則為他人之申請案。

§12.3　在美國境內秘密使用一製程且公開銷售製得之產品

美國專利法 35 USC §102(a)所謂的已為他人使用乃指知悉及使用是為大眾可取得者，故在美國境內秘密地使用一製程並非是大眾可取得的，除非大眾可由所銷售之產品知悉該製程才構成 35 USC §102(a)之已為他人所知及使用。

MPEP 2132

"The statutory language 'known or used by others in this country' (35 U.S.C. §102(a)), means knowledge or use which is accessible to the public." The knowledge or use is accessible to the public if there has been no deliberate attempt to keep it secret.

"The nonsecret use of a claimed process in the usual course of producing articles for commercial purposes is a public use." But a secret use of the process coupled with the sale of the product does not result in a public use of the process unless the public could learn the claimed process by examining the product.　Therefore, secret use of a process by another, even if the product is commercially sold, cannot result in a rejection under 35 U.S.C. 102(a)if an examination of the product would not reveal the process.

§12.4　請求項被預見

所謂請求項被預見（anticipated）乃指請求項中記載之每一元件均被明顯地，固有地敘述在單一的先前技術之中，此時審查委員就可引用 35USC§102 以預見（anticipation）核駁該請求項。當一請求項涵蓋有數種結構或組成，則如果在請求項範圍內之任何一種結構或組成從先前技術中是習知的，則此請求項被視為是預見的。

MPEP2131

"A claim is anticipated only if each and every element as set forth in the claim is found, either expressly or inherently described, in a single prior art reference. "When a claim covers several structures or compositions, either generically or as alternatives, the claim is deemed anticipated if any of the structures or compositions within the scope of the claim is known in the prior art."(claim to a system for setting a computer clock to an offset time to address the Year 2000 (Y2K) problem, applicable to records with year date data in "at least one of two-digit, three-digit, or four-digit" representations, was held anticipated by a system that offsets year dates in only two-digit formats). See also MPEP § 2131.02.< "The identical invention must be shown in as complete detail as is contained in the…… claim." The elements must be arranged as required by the claim, but this is not an ipsissimis verbis test, i.e., identity of terminology is not required.

實例 1

一專利申請案之 claim 1 如下：

　　Claim 1: An alloy consisting of 70.5 to 77.5% of iron, 15.0 to 17.0% of cobalt; 0.5 to 1.0% of carbon; up to 2.15% chromium, and at least 7% tungsten.

審查委員引證了一刊物其發表日期乃在本案申請日之前，其揭露了 76.0% iron, up to 15.0% cobalt, 0.5% carbon and8.50%tungsten 則審查委員可否以 35 USC §102(a) 主張 anticipation 核駁 claim 1？

可以。該引證雖未揭示 chromium，但 claim 1 為 up to 2.5% chromium 表示亦可不包含 chromium; 故該引證已揭示了 claim 1 之所有元件且發表日在其申請日（推斷之發明日）之前，故可以 35 USC §102(a)核駁。

實例 2

A 提出一美國專利申請案，此專利申請案請求一氣體組合物包括一特定範圍量之氧氣。而審查委員則引證出一美國專利在本申請案之有效申請日之前二年發證。此先前技術揭露了與本案相同之氣體組合物，但是其氧氣的含量範圍較本案為大，且其說明書中所揭露之氧含量之實施例並未包含在本案之氧氣含量範圍之內。而本案之說明書不但揭露了該氣體組合物之不同用途亦揭露了該範圍內之氧氣含量所造成之不可預期效果。請問本案所請求之發明是否已喪失新穎性（anticipated）？

未喪失新穎性，因為該引證並未特定出本案之氧氣之含量範圍。

MPEP 2131.03

Prior art which teaches a range within, overlapping or touching the claimed range anticipates of the prior art range discloses the claimed range within sufficient specificity.

When the prior art discloses a range which touches, overlaps or is within the claimed range, but no specific example falling within the claimed range are disclosed, a case-by-case determination must be made as to anticipation.

§12.5　解密文件之先前技術日期

美國專利局之審查委員在引用解密文件作為先前技術以用 35USC§102(a)核駁一請求項時通常考慮該解密文件之印刷日期及公告日期。印刷日期通常會出現在文件之上且可被考慮為文件準備要作有限制之散佈之日期。而發行（publication）之日期為大眾可取得之日期。故審查委員以解密文件作為 35USC§102(a)之預見之核駁時，發行之日期為已見於刊物之日期，而印刷之日期則為他人知道之日期。

MPEP 707.05(f) Effective Dates of Declassified Printed Matter

In using declassified material as references there are usually two pertinent dates to be considered, namely, the printing date and the publication date. The printing date in some instances will appear on the material and may be considered as that date when the material was prepared for limited distribution. The publication date is the date of release when the material was made available to the public. See Ex parteHarris, 79 USPQ 439 (Comm'r Pat. 1948). If the date of release does not appear on the material, this date may be determined by reference to the Office of Technical Services, Department of Commerce.

In the use of any of the above noted material as an anticipatory publication, the date of release following declassification is the effective date of publication within the meaning of the statute.

For the purpose of anticipation predicated upon prior knowledge under 35 U.S.C. 102(a) the above noted declassified material may be taken as prima facie evidence of such prior knowledge as of its printing date even though such material was classified at that time. When so used the material does not constitute an absolute statutory bar and its printing date may be antedated by an affidavit or declaration under 37 CFR 1.131.

實例

專利申請人收到一審定書（Office Action），審查委員用一已公告且已解密之文件以 35 USC §102(a)核駁本案之 claims，該引證之文件乃在本申請案之申請日前三個月印刷，且在印刷時列為保密文件（classified），但在本案申請日之後的 8 個月解密（declassified）並發行，且該引證文件揭露了本申請案 claim 中所有元件，請問審查委員以此解密公告文件作為引證以 35 USC §102(a)核駁是否適當？

適當，對於秘密文件，其解密、發行之日期為 publication 之日期，但是其印刷之日期可作為他人知道之日期。

§12.6　刊物之定義，包括之類型及其作為先前技術之日期

35 USC §102 中所謂的刊物作為先前技術引證（prior art reference）者乃指一文件已散布於大眾或大眾經合理之努力即可取得該文件即可作為 35 USC §102 之引

證。而該刊物並不限於在哪一地區發行之刊物或以哪一種語言發行，或者為手寫或者儲存於微影影片、磁碟片、光碟片，只要是有足夠量之影本已散布或者已在圖書館中編上目錄而使得大眾可取得（access）就可作為 35USC§102 之 prior art reference。另外，在網路上或以電子媒體之方式的發表亦構成 35 USC §102 之 prior art reference。在一研討會上以口頭發表同時以書面方式作無限制之散布亦構成 35 USC §102 之 prior art reference。但是在公司（組織）內部作散布而此散布乃是以密件方式進行則不構成 35 USC §102 之 prior art reference。

刊物作為 35 USC §102 之引證資料的日期如下：

1. 如為圖書館之刊物則以其編上目錄分類之日期為先前技術之日期。

2. 如果以郵件散布之刊物則以至少一大眾人士收到之日期為先前技術之日期。

3. 如為在網路或電子媒體上公布之資訊，則以在網路或媒體上公布之日期為先前技術之日期。

MPEP 2128

A reference is proven to be a "printed publication" "upon a satisfactory showing that such document has been disseminated or otherwise made available to the extent that persons interested and ordinarily skilled in the subject matter or art, exercising reasonable diligence, can locate it."

An electronic publication, including an on-line database or Internet publication, is considered to be a "printed publication" within the meaning of 35 U.S.C. 102(a) and (b) provided the publication was accessible to persons concerned with the art to which the document relates.

MPEP 2128.01

A paper which is orally presented in a forum open to all interested persons constitutes a "printed publication" if written copies are disseminated without restriction. Massachusetts Institute of Technology v. AB Fortia, 774 F.2d 1104, 1109, 227 USPQ 428, 432 (Fed. Cir. 1985) (Paper orally presented to between 50 and 500 persons at a scientific meeting open to all persons interested in the subject matter, with written copies distributed without restriction to all who requested, is a printed publication. Six persons requested and obtained copies.).

Documents and items only distributed internally within an organization which are intended to remain confidential are not "printed publications" no matter how many copies are distributed.

§12.7　引用二份或以上之先前技術核駁請求項

在引用 35 USC §102 核駁一請求項時通常只能引用單一引證資料而此單一引證資料需揭示被核駁的請求項的所有元件,且此單一引證資料需是可據以實施的(enabling or operable)。但在下列情形亦可引用另一輔助引證:

1. 在需證明主要引證包含了可實施之揭露時;
2. 需解釋主要引證中之用語之意義時;
3. 需證明在引證中未揭露之特徵為固有的時。

MPEP 2131.01

Normally, only one reference should be used in making a rejection under 35 U.S.C. 102. However, a 35 U.S.C. 102 rejection over multiple references has been held to be proper when the extra references are cited to:

(A) Prove the primary reference contains an "enabled disclosure;"

(B) Explain the meaning of a term used in the primary reference; or

(C) Show that a characteristic not disclosed in the reference is inherent.

§12.8　引證案之揭露不符合據以實施要件

當申請人認為以 35USC§102 核駁之先前技術不能實施時,申請人可提出事實證據證明該引證資料不能實施。以 35 USC §102 之引證只要揭示了申請案之請求項的所有元件就被假定是可實施的(operable),因而舉證責任就轉至申請人。另外,當引證資料為專利文件時,申請人雖可以宣誓書之方式爭論引證資料不能實施,但需提出優勢之證據佐證。(preponderance of the evidence)。

MPEP 2121

When the reference relied on expressly anticipates or makes obvious all of the elements of the claimed invention, the reference is presumed to be operable.　Once such a reference is found, the burden is on applicant to provide facts rebutting the presumption of operability.

MPEP 716.07

Since every patent is presumed valid (35 U.S.C. 282), and since that presumption includes the presumption of operability, examiners should not express any opinion on the operability of a patent.　Affidavits or declarations attacking the operability of a patent cited as a reference must rebut the presumption of operability by a preponderance of the evidence.

§12.9　已取得專利之定義及作為先前技術之日期

35USC§102 中所謂的"已取得專利"乃指一專利申請案賦予一專利且專利權是可執行的。但是美國專利局要引用一美國專利或外國之專利作為 35USC§102 之先前技術時，該專利必須是大眾可取得的（accessible），例如以刊物之形式公告。而，如果已被賦予專利，但是保持秘密未公告（秘密專利）則不可作為 35USC§102 之引證資料。而其作為先前技術之日期乃為賦予專利之日期。

MPEP 2126

Even if a patent grants an exclusionary right (is enforceable), it is not available as prior art under 35 U.S.C. 102(a)or(b) if it is secret or private.　In re Carlson, 983 F.2d 1032, 1037, 25 USPQ2d 1207, 1211 (Fed. Cir. 1992).　The document must be at least minimally available to the public to constitute prior art.

MPEP 2126.01

The date the patent is available as a reference is generally the date that the patent becomes enforceable.　This date is the date the sovereign formally bestows patents rights to the applicant.

§12.10 已公開使用之定義及實驗使用

　　35 USC §102(b)之公開使用乃指將發明於公開場合使用而言，而不論大眾是否可得知發明的內容，例如發明人將包含發明之裝置之機器或物品於公開場合展示或銷售，雖然其發明是隱藏起來，參觀者並不知曉其發明之裝置之操作也不知其發明之技術複雜性，則此使用仍為 35 USC §102(b)之公開使用，另外，如果發明人允許第三者無限制且在無保密狀況下使用其發明亦構成 35 USC §102(b)之公開使用。但是實驗使用（experimental use）即使是在公開場合，如果其使用的目的是為了實驗發明之可操作性，亦即為了付諸實施之實驗性質而非為了獲得使用之利潤時，則並非 35 USC §102(b)之已公開使用。

MPEP 2133.03(a)

Even If the Invention Is Hidden, Inventor Who Puts Machine or Article Embodying the Invention in Public View Is Barred from Obtaining a Patent as the Invention Is in Public Use

When the inventor or someone connected to the inventor puts the invention on display or sells it, there is a "public use" within the meaning of 35 U.S.C. 102(b) even though by its very nature an invention is completely hidden from view as part of a larger machine or article, if the invention is otherwise used in its natural and intended way and the larger machine or article is accessible to the public.

An Invention Is in Public Use If the Inventor Allows Another To Use the Invention Without Restriction or Obligation of Secrecy

"Public use" of a claimed invention under 35 U.S.C. 102(b) occurs when the inventor allows another person to use the invention without limitation, restriction or obligation of secrecy to the inventor."

MPEP 2133.03(e)

The question posed by the experimental use doctrine is "whether the primary purpose of the inventor at the time of the sale, as determined from an objective evaluation of the facts surrounding the transaction, was to conduct experimentation."

Experimentation must be the primary purpose and any commercial exploitation must be incidental.

If the use or sale was experimental, there is no bar under 35 U.S.C. 102(b). "A use or sale is experimental for purposes of section 102(b) if it represents a bona fide effort to perfect the invention or to ascertain whether it will answer its intended purpose....If any commercial exploitation does occur, it must be merely incidental to the primary purpose of the experimentation to perfect the invention."

"The experimental use exception...does not include market testing where the inventor is attempting to gauge consumer demand for his claimed invention. The purpose of such activities is commercial exploitation and not experimentation."

實例

發明人 X 在美國加州擁有一超級市場，但其超商之地板常因客人將口香糖丟至地板上而不易清理，X 於 2006 年 2 月 1 日完成了一種地板蠟並於 2006 年 4 月 1 日起將發明之地板蠟施打於超商之地板上，經過一年的實驗發現確實不易黏附口香糖因而清理地板變得較容易。發明人 X 於是於 2007 年 6 月 1 日提出美國專利申請，發明人 X 之公開使用其發明之地板臘是否會以已公開使用被 35 USC §102(b)核駁？

不會，因為其將發明之地板蠟施於地板上使用雖為公開之使用但是其乃為了實驗的目的，故自 2006 年 4 月 1 日至 2007 年 4 月 1 日乃為實驗使用，故於 2007 年 6 月 1 日提出專利申請雖已超過一年但並非 35 USC §102(b)之適用。但如果自 2008 年 4 月 1 日以後提出申請則構成公開使用。

§12.11 已銷售之定義

35 USC §102(b)所謂之銷售乃指在一專利申請案之有效申請日一年之前之銷售，要約銷售（Offer for sell）發生，而且銷售或要約銷售之標的與所請求之發明完全相同（fully anticipated）或使請求之標的為顯而易見的。此銷售或要約銷售可為有條件的銷售，秘密的銷售，對國外之銷售，被拒絕的要約銷售，單一銷售，無發明人同意之銷售等。另外要注意的是，要以"已銷售"為由用 35 USC §102(b)不予專利需下列二條件均於"關鍵日期（Critical date,申請日前一年）"之前存在：

(1)　專利申請所請求之產品即為銷售或要約銷售之標的。

(2)　所請求之發明已準備好申請專利（實際已付諸實施或發明人已準備完成足
　　　夠使此技藝人士可實施此發明之圖式或相關說明）。

MPEP 2133.03(b)

An impermissible sale has occurred if there was a definite sale, or offer to sell, more than 1 year before the effective filing date of the U.S. application and the subject matter of the sale, or offer to sell, fully anticipated the claimed invention or would have rendered the claimed invention obvious by its addition to the prior art.

MPEP 2133.03(c)

"[T]he on-sale bar applies when two conditions are satisfied before the critical date [more than one year before the effective filing date of the U.S. application]. First, the product must be the subject of a commercial offer for sale…. Second, the invention must be ready for patenting." Id. at 67, 119 S.Ct. at 311-12, 48 USPQ2d at 1646-47. "Ready for patenting," the second prong of the Pfaff test, "may be satisfied in at least two ways: by proof of reduction to practice before the critical date; or by proof that prior to the critical date the inventor had prepared drawings or other descriptions of the invention that were sufficiently specific to enable a person skilled in the art to practice the invention."

§12.12 發明人放棄了發明

35 USC §102(c)之發明人放棄了該發明乃指發明人有放棄該發明之意圖（intent），而且發明人要放棄該發明之意圖需從其行為可看出才是放棄了該發明。例如，發明人完成了一發明之後，過了相當長的時間並未提出專利申請並不一定是放棄了該發明。但是如果之後，見到他人欲發展相同之發明就取笑之，但當看到別人的發明成功時才想繼續發展該發明，則為放棄了該發明。

另外，發明人之專利中若揭露但未請求之發明（實施例）將被認為是放棄了該發明，但若於核准前公告之後的一年內另申請一專利請求該發明，或於取得專利之二年之內申請再發證（reissue）以擴大原專利之範圍，則可視為並未放棄該發明。

MPEP 2134

Actual abandonment under 35 U.S.C. 102(c) requires that the inventor intend to abandon the invention, and intent can be implied from the inventor's conduct with respect to the invention.

DELAY IN MAKING FIRST APPLICATION

Abandonment under 35 U.S.C. 102(c) requires a deliberate, though not necessarily express, surrender of any rights to a patent. To abandon the invention the inventor must intend a dedication to the public. Such dedication may be either express or implied, by actions or inactions of the inventor. Delay alone is not sufficient to infer the requisite intent to abandon. (Where the inventor does nothing over a period of time to develop or patent his invention, ridicules the attempts of another to develop that invention and begins to show active interest in promoting and developing his invention only after successful marketing by another of a device embodying that invention, the inventor has abandoned his invention under 35 U.S.C. 102(c).).

§12.13 符合 35USC102(d)之四要件

一美國專利申請案如果符合下列 4 條件則違反 35USC§102(d)之規定

(1)　外國之申請案之申請日是在美國專利申請案有效申請日一年之前。

(2)　外國申請案與美國申請案之申請人相同或者為其法律代表人或其受讓人。

(3)　外國申請案在美國申請案之有效申請日之前已核准（granted）〔所請核准乃指該專利可執行（enforceable）〕。

(4)　外國申請案與美國申請案乃有關於相同之發明。

上述第 4 項，外國申請案與美國專利申請案乃關於相同之發明，並不一定要在請求項中均請求相同之發明，只要美國專利申請案所請求之發明在外國申請案中有揭露即適用。

MPEP 2135.01 The Four Requirements of 35 U.S.C. 102(d)

Under 35 U.S.C. 102(d), the "invention…… patented" in the foreign country must be the same as the invention sought to be patented in the U.S. When the foreign patent contains the same claims as the U.S. application, there is no question that "the invention was first patented…… in a foreign country." However, the claims need not be identical or even within the same statutory class. If applicant is granted a foreign patent which fully discloses the invention and which gives applicant a number of different claiming options in the U.S., the reference in 35 U.S.C. 102(d) to "'invention…… patented' necessarily includes all the disclosed aspects of the invention. Thus, the section 102(d) bar applies regardless whether the foreign patent contains claims to less than all aspects of the invention." In essence, a 35 U.S.C. 102(d) rejection applies if applicant's foreign application supports the subject matter of the U.S. claims. (Applicant was granted a Spanish patent claiming a method of making a composition. The patent disclosed compounds, methods of use and processes of making the compounds. After the Spanish patent was granted, the applicant filed a U.S. application with claims directed to the compound but not the process of making it. The Federal Circuit held that it did not matter that the claims in the U.S. application were directed to the composition instead of the process because the foreign specification would have supported claims to the composition. It was immaterial that the formulations were unpatentable pharmaceutical compositions in Spain.).

實例

　　發明人 X 於台灣申請了一發明專利，專利申請書中揭露了一化合物，及製造此化合物之方法與用途，但只請求了此化合物之製法，本發明於申請之後 8 個月就核准取得專利，之後發明人 X 想針對此案申請美國專利但不請求該化合物之製法，而請求該化合物，然而要提申時已自台灣申請日超過一年，請問此美國專利申請案是否符合 35USC§102(d)之規定？

　　此美國專利申請案不符合 35USC 102(d)之規定。

§12.14 35USC102(e)之先前技術

　　美國專利局之審查委員用來作為 35USC§102(e)之先前技術之引證資料必需包括下列要件：

(1)　需為下列 5 種文件之一者：

　①　依照 35USC§122 核准前公告之美國專利申請案之公報；

　②　在 2000 年 11 月 29 日當日或之後申請之且指定美國之 PCT 申請案，且由 WIPO 以英文公告之公報；

③　在 2000 年 11 月 29 日當日或之後申請且指定美國，且由 WIPO 以英文公告之 PCT 申請案，其在進入美國國家階段後由美國專利局作核准前公告之公報；

④　由美國專利申請案核准之美國專利公報；

⑤　由 2000 年 11 月 29 日當日或之後申請的 PCT 申請案進入美國國家階段而核准之美國專利公報。

(2)　上述 5 種文件之發明人中有一人與專利申請案之發明人不相同；

(3)　用以核駁之揭露必需揭示在上述 5 種文件之中；

(4)　上述 5 種文件之先前技術日期必需在專利申請案之美國專利申請日之前。

MPEP2136.02 Content of the Prior Art Available Against the Claims [R-3]

I.< A 35 U.S.C. 102(e) REJECTION MAY RELY ON ANY PART OF THE PATENT OR APPLICATION PUBLICATION DIS-CLOSURE

Under 35 U.S.C. 102(e), the entire disclosure of a U.S. patent, a U.S. patent application publication, or an international application publication having an earlier effective U.S. filing date (which will include certain international filing dates) can be relied on to reject the claims. Sun Studs, Inc. v. ATA Equip. Leasing, Inc., 872 F.2d 978, 983, 10 USPQ2d 1338, 1342 (Fed. Cir. 1989). See MPEP § 706.02(a).

＊＊＊＊＊＊＊＊＊＊＊

II. < REFERENCE MUST ITSELF CONTAIN THE SUBJECT MATTER RELIED ON IN THE REJECTION

When a U.S. patent, a U.S. patent application publication, or an international application publication is used to reject claims under 35 U.S.C. 102(e), the disclosure relied on in the rejection must be present in the issued patent or application publication.

＊＊＊＊＊＊＊＊＊＊＊

MPEP 2136.04 Different Inventive Entity; Meaning of "By Another" [R-1]

IF THERE IS ANY DIFFERENCE IN THE INVENTIVE ENTITY, THE REFERENCE IS "BY ANOTHER"

＊＊＊＊＊＊＊＊＊＊＊

A DIFFERENT INVENTIVE ENTITY IS PRIMA FACIE EVIDENCE THAT THE REFERENCE IS "BY ANOTHER"

* * * * * * * * * * * *

§12.15 35USC102(e)之先前技術之日期

作為 35USC102(e)之先前技術之 5 種文件之中，如為美國專利或美國專利申請案之核准前公告公報，則先前技術日期為美國專利申請案之有效申請日。如為 PCT 專利申請案，則其先前技術日期為 PCT 專利申請日或所主張之美國專利申請案之有效申請日。

另外，上述 5 種文件之國外優先權日並不能作為 35USC§102(e)之先前技術日期。然而，其所主張優先權之暫時申請案之申請日可作為 35USC§102(e)之先前技術之日期，但是其作為核駁之內容需揭露在暫時申請案之中。

再者，CIP 專利申請案之母案之申請日只有當母案揭露了 CIP 專利申請案用來核駁之內容時，才可作為先前技術之日期。此外，作為 35USC§102(e)之先前技術之發明構想日及付諸實施日均不能作為 35USC§102(e)之先前技術之日期。

MPEP2136.03 Critical Reference Date [R-6]

I.FOREIGN PRIORITY DATE

Reference's Foreign Priority Date Under 35 U.S.C. 119(a)-(d) and (f) Cannot Be Used as the 35 U.S.C. 102(e) Reference Date

35 U.S.C. 102(e) is explicitly limited to certain references "filed in the United States before the invention thereof by the applicant" (emphasis added). Foreign applications' filing dates that are claimed (via 35 U.S.C. 119(a)-(d),(f)or 365(a)) in applications, which have been published as U.S. or WIPO application publications or patented in the U.S., may not be used as 35 U.S.C.102(e) dates for prior art purposes.

* * * * * * * * * * * *

II. INTERNATIONAL (PCT) APPLICA-TIONS; INTERNATIONAL APPLICA-TION PUBLICATIONS

If the potential reference resulted from, or claimed the benefit of, an international application, the following must be determined:

(A) If the international application meets the following three conditions:

(1) an international filing date on or after November 29, 2000;

(2) designated the United States; and

(3) published under PCT Article 21(2) in English,

the international filing date is a U.S. filing date for prior art purposes under 35 U.S.C. 102(e).

* * * * * * * * * * *

III. PRIORITY FROM PROVISIONAL APPLICATION UNDER 35 U.S.C. 119(e)

The 35 U.S.C. 102(e) critical reference date of a U.S. patent or U.S. application publications and certain international application publications entitled to the benefit of the filing date of a provisional application under 35 U.S.C. 119(e) is the filing date of the provisional application with certain exceptions if the provisional application(s) properly supports the subject matter relied upon to make the rejection in compliance with 35 U.S.C. 112, first paragraph.

* * * * * * * * * * *

IV. PARENT'S FILING DATE WHEN REFERENCE IS A CONTINUATION-IN-PART OF THE PARENT

Filing Date of U.S. Parent Application Can Only Be Used as the 35 U.S.C. 102(e) Date If It Supports the **>Subject Matter Relied Upon in the< Child

* * * * * * * * * * *

V. DATE OF CONCEPTION OR REDUCTION TO PRACTICE

35 U.S.C. 102(e) Reference Date Is the Filing Date Not Date of Inventor's Conception or Reduction to Practice

實例 1

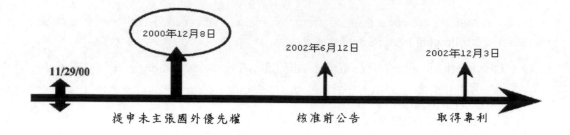

核准前公告作為引證的 35USC§102(e)(1)先前技術之日期為 2000 年 12 月 8 日。
專利作為引證資料之 35USC§102(e)(2)之先前技術日期為 2000 年 12 月 8 日。

實例 2

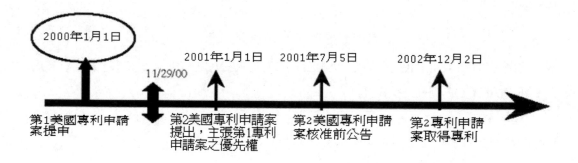

第 2 美國專利申請案之核准前公告作為引證的 35USC§102(e)(1)先前技術日期為 2000 年 1 月 1 日。

第 2 專利申請案之專利作為引證的 35USC§102(e)(2)的先前技術日期為 2000 年 1 月 1 日。

實例 3

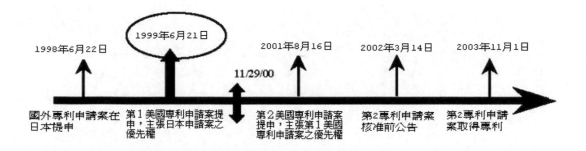

　　第 2 美國專利申請案之核准前公告之作為引證的 35USC§102(e)(1)先前技術日期為 1999 年 6 月 21 日。

　　第 2 專利申請案之專利作為引證的 35USC§102(e)(2)的先前技術日期為 1999 年 6 月 21 日。

實例 4

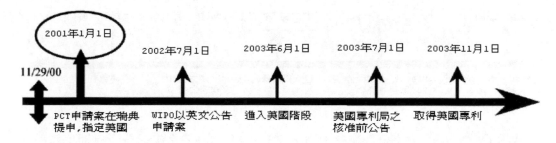

　WIPO 之英文公告作為引證之 35USC§102(e)(1)之先前技術日期為 2001 年 1 月 1 日。

　美國專利局之核准前公告作為引證之 35USC§102(e)(1)之先前技術日期為 2010 年 1 月 1 日。

　美國專利作為引證之 35USC§102(e)(1)之先前技術之日期為 2001 年 1 月 1 日。

實例 5

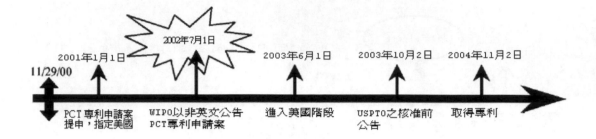

WIPO 之英文公告作為引證之 35USC§102(e)(1)之先前技術日期：無。

USPTO 之核准前公告作為引證之 35USC§102(e)(1)之先前技術日期：無。

美國專利作為引證之 35USC§102(e)(1)之先前技術日期：無。

實例 6

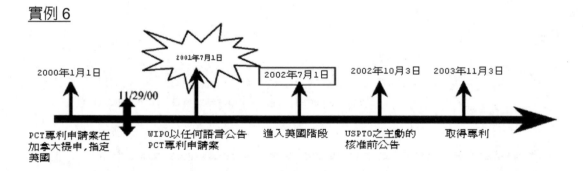

WIPO 之任何語言之公告作為引證之 35USC§102(e)(1)之先前技術日期：無。

USPTO 之主動核准前公告作為引證之 35USC§102(e)(1)之先前技術日期：無。

美國專利作為引證之 35USC§102(e)(1)之先前技術日期：2002 年 7 月 1 日。

實例 7

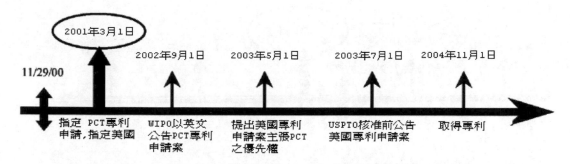

　　WIPO 之英文公告作為引證之 35USC§102(e)(1)之先前技術日期：2001 年 3 月 1 日。

　　USPTO 之核准前公告作為引證之 35USC§102(e)(1)之先前技術日期：2001 年 3 月 1 日。

　　美國專利作為引證 35USC§102(e)(1)之先前技術日期：2001 年 3 月 1 日。

實例 8

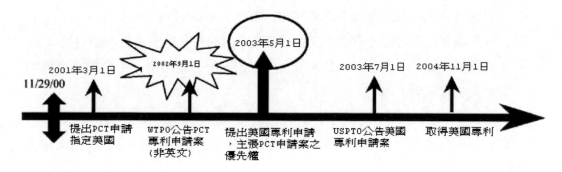

　　WIPO 之 PCT 專利申請案之公告作為引證之 35USC§102(e)(1)之先前技術日期：無。

　　美國專利局之公告作為引證之 35USC§102(e)(1)之先前技術日期：2003 年 5 月 1 日。

　　美國專利作為引證 35USC§102(e)(1)之先前技術日期：2003 年 5 月 1 日。

實例 9

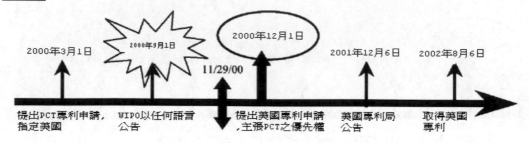

　　WIPO 之公告作為引證之 35USC§102(e)(1) 之先前技術日期：無。

　　美國專利局之公告作為引證之 35USC§102(e)(1) 之先前技術日期：2000 年 12 月 1 日。

　　美國專利作為引證 35USC§102(e)(1) 之先前技術日期：2000 年 12 月 1 日。

§12.16 35USC102(e)之暫時核駁

　　如果審查委員發現有一美國專利申請案其符合下列所有條件：

①　與一被審查之美國專利申請案共同審查中（co-pending）；

②　該美國專利申請案之申請日較該被審查之美國專利申請案之申請日為早；

③　揭露了一標的（內容）可用來預見（anticipate）該被審查中之美國專利申請案之請求項；

④　與被審查之美國專利申請案有不同之發明人實體（different inventive entity）；

　　則審查委員可能引證該美國專利申請案作為 35USC102(e) 之先前技術，暫時核駁被審查中之申請案之請求項。

MPEP2136.03 Critical Reference Date [R-6]

　　706.02(f)(2) Provisional Rejections Under 35 U.S.C. 102(e); Reference Is a Copending U.S. Patent Application [R-3]

　　If an earlier filed, copending, and unpublished U.S. patent application discloses subject matter which would anticipate the claims in a later filed pending U.S. application which has a different inventive entity, thc examiner should determine whether a provisional 35 U.S.C. 102(e) rejection of the later filed application can be made.

＊＊＊＊＊＊＊＊＊＊＊

I. COPENDING U.S. APPLICATIONS HAVING AT LEAST ONE COMMON INVENTOR OR ARE COMMONLY ASSIGNED

If (1) at least one common inventor exists between the applications or the applications are commonly assigned and (2) the effective filing dates are different, then a provisional rejection of the later filed application should be made. The provisional rejection is appropriate in circumstances where if the earlier filed application is published or becomes a patent it would constitute actual prior art under 35 U.S.C. 102. Since the earlier-filed application is not published at the time of the rejection, the rejection must be provisionally made under 35 U.S.C. 102(e).

A provisional rejection under 35 U.S.C. 102(e) can be overcome in the same manner that a 35 U.S.C. 102(e) rejection can be overcome. See MPEP § 706.02(b). The provisional rejection can also be overcome by abandoning the applications and filing a new application containing the subject matter of both.

＊＊＊＊＊＊＊＊＊＊＊

§12.17 克服 35USC102(e)之核駁

當審查委員用來核駁一請求項之 35USC102(e)之先前技術之美國專利，公告之美國專利申請案或公告之 PCT 申請案並不是法定上禁止事由（statutory bar，例如 102(b)，102(d)之先前技術）時，申請人可以呈送 37CFR 1.131 之宣誓書證明其發明日早於（antedate）該先前技術之日期以克服此核駁。或者亦可以 37CFR 1.132 宣誓書證明該先前技術之相關揭露事實上為申請人自己之作品（發明）以克服此核駁。或者亦可宣誓先前技術之相關揭露，事實上是敘述申請人自己先前之工作（例如，證明先前技術之發明人與申請人有工作上之關係而自申請人處知道申請人之發明）而克服此核駁。

MPEP 2136.05 Overcoming a Rejection Under 35 U.S.C. 102(e) [R-1]

A 35 U.S.C. 102(e) REJECTION CAN BE OVERCOME BY ANTEDATING THE FILING DATE OR SHOWING THAT DISCLOSURE RELIED ON IS APPLICANT"S OWN WORK

When a prior U.S. patent, ** U.S. patent application publication>,< or international application publication* is not a statutory bar, a 35 U.S.C. 102(e) rejection can be overcome by antedating the filing date (see MPEP § 2136.03 regarding critical reference date of 35 U.S.C. 102(e) prior art) of the reference by submitting an affidavit or declaration under 37 CFR 1.131 or by submitting an affidavit or declaration under 37 CFR 1.132 establishing that the relevant disclosure is applicant's own work.

＊＊＊＊＊＊＊＊＊＊＊

A 35 U.S.C. 102(e) REJECTION CAN BE OVERCOME BY SHOWING THE REFERENCE IS DESCRIBING APPLICANT'S OWN WORK

＊＊＊＊＊＊＊＊＊＊＊

Therefore, when the unclaimed subject matter of a reference is applicant's own invention, applicant may overcome a prima facie case based on the patent, ** U.S. patent application publication>,< or international application publication, by showing that the disclosure is a description of applicant's own previous work. Such a showing can be made by proving that the patentee, or ** the inventor(s) of the U.S. patent application publication or the international application publication, was associated with applicant (e.g. worked for the same company) and learned of applicant's invention from applicant.

＊＊＊＊＊＊＊＊＊＊＊

實例

　　發明人 A 在 2007 年 1 月 4 日在台灣構思了一製造半導體裝置的方法，並於 2007 年 10 月 1 日實施了此發明，並於 2008 年 4 月 1 日於美國提出專利申請。美國專利申請案包括了二項請求項，Claim 1 及 Claim 2。在審查過程中，審查委員引證了一發明人 R1 之英國專利，（申請日 2005 年 3 月 1 日，發證日 2007 年 3 月 22 日）以 35USC102(b)核駁 Claim 1，又引證一發明人 R2 的美國專利（申請日 2007 年 2 月 1 日，發證日 2009 年 1 月 4 日）以 35U2C102(e)核駁 Claim 2。請問發明人 A 需如何克服此 102(b)及 102(e)之核駁。

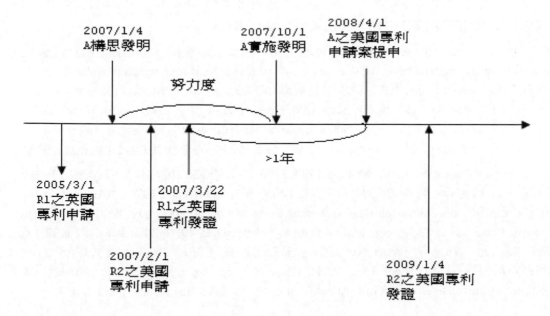

　　發明人 A 應刪除 Claim 1，因為 R1 之英國專利在發明人 A 之申請日的一年之前發證，故為 statutory bar。發明人可提出 37CFR 1.131 之宣誓書，指出其構思日為 2007 年 1 月 4 日，在引證 R2 之申請日之前，而且其自構思日至實施日（2007 年 10 月 1 日）之間有努力度（diligence）故發明日為 2007 年 1 月 4 日，早於引證 R2 之美國專利申請案之申請日。

§12.18 發明衍生自他人

　　如果一專利申請人之發明乃衍生自他人，則非專利申請人自己發明了要申請專利之標的。所謂衍生乃指構思完全來自他人，而藉由任何方式（方法）傳達至專利申請人。而傳達構思乃指傳達足夠使此技藝人士建造及成功地操作該發明。

MPEP 2137

Where it can be shown that an applicant "derived" an invention from another, a rejection under 35 U.S.C. 102(f) is proper.

＊＊＊＊＊＊＊＊＊＊＊

DERIVATION REQUIRES COMPLETE CONCEPTION BY ANOTHER AND COMMUNICATION TO THE ALLEGED DERIVER

"The mere fact that a claim recites the use of various components, each of which can be argumentatively assumed to be old, does not provide a proper basis for a rejection under 35 U.S.C. 102(f)." Ex parte Billottet, 192 USPQ 413, 415 (Bd. App. 1976). Derivation requires complete conception by another and communication of that conception by any means to the party charged with derivation prior to any date on which it can be shown that the one charged with derivation possessed knowledge of the invention. Kilbey v. Thiele, 199 USPQ 290, 294 (Bd. Pat. Inter. 1978).

See also Price v. Symsek, 988 F.2d 1187, 1190, 26 USPQ2d 1031, 1033 (Fed. Cir. 1993); Hedgewick v. Akers, 497 F.2d 905, 908, 182 USPQ 167, 169 (CCPA 1974). "Communication of a complete conception must be sufficient to enable one of ordinary skill in the art to construct and successfully operate the invention." Hedgewick, 497 F.2d at 908, 182 USPQ at 169. See also Gambro Lundia AB v. Baxter Healthcare Corp., 110 F.3d 1573, 1577, 42 USPQ2d 1378, 1383 (Fed. Cir. 1997) (Issue in proving derivation is "whether the communication enabled one of ordinary skill in the art to make the patented invention.").

實例

發明人 A 於 2005 年 12 月 15 日發明了一裝置 X，但其試作之結果卻不能作動(not operative)。發明人 A 於是於 2006 年 1 月 20 日將其發明內容及試驗結果揭露給 B，B 於是針對裝置 X 作研發並加上一新的特徵元件，而使得裝置 X 可作動。B 於是於 2006 年 4 月 10 日將改良之裝置 X 揭示給發明人 A，發明人 A 於 2006 年 5 月 20 日以自己為發明人提出一美國發明專利請求該經改良之裝置 X。請問，發明人 A 之美國專利申請案是否違反 35 USC§102(f)之規定？

發明人 A 之美國專利申請案違反 35 USC §102(f)之規定，所謂發明人乃是對所請求之發明（claimed invention）之構思上有貢獻之人，如果發明之標的乃衍生自他人（發明之構思來自他人並傳達至專利申請人）則此專利申請人並非自己發明了申請專利之標的。

§12.19 發明優先權

依照 35 USC §102(g)(2)，如果有一發明人在發明完成之後，隱藏（不發表）其發明，亦即並未積極提出發明專利之申請，如果有其他之發明人有同樣的發明，且積極準備專利申請且較該先發明人先提出後發明之專利申請，則該後發明人具有發明優先權，亦即應取得該發明之專利權。

實例

　　發明人 A 於 1995 年 10 月於美國發明了一新穎的馬鈴薯切割器，其可將馬鈴薯切割成斷面而為星形之馬鈴薯片，此種形狀之馬鈴薯片當油炸時可炸的更均勻。發明人 A 認為其發明相當有價值，就將該馬鈴薯切割器之設計圖交由其二位朋友簽名作證其為發明人並要求其二位朋友保密，隨後將設計圖鎖於保險箱之中達 12 年之久，而在此 12 年之中發明人 A 及其二位朋友均未將此發明揭露給任何人。在 2007 年 12 月另一發明人 B 設計出一相同之馬鈴薯切割器可將馬鈴薯切割成斷面為星形之馬鈴薯片。發明人 B 於是在 2008 年 6 月 1 日提出美國專利申請。發明人 A 見到發明人 B 之產品相當成功亦於 2008 年 6 月 1 日同一天提出專利申請。美國專利局之審查委員並未檢索到任何相關之先前技術，請問美國專利局應該給發明人 A 或發明人 B 專利？

　　發明人 A 雖較發明人 B 先發明，但是發明人 A 隱藏其發明及自發明日起 12 年均未提出專利申請，故發明人 B 應獲得專利。

---

35 USC §102(g)(2)

A person shall be entitled to a patent unless –

(g)(1) during the course of an interference conducted under section　135 or section 291, another inventor involved therein establishes, to the extent permitted in section 104, that before such person's invention thereof the invention was made by such other inventor and not abandoned, suppressed, or concealed, or (2) before such person's invention thereof, the invention was made in this country by another inventor who had not abandoned, suppressed, or concealed it. In determining priority of invention under this subsection, there shall be considered not only the respective dates of conception and reduction to practice of the invention, but also the reasonable　diligence of one who was first to conceive and last to reduce of practice, from a time prior to conception by the other.

§12.20 專利執業者準備專利說明書及提出專利申請之努力度

專利的申請被認為是推斷的付諸實施，在發明人完成發明之構思後將其發明內容交由專利執業者（專利律師，專利代理人，專利工程師）準備專利說明書及提出申請，如果專利執業者延遲此專利說明書撰寫（準備）及提出申請，則亦會被認為未有合理之努力度。但是，專利代理人在該期間內努力撰寫說明書或者專利執業者有其他不相關之案件在處理，但是依先後次序快速處理這些案件，或者專利執業者處理相關之案件而這些案件與專利申請案是有關的，都可被視為專利執業者具有合理之努力度。

MPEP 2138.06

DILIGENCE REQUIRED IN PREPARING AND FILING PATENT APPLICATION

The diligence of attorney in preparing and filing patent application inures to the benefit of the inventor. Conception was established at least as early as the date a draft of a patent application was finished by a patent attorney on behalf of the inventor. Conception is less a matter of signature than it is one of disclosure. Attorney does not prepare a patent application on behalf of particular named persons, but on behalf of the true inventive entity. Six days to execute and file application is acceptable. Haskell v. Coleburne, 671 F.2d 1362, 213 USPQ 192, 195 (CCPA 1982). See also Bey v. Kollonitsch, 866 F.2d 1024, 231 USPQ 967 (Fed. Cir. 1986) (Reasonable diligence is all that is required of the attorney. Reasonable diligence is established if attorney worked reasonably hard on the application during the continuous critical period. If the attorney has a reasonable backlog of unrelated cases which he takes up in chronological order and carries out expeditiously, that is sufficient. Work on a related case(s) that contributed substantially to the ultimate preparation of an application can be credited as diligence.).

第 13 章　非顯而易見性

目　錄

§13.1　非顯而易見性之規定

　　美國專利法有關非顯而易見性之規定乃訂立在 35 USC §103，其包括 103(a), 103(b)及 103(c)三部份。

　　103(a)：雖然發明並非如§102 所載被完全地揭露或敘述在引證之先前技術中，但是如果該發明之請求標的與先前技術之差異整體觀之，在發明之前對於熟習此技藝人士是顯而易見的話，則仍不能取得專利；專利性不應被發明之形式所否定。

　　103(b)：如果所請求之生物技術方法乃是使用一具有新穎性且非顯而易見之物質（組合物）之方法或是製造出一具有新穎且非顯而易見之物質（組合物）之製法，則雖構成該方法（製法）之步驟是習知的；但此生物技術方法（製法）將被認為是非顯而易見的，但是需符合下列條件：

a) 該生物技術方法及物質（組合物）之專利之請求項需包含在同一申請案中或者是具有相同之有效申請日之不同申請案中；

b) 此組合物及生物技術方法，在其發明完成時，乃由同一人所擁有或有義務讓渡給同一人；

c) 如果該生物技術方法及物質之請求項乃包括在不同之申請案中，此二專利之專利權應於同一天結束。

該所謂生物技術方法包括下列三種：

a) 一基因工程方法，用以改變或引導一單一或多細胞生物，以表現一外源性核苷酸序列，以抑制、除去、增大或改變一內源性核苷酸序列之表現；或表現一與該生物非自然相關聯之生理特性；

b) 一種細胞融合的方法，其用以產生一能夠表現特定蛋白質如單株抗體之細胞株；以及

c) 一種使用上述(a)及(b)式其組合之方法所製成之產品之方法。

103(c):

(1) 一由他發明人所發明之專利標的，其為§102(e), (f), (g)條款所定義之先前技術者，如果與一申請案所請求之發明在其發明完成時乃屬於同一人所有或有義務讓渡給同一人時不應排除其依本款之專利性。

(2) (A)如果所請求之發明乃由或為了在所請求之發明完成之前所簽署聯合開發合約者所完成者；(B)請求之發明是基於該聯合開發合約下所進行發明活動的結果；及(C)所請求之發明之專利申請案揭露了或修正後揭露了聯合開發合約參與者之名字時；則由其他人所發展之標的及所請求之標的應為視為第(1)項之屬於同一人所有或有義務讓渡給同一人。

(3) 第(2)項所謂之聯合開發合約乃指二個或以上的人或團體為了在所請求之發明的領域上所作的實驗、發展或研究工作上的合約、授權合約或合作協約而言。

35 USC §103:

35 U.S.C. 103Conditions for patentability; non-obvious subject matter.

(a)A patent may not be obtained though the invention is not identically disclosed or described as set forth in section 102 of this title, if the differences between the subject matter sought to be patented and the prior art are such that the subject matter as a whole would have been obvious at the time the invention was made to a person having ordinary skill in the art to which said subject matter pertains. Patentability shall not be negatived by the manner in which the invention was made.

(b)(1) Notwithstanding subsection (a), and upon timely election by the applicant for patent to proceed under this subsection, a biotechnological process using or resulting in a composition of matter that is novel under section 102 and nonobvious under subsection (a) of this section shall be considered nonobvious if-

(A)claims to the process and the composition of matter are contained in either the same application for patent or in separate applications having the same effective filing date; and

(B)the composition of matter, and the process at the time it was invented, were owned by the same person or subject to an obligation of assignment to the same person.

(2)A patent issued on a process under paragraph (1)-

(A)shall also contain the claims to the composition of matter used in or made by that process, or

(B)shall, if such composition of matter is claimed in another patent, be set to expire on the same date as such other patent, notwithstanding section 154.

(3)For purposes of paragraph (1), the term "biotechnological process" means-

(A)a process of genetically altering or otherwise inducing a single- or multi-celled organism to-

(i)express an exogenous nucleotide sequence,

(ii)inhibit, eliminate, augment, or alter expression of an endogenous nucleotide sequence, or

(iii)express a specific physiological characteristic not naturally associated with said organism;

(B)cell fusion procedures yielding a cell line that expresses a specific protein, such as a monoclonal antibody; and

(C)a method of using a product produced by a process defined by subparagraph (A) or (B), or a combination of subparagraphs (A) and (B).

(c)(1) Subject matter developed by another person, which qualifies as prior art only under one or more of subsections (e), (f), and (g) of section 102 of this title, shall not preclude patentability under this section where the subject matter and the claimed invention were, at the time the claimed invention was made, owned by the same person or subject to an obligation of assignment to the same person.

(2)For purposes of this subsection, subject matter developed by another person and a claimed invention shall be deemed to have been owned by the same person or subject to an obligation of assignment to the same person if —

(A)the claimed invention was made by or on behalf of parties to a joint research agreement that was in effect on or before the date the claimed invention was made;

(B)the claimed invention was made as a result of activities undertaken within the scope of the joint research agreement; and

(C)the application for patent for the claimed invention discloses or is amended to disclose the names of the parties to the joint research agreement.

(3)For purposes of paragraph (2), the term "joint research agreement" means a written contract, grant, or cooperative agreement entered into by two or more persons or entities for the performance of experimental, developmental, or research work in the field of the claimed invention.

§13.2　解讀請求項

審查委員在審查一申請案之過程中乃依據請求項所賦予之最廣且合理的解釋去解讀請求項用語以檢索先前技術。已核准之專利其請求項之範圍乃是基於說明書、審查過程檔案、先前技術以及其他請求項去解讀去定義，但是審查委員在審查一專利申請案時卻是以請求項用語可允許的最廣的解釋去解讀請求項並據此檢索先前技術。亦即，一審查委員會以一般之意義（plain meaning）去解釋請求項用語，除非在說明書另有定義。

所謂一般之意義乃指熟習此技藝人士所認知之意義。例如請求項中敘及一銅基質（copper substrate），則審查委員會將其解釋為元素銅及銅合金，但不會包括銅化合物。但是若請求項中一元件以裝置加功能用語形式撰寫，則審查委員會將其解讀成涵蓋"所有揭露在說明書中用來執行所述之功能（function）之元件及其均等元件以作前案檢索。

MPEP 2111.01 Plain Meaning

While the ** claims of issued patents are interpreted in light of the specification, prosecution history, prior art and other claims, this is not the mode of claim interpretation to be applied during examination. During examination, the claims must be interpreted as broadly as their terms reasonably allow. This means that the words of the claim must be given their plain meaning unless applicant has provided a clear definition in the specification. There is one exception, and that is when an element is claimed using language falling under the scope of 35 U.S.C. 112, 6th paragraph (often broadly referred to as means or step plus function language). In that case, the specification must be consulted to determine the structure, material, or acts corresponding to the function recited in the claim. In re Donaldson, 16 F.3d 1189, 29 USPQ2d 1845 (Fed. Cir. 1994) (see MPEP §2181-§2186).

When not defined by applicant in the specification, the words of a claim must be given their plain meaning. In other words, they must be read as they would be interpreted by those of ordinary skill in the art.

＊＊＊＊＊＊＊＊＊＊＊

§13.3 以 35USC103 核駁之先前技術

所有可作為 35 USC §102 之先前技術均可作為 35 USC §103 之先前技術，另外申請人自己承認的先前技術亦可作為 35 USC §103 之先前技術。申請人自己承認的先前技術包括在說明書中標示為 prior art 者，在專利審查過程中所有申請人承認為先前技術者。另外，請求項若為吉普森形式，則此吉普森請求項之前言部份亦可被作為審查委員之先前技術。

MPEP 2141.01 Scope and Content of the Prior Art [R-6]

* * * * * * * * * * * *

Subject matter that is prior art under 35 U.S.C. 102 can be used to support a rejection under section 103.

A 35 U.S.C. 103 rejection is based on 35 U.S.C. 102(a), 102(b), 102(e), etc. depending on the type of prior art reference used and its publication or issue date. For instance, an obviousness rejection over a U.S. patent which was issued more than 1 year before the filing date of the application is said to be a statutory bar just as if it anticipated the claims under 35 U.S.C. 102(b). Analogously, an obviousness rejection based on a publication which would be applied under 102(a) if it anticipated the claims can be overcome by swearing behind the publication date of the reference by filing an affidavit or declaration under 37 CFR 1.131.

* * * * * * * * * * * *

實例

發明人 A 之專利申請案在審查過程中引證了一 X 之專利作為先前技術。發明人 A 之代理人在答辯時承認與其所有的請求項最相關的先前技術就是審查委員所引證之 X 之專利並修改請求項之後獲得專利。在獲得專利之後一年；未滿二年時發明人 A 發現其在該 X 之專利之申請日之前已構思且付諸實施其自己的發明並有相當之證據，故欲申請再發證（reissue）來擴大其專利之範圍，請問發明人 A 之再發證申請案是否可克服引證的 X 專利之先前技術？

無法克服 X 專利之先前技術，因為發明人在審查過程中已明白地承認 X 專利為先前技術。

MPEP 2129

ADMISSIONS BY APPLICANT CONSTITUTE PRIOR ART

When applicant states that something is prior art, if is taken as being available as prior art against the claims. Admitted prior art can be used in obviousness rejection.　In re Nomiya,

509 F.2d 566, 184 USPQ 607, *>611< (CCPA 1975) (Figures in the application labeled "prior art" held to be an admission that what was pictured was prior art relative to applicant's invention.).

A JEPSON CLAIM RESULTS IN AN IMPLIED ADMISSION THAT PREAMBLE IS PRIOR ART

The preamble elements in a Jepson-type claim (i.e., a claim of the type discussed in 37 CFR 1.75(e); see MPEP §608.01(m)) "are impliedly admitted to be old in the art,... but it is only an implied admission."

＊＊＊＊＊＊＊＊＊＊＊

§13.4　判定為 35USC§103 之顯而易見之基準

美國專利局之 MPEP2141 記載了審查委員判定為 35USC§103 之顯而易見之基準（guidelines）。此審查基準乃是美國專利局依據目前對於專利法了解且相信與最高法院之 KSR v. Teleflex（KSR）判例結果完全一致，此基準說明如下：

Ⅰ. KSR 判例之決定及顯而易見法律之原則

在最高法院之 KSR 判例中，最高法院確認了 Graham V. John Deere Co 之判例中判定顯而易見之架構步驟仍適用，但認為在適用 TSM 測試（Teaching-Suggestion-Motivation）決定是否為顯而易見時太無彈性（overly rigid）及形式化（formalistic way），詳而言之，最高法院認為聯邦巡迴法庭及美國專利局之審查委員在判定是否為顯而易見時有下列 4 種錯誤：

(1) 只針對專利申請人（專利權人）想要解決之問題作審查；
(2) 想要解決問題之此技藝人士將只被引導至設計來解決相同問題之先前技術之那些元件；
(3) 不能僅僅憑顯示元件之組合是 "顯然會嘗試"判定一請求項是顯而易見；
(4) 過分強調 "後見之明"之偏見之風險，因而造成判定顯而易見時太無彈性。

II. Graham v. John Deere Co.之基本事實判定

審查委員需依據下列順序對事實作判定：

(A) 先前技術所揭露之內容及範圍；

(B) 決定專利申請案所請求之發明標的與先前技術之差異；

(C) 熟習此技藝人士在此相關領域之技術水平；

另外，當需評估第二層考量（Secondary consideration）之客觀證據，其包括：

a) 商業上之成功　　　　　　b) 長期但未解決之需要

c) 其他人之失敗　　　　　　d) 發現先前技術之問題所在

e) 競爭者之仿冒　　　　　　f) 授權給他人使用

g) 不可預期之效果　　　　　h) 業界之質疑可解決先前技術之問題

III. 支持 35USC§103（顯而易見）核駁之理論基礎

　　一旦上述 Graham 事實判定完成，審查委員就需對於所請求之發明是否為顯而易見作決定。以下為審查委員在作決定時之基準。

　　首先，先前技術並不限定於引用之參考文獻，亦應包括此技藝人士所了解的先前技術。先前技術並不一定要教示或建議了所有請求項之限制條件，但是審查委員需解釋為何先前技術與所請求之發明之差異，對此技藝人士是顯而易見的。亦即，先前技術與所請求之發明存在有差異並不能確立發明是非顯而易見的，因為，先前技術與請求之發明之間隙可能並非大到可使此技藝人士認為是非顯而易見的。在判定非顯而易見時，特定的動機去製造所請求之發明及發明人正想要解決之問題均不是決定的因素。適當的分析為"在考量所有的事實後，所請求之發明是否對此技藝人士是顯而易見的"。可能有引證之先前技術之揭露以外之因素可讓此技藝人士將該間隙連接起來。

　　如果審查委員之先前技術的檢索及 Graham 之事實的判定顯示可使用 TSM 測試作為理論基礎以決定為顯而易見時，則可作出顯而易見之決定，雖然最高法院警告在應用 TSM 測試時不能太沒彈性。亦即，TSM 測試對於判定是否為顯而易見是有幫助的，但不是唯一的測試方法。

　　下列七種情況是美國專利局列舉可用來支持顯而易見核駁之理論基礎：

(A) 依照習知之方法結合先前技術之元件以產生可預測之結果;

(B) 單純用另一元件取代一習知之元件以得到可預測之結果;

(C) 使用習知之技術以相同之方式改良類似之裝置,方法或產品;

(D) 將一習知之技術應用至一準備作改良之習知裝置,方法或產品以產生可預期之結果;

(E) "顯然會嘗試",從有限數目之特定的可預測的解決方案中作選擇,且具有合理之成功機會;

(F) 在一技術領域之習知工作,可促使其改變以因應設計之需要或市場因素用在相同領域或不同領域,如果此改變對於此技藝人士是可預測的;

(G) 在先前技術之某些教示,建議或動機,其可引導此技藝人士改良先前技術或結合先前技術之教示以達到所請求之發明者。

Ⅳ. 申請人之答辯

一旦審查委員完成了 Graham 之事實判定,並下結論所請求的發明是顯而易見的話,舉證責任就轉換到申請人,申請人必需(A)顯示美國專利局在事實判定上之錯誤,或(B)提供證據顯示所請求的標的為非顯而易見。依照 37CFR 1.111(b),申請人必需清楚且特定地指出審定書之錯誤,且針對每一核駁理由作答辯。答辯必需指出可使請求項具有專利性之與先前技術之區別。通常只陳述(爭論)審定書並未確立是推斷的顯而易見或審查委員據以核駁之一般知識並無書面證據等理由並不是有效的答辯。

Ⅴ. 考慮申請人之答辯證據

美國專利局之審查委員在作顯而易見之判定時應考慮申請人提出之答辯證據。答辯證據可包括第二層考量之證據,例如:商業上之成功,長期未解決之需要,其他人失敗,及不可預期之結果等。例如,當一請求項為一組合時,申請人可提供證據或答辯顯示:

(A) 此技藝人士無法藉由習知之方法將所請求之元件組合起來(技術上有困難);

(B) 組合中之元件並非單純地只執行各元件分別執行之功能;或

(C) 其完成所請求之組合,此組合是不可預期的。

MPEP 2129

2141 >Examination Guidelines for Determining Obviousness Under< 35 U.S.C. 103** [R-6]

I.The KSR Decision and Principles of the Law of Obviousness

The Supreme Court in KSR reaffirmed the familiar framework for determining obviousness as set forth in Graham v. John Deere Co. (383 U.S. 1, 148 USPQ 459 (1966)), but stated that the Federal Circuit had erred by applying the teaching-suggestion-motivation (TSM) test in an overly rigid and formalistic way. KSR, 550 U.S. at ___, 82 USPQ2d at 1391. Specifically, the Supreme Court stated that the Federal Circuit had erred in four ways: (1) "by holding that courts and patent examiners should look only to the problem the patentee was trying to solve " (Id. at ___, 82 USPQ2d at 1397); (2) by assuming "that a person of ordinary skill attempting to solve a problem will be led only to those elements of prior art designed to solve the same problem" (Id.); (3) by concluding "that a patent claim cannot be proved obvious merely by showing that the combination of elements was 'obvious to try'" (Id.); and (4) by overemphasizing "the risk of courts and patent examiners falling prey to hindsight bias" and as a result applying "[r]igid preventative rules that deny factfinders recourse to common sense" (Id.).

＊＊＊＊＊＊＊＊＊＊＊

II.The Basic Factual Inquiries of Graham v. John Deere Co.

＊＊＊＊＊＊＊＊＊＊＊

A.Determining the Scope and Content of the Prior Art

＊＊＊＊＊＊＊＊＊＊＊

B.Ascertaining the Differences Between the Claimed Invention and the Prior Art

＊＊＊＊＊＊＊＊＊＊＊

C.Resolving the Level of Ordinary Skill in the Art

＊＊＊＊＊＊＊＊＊＊＊

III.RATIONALES TO SUPPORT REJECTIONS UNDER 35 U.S.C.103

Once the Graham factual inquiries are resolved, Office personnel must determine whether the claimed invention would have been obvious to one of ordinary skill in the art.

* * * * * * * * * *

Prior art is not limited just to the references being applied, but includes the understanding of one of ordinary skill in the art. The prior art reference (or references when combined) need not teach or suggest all the claim limitations, however, Office personnel must explain why the difference(s) between the prior art and the claimed invention would have been obvious to one of ordinary skill in the art. The "mere existence of differences between the prior art and an invention does not establish the invention's nonobviousness." Dann v. Johnston, 425 U.S. 219, 230, 189 USPQ 257, 261 (1976). The gap between the prior art and the claimed invention may not be "so great as to render the [claim] nonobvious to one reasonably skilled in the art."Id. In determining obviousness, neither the particular motivation to make the claimed invention nor the problem the inventor is solving controls. The proper analysis is whether the claimed invention would have been obvious to one of ordinary skill in the art after consideration of all the facts. See 35 U.S.C. 103(a). Factors other than the disclosures of the cited prior art may provide a basis for concluding that it would have been obvious to one of ordinary skill in the art to bridge the gap. The rationales discussed below outline reasoning that may be applied to find obviousness in such cases.

If the search of the prior art and the resolution of the Graham factual inquiries reveal that an obviousness rejection may be made using the familiar teaching-suggestion-motivation (TSM) rationale, then such a rejection should be made

* * * * * * * * * *

Exemplary rationales that may support a conclusion of obviousness include:

(A) Combining prior art elements according to known methods to yield predictable results;

(B) Simple substitution of one known element for another to obtain predictable results;

(C) Use of known technique to improve similar devices (methods, or products) in the same way;

(D) Applying a known technique to a known device (method, or product) ready for improvement to yield predictable results;

(E) "Obvious to try" - choosing from a finite number of identified, predictable solutions, with a reasonable expectation of success;

(F) Known work in one field of endeavor may prompt variations of it for use in either the same field or a different one based on design incentives or other market forces if the variations are predictable to one of ordinary skill in the art;

(G) Some teaching, suggestion, or motivation in the prior art that would have led one of ordinary skill to modify the prior art reference or to combine prior art reference teachings to arrive at the claimed invention.

＊＊＊＊＊＊＊＊＊＊＊

IV.APPLICANT's REPLY

Once Office personnel have established the Graham factual findings and concluded that the claimed invention would have been obvious, the burden then shifts to the applicant to (A) show that the Office erred in these findings or (B) provide other evidence to show that the claimed subject matter would have been nonobvious. 37 CFR 1.111(b) requires applicant to distinctly and specifically point out the supposed errors in the Office's action and reply to every ground of objection and rejection in the Office action. The reply must present arguments pointing out the specific distinction believed to render the claims patentable over any applied references.

If an applicant disagrees with any factual findings by the Office, an effective traverse of a rejection based wholly or partially on such findings must include a reasoned statement explaining why the applicant believes the Office has erred substantively as to the factual findings. A mere statement or argument that the Office has not established a prima facie case of obviousness or that the Office's reliance on common knowledge is unsupported by documentary evidence will not be considered substantively adequate to rebut the rejection or an effective traverse of the rejection under 37 CFR 1.111(b).

＊＊＊＊＊＊＊＊＊＊＊

V.CONSIDERATION OF APPLICANT'S REBUTTAL EVIDENCE

Office personnel should consider all rebuttal evidence that is timely presented by the applicants when reevaluating any obviousness determination. Rebuttal evidence may include evidence of "secondary considerations," such as "commercial success, long felt but unsolved needs, [and] failure of others"(Graham v. John Deere Co., 383 U.S. at 17, 148 USPQ at 467), and may also include evidence of unexpected results. As set forth above, Office personnel must articulate findings of fact that support the rationale relied upon in an obviousness rejection. As a result, applicants are likely to submit evidence to rebut the fact finding made by Office personnel. For example, in the case of a claim to a combination, applicants may submit evidence or argument to demonstrate that:

(A) one of ordinary skill in the art could not have combined the claimed elements by known methods (e.g., due to technological difficulties);

(B) the elements in combination do not merely perform the function that each element performs separately; or

(C) he results of the claimed combination were unexpected.

＊＊＊＊＊＊＊＊＊＊＊

§13.5　推斷的顯而易見核駁

美國專利局之審查委員要作推斷的顯而易見（Prima Facie obviousness）核駁需負最先的舉證責任。如果審查委員未建立推斷的顯而易見核駁，則申請人並無義務提供非顯而易見之證據。但是，一旦審查委員提出事實之證據，建立了推斷的顯而易見的核駁，則舉證責任轉移至申請人。申請人可提出另外的證據，例如所請求之發明具有不可預期的改良的性質之比較試驗資料。

最高法院在 KSR 之判決中指出支持 35USC§103 之顯而易見核駁之分析必需是很明顯的。顯而易見的核駁不能只有結論式的陳述，相對地，必需有清楚的理由及一些理論基礎以支持顯而易見之結論。

如果審查委員發現有事實上之支持可以 35USC§103 顯而易見性核駁所請求之發明，審查委員需再考慮申請人提供之所有支持專利性之證據。最終專利性之決定必需依據整個記錄，有占多數之證據，且合理考量答辯之說服性及第二層考量之證據。

MPEP 2142 Legal Concept of Prima Facie Obviousness [R-6]

＊＊＊＊＊＊＊＊＊＊＊

The examiner bears the initial burden of factually supporting any prima facie conclusion of obviousness. If the examiner does not produce a prima facie case, the applicant is under no obligation to submit evidence of nonobviousness. If, however, the examiner does produce a prima facie case, the burden of coming forward with evidence or arguments shifts to the applicant who may submit additional evidence of nonobviousness, such as comparative test data showing that the claimed invention possesses improved properties not expected by the prior art. The initial evaluation of prima facie obviousness thus relieves both the examiner and applicant from evaluating evidence beyond the prior art and the evidence in the specification as filed until the art has been shown to *>render obvious< the claimed invention.

＊＊＊＊＊＊＊＊＊＊＊

ESTABLISHING A PRIMA FACIE CASE OF OBVIOUSNESS

**>The key to supporting any rejection under 35 U.S.C. 103 is the clear articulation of the reason(s) why the claimed invention would have been obvious. The Supreme Court in KSR International Co. v. Teleflex Inc., 550 U.S. ___, ___, 82 USPQ2d 1385, 1396 (2007) noted that the analysis supporting a rejection under 35 U.S.C. 103 should be made explicit. The Federal Circuit has stated that "rejections on obviousness cannot be sustained with mere conclusory statements; instead, there must be some articulated reasoning with some rational underpinning to support the legal conclusion of obviousness." In re Kahn, 441 F.3d 977, 988, 78 USPQ2d 1329, 1336 (Fed. Cir. 2006). See also KSR, 550 U.S. at ___ , 82 USPQ2d at 1396 (quoting Federal Circuit statement with approval). <

If the examiner determines there is factual support for rejecting the claimed invention under 35 U.S.C. 103, the examiner must then consider any evidence supporting the patentability of the claimed invention, such as any evidence in the specification or any other evidence submitted by the applicant. The ultimate determination of patentability is based on the entire record, by a preponderance of evidence, with due consideration to the persuasiveness of any arguments and any secondary evidence. In re Oetiker, 977 F.2d 1443, 24 USPQ2d 1443 (Fed. Cir. 1992). The legal standard of "a preponderance of evidence" requires the evidence to be more convincing than the evidence which is offered in opposition to it. With regard to rejections under 35 U.S.C. 103, the examiner must provide evidence which as a whole shows that the legal determination sought to be proved (i.e., the reference teachings establish a prima facie case of obviousness) is more probable than not.

實例 1

Combining Prior Art Elements According to Known Methods To Yield Predictable Results

請求的發明為一舖路機，其將幾種習知之元件組合在單一的底盤之上。習知的舖路機，將分佈及成形瀝青之設備結合在一底盤之中。請求項包括了一輻射熱燃燒器之元件，其結合至舖路機之側邊以防止連續舖路時瀝青冷卻，習知的舖路機亦使用輻射熱來軟化瀝青，但並未使用燃燒器，請求之發明之元件都是習知的，與習知技術不同的是將舊的元件組合在單一底盤之上。

此發明乃為推斷的顯而易見，因為此發明乃將舊的元件（燃燒器）用習知的方法組裝在底盤之上，且未產出任何新的或不同的功能。

實例 2

Simple Substitution of One Known Element for Another To Obtain Predictable Results

所請求之發明是有關於將咖啡或茶去除咖啡因的方法。先前技術 A 之除去咖啡因的方法是將植物之咖啡因先吸收在油中，再用萃取將咖啡因移去。請求之發明與先前技術 A 之方法不同之處在於使用蒸發（蒸餾）將咖啡因從油中去除。但是另一先前技術 B 之方法是將咖啡懸浮在油中，然後直接用蒸餾除去咖啡因。

所請求之發明為推斷的顯而易見，因為其只用一習知步驟來取代先前技術之步驟以得到可預測之結果。

實例 3

Use of Known Technique To Improve Similar Devices (Methods, or Products) in the Same Way

所請求之發明是有關於一裝置可在當從變流器輸出之電流超過某一預設的極限值，一非常短的時間時，將此變流器停止作動以防止變流器損害，先前技術則敘述到一裝置其藉由控制裝置來保護變流器之電路。此先前技術亦提到當變流器在高負載電流狀況下時變流器會停止作動，以保護變流器之電路。亦即，先前技術藉由一切斷開關使變流器不作動以保護變流器。

所請求的發明是推斷的顯而易見的，因為所請求的發明使用習知之技術以相同之方式改良類似之裝置。

實例 4

Applying a Known Technique to a Known Device (Method, or Product) Ready for Improvement To Yield Predictable Results

所請求之發明乃為一用來自動記錄銀行支票或存款之系統，依照此系統，顧客在各銀行支票或存款條上給予一數字分類碼，此系統就會以磁性墨水記錄下此數字分類碼，如此，則銀行就可提供顧客對帳單（statement），而此帳單是依不同之交易分類以小計（subtotal）來呈現。而習知之銀行的資料處理系統，電腦讀取用來確認帳單磁性墨水以便產生交易之記錄。亦即，習知之系統各帳號來追蹤客戶項目並提供交易之記錄，所請求之發明只是再加一分類碼以能將交易細分類而提供交易之細目（breakdown）。

所請求之發明為推斷的顯而易見，因其將習知之技術應用至習知之裝置（一般電腦）以產生可預期之結果。

實例 5

Obvious To Try" - Choosing From a Finite Number of Identified, Predictable Solutions, With a Reasonable Expectation of Success.

所請求之發明乃有關於一種習知藥物 oxybutynin 之緩釋配方。此配方可將 oxybutynin 在 24 小時之期間內以特定之速率釋放出來。oxybutynin 為人習知是一種高水溶性之藥物。

一先前技術教示了高水溶性藥物之稀釋配方，其例子為 morphine 之緩釋配方，而且亦提到 oxybutynin 屬於高水溶性藥物。另一先前技術教示了 oxybutynin 之緩釋配方，但其緩釋速率與所請求之發明不相同。又有一先前技術提到在 24 小時之期間將藥物釋出之一般方法。

所請求之發明為推斷的顯而易見，因為該發明乃從有限的，可預測的解決方法中作選擇，且有合理的成功機會。

實例 6

Known Work in One Field of Endeavor May Prompt Variations of It for Use in Either the Same Field or a Different One Based on Design Incentives or Other Market Forces if the Variations Are Predictable to One of Ordinary Skill in the Art

所請求之發明是有關於使用生物識別（bioauthentication）之消費性電子裝置以讓經授權之使用者經由通訊網路下訂單購買物品或服務。一習知的消費性電子裝置與所請求之發明類似，不同的是使用密碼識別裝置（password authentication device）。另一先前技術揭示了在消費性電子裝置（搖控器）上使用生物識別裝置（指紋感知器）以提供生物識別資訊。另一先前技術亦揭示了用生物識別取代 PIN 識別以讓使用者經由消費電子裝置去取得貸款（credit）。

所請求的發明是推斷的顯而易知，因為改變為生物識別是此技藝人士可預測的。

§13.6　以 TSM 測試作出推斷的顯而易見核駁

當審查委員在發現先前技術中之某些教示，建議或動機，其可引導此技藝人士改良先前技術或結合先前技術以達到所請求之發明（TSM 測試），而要以此作為理論基礎（rational）作出推斷的顯而易見核駁時，審查委員必需先進行 Graham 之事實判定，然後確定下列事項：

(1)　在引證之先前技術本身或此技藝人士一般可得之知識中，有一些教示，建議、動機去改良先前技術或結合先前技術之教示；

(2)　有合理成功之期待；

(3)　可能需要任何其他根據 Graham 事實判定之發現，以解釋顯而易見之判定。

此 TSM 測試是很彈性的，並不一定需要有很明顯的教示去結合先前技術，教示建議動機可能是不明顯的，而可存在於此技藝人士之知識或在要解決之問題之本質中。此外，當二件或以上之先前技術中之教示互相抵觸時，審查委員必需衡量哪一件先前技術占有較重要的建議性。此外，單純先前技術可被結合或改良並不一定使所得之結合是顯而易見的，除非所得到之結合對此技藝人士是可預測的，再者，審查委員亦不能只作出" 將先前技術改良以達到所請求之發明是在此技藝人士之能力之內，因為引證之先前技術教示所請求之發明的所有特徵"之陳述，就作出推斷的顯而易見之判定。而必需指出一些客觀的理由支持此技藝人士會結合先前技術之教示。

另外，如果提議之改良會使得先前技術的發明被改良後不能滿足所希望的目的的話，則並無教示或動機或使此技藝人士作該提議的改良。亦即，不能根據此提議的改良來判定是推斷的顯而易見。同樣的，如果先前技術之改良或結合會改變被結合（改良）之先前技術之發明之動作原理的話，先前技術之教示不足夠作推斷的顯而易見的判定。

在進行 TSM 測試時，還需該結合或改良先前技術有合理的成功期待才行。但此成功的預測並不必是絕對的，只要至少有某些程度的成功預測即可。但是申請人如果提出無合理成功之期待的證據，就可能支持其非顯而易見的結論。上述結合或改良成功之預測性之決定的時點是以所請求的發明完成時之先前技術水平作判定。

最後，在作推斷的顯而易見判定時，所有請求項之限制條件均必需考慮，即使限制條件是界定不清楚的或是在說明書中未有支持的，均需考慮。

MPEP 2143 >Examples of< Basic Requirements of a Prima Facie Case of Obviousness

* * * * * * * * * * * *

(G) Some Teaching, Suggestion, or Motivation in the Prior Art That Would Have Led One of Ordinary Skill To Modify the Prior Art Reference or To Combine Prior Art Reference Teachings To Arrive at the Claimed Invention

To reject a claim based on this rationale, Office personnel must resolve the Graham factual inquiries. Then, Office personnel must articulate the following:

(1) a finding that there was some teaching, suggestion, or motivation, either in the references themselves or in the knowledge generally available to one of ordinary skill in the art, to modify the reference or to combine reference teachings;

(2) a finding that there was reasonable expectation of success; and

(3) whatever additional findings based on the Graham factual inquiries may be necessary, in view of the facts of the case under consideration, to explain a conclusion of obviousness.

The Courts have made clear that the teaching, suggestion, or motivation test is flexible and an explicit suggestion to combine the prior art is not necessary. The motivation to combine may be implicit and may be found in the knowledge of one of ordinary skill in the art, or, in some cases, from the nature of the problem to be solved.

* * * * * * * * * * *

MPEP 2143.01 Suggestion or Motivation To Modify the References [R-6]

I. *PRIOR ART **>SUGGESTION OF< THE DESIRABILITY OF THE CLAIMED INVENTION

* * * * * * * * * * *

Obviousness can * be established by combining or modifying the teachings of the prior art to produce the claimed invention where there is some teaching, suggestion, or motivation to do so.

* * * * * * * * * * *

II. WHERE THE TEACHINGS OF THE PRIOR ART CONFLICT, THE EXAMINER MUST WEIGH THE SUGGESTIVE POWER OF EACH REFERENCE

The test for obviousness is what the combined teachings of the references would have suggested to one of ordinary skill in the art, and all teachings in the prior art must be considered to the extent that they are in analogous arts. Where the teachings of two or more prior art references conflict, the examiner must weigh the power of each reference to suggest solutions to one of ordinary skill in the art, considering the degree to which one reference might accurately discredit another.

＊＊＊＊＊＊＊＊＊＊＊

III. FACT THAT REFERENCES CAN BE COMBINED OR MODIFIED **>MAY NOT BE< SUFFICIENT TO ESTABLISH PRIMA FACIE OBVIOUSNESS

The mere fact that references can be combined or modified does not render the resultant combination obvious unless **>the results would have been predictable to one of ordinary skill in the art.

＊＊＊＊＊＊＊＊＊＊＊

IV. *>MERE STATEMENT< THAT THE CLAIMED INVENTION IS WITHIN THE CAPABILITIES OF ONE OF ORDINARY SKILL IN THE ART IS NOT SUFFICIENT BY ITSELF TO ESTABLISH PRIMA FACIE OBVIOUSNESS

A statement that modifications of the prior art to meet the claimed invention would have been "'well within the ordinary skill of the art at the time the claimed invention was made'" because the references relied upon teach that all aspects of the claimed invention were individually known in the art is not sufficient to establish a prima facie case of obviousness without some objective reason to combine the teachings of the references.

＊＊＊＊＊＊＊＊＊＊＊

V. THE PROPOSED MODIFICATION CANNOT RENDER THE PRIOR ART UNSATISFACTORY FOR ITS INTENDED PURPOSE

If proposed modification would render the prior art invention being modified unsatisfactory for its intended purpose, then there is no suggestion or motivation to make the proposed modification.

＊＊＊＊＊＊＊＊＊＊＊

VI. THE PROPOSED MODIFICATION CANNOT CHANGE THE PRINCIPLE OF OPERATION OF A REFERENCE

If the proposed modification or combination of the prior art would change the principle of operation of the prior art invention being modified, then the teachings of the references are not sufficient to render the claims prima facie obvious.

＊＊＊＊＊＊＊＊＊＊＊

2143.02 Reasonable Expectation of Success Is Required [R-6]

＊＊＊＊＊＊＊＊＊＊

I. < OBVIOUSNESS REQUIRES ONLY A REASONABLE EXPECTATION OF SUCCESS

The prior art can be modified or combined to reject claims as prima facie obvious as long as there is a reasonable expectation of success.

＊＊＊＊＊＊＊＊＊＊＊

II. < AT LEAST SOME DEGREE OF PREDICTABILITY IS REQUIRED; APPLICANTS MAY PRESENT EVIDENCE SHOWING THERE WAS NO REASONABLE EXPECTATION OF SUCCESS

Obviousness does not require absolute predictability, however, at least some degree of predictability is required. Evidence showing there was no reasonable expectation of success may support a conclusion of nonobviousness.

＊＊＊＊＊＊＊＊＊＊＊

III. < PREDICTABILITY IS DETERMINED AT THE TIME THE INVENTION WAS MADE

Whether an art is predictable or whether the proposed modification or combination of the prior art has a reasonable expectation of success is determined at the time the invention was made.

＊＊＊＊＊＊＊＊＊＊＊

MPEP 2143.03 All Claim Limitations Must Be **>Considered< [R-6]

** "All words in a claim must be considered in judging the patentability of that claim against the prior art." In re Wilson, 424 F.2d 1382, 1385, 165 USPQ 494, 496 (CCPA 1970). If an independent claim is nonobvious under 35 U.S.C. 103, then any claim depending therefrom is nonobvious. In re Fine, 837 F.2d 1071, 5 USPQ2d 1596 (Fed. Cir. 1988).

I. < INDEFINITE LIMITATIONS MUST BE CONSIDERED

A claim limitation which is considered indefinite cannot be disregarded. If a claim is subject to more than one interpretation, at least one of which would render the claim unpatentable over the prior art, the examiner should reject the claim as indefinite under 35 U.S.C. 112, second paragraph (see MPEP § 706.03(d)) and should reject the claim over the prior art based on the interpretation of the claim that renders the prior art applicable.

＊＊＊＊＊＊＊＊＊＊＊

II. < LIMITATIONS WHICH DO NOT FIND SUPPORT IN THE ORIGINAL SPECIFICATION MUST BE CONSIDERED

＊＊＊＊＊＊＊＊＊＊＊

§13.7 依據一般知識作顯而易見之核駁

在某些情況下，美國專利局之審查委員可以未附上書面證據之支持，就以官方通知依據一般知識作出 35USC103 顯而易見核駁。但是，此核駁必需依照以下程序：

A. 決定是否在無支持之證據下以官方通知為一般知識是適當的

審查委員只有在可立即及明顯地顯示請求項之發明為一般知識或習知之技術時，才可不附上何支持之證據以官方通知核駁請求項。

B. 官方通知之技術上之理由必需是很清楚的且無誤的

審查委員如果無任何支持書面引證資料以官方通知核駁一請求項，則審查委員必需提供清楚、無誤的，特定的科學與技術上之事實理由來支持其主張為一般知識之核駁。

C. 如果申請人對審查委員之一般知識之核駁提出挑戰，則審查委員必需提出證據以支持其核駁理由

申請人對於審查委員之一般知識之核駁之反駁必需清楚地指出審查委員之錯誤。亦即，需包括為何不是一般知識或習知之技術之聲明。如果申請人充份地反駁了審查委員之主張，則審查委員在下一審定書中必需提供書面之證據以維持其核駁之主張。

如果申請人未反駁審查委員之一般知識之核駁或反駁之理由不完全，則審查委員會在下一審定書中清楚地指出一般知識及習知技術為申請人承認之先前技術（admitted prior art）。

D. 決定下次之審定書是否為最終審定書

申請人答辯後，如果審查委員在下一審定書中加入書面之引證資料，而此引證資料為支持其一般知之核駁理由的話，則此一審定書可為最終審定書，如果此加入之書面引證資料並非支持其一般知識之核駁，則此審定書可能不是最終審定書。

MPEP 2144.03 Reliance on Common Knowledge in the Art or "Well Known" Prior Art [R-6]

In *>certain< circumstances >where appropriate<, ** an examiner *>may< take official notice of facts not in the record or * rely on "common knowledge" in making a rejection, however such rejections should be judiciously applied.

＊＊＊＊＊＊＊＊＊＊＊

In light of recent Federal Circuit decisions as discussed below and the substantial evidence standard of review now applied to USPTO Board decisions, the following guidance is provided in order to assist the examiners in determining when it is appropriate to take official notice of facts without supporting documentary evidence or to rely on common knowledge in the art in making a rejection, and if such official notice is taken, what evidence is necessary to support the examiner's conclusion of common knowledge in the art.

* * * * * * * * * * *

A.Determine When It Is Appropriate To Take Official Notice Without Documentary Evidence To Support the Examiner's Conclusion

* * * * * * * * * * *

Official notice unsupported by documentary evidence should only be taken by the examiner where the facts asserted to be well-known, or to be common knowledge in the art are capable of instant and unquestionable demonstration as being well-known.

* * * * * * * * * * *

B.If Official Notice Is Taken of a Fact, Unsupported by Documentary Evidence, the Technical Line of Reasoning Underlying a Decision To Take Such Notice Must Be Clear and Unmistakable

* * * * * * * * * * *

The examiner must provide specific factual findings predicated on sound technical and scientific reasoning to support his or her conclusion of common knowledge.

* * * * * * * * * * *

C.If Applicant Challenges a Factual Assertion as Not Properly Officially Noticed or Not Properly Based Upon Common Knowledge, the Examiner Must Support the Finding With Adequate Evidence

To adequately traverse such a finding, an applicant must specifically point out the supposed errors in the examiner's action, which would include stating why the noticed fact is not considered to be common knowledge or well-known in the art.

* * * * * * * * * * *

applicant adequately traverses the examiner's assertion of official notice, the examiner must provide documentary evidence in the next Office action if the rejection is to be maintained.

* * * * * * * * * * *

If applicant does not traverse the examiner's assertion of official notice or applicant's traverse is not adequate, the examiner should clearly indicate in the next Office action that the common knowledge or well-known in the art statement is taken to be admitted prior art because applicant either failed to traverse the examiner's assertion of official notice or that the traverse was inadequate. If the traverse was inadequate, the examiner should include an explanation as to why it was inadequate.

D.Determine Whether the Next Office Action Should Be Made Final

If the examiner adds a reference in the next Office action after applicant's rebuttal, and the newly added reference is added only as directly corresponding evidence to support the prior common knowledge finding, and it does not result in a new issue or constitute a new ground of rejection, the Office action may be made final. If no amendments are made to the claims, the examiner must not rely on any other teachings in the reference if the rejection is made final. If the newly cited reference is added for reasons other than to support the prior common knowledge statement and a new ground of rejection is introduced by the examiner that is not necessitated by applicant's amendment of the claims, the rejection may not be made final. See MPEP § 706.07(a).

＊＊＊＊＊＊＊＊＊＊＊

§13.8　以判例作顯而易見核駁之依據

如果被審查中之專而案件與判例中之案件相當近似，美國專利局之審查委員亦可以法院之類似判例作依據以顯而易見核駁專利申請案，其可分為以下數種：

Ⅰ.屬於美感之設計改變（aesthetic design changes）

如果發明之特徵只在於裝飾上而無機械上之功能則為顯而易見。

Ⅱ.單純省去一步驟或元件及其功能之發明（elimination of a step or an element and its function）

A：一發明為省去一元件及其功能，而此元件之功能非所希望者，則此發明為顯而易見；

例如一發明乃是有關於使用一組合物來防止金屬表面腐蝕的方法，此組合物乃由 epoxy resin, petroleum sulfonate 及 hydrocarbon diluent 所組成（consisting of）。審查委員引證了一主要先前技術其使用一組合物包括：epoxy resin, hydrocarbon diluent 及 polybasic acid salt 作金屬之防蝕，其中說明該 polybasic acid salt 當使用在 freshwater 環境中對防蝕特別有利。另一次要先前技術則教示加入 petroleum sulfonate 可防蝕。

在此情形此發明乃為顯而易見，因為本發明只省去了 polybasic acid salt，而此 salt 之功用（function）（在 freshwater 防蝕效果增加）並非本發明所要求或希望者。

B：發明為省去一元件但保留此元件之功能則此發明為非顯而易見之發明，例如一請求之發明為一印有標記之一片狀物，其上直接粘合有一可擦去的金屬薄層，而先前技術揭示了一類似的印有標記的片狀物其更包括了一中間透明防擦去的保護層。則雖然該發明省去了先前技術之透明層，但此先前技術之透明層之功能在該發明中被保留，因為金屬薄層可被擦去但不會擦去所印之標記。因此，該發明為非顯而易見的。

III.將人力活動自動化為顯而易知

例如一請求之發明乃為一耐久的鑄模裝置用來成型一卡車之汽缸，此裝置結合了習知的耐久鑄模結構及一計時器與電磁閥，藉由此計時器及電磁閥來自動作動壓力閥系統以於一特定時間之後釋出模芯。此自動化系統事實上只是用機械裝置來取代人力活動以達成相同之結果，故為顯而易知。

IV.僅改變尺寸、改變形狀、或改變步驟之順序，為顯而易見

改變尺寸（大小）、比例及步驟之順序被認為僅是一種選擇而對此技藝人士是顯而易見的，除非有證據顯示改變尺寸、形狀是重要的改變實行方法之步驟順序會有新的或不可預期之效果。

V.僅將裝置改為可攜帶的、一體成型的、可分離的、可調整的或將方法改為連續的是為顯而易見

VI.將一裝置之構件反轉、複製或重新安排可為顯而易見。除非可達成新的，不可預期之效果

VII.純化之舊的產物

在考慮舊的產物（化合物、組合物等）的純化形式是否為顯而易見的因素包括是否所請求之純化的產物與先前技術中極相關之物質有相同之用途，以及是否先前技術建議了所請求之純化產物之特定形式或構造，或者建議了得到所請求之純化產物之適當方法而定。例如，所請求之化合物為一已知化合物之可自由流動的結晶形式，而先前技術所揭示的是該化合物之粘稠液體形式。因為先前技術並未揭示結晶形式之該化合物亦未揭示如何得到此結晶，故該化合物是為非顯而易見的。

MPEP 2144.04 Legal Precedent as Source of Supporting Rationale [R-6]

As discussed in MPEP § 2144, if the facts in a prior legal decision are sufficiently similar to those in an application under examination, the examiner may use the rationale used by the court. Examples directed to various common practices which the court has held normally require only ordinary skill in the art and hence are considered routine expedients are discussed below. If the applicant has demonstrated the criticality of a specific limitation, it would not be appropriate to rely solely on case law as the rationale to support an obviousness rejection.

I.AESTHETIC DESIGN CHANGES

＊＊＊＊＊＊＊＊＊＊＊

II.ELIMINATION OF A STEP OR AN ELEMENT AND ITS FUNCTION

A.Omission of an Element and Its Function Is Obvious if the Function of the Element Is Not Desired

＊＊＊＊＊＊＊＊＊＊＊

B.Omission of an Element with Retention of the Element"s Function Is an Indicia of Unobviousness

＊＊＊＊＊＊＊＊＊＊＊

III.AUTOMATING A MANUAL ACTIVITY

＊＊＊＊＊＊＊＊＊＊＊

IV.CHANGES IN SIZE, SHAPE, OR SEQUENCE OF ADDING INGREDIENTS

＊＊＊＊＊＊＊＊＊＊＊

V. MAKING PORTABLE, INTEGRAL, SEPARABLE, ADJUSTABLE, OR CONTINUOUS

＊＊＊＊＊＊＊＊＊＊＊

VI.REVERSAL, DUPLICATION, OR REARRANGEMENT OF PARTS

＊＊＊＊＊＊＊＊＊＊＊

VII.PURIFYING AN OLD PRODUCT

§13.9 數值範圍之顯而易見性

　　所請求之數值範圍如果與先前技術所揭示之數值範圍重疊（overlap）或者落入該範圍之內就為推斷的顯而易見。此外，濃度或溫度等數值與先前技術不同並不能支持其具有專利性，除非提出證據證明此溫度或濃度等數值是重要的（critical）。另外，藉由日常的實驗發現溫度或濃度等變數(數值)之最佳範圍或可實行範圍（optimum or workable range）亦是不具發明性的。例如所請求之製程是在 40℃ 及 80℃ 之範圍及酸濃度為 25% 至 70% 之範圍內進行，而先前技術之製程乃在 100℃ 及酸濃度為 10% 進行時，則所請求之製程就是推斷的顯而易見，但是要判定一變數（數值）是最佳範圍或可實行之範圍需先有一前提，就是此變數（數值）是一種用來達成一已被認知之結果（recognized result）之結果－有效（result-effective varible）之變數者。例如所請求之發明為一廢水處理槽，處理槽之體積與收縮口之比率為 0.12 gal/sgft。但是先前技術並不知道廢水處理槽之處理容量（treatment capacity）乃是槽體積與收縮口之比率之函數，則此比率就不是一結果－有效變數，因此找出槽體積與收縮口比率之最佳範圍就不是推斷的顯而易見。

MPEP2144.05 Obviousness of Ranges [R-5]

See MPEP § 2131.03 for case law pertaining to rejections based on the anticipation of ranges under 35 U.S.C. 102 and 35 U.S.C. 102/ 103.

I.OVERLAP OF RANGES

In the case where the claimed ranges "overlap or lie inside ranges disclosed by the prior art" a prima facie case of obviousness exists.

＊＊＊＊＊＊＊＊＊＊＊

II.OPTIMIZATION OF RANGES

A.Optimization Within Prior Art Conditions or Through Routine Experimentation

Generally, differences in concentration or temperature will not support the patentability of subject matter encompassed by the prior art unless there is evidence indicating such concentration or temperature is critical. "[W]here the general conditions of a claim are disclosed in the prior art, it is not inventive to discover the optimum or workable ranges by routine experimentation." In re Aller, 220 F.2d 454, 456, 105 USPQ 233, 235 (CCPA 1955) (Claimed process which was performed at a temperature between 40°C and 80°C and an acid concentration between 25% and 70% was held to be prima facie obvious over a reference process which differed from the claims only in that the reference process was performed at a temperature of 100°C and an acid concentration of 10%.);

＊＊＊＊＊＊＊＊＊＊＊

B.Only Result-Effective Variables Can Be Optimized

A particular parameter must first be recognized as a result-effective variable, i.e., a variable which achieves a recognized result, before the determination of the optimum or workable ranges of said variable might be characterized as routine experimentation. In re Antonie, 559 F.2d 618, 195 USPQ 6 (CCPA 1977) (The claimed wastewater treatment device had a tank volume to contractor area of 0.12 gal./sq. ft. The prior art did not recognize that treatment capacity is a function of the tank volume to contractor ratio, and therefore the parameter optimized was not recognized in the art to be a result- effective variable.). See also In re Boesch, 617 F.2d 272, 205 USPQ 215 (CCPA 1980) (prior art suggested proportional balancing to achieve desired results in the formation of an alloy).

＊＊＊＊＊＊＊＊＊＊＊

§13.10 物種請求項之顯而易見性

當一請求項（claim）請求一物種（species），例如一化合物或一次群組（subgenus），而當審查委員找到一單一先前技術揭露了一群組（genus），例如一通式表示之化合物，此群組涵蓋了所請求之物種或次群組，但並未特別揭示出該物種（群組），此時審查委員如要以 35USC103 作顯而易見之核駁，則應努力再找出另外的先前技術以顯示整體觀之所請求之物種是顯而易見的。如果不能再找到之外的先前技術，而要以單一的先前技術作出推斷的以顯而易見核駁，需考慮以下因素：

所請求之物種（species）被單一的先前技術之群組（genus）所涵蓋之事實並不足以讓審查委員下推斷的顯而易見之判定。審查委員需先依照 Graham 之法則先決定先前技術之內容及範圍，例如需先確定(A)先前技術之群組之結構及說明書中敘述到之物種及次群組，(B)說明書中所揭示之群組之物理、化學性質及用途及使用此群組之用途上的限制及問題，(C)此技術領域之可預測性；及(D)所有可考慮到之由此群組涵蓋之物種之數目。接著，確定先前技術之群組中有揭露之最接近的物種以及與所請求之發明的物種之差異。再接下來，需決定此技藝中技術之水平。最後再決定此技藝人士是否會有動機去選擇所請求之物種。在決定此技藝人士是否會有動機去選擇所請求之物種時，審查委員需考慮所有相關的先前技術之教示，特別需將焦點放在以下數點：

(a)考慮群組之大小（數目）

群組之大小與顯而易見之結論並沒有絕對之關係，亦即，單純因為群組之數目小，不一定就是顯而易見的，但是如果群組太小則此群組可能預見（anticipate）所請求之物種。例如在 In re Petering 之判例中，先前技術之群組只包括 20 個化合物，則此技藝人士可直接就想到該物種。

(b)考慮明顯的教示

如果引證之先前技術明顯地教示了一特定的理由去選擇所請求之物種（次群組），則審查委員需指出該揭露，並解擇為何此技藝人士會去選擇所請求之物種。

(c)考慮結構類似性之教示

　　需考慮任何先前技術所揭露之群組中典型的，最佳的，較佳的物種或次群組之教示。如果此種物種或次群組在結構上類似於所請求之物種，則此技藝人士就可能會從群組中去選擇所請求之物種，因為結構類似之物種通常具有類似之性質。結構上之類似會提供此技藝人士動機去改良習知之化合物以得到新的化合物。例如習知之化合物會建議其同系物（homologs），因為同系物時常具有類似之性質，因而此技藝之化學家通常會想去製造以得到具改良性質的化合物。

(d)考慮類似性質或用途之教示

　　類似的性質或用途通常會提供給此技藝人士動機去製造與先前技術中結構類似之物種，相對地，如果每一物種缺乏任何有用的性質，則此技藝人士不會有動機去製造或選擇該物種。

(e)考慮該技術領域之可預測性

　　如果該技術領域是不可預期的，則不太可能先前技術中之類似結構的物種會使所請求物種顯而易見。例如，在 In re Schechter 之判例中指出"在殺蟲劑領域中之不可預測性高，所以結構上之類似性是判定顯而易見之結論之不利因素。但是，顯而易見並不要求絕對的可預測性，僅僅有得到類似性質成功之期待就可能作為判定顯而易見之因素。

(f)考慮其他支持去選擇所請求物種之因素。

MPEP 2144.08 Obviousness of Species When Prior Art Teaches Genus [R-6]

I.** EXAMINATION OF CLAIMS DIRECTED TO SPECIES OF CHEMICAL COMPOSITIONS BASED UPON A SINGLE PRIOR ART REFERENCE

　　**>When< a single prior art reference which discloses a genus encompassing the claimed species or subgenus but does not expressly disclose the particular claimed species or subgenus*>,< Office personnel should attempt to find additional prior art to show that the differences between the prior art primary reference and the claimed invention as a whole would have been obvious. Where such additional prior art is not found, Office personnel should **>consider the factors discussed below< to determine whether a single reference 35 U.S.C. 103 rejection would be appropriate. **

1.Determine the Scope and Content of the Prior Art

＊＊＊＊＊＊＊＊＊＊

In the case of a prior art reference disclosing a genus, Office personnel should make findings as to:

(A) the structure of the disclosed prior art genus and that of any expressly described species or subgenus within the genus;

(B) any physical or chemical properties and utilities disclosed for the genus, as well as any suggested limitations on the usefulness of the genus, and any problems alleged to be addressed by the genus;

(C) the predictability of the technology; and

(D) the number of species encompassed by the genus taking into consideration all of the variables possible.

2.Ascertain the Differences Between the Closest Disclosed Prior Art Species or Subgenus of Record and the Claimed Species or Subgenus

＊＊＊＊＊＊＊＊＊＊

3.Determine the Level of Skill in the Art

Office personnel should evaluate the prior art from the standpoint of the hypothetical person having ordinary skill in the art at the time the claimed invention was made.

＊＊＊＊＊＊＊＊＊＊

4.Determine Whether One of Ordinary Skill in the Art Would Have Been Motivated To Select the Claimed Species or Subgenus

In light of the findings made relating to the three Graham factors, Office personnel should determine whether >it would have been obvious to< one of ordinary skill in the relevant art ** to make the claimed invention as a whole, i.e., to select the claimed species or subgenus from the disclosed prior art genus. ** To address this key issue, Office personnel should consider all relevant prior art teachings, focusing on the following, where present.

(a)Consider the Size of the Genus

Consider the size of the prior art genus, bearing in mind that size alone cannot support an obviousness rejection. See, e.g., Baird, 16 F.3d at 383, 29 USPQ2d at 1552 (observing that "it is not the mere number of compounds in this limited class which is significant here but, rather, the total circumstances involved"). There is no absolute correlation between the size of the prior art genus and a conclusion of obviousness. Id. Thus, the mere fact that a prior art genus contains a small number of members does not create a per se rule of obviousness. ** However, a genus may be so small that, when considered in light of the totality of the circumstances, it would anticipate the claimed species or subgenus. For example, it has been held that a prior art genus containing only 20 compounds and a limited number of variations in the generic chemical formula inherently anticipated a claimed species within the genus because "one skilled in [the] art would...... envisage each member" of the genus. In re Petering, 301 F.2d 676, 681, 133 USPQ 275, 280 (CCPA 1962) (emphasis in original).

＊＊＊＊＊＊＊＊＊＊＊

(b)Consider the Express Teachings

If the prior art reference expressly teaches a particular reason to select the claimed species or subgenus, Office personnel should point out the express disclosure **>and explain why it would have been obvious to< one of ordinary skill in the art to select the claimed invention.

＊＊＊＊＊＊＊＊＊＊＊

(c)Consider the Teachings of Structural Similarity

Consider any teachings of a "typical," "preferred," or "optimum" species or subgenus within the disclosed genus. If such a species or subgenus is structurally similar to that claimed, its disclosure may *>provide a reason for< one of ordinary skill in the art to choose the claimed species or subgenus from the genus, based on the reasonable expectation that structurally similar species usually have similar properties. See, e.g., Dillon, 919 F.2d at 693, 696, 16 USPQ2d at 1901, 1904. See also Deuel, 51 F.3d at 1558, 34 USPQ2d at 1214 ("Structural relationships may provide the requisite motivation or suggestion to modify known compounds to obtain new compounds. For example, a prior art compound may suggest its homologs because homologs often have similar properties and therefore chemists of ordinary skill would ordinarily contemplate making them to try to obtain compounds with improved properties."). **

＊＊＊＊＊＊＊＊＊＊＊

(d)Consider the Teachings of Similar Properties or Uses

Consider the properties and utilities of the structurally similar prior art species or subgenus. It is the properties and utilities that provide real world motivation for a person of ordinary skill to make species structurally similar to those in the prior art. Dillon, 919 F.2d at 697, 16 USPQ2d at 1905; In re Stemniski, 444 F.2d 581, 586, 170 USPQ 343, 348 (CCPA 1971). Conversely, lack of any known useful properties weighs against a finding of motivation to make or select a species or subgenus.

* * * * * * * * * * *

(e)Consider the Predictability of the Technology

* * * * * * * * * * *

(e)Consider the Predictability of the Technology

Consider the predictability of the technology. See, e.g., Dillon, 919 F.2d at 692-97, 16 USPQ2d at 1901-05; In re Grabiak, 769 F.2d 729, 732-33, 226 USPQ 870, 872 (Fed. Cir. 1985). If the technology is unpredictable, it is less likely that structurally similar species will render a claimed species obvious because it may not be reasonable to infer that they would share similar properties.

* * * * * * * * * * *

In re Schechter, 205 F.2d 185, 191, 98 USPQ 144, 150 (CCPA 1953) (unpredictability in the insecticide field, with homologs, isomers and analogs of known effective insecticides having proven ineffective as insecticides, was considered as a factor weighing against a conclusion of obviousness of the claimed compounds). However, obviousness does not require absolute predictability, only a reasonable expectation of success, i.e., a reasonable expectation of obtaining similar properties. See, e.g., In re O'Farrell, 853 F.2d 894, 903, 7 USPQ2d 1673, 1681 (Fed. Cir. 1988).

* * * * * * * * * * *

(f)Consider Any Other Teaching To Support the Selection of the Species or Subgenus

* * * * * * * * * * *

§13.11 同系物、類似物、異構物之顯而易見性

當所請求之化合物與先前技術之化合物有非常接近之結構類似性時，審查委員可能會作出推斷的顯而易見之核駁，因為此技藝人士在期待化合物在結構上類似會

有類似之性質下,會有動機去製造所請求之化合物。但是,並非引證之先前技術所揭示之化合物與所請求之化合物是異構物或同系物就足以判定是顯而易見的,雖然位置異構物與同系物通常有足夠近似之結構,因而存在有一假定之期望,會有類似之性質。雖為同系物如果與鄰近之同系物相差甚大,亦可能無法期待具有類似之性質。所以在判定具有接近之結構類似性之同系物,異構物的顯而易見性時,尚需考慮其他相關之事實。亦即,先前技術揭示之化合物並不一定需要是所請求之化合物的異構物或同系物。如果其在結構與性質上相當接近,可能使此技藝人士有動機去製造所請求的化合物。

另外,如果先前技術揭示了與所請求之化合物在結構及性質上類似之化合物,但是並未揭示、未建議製造所請求之化合物之方法,則所請求之化合物對此技藝人士是顯而易見之假定可能會被排除。

再者,如果所引證之先前技術之化合物在結構上與所請求之化合物類似,但申請人可證實他們之間並無類似性質之合理期待時,顯而易見之期待亦有可能被排除。

另外,如果先前技術所揭示之類似化合物不具有用途,或其用途只是作為中間產物,則亦有可能所請求化合物並不是顯而易見的。

如果審查委員因結構類似而發下推斷的顯而易見核駁時,申請人亦可藉由37CFR 1.132 宣誓書主張所請求之化合物有不可預期之效果而克服之。例如在 In Re Papesch 之判例中,所請求之三乙基化合物(triethylated compounds)具有抗發炎之活性,而引證之類似物,三甲基化合物(trimethylated compounds)並無抗發炎之活性,申請人以宣誓書之方式提作答辯而克服了此推斷的顯而易見之核駁。

MPEP 2144.09 Close Structural Similarity Between Chemical Compounds (Homologs, Analogues, Isomers) [R-6]

I. < REJECTION BASED ON CLOSE STRUCTURAL SIMILARITY IS FOUNDED ON THE EXPECTATION THAT COMPOUNDS SIMILAR IN STRUCTURE WILL HAVE SIMILAR PROPERTIES

A prima facie case of obviousness may be made when chemical compounds have very close structural similarities and similar utilities. "An obviousness rejection based on similarity in chemical structure and function entails the motivation of one skilled in the art to make a claimed compound, in the expectation that compounds similar in structure will have similar properties."

* * * * * * * * * * * *

II. < HOMOLOGY AND ISOMERISM ARE FACTS WHICH MUST BE CONSIDERED WITH ALL OTHER RELEVANT FACTS IN DETERMINING OBVIOUSNESS

* * * * * * * * * * *

Compounds which are position isomers (compounds having the same radicals in physically different positions on the same nucleus) or homologs (compounds differing regularly by the successive addition of the same chemical group, e.g., by -CH2- groups) are generally of sufficiently close structural similarity that there is a presumed expectation that such compounds possess similar properties. In re Wilder, 563 F.2d 457, 195 USPQ 426 (CCPA 1977). See also In re May, 574 F.2d 1082, 197 USPQ 601 (CCPA 1978) (stereoisomers prima facie obvious).

Isomers having the same empirical formula but different structures are not necessarily considered equivalent by chemists skilled in the art and therefore are not necessarily suggestive of each other. Ex parte Mowry, 91 USPQ 219 (Bd. App. 1950) (claimed cyclohexylstyrene not prima facie obvious over prior art isohexylstyrene). Similarly, homologs which are far removed from adjacent homologs may not be expected to have similar properties. In re Mills, 281 F.2d 218, 126 USPQ 513 (CCPA 1960) (prior art disclosure of C8 to C12 alkyl sulfates was not sufficient to render prima facie obvious claimed C1 alkyl sulfate).

Homology and isomerism involve close structural similarity which must be considered with all other relevant facts in determining the issue of obviousness.

* * * * * * * * * * * *

Homology and isomerism involve close structural similarity which must be considered with all other relevant facts in determining the issue of obviousness.

* * * * * * * * * * *

III. <PRESENCE OF A TRUE HOMOLOGOUS OR ISOMERIC RELATIONSHIP IS NOT CONTROLLING

* * * * * * * * * * *

The court held that although the prior art compounds were not true homologs or isomers of the claimed compounds, the similarity between the chemical structures and properties is sufficiently close that one of ordinary skill in the art would have been motivated to make the claimed compounds in searching for new pesticides.).

＊＊＊＊＊＊＊＊＊＊＊

IV.　< PRESENCE OR ABSENCE OF PRIOR ART SUGGESTION OF METHOD OF MAKING A CLAIMED COMPOUND MAY BE RELEVANT IN DETERMINING PRIMA FACIE OBVIOUSNESS

"[T]he presence-or absence-of a suitably operative, obvious process for making a composition of matter may have an ultimate bearing on whether that composition is obvious-or nonobvious-under 35 U.S.C. 103." In re Maloney, 411 F.2d 1321, 1323, 162 USPQ 98, 100 (CCPA 1969).

"[I]f the prior art of record fails to disclose or render obvious a method for making a claimed compound, at the time the invention was made, it may not be legally concluded that the compound itself is in the possession of the public. In this context, we say that the absence of a known or obvious process for making the claimed compounds overcomes a presumption that the compounds are obvious, based on the close relationships between their structures and those of prior art compounds." In re Hoeksema, 399 F.2d 269, 274-75, 158 USPQ 597, 601 (CCPA 1968).

＊＊＊＊＊＊＊＊＊＊＊

V.　<PRESUMPTION OF OBVIOUSNESS BASED ON STRUCTURAL SIMILARITY IS OVERCOME WHERE THERE IS NO REASONABLE EXPECTATION OF SIMILAR PROPERTIES

The presumption of obviousness based on a reference disclosing structurally similar compounds may be overcome where there is evidence showing there is no reasonable expectation of similar properties in structurally similar compounds.

VI.　<IF PRIOR ART COMPOUNDS HAVE NO UTILITY, OR UTILITY ONLY AS INTERMEDIATES, CLAIMED STRUCTURALLY SIMILAR COMPOUNDS MAY NOT BE PRIMA FACIE OBVIOUS OVER THE PRIOR ART

If the prior art does not teach any specific or significant utility for the disclosed compounds, then the prior art is **>unlikely< to render structurally similar claims prima facie obvious **>in the absence of any reason< for one of ordinary skill in the art to make the reference compounds **>or< any structurally related compounds. In re Stemniski, 444 F.2d 581, 170 USPQ 343 (CCPA 1971).

＊＊＊＊＊＊＊＊＊＊＊

VII. < PRIMA FACIE CASE REBUTTABLE BY EVIDENCE OF SUPERIOR OR UNEXPECTED RESULTS

A prima facie case of obviousness based on structural similarity is rebuttable by proof that the claimed compounds possess unexpectedly advantageous or superior properties. In re Papesch, 315 F.2d 381, 137 USPQ 43 (CCPA 1963) (Affidavit evidence which showed that claimed triethylated compounds possessed anti-inflammatory activity whereas prior art trimethylated compounds did not was sufficient to overcome obviousness rejection based on the homologous relationship between the prior art and claimed compounds.);

****

§13.12 類似與非類似之技藝

審查委員要用來引證之先前技術必須是屬於類似技藝（analogous art）領域，而所謂類似的技藝是屬於發明人努力的領域或者必須合理地與發明所要處理之問題有關。因此美國專利局之審查委員在尋找先前技術大都會依照美國專利分類來進行。但是審查委員認為在結構及功能上類似與差異在判斷是否為類似技藝上需占更大的比重。例如申請人之發明是有關於人行道地板格柵（pedestrian floor gratings），而審查委員發現二引證，一是結構格柵（structural gratings），一是鞋子刮刀，雖非人行道地板用格柵，但其結構及功能均類似，則此二引證仍是屬於類似技藝。

MPEP 2141.01(a) Analogous and Nonanalogous Art [R-6]

I. TO RELY ON A REFERENCE UNDER 35 U.S.C. 103, IT MUST BE ANALOGOUS PRIOR ART

The examiner must determine what is "analogous prior art" for the purpose of analyzing the obviousness of the subject matter at issue. "In order to rely on a reference as a basis for rejection of an applicant's invention, the reference must either be in the field of applicant's endeavor or, if not, then be reasonably pertinent to the particular problem with which the inventor was concerned." In re Oetiker, 911 F.2d 1443, 1446, 24 USPQ2d 1443, 1445 (Fed. Cir. 1992). See also In re Deminsk, 796 F.2d 436, 230 USPQ 313 (Fed. Cir 1986); In re Clay, 966 F.2d 656, 659, 23 USPQ2d 1058, 1060-61 (Fed. Cir 1992) ("A reference is reasonably pertinent if, even though it may be in a different field from that of the inventor's endeavor, it is one which, because of the matter with which it deals, logically would have commended itself to an inventor's attention in considering his problem."); and Wang Laboratories Inc. v. Toshiba Corp., 993 F.2d 858, 26 USPQ2d 1767 (Fed. Cir. 1993).

II. PTO CLASSIFICATION IS SOME EVIDENCE OF ANALOGY, BUT SIMILARITIES AND DIFFERENCES IN STRUCTURE AND FUNCTION CARRY MORE WEIGHT

While Patent Office classification of references and the cross-references in the official search notes of the class definitions are some evidence of "nonanalogy" or "analogy" respectively, the court has found "the similarities and differences in structure and function of the inventions to carry far greater weight." In re Ellis, 476 F.2d 1370, 1372, 177 USPQ 526, 527 (CCPA 1973) (The structural similarities and functional overlap between the structural gratings shown by one reference and the shoe scrapers of the type shown by another reference were readily apparent, and therefore the arts to which the reference patents belonged were reasonably pertinent to the art with which appellant's invention dealt (pedestrian floor gratings).).

＊＊＊＊＊＊＊＊＊＊

§13.13 35USC103(b)法條之意義

35 USC §103(b)乃是有關於生物技術方法之非顯而易見性，其規定如果生物技術方法所用之原料或製出之最終產品是具有專利性（新穎性及非顯而易見性）的話，則雖然該生物技術方法之各步驟是一般業界使用之製程，該生物技術方法仍是非顯而易見的。然而，在實務上 35 USC §103(b)之法條很少或從未被審查委員使用過，因為在 1995 年 In re Ochini, 71F.,3d. 1565, 37 USPQ2d 1127(Fed. Cir. 1995)及 In re Brouwer, 77.F3d 422; 37 USPQ2d 1633 (Fed. Cir. 1996)二個判例中，法官指出在判定一製程（方法）（process or method）之發明時，需考慮所有的限制條件，亦即除了考慮製程之步驟是否顯而易見外，尚需考慮製出之產品或使用之原料是否為顯而易知，因而所引證之先前技術必須顯示此技藝人士有動機（motivation）去使用或製造一非顯而易見之產品或原料方可為顯而易見之核駁。亦即，由 Ochiiai 及 Brouwer 二判例，如果一製法所用之原料或製出之產品是具有專利性的，雖其製法步驟是習知的，該製法仍可被視為是非顯而易見的，上述之製法自然包括了生物技術之方法，因而 35 USC §103(b)已不適用。

MPEP 2116.01 Novel, Unobvious Starting Material or End Product [R-6]

All the limitations of a claim must be considered when weighing the differences between the claimed invention and the prior art in determining the obviousness of a process or method claim. See MPEP § 2143.03.

In re Ochiai, 71 F.3d 1565, 37 USPQ2d 1127 (Fed. Cir. 1995) and In re Brouwer, 77 F.3d 422, 37 USPQ2d 1663 (Fed. Cir. 1996) addressed the issue of whether an otherwise conventional process could be patented if it were limited to making or using a nonobvious product. In both cases, the Federal Circuit held that the use of per se rules is improper in applying the test for obviousness under 35 U.S.C. 103. Rather, 35 U.S.C. 103 requires a highly fact-dependent analysis involving taking the claimed subject matter as a whole and comparing it to the prior art. "A process yielding a novel and nonobvious product may nonetheless be obvious; conversely, a process yielding a well-known product may yet be nonobvious." TorPharm, Inc. v. Ranbaxy Pharmaceuticals, Inc., 336 F.3d 1322, 1327, 67 USPQ2d 1511, 1514 (Fed. Cir. 2003). **

＊＊＊＊＊＊＊＊＊＊＊＊

§13.14 35USC103(c)之規定與運用

至 1994 年 11 月 29 日起，35USC102(e)之先前技術之標的不能再用來作為依據 35USC103(a)之顯而易見核駁之先前技術，如果該標的與所請求發明在發明完成時是屬於同一人所有或有義務讓給同一人。之後 2004 年之 CREATE ACT 又將 35USC103(c)修正成：由他人發展出之標的，如果符合下列三條件，亦被視為屬於同一人所有或有義務讓渡給同一人：

(A)如果所請求之發明乃由或為了在所請求之發明完成之前所簽署聯合開發合約者所完成者；

(B)請求之發明是基於該聯合開發合約下所進行發明活動的結果；及

(C)所請求之發明之專利申請案揭露了或修正後揭露了聯合開發合約參與者之名字時；則由其他人所發展之標的及所請求之標的應為視為第(1)項之屬於同一人所有或有義務讓渡給同一人。

此 35USC103(c)之修正適用於 2004 年 12 月 1 日當日或之後發證之所有專利（包括再發證專利），而在 2004 年 12 月 1 日當天或之後仍為審查中之案件均適用 35USC103(c)之修正。

實例

其公司研發部門之發明人 A 發明了一晶圓研磨組合物（2004 年 4 月 1 日）包括成份 X,Y,及一 pH 調整劑（特徵），提出美國專利申請（申請日為 2005 年 1 月 1 日），主張 TW 優先權（2004 年 7 月 1 日）），此美國專利申請案於 2006 年 1 月 2 日於美國早期公開（publication），而該公司製造部門之另一發明人 B 參考了發明人 A 之發明，改良了此晶圓研磨組合物（2004 年 12 月 1 日），包括成份 X,Y,Z,一 pH 調整劑及一防止沉澱劑（特徵）並於 2005 年 7 月 1 日於美國提出申請，審查委員引證了發明人 A 之早期公開公報〔102(e)之引證〕及另一美國專利 C 其揭露了在磁鐵之研磨組合物上使用了該防止沉澱劑，而以 35 USC §103(a)核駁發明人 B 之發明。請問在此情況下可否適用 35 USC §103(c)主張非顯而易見而克服核駁？

由於在發明人 B 發明之時（2004 年 12 月 1 日），此二專利之發明人 A 與 B 均有義務將其專利讓予（或已讓予）該公司，故可主張 35 USC §103(c)並非顯而易見。

MPEP 2146 35 U.S.C. 103(c) [R-3]

＊＊＊＊＊＊＊＊＊＊＊

was made, owned by the same person or subject to an obligation of assignment to the same person." This amendment to 35 U.S.C. 103(c) was made pursuant to section 4807 of the American Inventors Protection Act of 1999 (AIPA); see Pub. L. 106-113, 113 Stat. 1501, 1501A-591 (1999). The changes to 35 U.S.C. 102(e) in the Intellectual Property and High Technology Technical Amendments Act of 2002 (Pub. L. 107-273, 116 Stat. 1758 (2002)) did not affect the exclusion under 35 U.S.C. 103(c) as amended on November 29, 1999. Subsequently, the Cooperative Research and Technology Enhancement Act of 2004 (CREATE Act) (Pub. L. 108-453, 118 Stat. 3596 (2004)) further amended 35 U.S.C. 103(c) to provide that subject matter developed by another person shall be treated as owned by the same person or subject to an obligation of assignment to the same person for purposes of determining obviousness if three conditions are met:

(A) the claimed invention was made by or on behalf of parties to a joint research agreement that was in effect on or before the date the claimed invention was made;

(B) the claimed invention was made as a result of activities undertaken within the scope of the joint research agreement; and

(C) the application for patent for the claimed invention discloses or is amended to disclose the names of the parties to the joint research agreement (hereinafter "joint research agreement disqualification").

These changes to 35 U.S.C. 103(c) apply to all patents (including reissue patents) granted on or after December 10, 2004. The amendment to 35 U.S.C. 103(c) made by the AIPA to change "subsection (f) or (g)" to "one of more of subsections (e), (f), or (g)" applies to applications filed on or after November 29, 1999. It is to be noted that, for all applications (including reissue applications), if the application is pending on or after December 10, 2004, the 2004 changes to 35 U.S.C. 103(c), which effectively include the 1999 changes, apply; thus, the November 29, 1999 date of the prior revision to 35 U.S.C. 103(c) is no longer relevant.

第 14 章　可專利之標的與實用性

目　錄

§14.1　可專利之標的

依照美國專利法 35USC§101，可授予實用專利之標的為：方法、機械、物品或組合物。依照 35USC§171，可授予新式樣專利之標的為：物品之新穎的，原創的、裝飾性的設計。依照 35USC§161，可授予植物專利之標的為：以無性繁殖方式複製之獨特且新穎的植物新品種，其為非球根繁殖之植物或不是在非栽培狀態下發現之植物。

---

35 U.S.C. 101 Inventions patentable.

Whoever invents or discovers any new and useful process, machine, manufacture, or composition of matter, or any new and useful improvement thereof, may obtain a patent therefor, subject to the conditions and requirements of this title.

35 U.S.C. 171 Patents for designs.

Whoever invents any new, original, and ornamental design for an article of manufacture may obtain a patent therefor, subject to the conditions and requirements of this title.

The provisions of this title relating to patents for inventions shall apply to patents for designs, except as otherwise provided.

35 U.S.C. 161 Patents for plants.

Whoever invents or discovers and asexually reproduces any distinct and new variety of plant, including cultivated sports, mutants, hybrids, and newly found seedlings, other than a tuber propagated plant or a plant found in an uncultivated state, may obtain a patent therefor, subject to the conditions and requirements of this title.

The provisions of this title relating to patents for inventions shall apply to patents for plants, except as otherwise provided.

§14.2　活的標的之可專利性

　　1980 年美國最高法院在 *Court in* Diamond v. Chakrabarty 之判例中判定藉由遺傳工程（genetic engineering）產出之微生物是可專利之標的。因此活的標的（Living Subject Matter）是美國專利法上可予專利的標的。從此最高法院之判例中可看出，涵蓋活的標的物之發明和與專利性之議題是無關的，最高法院之是否為可予專利之標的之測試乃在於此活的標的物是否有人類的創造力介入。

　　最高法院在該判例之測試中指出下列要點：

(A) 自然法則，物理現象及抽象的觀念是不可專利之標的。

(B) 非天然產出之製品、組合物（人類創造力之產物），其具有特定之性質，用途者是可專利之標的。

(C) 在地球上發現之新的礦物或在野地上發現的新的植物都不是可予專利的標的。

(D) 從原料藉由手工或機器，給這些原料新的形式、特性，性質而得到之產物均是可予專利之標的。

　　在 *J.E.M. Ag Supply, Inc. v. Pioneer Hi-Bred Int' l, Inc.* 之判例中最高法院亦決定新的植物品種（newly developed plant breeds）是 35USC§101 中所述之可予專利的標的，雖然植物專利及植物品種保護法已經對於植物給予保護。此外，在 1987 年 *In Ex parte Allen* 之判例中，CAFC（聯邦巡迴法庭）亦決定一種多倍體牡蠣為 35USC§101 之可予專利之標的，只要其滿足其他的專利要件。之後不久，美國專利局就公告非天然產生，非人類多細胞生物是 35USC§101 之可予專利標的。

　　因而，除了人類不是 35USC§101 可予專利之標的外，其他動物、植物、微生物只要是非天然產生，有人類的創造力介入者均是可予專利之標的。

MPEP 2105 Patentable Subject Matter - Living Subject Matter [R-1]

　　The decision of the Supreme Court in Diamond v. Chakrabarty, 447 U.S. 303, 206 USPQ 193 (1980), held that microorganisms produced by genetic engineering are not excluded from patent protection by 35 U.S.C. 101. It is clear from the Supreme Court decision and opinion that the question of whether or not an invention embraces living matter is irrelevant to the issue of patentability. The test set down by the Court for patentable subject matter in this area is whether the living matter is the result of human intervention.

* * * * * * * * * * *

The tests set forth by the Court are (note especially the italicized portions):

(A) "The laws of nature, physical phenomena and abstract ideas" are not patentable subject matter.

(B) A "nonnaturally occurring manufacture or composition of matter - a product of human ingenuity -having a distinctive name, character, [and] use" is patentable subject matter.

(C) "[A] new mineral discovered in the earth or a new plant found in the wild is not patentable subject matter. Likewise, Einstein could not patent his celebrated E=mc2; nor could Newton have patented the law of gravity. Such discoveries are 'manifestations of...... nature, free to all men and reserved exclusively to none.'"

(D) "[T]he production of articles for use from raw materials prepared by giving to these materials *new forms, qualities, properties, or combinations whether by hand labor or by machinery"* [emphasis added] is a "manufacture" under 35 U.S.C. 101.

* * * * * * * * * * *

**>In another case addressing< the scope of 35 U.S.C. 101, the **>Supreme Court< held that patentable subject matter under 35 U.S.C. 101 includes **>newly developed plant breeds<, even though plant protection is also available under the Plant Patent Act (35 U.S.C. 161-164) and the Plant Variety Protection Act (7 U.S.C. 2321 *et. seq.*). **> *J.E.M. Ag Supply, Inc. v. Pioneer Hi-Bred Int' l, Inc.,* 534 U.S. 124, 143-46, 122 S.Ct. 593, 605-06, 60 USPQ2d 1865, 1874 (2001) (The scope of coverage of 35 U.S.C.101 is not limited by the Plant Patent Act or the Plant Variety Protection Act; each statute can be regarded as effective because of its different requirements and protections)

* * * * * * * * * * *

In Ex parte Allen, 2 USPQ2d 1425 (Bd. Pat. App. & Inter. 1987), the Board decided that a polyploid Pacific coast oyster could have been the proper subject of a patent under 35 U.S.C. 101 if all the criteria for patentability were satisfied. Shortly after the *Allen* decision, the Commissioner of Patents and Trademarks issued a notice (Animals - Patentability, 1077 O.G. 24, April 21, 1987) that the Patent and Trademark Office would now consider nonnaturally occurring, nonhuman multicellular living organisms, including animals, to be patentable subject matter within the scope of 35U.S.C. 101.

If the broadest reasonable interpretation of the claimed invention as a whole encompasses a human being, then a rejection under 35 U.S.C. 101 must be made indicating that the claimed invention is directed to nonstatutory subject matter.

§14.3　決定標的是否為可專利之標的之步驟

美國專利局依下列程序（步驟）決定所請求之發明是否為可予專利（適格）之標的（Patent subject matter eligibility）。

Ⅰ．決定申請人所發明的是什麼，及要尋求專利保護的是什麼？

(a)　確認及了解發明之任何用途及／或實際之應用

(b)　研讀發明之詳細揭露及特定之實施例

(c)　研讀請求項

Ⅱ．進行先前技術之完整檢索。

Ⅲ．決定所請求之發明是否符合 35USC§101 之合格專利標的之專利要件：

(a)　所請求之發明是否為法定之可予專利標的之範疇（亦即方法、機械、物品及組合物）；

(b)　所請求之發明是否屬於 35USC§101 之不予專利之標的：自然法則，自然現象，或抽象概念；

(c)　所請求之發明是涵蓋 35USC§101 之不予專利之標的或為其實際應用：①藉由物理轉換之實際應用？②可產生有用、有形及具體之結果之實際應用？

(d)　所請求之發明是否占用（preempt）35USC§101 不予專利標的（自然法則，自然現象及抽象之概念）。

(e)　建立推斷的實例（prima facie case）之記錄。

Ⅳ. 評估申請案是否符合 35USC§101（實用性）及 35USC§112 之規定。

Ⅴ. 決定所請求之發明是否符合 35USC§102 及 35USC§103 之規定。

Ⅵ. 清楚傳達不予專利之判定，結論及基礎，檢閱所有擬核駁理由及確認無專利性之推斷的決定（Prima Facie determination）。

在上述步驟(Ⅲ)(a)中決定所請求之發明是否為法定可予專利之標的時，需注意請求項是否為裝置與該裝置所執行之方法步驟之組合，如果請求項同時記載了結構及步驟作為元件則為混合請求項（hybrid claim），不符合 35USC§101 之規定。但是如果裝置請求項中只使用了功能來作為結構元件之限制條件（定義或修飾元件）則並非混合請求項，符合 35USC§101 之規定。

在步驟(Ⅲ)(a)中決定了所請求之發明是屬於可予專利之四種標的之後，此分析並未結束，因為可能所請求之發明實際上是自然法則，自然現象或抽象之概念，故需再進行步驟(Ⅲ)(b)步驟，進行此步驟之分析時需整體來看所請求之發明是否為自然法則、自然現象、抽象概念或是這些法則、現象、概念之實際應用。

在步驟(Ⅲ)(c)中，如果所請求的發現是①將一物品或有形體轉換成不同之狀態或物件，或②可產生有用、有形及具體之結果的話，則是自然法則，自然現象及抽象概念之實際應用，是可予專利之標的。故在步驟Ⅲ(c)之中，美國專利局審查委員如果認為所請求之發明有對物體作轉換（transformation or reduction）則將終止此分析而判定所請求之發明為可予專利之標的，如果未發現有轉換，則再進行分析看所請求之發明是否可產生有用、有形且具體之結果。

（Ⅰ）有用的結果：指所請求之發明之用途（Utility）必需是特定的（Specific），實質的（substantial）及可信的（credible）。

（Ⅱ）有形的結果：所謂有形的結果並非請求項一定要與一裝置或機械連接，但是請求項不能只記載自然法則，自然現象或抽象的概念。

（Ⅲ）具體的結果：所謂具體之結果是指所請求之發明所產生之結果是可確定的，例如方法所產生之結果必需是可重複的（具再現性的）。

　　在步驟(Ⅲ)(d)中，即使一請求項看起來是應用自然法則、自然現象或抽象之概念，是可予專利之標的，美國專利還需再分析是否所請求之發明實際上占有了（preempts）該自然法則，自然現象或抽象概念。例如在 *Benson* 之案例中，請求項記載了一電腦其只計算一數學式子，以防止他人使用該數學式子。或者請求發明為一光碟片其只貯存了數學式子。上述二例子均實際上占用了自然法則、自然現象或抽象之概念。

MPEP2106 Patent Subject Matter Eligibility [R-6]

GUIDELINES FLOWCHART

DETERMINE WHAT APPLICANT HAS INVENTED AND IS SEEKING TO PATENT

- Identify and Understand Any Utility and/or Practical Application Asserted for the Invention
- Review the Detailed Disclosure and Specific Embodiments of the Invention
- Review the Claims

↓

CONDUCT A THOROUGH SEARCH OF THE PRIOR ART

↓

DETERMINE WHETHER THE CLAIMED INVENTION COMPLIES WITH THE SUBJECT MATTER ELIGIBILITY REQUIREMENT OF 35 U.S.C. 101

- Does the Claimed Invention Fall Within an Enumerated Statutory Category?
- Does the Claimed Invention Fall *>Within< a 35 U.S.C. 101 Judicial Exception – Law of Nature, Natural Phenomena or Abstract Idea?
- Does the Claimed Invention Cover a 35 U.S.C. 101 Judicial Exception, or a Practical Application of a 35 U.S.C. 101 Judicial Exception?
  - Practical Application by Physical Transformation?
  - Practical Application That Produces a Useful (** 35 U.S.C. 101 utility), Tangible, and Concrete Result?
- Does the Claimed Invention Preempt a 35 U.S.C. 101 Judicial Exception (Abstract Idea, Law of Nature, or Natural Phenomenon)?
- Establish on the Record a Prima Facie Case

↓

EVALUATE APPLICATION FOR COMPLIANCE WITH 35 U.S.C. 101 (UTILITY) AND 112

↓

DETERMINE WHETHER THE CLAIMED INVENTION COMPLIES WITH

35 U.S.C. 102 AND 103

↓

CLEARLY COMMUNICATE FINDINGS, CONCLUSIONS AND THEIR BASES
- Review all the proposed rejections and their bases to confirm any prima facie determination of unpatentability.

---

* * * * * * * * * * *

For example, a claimed invention may be a combination of devices that appear to be directed to a machine and one or more steps of the functions performed by the machine. Such instances of mixed attributes, although potentially confusing as to which category of patentable subject matter the claim belongs, does not affect the analysis to be performed by USPTO personnel. Note that an apparatus claim with process steps is not classified as a "hybrid" claim; instead, it is simply an apparatus claim including functional limitations. See, e.g., R.A.C.C. Indus. v. Stun-Tech, Inc., 178 F.3d 1309 (Fed. Cir. 1998) (unpublished).

* * * * * * * * * * *

C. Determine Whether the Claimed Invention Falls Within 35 U.S.C. 101 Judicial Exceptions - Laws of Nature, Natural Phenomena and Abstract Ideas

While abstract ideas, natural phenomena, and laws of nature are not eligible for patenting, methods and products employing abstract ideas, natural phenomena, and laws of nature to perform a real-world function may well be. In evaluating whether a claim meets the requirements of section 101, the claim must be considered as a whole to determine whether it is for a particular application of an abstract idea, natural phenomenon, or law of nature, and not for the abstract idea, natural phenomenon, or law of nature itself.

* * * * * * * * * * *

2. Determine Whether the Claimed Invention is a Practical Application of an Abstract Idea, Law of Nature, or Natural Phenomenon (35 U.S.C. 101 Judicial Exceptions)

* * * * * * * * * * *

A claimed invention is directed to a practical application of a 35 U.S.C. 101 judicial exception when it:

(A) "transforms" an article or physical object to a different state or thing; or

(B) otherwise produces a useful, concrete and tangible result, based on the factors discussed below.

(1)Practical Application by Physical Transformation

USPTO personnel first shall review the claim and determine if it provides a transformation or reduction of an article to a different state or thing. If USPTO personnel find such a transformation or reduction, USPTO personnel shall end the inquiry and find that the claim meets the statutory requirement of 35 U.S.C. 101. If USPTO personnel do not find such a transformation or reduction, they must determine whether the claimed invention produces a useful, concrete, and tangible result.

(2)Practical Application That Produces a Useful, Concrete, and Tangible Result

＊＊＊＊＊＊＊＊＊＊＊

a)"USEFUL RESULT"

For an invention to be "useful" it must satisfy the utility requirement of section 101. The USPTO's official interpretation of the utility requirement provides that the utility of an invention has to be (i) specific, (ii) substantial and (iii) credible.

＊＊＊＊＊＊＊＊＊＊＊

b) "TANGIBLE RESULT"

The tangible requirement does not necessarily mean that a claim must either be tied to a particular machine or apparatus or must operate to change articles or materials to a different state or thing. However, the tangible requirement does require that the claim must recite more than a 35 U.S.C. 101 judicial exception, in that the process claim must set forth a practical application of that judicial exception to produce a real-world result.

＊＊＊＊＊＊＊＊＊＊＊

c)"CONCRETE RESULT"

Another consideration is whether the invention produces a "concrete" result. Usually, this question arises when a result cannot be assured. In other words, the process must have a result that can be substantially repeatable or the process must substantially produce the same result again.

＊＊＊＊＊＊＊＊＊＊＊

3.Determine Whether the Claimed Invention Preempts a 35 U.S.C. 101 Judicial Exception (Abstract Idea, Law of Nature, or Natural Phenomenon)

Even when a claim applies a mathematical formula, for example, as part of a seemingly patentable process, USPTO personnel must ensure that it does not in reality "seek[] patent protection for that formula in the abstract.

＊＊＊＊＊＊＊＊＊＊＊

Thus, a claim that recites a computer that solely calculates a mathematical formula (see Benson) or a computer disk that solely stores a mathematical formula is not directed to the type of subject matter eligible for patent protection. If USPTO personnel determine that the claimed invention preempts a 35 U.S.C.101judicial exception, they must identify the abstraction, law of nature, or natural phcnomcnon and explain why the claim covers every substantial practical application thereof.

$$* * * * * * * * * * * *$$

§14.4　非法定可專利之標的之電腦相關發明

　　敘述性資料（descriptive material）通常可分為功能性敘述資料（functional descriptive material)及非功能性敘述資料（nonfunctional descriptive material）。所謂功能性敘述資料，例如資料結構（data structures）及當使用作為電腦構件時賦予功能之電腦程式。而非功能性敘述資料，例如音樂或文學作品，及資料之編纂及安排。當請求項單純請求敘述資料，不論是功能性或非功能性時均非法定可予專利之標的。但是如果所請求的是紀錄（儲存）在電腦可閱讀之媒體上之功能性敘述資料時，而變成與媒體有結構及功能上之關係時，其可為法定可予專利之標的。但是如果所請求的是記錄在媒體上之非功能性敘述資料時其仍不是法定可予專利之標的，因為其並未呈現功能性以滿足實際應用之要件。

MPEP 2106.01 Computer-Related Nonstatutory Subject Matter [R-6]

　　Descriptive material can be characterized as either "functional descriptive material" or "nonfunctional descriptive material." In this context, "functional descriptive material" consists of data structures and computer programs which impart functionality when employed as a computer component. (The definition of "data structure" is "a physical or logical relationship among data elements, designed to support specific data manipulation functions." The New IEEE Standard Dictionary of Electrical and Electronics Terms 308 (5th ed. 1993).) "Nonfunctional descriptive material" includes but is not limited to music, literary works, and a compilation or mere arrangement of data.

　　Both types of "descriptive material" are nonstatutory when claimed as descriptive material per se, 33 F.3d at 1360, 31 USPQ2d at 1759. When functional descriptive material is recorded on some computer-readable medium, it becomes structurally and functionally interrelated to the medium and will be statutory in most cases since use of technology permits the function of the descriptive material to be realized.

* * * * * * * * * * *

When nonfunctional descriptive material is recorded on some computer-readable medium, in a computer or on an electromagnetic carrier signal, it is not statutory since no requisite functionality is present to satisfy the practical application requirement.

* * * * * * * * * * *

§14.5 請求方法作為 35USC 101 之適格標的之限制

方法是 35USC§101 中所敘及 4 種可予專利之標的，方法、機器、物品及組合物之一。但是，所請求之方法是否為一合格之可予專利標的，則需依照美國專利局之臨時審查基準作判斷，此臨時審查基準乃是美國專利局回應美國聯邦巡迴法院（CAFC）對 In re Bilski 案作判決所訂立之審查基準以審查方法請求項，特別是商業方法，電腦軟體是否為適格之可予專利標的之基準。該 Bilski 案（美國專利申請案 08/833,892）之請求項如下：

A method for managing the consumption risk costs of a commodity sold by a commodity provider at a fixed price,comprising the steps of:

(a) initiating a series of transactions between said commodity provider and consumers of said commodity wherein said consumers purchase said commodity at a fixed rate based upon historical averages, said fixed rate correspondimg to a risk position of said consumer;

(b) identifying market participants for said commodity having a counter-risk position to said consumers; and

(c) initiating a series of transactions between said commodity provider and said market participants at a second fixed rate such that said series of market participant transactions balances the risk position of said series of consumer transactions.

此請求項所請求之方法被美國專利局之專利上訴暨抵觸委員會（BPAI）判定為非適格之可予專利標的。Bilski 不服，上訴至 CAFC，CAFC 之法官使用機器－或-

轉換測試（Machine-or-transformation test），判定此方法並未連接至一機器（tied to a machine），或將一標的轉換成不同之狀態或事務，故判決 Bilski 之方法非 35USC§101 之專利適格標的。

依照此臨時審查基準，所請求之方法必需：(1)與另三種可予專利之標的之一結合（亦即需與機器、製品、組合物之一種或以上之結合）；或者(2)此方法將該另一種標的（如組合物、物品）轉換成不同之狀態或轉換成不同之事物。如果所請求之方法並非上述二情形者，則將被判定為非法定可予專利之方法。舉例來說，如果一請求項請求單純心智之步驟（purely mental step），則非合格之法定可予專利標的。因此，請求項必需正面地記載其連接之其他可予專利之標的。例如，在請求項中將執行方法步驟之裝置特別指出（identified），或者正面地指出被轉換之材料或被變換成何種狀態。上述之判定方法被稱為"機器－或－轉換測試"（machine-or-transformation test）。美國專利局以二種方式來進行"機器-或-轉換測試"。首先，加入單純的使用領域（field of use）之限制條件，通常不能使方法變成合格的可予專利標的。亦即，機器或轉換必需對於方法請求項之範圍賦予有意義之限制才可通過此測試。此外，不重要的額外解決之活動（extra-solution activity）不能將不可專利之方法轉換成可專利之方法。此表示在一不重要的步驟中，例如資料蒐集或輸出之步驟中，記載一特定的機器或一特定物品之特定轉換作為限制條件，並不能通過此測試。

如果經過上述測試之後，所請求之方法是合格的法定可予專利標的，需再決定的請求之方法是否為屬於法定不可予專利之標的：自然法則，自然現象或抽象概念。以及所請求之發明是否為 35USC§101 之不可專利標的之實際應用，以及所請求之發明是否占用了 35USC§101 之不可專利標的。

§14.6　實用性，實用性要件之審查基準

所謂實用性（具有用途（utility, use）），依照 35USC§101，請求專利之標的之方法、機器、製品及組合物必需是新的，有用的（useful）才是可予專利的。另外依照 35USC§102 第 1 段，說明書需包含發明之書面敘述以使此技藝人士可製造及使用

此發明。美國專利局對於一專利申請案是否符合實用性之要件（utility requirement）之審查基準如下：

(A) 請求項及支持請求項之書面敘述顯示具有確立之用途：

如果在案件審查過程中很明顯地所請求之發明從所揭露之內容來看具有確立的（well-established）用途，則不發出缺乏實用性之核駁。所謂確立的用途（well-established utility）乃指：(i)此技藝人士基於發明之特性（如產品或方法之性質或應用）可立即知道為何該發明是有用的（具有用途）及(ii)此用途是特定的、實質的及可信的。

(B) 請求項及支持請求項之書面敘述中主張特定的、實質的及可信的用途。

(1) 如果申請人有主張所請求的發明作為特定的，實用目的是有用的（即具有特定的及實質的用途），而且此主張被此技藝人士認為是可信的，則不發下缺乏實用性之核駁。

(2) 如果在支持之書面敘述中，申請人未主張任何特定的、實質的之可信的用途的話，則請求之發明被視為不具有確立的用途，會以缺乏實用性用35USC§101 核駁，同時以不符合據以實施要件（未教示如何使用）用35USC§112 第 1 段核駁。

(3) 當請求項被以缺乏用途核駁時，舉證責任就轉移至申請人，申請人必需：

(i) 明確地指出所請求之發明之特定及實質之用途，及

(ii) 提供證據證明在專利申請時此技藝人士已知道所指出之特定及實質用途在申請時已確立。

(C) 以缺乏實用性核駁時需包括有詳細的解釋。

審查委員，如果可能，要附上書面之證據（不論公告日之科技期利刊等）支持其推斷的（prima facie）缺乏實用性之核駁，如果沒有書面之證據，則審查委員需解釋其結論之科學依據。

(D) 申請人所主張之發明的用途被認為是特定的、實質的及可信的時，缺乏實用性之核駁應撤回，除非審查委員有證據可反駁申請人所主張的用途。則審查委員應相信申請人之主張實用性之陳述（statement）是真的。同樣的，如果由合格之專家提出證據主張具有實用性，則審查委員應接受此專家之意見。

MPEP 2107 Guidelines for Examination of Applications for Compliance with the Utility Requirement

＊＊＊＊＊＊＊＊＊＊＊

II.EXAMINATION GUIDELINES FOR THE UTILITY REQUIREMENT

Office personnel are to adhere to the following procedures when reviewing patent applications for compliance with the "useful invention" ("utility") requirement of 35 U.S.C. 101 and 112, first paragraph.

(A) Read the claims and the supporting written description.

(1) Determine what the applicant has claimed, noting any specific embodiments of the invention.

(2) Ensure that the claims define statutory subject matter (i.e., a process, machine, manufacture, composition of matter, or improvement thereof).

(3) If at any time during the examination, it becomes readily apparent that the claimed invention has a well-established utility, do not impose a rejection based on lack of utility. An invention has a well-established utility if (i) a person of ordinary skill in the art would immediately appreciate why the invention is useful based on the characteristics of the invention (e.g., properties or applications of a product or process), and (ii) the utility is specific, substantial, and credible.

(B) Review the claims and the supporting written description to determine if the applicant has asserted for the claimed invention any specific and substantial utility that is credible:

(1) If the applicant has asserted that the claimed invention is useful for any particular practical purpose (i.e., it has a "specific and substantial utility") and the assertion would be considered credible by a person of ordinary skill in the art, do not impose a rejection based on lack of utility.

＊＊＊＊＊＊＊＊＊＊＊

(2) If no assertion of specific and substantial utility for the claimed invention made by the applicant is credible, and the claimed invention does not have a readily apparent well-established utility, reject the claim(s) under 35 U.S.C. 101 on the grounds that the invention as claimed lacks utility. Also reject the claims under 35 U.S.C. 112, first paragraph, on the basis that the disclosure fails to teach how to use the invention as claimed. The 35

U.S.C. 112, first paragraph, rejection imposed in conjunction with a 35 U.S.C. 101 rejection should incorporate by reference the grounds of the corresponding 35 U.S.C. 101 rejection.

(3) If the applicant has not asserted any specific and substantial utility for the claimed invention and it does not have a readily apparent well-established utility, impose a rejection under 35 U.S.C. 101, emphasizing that the applicant has not disclosed a specific and substantial utility for the invention. Also impose a separate rejection under 35 U.S.C. 112, first paragraph, on the basis that the applicant has not disclosed how to use the invention due to the lack of a specific and substantial utility. The 35 U.S.C. 101 and 112 rejections shift the burden of coming forward with evidence to the applicant to:

(i) Explicitly identify a specific and substantial utility for the claimed invention; and

(ii) Provide evidence that one of ordinary skill in the art would have recognized that the identified specific and substantial utility was well-established at the time of filing. The examiner should review any subsequently submitted evidence of utility using the criteria outlined above. The examiner should also ensure that there is an adequate nexus between the evidence and the properties of the now claimed subject matter as disclosed in the application as filed. That is, the applicant has the burden to establish a probative relation between the submitted evidence and the originally disclosed properties of the claimed invention.

* * * * * * * * * * *

(C) Any rejection based on lack of utility should include a detailed explanation why the claimed invention has no specific and substantial credible utility. Whenever possible, the examiner should provide documentary evidence regardless of publication date (e.g., scientific or technical journals, excerpts from treatises or books, or U.S. or foreign patents) to support the factual basis for the prima facie showing of no specific and substantial credible utility. If documentary evidence is not available, the examiner should specifically explain the scientific basis for his or her factual conclusions.

(D) A rejection based on lack of utility should not be maintained if an asserted utility for the claimed invention would be considered specific, substantial, and credible by a person of ordinary skill in the art in view of all evidence of record.

Office personnel are reminded that they must treat as true a statement of fact made by an applicant in relation to an asserted utility, unless countervailing evidence can be provided that shows that one of ordinary skill in the art would have a legitimate basis to doubt the credibility of such a statement. Similarly, Office personnel must accept an opinion from a qualified expert that is based upon relevant facts whose accuracy is not being questioned; it is improper to disregard the opinion solely because of a disagreement over the significance or meaning of the facts offered.

* * * * * * * * * * *

§14.7　缺乏實用性之發明

　　通常所請求之發明缺乏實用性有二種情形，第 1 種情況是申請人在說明書中未指出發明之任何特定（specific）及實質的（substantial）用途，或者未揭露足夠的資訊，以讓技藝人士可立即了解此發明之用途（實用性）。另一種情形是申請人在說明書中所主張之用途是不可信的。以下說明之：

I.　特定及實質的用途

(A)　特定的用途（specific utility）

　　所謂特定的用途乃指此用途對於所請求之標的是特定的，而且可提供大眾明確及特殊的利益而言。例如所請求者為一醫藥用之化合物，而申請人只指出此醫藥化合物可治療癌症，但未指出哪一種癌症，或只指出此醫藥化合物之有用的生物學性質（biological properties），則被認為是不具有特定的用途。又例如所請求之發明為一多核苷酸（polynucleotide），在說明書中只揭露可作為基因探針或染色體標誌，但未揭露特定的 DNA 標靶（DNA target），則會被認為不具有特定之用途。

(B)　實質的用途（substantial utility）

　　所謂實質的用途乃指所請求之發明以其所揭露的現有之形態（current form）對大眾是有用的。如果要再經研究之後在將來可被證明是有用的，則不是實質的用途。簡單來說，就是所主張之用途必需顯示所請求發明對大眾具有明顯且立即可得之利益。亦即，實質的用途應是"真實世界"之用途。例如下列數種發明將被認為不具實質的用途：

(a)　基礎研究，例如，所請求之產品本身之性質或者機轉(mechanisms)之研究。

(b)　治療一未特定之疾病或症狀之方法。

(c)　分析或確認一物質之方法，但是該物質本身並無特定及/或實質之用途。

(d)　製造一物質之方法，但該物質本身並無特定、實質上及可信之用途。

(e) 中間產物其用來製造一最終產品，但此最終產品本身並無特定、實質及可信之用途。

II. 完全不能操作之發明：不可信之用途

一個發明，如果是不能操作的（亦即不能操作以產生所請求之結果）則不是有用之發明。但是要完全不能達成有用的結果才不是有用的發明，發明只要有很小程度上之用途就是有用的發明。因此，不能因為說明書中之某些實施例不完全或操作上不順暢，就認為不是有用的發明。一發明亦不必要能完成所有希望的功能才是有用的發明。亦即，一發明只要能夠在達成有用的結果上可部份成功，即是有用之發明。

III. 治療或醫藥上之用途

一發明其主張在治療人類及動物上具有（用途）者，在審查是否具有實用性乃採用與其他發明相同之法律標準。只要藥物或治療上之發明可對大眾提供任何立即之利益的話，就算符合 35USC§101 關於實用性之要件。因此，只要單單指出一化合物之藥理活性（pharmacological activities），就被視為提供大眾立即的利益而滿足實用性要件。在 Nelson v. Bowler 之判例中，Nelson 請求了一合成的 prostaglandin，其在說明書中揭露了測試以天然產出之 prostaglandin 和合成之 prostaglandin 的生物活性之結果，法院判定此揭露已符合了 35USC§101 實用性之要件。法院同時指出專利法上之治療用途要件並不應與食品藥物管理局（FDA）之有關藥物之安全性及有效性之要件混淆。

如果一發明不能滿足 35USC§101 之實用性之要件（無用途），則同時不能滿足 35USC§112 第 1 段之據以實施要件，因為如果一發明沒有用途，則此技藝人士無法使用該發明。

但是，申請人揭露了一發明之特定且實質之用途，也提供了可信的證據來支持此特定及實質之用途，並不表示所請求之發明符合 35USC§112 第 1 段之據以實施要件。例如，申請人請求了以某種化合物治療某種癌症或症狀之方法，且提供了可信之證據證明此化合物是有用的（useful），但是此技藝人士在實際上實施此發明時卻必需進行不適當的實驗才可能使用此發明，則此發明是不符 35USC§112 第 1 段之據以實施要件。

MPEP 2107.01 General Principles Governing Utility Rejections [R-5]

＊＊＊＊＊＊＊＊＊＊＊

Deficiencies under the "useful invention" requirement of 35 U.S.C. 101 will arise in one of two forms. The first is where it is not apparent why the invention is "useful." This can occur when an applicant fails to identify any specific and substantial utility for the invention or fails to disclose enough information about the invention to make its usefulness immediately apparent to those familiar with the technological field of the invention.

＊＊＊＊＊＊＊＊＊＊＊

I.SPECIFIC AND SUBSTANTIAL REQUIREMENTS

＊＊＊＊＊＊＊＊＊＊＊

A.Specific Utility

A "specific utility" is specific to the subject matter claimed >and can "provide a well-defined and particular benefit to the public.

＊＊＊＊＊＊＊＊＊＊＊

For example, indicating that a compound may be useful in treating unspecified disorders, or that the compound has "useful biological" properties, would not be sufficient to define a specific utility for the compound. >See, e.g., In re Kirk, 376 F.2d 936, 153 USPQ 48 (CCPA 1967); In re Joly, 376 F.2d 906, 153 USPQ 45 (CCPA 1967).< Similarly, a claim to a polynucleotide whose use is disclosed simply as a "gene probe" or "chromosome marker" would not be considered to be specific in the absence of a disclosure of a specific DNA target. >See In re Fisher, 421 F.3d at 1374, 76 USPQ2d at 1232

＊＊＊＊＊＊＊＊＊＊＊

B.Substantial Utility

*>"[A]n application must show that an invention is useful to the public as disclosed in its current form, not that it may prove useful at some future date after further research. Simply put, to satisfy the 'substantial' utility requirement, an asserted use must show that the claimed invention has a significant and presently available benefit to the public."

＊＊＊＊＊＊＊＊＊＊＊

On the other hand, the following are examples of situations that require or constitute carrying out further research to identify or reasonably confirm a "real world" context of use and, therefore, do not define "substantial utilities":

(A) Basic research such as studying the properties of the claimed product itself or the mechanisms in which the material is involved;

(B) A method of treating an unspecified disease or condition;

(C) A method of assaying for or identifying a material that itself has no specific and/or substantial utility;

(D) A method of making a material that itself has no specific, substantial, and credible utility; and

(E) A claim to an intermediate product for use in making a final product that has no specific, substantial and credible utility.

＊＊＊＊＊＊＊＊＊＊＊

II.WHOLLY INOPERATIVE INVENTIONS; "INCREDIBLE" UTILITY

An invention that is "inoperative" (i.e., it does not operate to produce the results claimed by the patent applicant) is not a "useful" invention in the meaning of the patent law. See, e.g., Newman v. Quigg, 877 F.2d 1575, 1581, 11 USPQ2d 1340, 1345 (Fed. Cir. 1989); In re Harwood, 390 F.2d 985, 989, 156 USPQ 673, 676 (CCPA 1968) ("An inoperative invention, of course, does not satisfy the requirement of 35 U.S.C. 101 that an invention be useful."). However, as the Federal Circuit has stated, "[t]o violate [35 U.S.C.] 101 the claimed device must be totally incapable of achieving a useful result." Brooktree Corp. v. Advanced Micro Devices, Inc., 977 F.2d 1555, 1571, 24 USPQ2d 1401, 1412 (Fed. Cir. 1992) (emphasis added). See also E.I. du Pont De Nemours and Co. v. Berkley and Co., 620 F.2d 1247, 1260 n.17, 205 USPQ 1, 10 n.17 (8th Cir. 1980) ("A small degree of utility is sufficient . . . The claimed invention must only be capable of performing some beneficial function . . . An invention does not lack utility merely because the particular embodiment disclosed in the patent lacks perfection or performs crudely . . . A commercially successful product is not required . . . Nor is it essential that the invention accomplish all its intended functions . . . or operate under all conditions . . . partial success being sufficient to demonstrate patentable utility . . . In short, the defense of non-utility cannot be sustained without proof of total incapacity." If an invention is only partially successful in achieving a useful result, a rejection of the claimed invention as a whole based on a lack of utility is not appropriate.

＊＊＊＊＊＊＊＊＊＊＊

III.THERAPEUTIC OR PHARMACOLOGICAL UTILITY

Inventions asserted to have utility in the treatment of human or animal disorders are subject to the same legal requirements for utility as inventions in any other field of technology.

＊＊＊＊＊＊＊＊＊＊＊

As such, pharmacological or therapeutic inventions that provide any "immediate benefit to the public" satisfy 35 U.S.C. 101.

＊＊＊＊＊＊＊＊＊＊＊

Courts have repeatedly found that the mere identification of a pharmacological activity of a compound that is relevant to an asserted pharmacological use provides an "immediate benefit to the public" and thus satisfies the utility requirement.

＊＊＊＊＊＊＊＊＊＊＊

Nelson had developed and claimed a class of synthetic prostaglandins modeled on naturally occurring prostaglandins.

＊＊＊＊＊＊＊＊＊＊＊

Nelson included in his application the results of tests demonstrating the bioactivity of his new substituted prostaglandins relative to the bioactivity of naturally occurring prostaglandins.

＊＊＊＊＊＊＊＊＊＊＊

The Federal Circuit has reiterated that therapeutic utility sufficient under the patent laws is not to be confused with the requirements of the FDA with regard to safety and efficacy of drugs to marketed in the United States.

＊＊＊＊＊＊＊＊＊＊＊

IV.RELATIONSHIP BETWEEN 35 U.S.C. 112, FIRST PARAGRAPH, AND 35 U.S.C. 101

A deficiency under >the utility prong of< 35 U.S.C. 101 also creates a deficiency under 35 U.S.C. 112, first paragraph.

＊＊＊＊＊＊＊＊＊＊＊

("The how to use prong of section 112 incorporates as a matter of law the requirement of 35 U.S.C. 101 that the specification disclose as a matter of fact a practical utility for the invention. ...... If the application fails as a matter of fact to satisfy 35 U.S.C. § 101, then the application also fails as a matter of law to enable one of ordinary skill in the art to use the invention under 35 U.S.C. § 112.");

＊＊＊＊＊＊＊＊＊＊＊

The fact that an applicant has disclosed a specific utility for an invention and provided a credible basis supporting that specific utility does not provide a basis for concluding that the claims comply with all the requirements of 35 U.S.C. 112, first paragraph. For example, if an applicant has claimed a process of treating a certain disease condition with a certain compound and provided a credible basis for asserting that the compound is useful in that regard, but to actually practice the invention as claimed a person skilled in the relevant art would have to engage in an undue amount of experimentation, the claim may be defective under 35 U.S.C. 112, but not 35 U.S.C. 101.

＊＊＊＊＊＊＊＊＊＊＊

§14.8　缺乏實用性核駁之程序

美國專利局以缺乏實用性核駁一請求項時，依照以下之程序進行：

I . 將焦點放在請求項所請求之發明以考量是否缺乏實用性

美國專利局之審查在審查是否具有實用性時乃針對每一請求項所請求之發明逐一作審查。當然，當一獨立項之發明被認為具有實用性時，其依附項，如果請求之標的相同，通常被認為是有實用性的。因此除非依附項所定義之發明中所指出之實用性與獨立項不同時，審查委員才需另外去審查依附項之實用性。

再者，當所請求之發明，特別當標的是產品時，有數種不同之用途（實用性）時，申請人只要主張其中之一種實用性是可信的，該發明就具有實用性。

II.在說明書中申請人是否有主張所請求之發明具有實用性或有確立用途

審查委員先首先檢查在說明書中是否有任何主張所請求之發明在達成某特定目的上是有用的陳述。完整的揭露應該包括指出特定的、實質的用途之陳述。所謂特定的及實質的實用性的陳述必需完整且清楚地解釋為何申請人相信所請求的發明是有用的（useful）。例如一化合物在治療一特定的症狀上被相信是有用的。通常來說，除了所請求之發明具有確立的用途（實用性）（well-established utility）之外，如果申請人不能在說明書中特定地指出為何所請求之發明相信是有用的（useful），就被認為所請求之發明不符 35USC§101 之實用性要件及 35USC§112 第 2 段之據以實施要件。

但是，說明書中雖沒有主張具有特定的、實質的用途之陳述，並不一定就表示所請求之說明不具有實用性。在此情況下，美國專利局之審查委員會再決定所請求發明是否有確立之用途。所請求之發明被認為是具有實用性的，如果(i)此技藝人士基於發明之特性（產品或方法之性質或應用）可立即了解為何發明是有用的，及(ii)此用途是特定的、實質的及可信的。例如一申請案是有關於選殖及特定化一習知的蛋白質，如胰島素之核苷酸序列。在說明書中雖未主張此發明具有用途，但此技藝人士均知胰島素具有確立之用途。

Ⅲ. 評估所主張之用途之可信度

多數的情況，申請人之用途的主張就形成了實用性要件已滿足之假定。亦即，申請人的實用性之主張是可信的，除非(A)此主張之邏輯性有嚴重之錯誤，或(B)主張之基礎之事實與主張邏輯不一致。所以美國專利局之審查委員如果要推翻此假定，需指出占優勢之書面證據。

Ⅳ. 推斷缺乏實用性需負舉證責任

美國專利局之審查如果要以缺乏實用性核駁，則需(A)建立推斷的案件顯示請求之發明缺乏實用性，及(B)提供足夠的證據以支持此推斷之案例。

Ⅴ. 審查委員可要求提出書面證據支持所主張之用途

在適當之情形下，例如，此技藝人士會認為方法在操作上不可能時，審查委員可要求申請人提供證據支持其所主張之用途。

VI. 考慮缺乏實用性之推斷核駁之答辯

如果申請人對缺乏實用性之推斷核駁作出答辯，則美國專利局會再審閱原有之說明書之揭露、審查委員以推斷核駁之證據、請求項之修正、及申請人所提供用以主張特定及實質的用途之證據。如果整體來說，所主張實用性不是特定、實質及可信的，則維持缺乏實用性之核駁。如果整體來看所主張的實用性由此技藝人士可考慮為可信的，則審查委員將不維持原缺乏實用性之核駁。

VII. 評估關於實用性（用途）之證據

申請人用來主張實用性之證據並沒有規定量要多少或者證據之特性需如何才可。需要用來支持所主張之實用性之證據之特性及量會依所請求之標的之不同而有所不同。另外，申請人不必提供足以確立所主張之實用性為超越合理懷疑之真實的證據。也不必提供統計上可確立為正確之證據以主張實用性。

MPEP 2107.02 Procedural Considerations Related to Rejections

I.THE CLAIMED INVENTION IS THE FOCUS OF THE UTILITY REQUIREMENT

The claimed invention is the focus of the assessment of whether an applicant has satisfied the utility requirement. Each claim (i.e., each "invention"), therefore, must be evaluated on its own merits for compliance with all statutory requirements. Generally speaking, however, a dependent claim will define an invention that has utility if the >independent< claim from which the dependent claim depends is drawn to the same statutory class of invention as the dependent claim and the independent claim defines< an invention having utility. An exception to this general rule is where the utility specified for the invention defined in a dependent claim differs from that indicated for the invention defined in the independent claim from which the dependent claim depends.

＊＊＊＊＊＊＊＊＊＊＊

It is common and sensible for an applicant to identify several specific utilities for an invention, particularly where the invention is a product (e.g., a machine, an article of manufacture or a composition of matter). However, regardless of the category of invention that is claimed (e.g., product or process), an applicant need only make one credible assertion of specific utility for the claimed invention to satisfy 35 U.S.C. 101

＊＊＊＊＊＊＊＊＊＊＊

II. IS THERE AN ASSERTED OR WELL-ESTABLISHED UTILITY FOR THE CLAIMED INVENTION?

Upon initial examination, the examiner should review the specification to determine if there are any statements asserting that the claimed invention is useful for any particular purpose. A complete disclosure should include a statement which identifies a specific and substantial utility for the invention.

A statement of specific and substantial utility should fully and clearly explain why the applicant believes the invention is useful. Such statements will usually explain the purpose of or how the invention may be used (e.g., a compound is believed to be useful in the treatment of a particular disorder). Regardless of the form of statement of utility, it must enable one ordinarily skilled in the art to understand why the applicant believes the claimed invention is useful.

Except where an invention has a well-established utility, the failure of an applicant to specifically identify why an invention is believed to be useful renders the claimed invention deficient under 35 U.S.C. 101 and 35 U.S.C. 112, first paragraph.

Occasionally, an applicant will not explicitly state in the specification or otherwise assert a specific and substantial utility for the claimed invention. If no statements can be found asserting a specific and substantial utility for the claimed invention in the specification, Office personnel should determine if the claimed invention has a well-established utility. An invention has a well-established utility if (i) a person of ordinary skill in the art would immediately appreciate why the invention is useful based on the characteristics of the invention (e.g., properties or applications of a product or process), and (ii) the utility is specific, substantial, and credible. If an invention has a well- established utility, rejections under 35 U.S.C. 101 and 35 U.S.C. 112, first paragraph, based on lack of utility should not be imposed. In re Folkers, 344 F.2d 970, 145 USPQ 390 (CCPA 1965). For example, if an application teaches the cloning and characterization of the nucleotide sequence of a well-known protein such as insulin, and those skilled in the art at the time of filing knew that insulin had a well-established use, it would be improper to reject the claimed invention as lacking utility solely because of the omitted statement of specific and substantial utility.

＊＊＊＊＊＊＊＊＊＊＊

III. EVALUATING THE CREDIBILITY OF AN ASSERTED UTILITY

＊＊＊＊＊＊＊＊＊＊＊

In most cases, an applicant's assertion of utility creates a presumption of utility that will be sufficient to satisfy the utility requirement of 35 U.S.C. 101.

＊＊＊＊＊＊＊＊＊＊＊

Thus, to overcome the presumption of truth that an assertion of utility by the applicant enjoys, Office personnel must establish that it is more likely than not that one of ordinary skill in the art would doubt (i.e., "question") the truth of the statement of utility. The evidentiary standard to be used throughout ex parte examination in setting forth a rejection is a preponderance of the totality of the evidence under consideration.

＊＊＊＊＊＊＊＊＊＊＊

IV. INITIAL BURDEN IS ON THE OFFICE TO ESTABLISH A PRIMA FACIE CASE AND PROVIDE EVIDENTIARY SUPPORT THEREOF

To properly reject a claimed invention under 35 U.S.C. 101, the Office must (A) make a prima facie showing that the claimed invention lacks utility, and (B) provide a sufficient evidentiary basis for factual assumptions relied upon in establishing the prima facie showing.

＊＊＊＊＊＊＊＊＊＊＊

V. EVIDENTIARY REQUESTS BY AN EXAMINER TO SUPPORT AN ASSERTED UTILITY

In appropriate situations the Office may require an applicant to substantiate an asserted utility for a claimed invention. See In re Pottier, 376 F.2d 328, 330, 153 USPQ 407, 408 (CCPA 1967) ("When the operativeness of any process would be deemed unlikely by one of ordinary skill in the art, it is not improper for the examiner to call for evidence of operativeness.").

＊＊＊＊＊＊＊＊＊＊＊

VI. CONSIDERATION OF A REPLY TO A PRIMA FACIE REJECTION FOR LACK OF UTILITY

＊＊＊＊＊＊＊＊＊＊＊

If the applicant responds to the prima facie rejection, Office personnel should review the original disclosure, any evidence relied upon in establishing the prima facie showing, any claim amendments, and any new reasoning or evidence provided by the applicant in support of an asserted specific and substantial credible utility. It is essential for Office personnel to recognize, fully consider and respond to each substantive element of any response to a rejection based on lack of utility. Only where the totality of the record continues to show that the asserted utility is not specific, substantial, and credible should a rejection based on lack of utility be maintained. If the record as a whole would make it more likely than not that the asserted utility for the claimed invention would be considered credible by a person of ordinary skill in the art, the Office cannot maintain the rejection. In re Rinehart, 531 F.2d 1048, 1052, 189 USPQ 143, 147 (CCPA 1976).

VII. EVALUATION OF EVIDENCE RELATED TO UTILITY

There is no predetermined amount or character of evidence that must be provided by an applicant to support an asserted utility, therapeutic or otherwise. Rather, the character and amount of evidence needed to support an asserted utility will vary depending on what is claimed (Ex parte Ferguson, 117 USPQ 229 (Bd. App. 1957)), and whether the asserted utility appears to contravene established scientific principles and beliefs. In re Gazave, 379 F.2d 973, 978, 154 USPQ 92, 96 (CCPA 1967); In re Chilowsky, 229 F.2d 457, 462, 108 USPQ 321, 325 (CCPA 1956). Furthermore, the applicant does not have to provide evidence sufficient to establish that an asserted utility is true "beyond a reasonable doubt." In re Irons, 340 F.2d 974, 978, 144 USPQ 351, 354 (CCPA 1965). Nor must an applicant provide evidence such that it establishes an asserted utility as a matter of statistical certainty.

＊＊＊＊＊＊＊＊＊＊＊＊

§14.9　具有治療及醫藥上之用途之發明的實用性

聯邦法院對於申請人所主張之治療及醫藥上的用途是否符合 35USC§101 之實用性的規定已有一致之決定，因此美國專利局在申請人主張其發明有治療及醫藥上之用途時，有一些特別的考量。說明如下：

Ⅰ. 證據與主張之用途有合理之相互關連即足夠

一般的原則為：一化合物之藥學或其他生物活性之證據是與醫療用途相關的，只要活性與所主張用途間有合理的相關連即可。申請人可藉由統計上之相關資料，書面上之證據，例如，科學期刊之論文來建立合理的相互關連，申請人不必提供特定的活性與一化合物所主張醫療用之關係，不必提供在治療人類疾病上真正成功之證據。

Ⅱ. 與具有確立之用途之化合物之結構相似性

如果申請人所請求之新的化合物之結構與一習知化合物類似。而此習知之化合物為人知道是具有治療及醫藥上之用途的話，美國專利局之審查委員在決定是否此技藝人士會認為所主張之用途是可信的時需給予相當之比重。但是審查委員不但需評估是否結構之類似性確實存在，還需檢視結構類似是否確實與所主張之用途相關。

Ⅲ. 體外（IN VITRO）或動物實驗之資料通常足夠支持醫療之用途

如果申請人提供體外試驗、或動物試驗、或兩者之資料，並說明為何提供之資料可支持所主張之用途時，美國專利局會決定此技藝人士是否會將所提供之資料及說明視為所主張之用途的合理推測。如果是，則此體外、動物資料就足以支持所主張之用途。

但是，申請人所提供之證據並不一定必需是對該特定疾病或症狀為業界作承認所接受之動物模式。只要此技藝人士可接受其可合理預測在人類具有用途，則其證據就被認為是足夠支持醫療上的用途。

Ⅳ. 人類臨床資料

目前並無判例或法律要求申請人需提供人類臨床資料來建立其發明在治療人類疾病上之用途，即使此技藝人士可能無法接受人類臨床試驗以外之其他證據時亦然。此外，如果申請人已經開始進行一醫療產品或方法之人類臨床試驗，則美國專利局就應該假定申請人已經建立了所請求之標的與所主張醫療用途之關係。

Ⅴ. 安全及藥效之考量

　　美國專利局在審查一所請求之發明之實用性（是否具有用途）時，只針對專利法上之要件作審查，而不管所請求之發明的安全、銷售、廣告是否符合其他法律之要求。因而，FDA（食品藥物管理局）之核准上市並非美國專利局在審查一化合物是否合乎專利法上有用（useful）之要件之先決條件。所以，在某些情況下，申請人需提供證據顯示其發明可如所請求地作運作（操作），但是美國專利局不會要求提供在治療人類疾病上之安全性及藥效程度之證據。

Ⅵ. 特定疾病（症狀）之治療

　　如果所請求之發明是治療某種特定疾病（症狀）之方法，而此種特定之病症在之前並未有任何成功治療（治癒）之例子時，並不能就下結論此發明缺乏實用性（具有用途）。美國專利局之審查委員仍需小心地依據所揭露之資訊決定所主張之用途是否是可信的，如果申請人可提供由此方面之專家指出有治療（治癒）此病症之合理的成功機率，並以相同之理由、證據支持的話，通常就被認為所主張之用途是可信的。

MPEP2107.03 Special Considerations for Asserted Therapeutic or Pharmacological Utilities

I.A REASONABLE CORRELATION BETWEEN THE EVIDENCE AND THE ASSERTED UTILITY IS SUFFICIENT

As a general matter, evidence of pharmacological or other biological activity of a compound will be relevant to an asserted therapeutic use if there is a reasonable correlation between the activity in question and the asserted utility. Cross v. Iizuka, 753 F.2d 1040, 224 USPQ 739 (Fed. Cir. 1985); In re Jolles, 628 F.2d 1322, 206 USPQ 885 (CCPA 1980); Nelson v. Bowler, 626 F.2d 853, 206 USPQ 881 (CCPA 1980). An applicant can establish this reasonable correlation by relying on statistically relevant data documenting the activity of a compound or composition, arguments or reasoning, documentary evidence (e.g., articles in scientific journals), or any combination thereof. The applicant does not have to prove that a correlation exists between a particular activity and an asserted therapeutic use of a compound as a matter of statistical certainty, nor does he or she have to provide actual evidence of success in treating humans where such a utility is asserted.

＊＊＊＊＊＊＊＊＊＊＊

II.STRUCTURAL SIMILARITY TO COMPOUNDS WITH ESTABLISHED UTILITY

Courts have routinely found evidence of structural similarity to a compound known to have a particular therapeutic or pharmacological utility as being supportive of an assertion of therapeutic utility for a new compound.

＊＊＊＊＊＊＊＊＊＊＊

Such evidence should be given appropriate weight in determining whether one skilled in the art would find the asserted utility credible. Office personnel should evaluate not only the existence of the structural relationship, but also the reasoning used by the applicant or a declarant to explain why that structural similarity is believed to be relevant to the applicant"s assertion of utility.

III.DATA FROM IN VITRO OR ANIMAL TESTING IS GENERALLY SUFFICIENT TO SUPPORT THERAPEUTIC UTILITY

＊＊＊＊＊＊＊＊＊＊＊

If an applicant provides data, whether from in vitro assays or animal tests or both, to support an asserted utility, and an explanation of why that data supports the asserted utility, the Office will determine if the data and the explanation would be viewed by one skilled in the art as being reasonably predictive of the asserted utility.

＊＊＊＊＊＊＊＊＊＊＊

Evidence does not have to be in the form of data from an art-recognized animal model for the particular disease or disease condition to which the asserted utility relates. Data from any test that the applicant reasonably correlates to the asserted utility should be evaluated substantively. Thus, an applicant may provide data generated using a particular animal model with an appropriate explanation as to why that data supports the asserted utility. The absence of a certification that the test in question is an industry-accepted model is not dispositive of whether data from an animal model is in fact relevant to the asserted utility. Thus, if one skilled in the art would accept the animal tests as being reasonably predictive of utility in humans, evidence from those tests should be considered sufficient to support the credibility of the asserted utility.

＊＊＊＊＊＊＊＊＊＊＊

IV.HUMAN CLINICAL DATA

Office personnel should not impose on applicants the unnecessary burden of providing evidence from human clinical trials. There is no decisional law that requires an applicant to provide data from human clinical trials to establish utility for an invention related to treatment of human disorders.

＊＊＊＊＊＊＊＊＊＊＊

(human clinical data is not required to demonstrate the utility of the claimed invention, even though those skilled in the art might not accept other evidence to establish the efficacy of the claimed therapeutic compositions and the operativeness of the claimed methods of treating humans).

＊＊＊＊＊＊＊＊＊＊＊

general rule, if an applicant has initiated human clinical trials for a therapeutic product or process, Office personnel should presume that the applicant has established that the subject matter of that trial is reasonably predictive of having the asserted therapeutic utility.

V.SAFETY AND EFFICACY CONSIDERA-TIONS

The Office must confine its review of patent applications to the statutory requirements of the patent law.

＊＊＊＊＊＊＊＊＊＊＊

Thus, in challenging utility, Office personnel must be able to carry their burden that there is no sound rationale for the asserted utility even though experts designated by Congress to decide the issue have come to an opposite conclusion. "FDA approval, however, is not a prerequisite for finding a compound useful within the meaning of the patent laws.

＊＊＊＊＊＊＊＊＊＊＊

Thus, while an applicant may on occasion need to provide evidence to show that an invention will work as claimed, it is improper for Office personnel to request evidence of safety in the treatment of humans, or regarding the degree of effectiveness.

＊＊＊＊＊＊＊＊＊＊＊

VI.TREATMENT OF SPECIFIC DISEASE CONDITIONS

Claims directed to a method of treating or curing a disease for which there have been no previously successful treatments or cures warrant careful review for compliance with 35 U.S.C. 101.

* * * * * * * * * * *

The fact that there is no known cure for a disease, however, cannot serve as the basis for a conclusion that such an invention lacks utility. Rather, Office personnel must determine if the asserted utility for the invention is credible based on the information disclosed in the application.

* * * * * * * * * * *

Thus, affidavit evidence from experts in the art indicating that there is a reasonable expectation of success, supported by sound reasoning, usually should be sufficient to establish that such a utility is credible.

第 15 章　答辯與修正

目　錄

§15.1　先前修正之內容及時間上之限制

　　申請人在專利案提出申請後如要對說明書、請求項、圖式作先前修正（preliminary amendment）通常需在審查委員發出第 1 次審定書（first office action）之前收到，才會被接受（entered），但是如果審查委員已在準備發出第 1 次審定書而此先前修正會影響到審定書之實質內容時（審查委員會考慮先前修正之內容及其第 1 次審定書完成之程度），此先前修正亦可能不被接受。然而，如果申請人之先前修正是在申請日起 3 個月內提出，則審查委員不能拒絕此先前修正。

　　申請人所作之先前修正內容必需在最初申請之專利說明書、圖式、請求項可支持者才可被接受。另外，如果提出專利申請時同時作先前修正而且申請時並未提出宣誓書，則此先前修正之內容可不必由說明書、圖式，請求項所支持，但是在後補之宣誓書中必需提及此先前修正之內容。再者，如果提出申請時未繳納申請規費亦可藉由先前修正將請求項刪除以減少超項費。

---

37 CFR§1.115 Preliminary amendments.

(a) A preliminary amendment is an amendment that is received in the Office (§1.6) on or before the mail date of the first Office action under §1.104.The patent application publication may include preliminary amendments (§1.215(a)).

(1)A preliminary amendment that is present on the filing date of an application is part of the original disclosure of the application.

(2) A preliminary amendment filed after the filing date of the application is not part of the original disclosure of the application.

(b) A preliminary amendment in compliance with § 1.121 will be entered unless disapproved by the Director.

(1) A preliminary amendment seeking cancellation of all the claims without presenting any new or substitute claims will be disapproved.

(2) A preliminary amendment may be disapproved if the preliminary amendment unduly interferes with the preparation of a first Office action in an application. Factors that will be considered in disapproving a preliminary amendment include:

(i) The state of preparation of a first Office action as of the date of receipt (§1.6) of the preliminary amendment by the Office; and

(ii) The nature of any changes to the specification or claims that would result from entry of the preliminary amendment.

(3) A preliminary amendment will not be disapproved under (b)(2) of this section if it is filed no later than:

(i) Three months from the filing date of an application under §1.3(b);

(ii)The filing date of a continued prosecution application under §1.53(d); or

(iii) Three months from the date the national stage is entered as set forth in §1.491in an international application.

(4) The time periods specified in paragraph (b)(3) of this section are not extendable.

§15.2 法定期限

在專利申請案之審查過程中答復美國專利局之官方函之期限有下列二種：(a)法定期限（statutory due dates）及(b)非法定期限（non-statutory due dates）對於美國專利局之審定書（office action,OA）之答復之期限為法定期限，均為自美國專利局之郵寄日期起算 6 個月為期限。但是，美國專利局之人員(審查委員)可自行訂定縮短的法定期限（shortened statutory period(SSP)）。申請人如果不能在 SSP 之內答復則需申請延期並繳延期費，美國專利局訂定之法定期限必需至少 30 天。

通常審查委員如果發下的審定書是限制/選擇之審定書時，其 SSP 會訂為 1 個月，可再延期（同時繳延期費）5 個月。而如果發下的審定書是針對專利要件者，則 SSP 通常訂為 3 個月可再延期 3 個月，而如果是 Ex Parte Quayle 之審定書專利申請案已完全能符合專利要件，僅需對一些程式（打字錯誤等）之錯誤作修正則 SSP 是二個月，可延期 4 個月。

而對於上述超過 SSP 期限之延期可不必在 SSP 期限前提出，只要在答復時提出請願（petition，申請）並繳交延期費即可。如果只繳完了延期費或授權美國專利局於帳戶中扣款但並未提出延期之請願，仍可被視為延期之請願（constructive petition for extension）。上述的自動延期申請需在審定書之郵寄日起 6 個月內提出，如果超過 6 個月之法定期限，除非提出正當理由，否則將被視為未答復而造成專利申請案之放棄。若 SSP 之期限當日為週六、週日或連續假日則期限為其次日（美國專利局上班之日）。

35 USC 133

Upon failure of the applicant to prosecute the application within six months after any action therein, of which notice has been given or mailed to the applicant, or within such shorter time, not less than thirty days, as fixed by the Director in such action, the application shall be regarded as abandoned by the parties thereto, unless it be shown to the satisfaction of the Director that such delay was unavoidable.

37 CFR 1.136

(a)(1) If an applicant is required to reply within a nonstatutory or shortened statutory time period, applicant may extend the time period for reply up to the earlier of the expiration of any maximum period set by statute or five months after the time period set for reply, if a petition for an extension of time and the fee set in §1.17(a) are filed, unless:

(i) Applicant is notified otherwise in an Office action;

(ii) The reply is a reply brief submitted pursuant to § 1.193(b);

(iii) The reply is a request for an oral hearing submitted pursuant to §1.194(b);

(iv) The reply is to a decision by the Board of Patent Appeals and Interferences pursuant to §1.196 §1.197 or §1.304; or

(v) The application is involved in a interference declared pursuant to §1.611; or

(2) The date on which the petition and the fee have been filed is the date for purposes of determining the period of extension and the corresponding amount of the fee. The expiration of the time period is determined by the amount of the fee paid. A reply must be filed prior to the expiration of the period of extension to avoid abandonment of the application (§1.135), but in no situation may an applicant reply later than the maximum time period set by statute, or be granted an extension of time under paragraph (b) of this section when the provisions of this paragraph are available. See §1.36 (b) for extensions of time to relating to proceedings pursuant to §§ 1.193(b), 1.194, 1.196 or 1.197 §1.304 for extensions of time to appeal to he U.S. Court of Appeals for the Federal Circuit or to commence a civil action; § 1.550(c) for extensions of time in ex parte reexamination partes,1.956 for extensions of time in inter partes reexamination proceedings; and § 1.645 for extensions of time in interference proceedings.

(3)A written request may be submitted in an application that is an authorization to treat any concurrent or future reply, requiring a petition for an extension of time under this paragraph for its timely submission, as incorporating a petition for extension of time for the appropriate length of time. An authorization to charge all required fees, fees under § 1.17, or all required extension of time fees will be treated as a constructive petition for an extension of time in any concurrent or future reply requiring a petition for an extension of time under this paragraph for its timely

submission. Submission of the fee set forth in § 1.17(a) will also be treated as a constructive petition for an extension of time in any concurrent reply requiring a petition for an extension of time under this paragraph for its timely submission.

(b) When a reply cannot be filed within the time period set for such reply and the provisions of paragraph (a) of this section are not available, the period for reply will be extended only for sufficient cause and for a reasonable time specified. Any request for an extension of time under this paragraph must be filed on or before the day on which such reply is due,but the mere filing of such a request will not affect any extension under this paragraph. In no situation

can any extension carry the date on which reply is due beyond the maximum time period set by statute. See § 1.304 for extensions of time to appeal to the U.S. Court of Appeals for the Federal Circuit or to commence a civil action; § 1.645 for extensions of time in ex parte reexamination proceedings; § 1.550(c) for extensions of time in inter partes reexamination proceedings; and § 1.956 for extensions of time in inter partes reexamination proceedings.

＊＊＊＊＊＊＊＊＊＊＊

37 CFR 1.7(a)

a) Whenever periods of time are specified in this part in days, calendar days are intended. When the day, or the last day fixed by statute or by or under this part for taking any action or paying any fee in the United States Patent and Trademark Office falls on Saturday, Sunday, or on a Federal holiday within the District of Columbia, the action may be taken, or the fee paid, on the next succeeding business day which is not a Saturday, Sunday, or a Federal holiday.

＊＊＊＊＊＊＊＊＊＊＊

實例 1

當審查委員對一專利申請案發下最初審定書，審定書之郵寄日為 6 月 30 日，並訂三個月作為 SSP，而該年之 12 月 30 日為週六，請問本案申請人最晚可於何日期前提出答辯？需繳幾個月的延期費？

最初審定書之郵寄日為 6 月 30 日，故答辯之最晚日應為 12 月 30 日，但因 12 月 30 日為週六，而翌年之 1 月 1 日為假日，故最晚可於翌年之 1 月 2 日前提出答辯，需繳三個月之延期費。

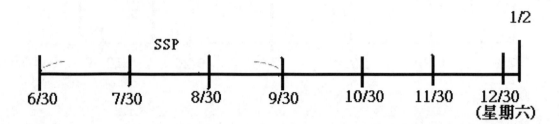

實例 2

申請人收到一最初審定書（first office action）其郵寄日為 10 月 31 日並設定 SSP 為三個月，請問申請人在何日期前提出答辯需繳一個月、二個月及三個月之延期費？

申請人如於翌年之 2 月 1 日至 2 月 28 日（閏年則為 2 月 29 日）之間提出答辯則需繳 1 個月之延期費，如於 3 月 1 日至 3 月 31 日之間提出答辯則需繳 2 個月之延期費，而如於 4 月 1 日至 4 月 30 日之間提出答辯則需繳 3 個月之延期費（4 月 30 日為最終期限不能延期）。

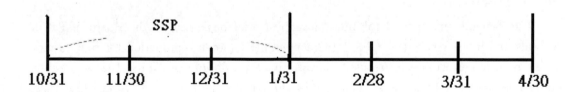

實例 3

審查委員於 5 月 20 日發下一最初審定書（first office action），申請人於 10 月 10 日提出答辯並申請延期但只繳了一個月的延期費（USD130），申請人後來於 10 月 25 日發現應繳二個月之延期費，故提出之答辯無效，故於 10 月 26 日再次提出答辯與延期申請，則請問申請人應繳多少金額之延期費？

第 3 個月的延期費為 USD 1110，故申請人只需繳交 USD 1110-130=USD 980 之延期費即可。

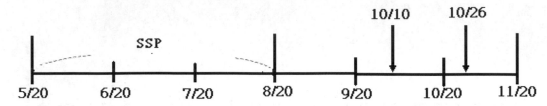

§15.3 非法定期限

美國專利局發出之 Notice to file missing parts（如缺宣誓書等文件），notice to file corrected application papers（如圖式不合規定），以及提出訴願聲明（notice of appeal），後補之訴願理由（appeal brief）等均非 35USC§133 所述之審定書（action），因此對這些通知書及決定書之答復期限為非法定期限，通常為郵寄日起二個月，但除非另有規定都可依 37CFR 1.136(a)延期 5 個月（亦即最長之回覆期限為 7 個月）。

MPEP 702.02(d)

The 2-month time period for filing an appeal brief on appeal to the Board of Patent Appeals and Interferences (37 CFR 1.192(a)) and the 1-month time period for filing a new appeal brief to correct the deficiencies in a defective appeal brief (37 CFR 1.192(d)) are time periods, but are not (shortened) statutory periods for reply set pursuant to 35 U.S.C. 133. Thus, these periods are, unless otherwise provided, extendable by up to 5 months under 37 CFR 1.136(a), and, in an exceptional situation, further extendable under 37 CFR 1.136(b) (i.e., these periods are not statutory periods subject to the 6-month maximum set in 35 U.S.C. 133). In addition, the failure to file an appeal brief (or a new appeal brief) within the time period set in 37 CFR 1.192(a) (or (d)) results in dismissal of the appeal. The dismissal of an appeal results in abandonment, unless there is any allowed claim(s) (see MPEP § 1215.04), in which case the examiner should cancel the nonallowed claims and allow the application.

The 2-month time period for reply to A Notice to File Missing Parts of an Application is not identified on the Notice as a statutory period subject to 35 U.S.C. 133. Thus, extensions of time of up to 5 months under 37 CFR 1.136(a), followed by additional time under 37 CFR 1.136(b), when appropriate, are permitted.

§15.4　最終審定書之答辯期限及延期

　　當發下最終審定書（final office action）時除了修正說明書及請求項（claim）上有所限制外，需注意的是提出完整答辯（fully responsive reply）後該最終審定之答辯期限並非就停止，除非在六個月之答辯期限內收到核准通知書，例如提出答辯後審查委員仍不願核准，而發下建議性審定書（advisory action），而此建議性審定書不會另設定一答辯期限，其答辯期限仍為原最終審定書之答辯期限（自最終審定書郵寄日起 6 個月）。另外，例如提出答辯後審查委員超過最終審定書郵寄日起算 6 個月後未發下任何審定書（action）或核准通知書，或撤回最終審定書時，申請人如未在 6 個月屆滿前提出訴願請求或延續案，該案將被放棄掉。此外，如果申請人在收到最終審定書後二個月內提出完整的答辯，同時審查委員對此答辯發下之建議性審定書（advisory action）是在最終審定書郵寄日之三個月後的話，則對該最終審定的到期日（due date）就自動轉移成美國專利局郵寄建議性審定書之日期或自最終審定書郵寄日之六個月，視哪一日期為早。但對此最終審定之答辯日絕不能超過最終審定郵寄日之六個月。

MPEP 706.007(f)

＊＊＊＊＊＊＊＊＊＊＊

　　All final rejections setting a 3-month shortened statutory period (SSP) for reply should contain on of form paragraphs 7.39, 7.40, 7.40.01, 7.41,7.41.03, or 7.42.09 advising applicant that if the reply is filed within 2 months of the date of the final Office action, the shortened statutory period will expire at 3 months from the date of the final rejection or on the date the advisory action is mailed, whichever is later. Thus, a variable reply period will be established. > If the last day of "2 months of the date of the final Office action" falls on Saturday, Sunday, or a Federal holiday within the District of Columbia, and a reply is filed on the next succeeding day which is not a Saturday, Sunday, or a Federal holiday, pursuant to 37 CFR 1.7(a), the reply is deemed to have been filed within the 2 months period and the shortened statutory period will expire at 3 months from the date of the final rejection or on the mailing date of the advisory action, whichever is later (see MPEP §710.05). <In no event can the statutory period for reply expire later than 6 months from the date of the final rejection.

＊＊＊＊＊＊＊＊＊＊＊＊

實例 1

美國專利局於 6 月 10 日對專利申請案發下最終審定書，設定之答辯期限（SSP）為三個月，申請人於 10 月 9 日提出答辯並請求延期一個月，但並未繳延期費，請問本案是否會放棄？

除非申請人於 12 月 10 日前補繳適當延期費，否則本申請案將在 9 月 11 日起被放棄。

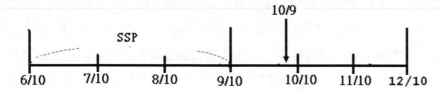

實例 2

美國專利局於 7 月 21 日發下一最終審定書，訂立之答辯期限（SSP）為三個月，申請人於 9 月 7 日提出答辯，但審查委員 11 月 4 日才發下建議性審定書（advisory action）。申請人於 12 月 4 日提出訴願聲明（notice of appeal），請問申請人需繳幾個月的延期費？

申請人只要繳 1 個月的延期費，因為申請人之答辯是於最終審定書之二個月內提出，且審查委員之建議性審定書乃在三個月後才發下，故延期費之起算日為發下建議性審定書之日期。

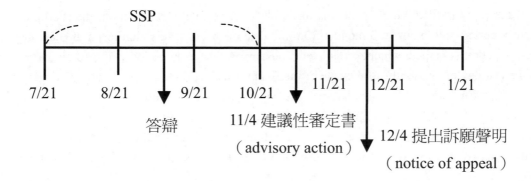

實例 3

　　美國專利局於 4 月 17 日發下一最終審定書，設定之答辯期限（SSP）為三個月，申請人二個月時（6 月 17 日）提出答辯，但審查委員於 11 月 2 日始發下建議性審定書，請問申請人最晚需提出訴願聲明以防止此案被放棄之日期為 10 月 17 日或 11 月 2 日？

　　最晚仍需於最終審定書之日期之六個月（10 月 17 日）提出訴願聲明，而非 11 月 2 日（建議性審定書發下日）。

日程圖

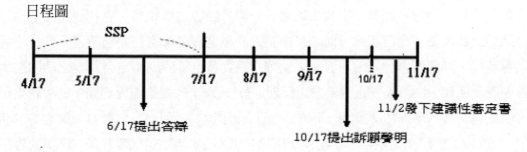

實例 4

　　美國專利局於 2002 年 7 月 16 日發下一 Ex Part Quayle 審定書指出對本案之專利要件之審查已結束，但有一些程式要件需作修正，並設定答辯期限（SSP）為二個月，申請人於 9 月 1 日提出修正，審查委員於 9 月 14 日發下建議性審定書指出所作之修正需再作檢索故不能接受（entered），請問申請人於何日期之前可再提出之修正而不會讓此案被放棄？

　　申請人應於 Ex Part Quayle OA 發下之 6 個月，2003 年 1 月 16 日前提出可被審查委員接受之修正，以免本案被放棄。

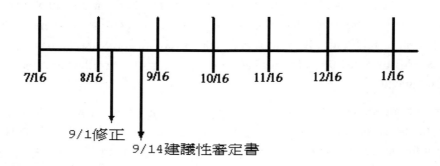

§15.5 審定書之引證資料不正確或審定書包含錯誤

如果審定書中之引證資料不正確或錯誤是很明顯的或微小的錯誤，申請人可自行取得正確之引證資料或不致於影響對審定書提出答辯及修正的話，可不必作回應。而如果此不正確或錯誤足以使申請人不能答辯或修正的話，例如引證資料不對，或對有些請求項未審查等，申請人如果在一個月內通知專利局，專利局將設定答辯日期從此錯誤更正之日起算，而如果申請人於審定書之郵寄日期起超過一個月才通知專利局，則美國專利局重新訂立之期限將只實質上等於原審定書設定之期間之剩餘期間。例如，申請人在原審定書之日期之後 5 週才告知美國專利局，則申請人可答辯之期間會只剩下約 2 個月，例如 9 週以後才通知專利局，仍會有一個的答辯期限。而如果申請人在超過三個月之答辯期（SSP）後才通知美國專利局，則美國專利局不會重新設定答辯期限，申請人需依原期限為基準繳交延期費。

MPEP 710.06 Situations When Reply Period Is Reset or Restarted

Where the citation of a reference is incorrect or an Office action contains some other error that affects applicant's ability to reply to the Office action and this error is called to the attention of the Office within 1 month of the mail date of the action, the Office will restart the previously set period for reply to run from the date the error is corrected, if requested to do so by applicant. If the error is brought to the attention of the Office within the period for reply set in the Office action but more than 1 month after the date of the Office action, the Office will set a new period for reply, if requested to do so by the applicant, to substantially equal the time remaining in the reply period. For example, if the error is brought to the attention of the Office 5 weeks after mailing the action, then the Office would set a new 2-month period for reply. The new period for reply must be at least 1 month and would run from the date the error is corrected. See MPEP § 707.05(g) for the manner of correcting the record where there has been an erroneous citation.

Where for any reason it becomes necessary to remail any action (MPEP § 707.13), the action should be correspondingly redated, as it is the remailing date that establishes the beginning of the period for reply,1924 C.D. 153,329 O.G. 536 (Comm'r Pat. 1924).

A supplementary action after a rejection explaining the references more explicitly or giving the reasons more fully, even though no further references are cited, establishes a new date from which the statutory period runs.

If the error in citation or other defective Office action is called to the attention of the Office after the expiration of the period for reply, the period will not be restarted and any appropriate extension fee will be required to render a reply timely. The Office letter correcting the error will note that the time period for reply remains as set forth in the previous Office action.

＊＊＊＊＊＊＊＊＊＊＊

§15.6　申請人延遲收到審定書之答辯期限

如果郵局之寄送延遲致使申請人延遲收到審定書，而情況符合下列三要件時，申請人可提出請願（petition）請求美國專利局所設定之期限自收到審定書起算：

(A) 申請人於收到審定書之二週之內提出請願。

(B) 收到審定書時答辯期間已過了相當久（例如至少一個月以上或二個月或三個月）。

(c) 請願書需包括(1)蓋記有郵戳以證明收到審定書之文件影本(2)一聲明（statement）說明上述延遲收到之事宜。

如果是由於美國專利局在審定書上印上日期後延遲交至美國郵局致使申請人延期收到審定書，如果符合下列三要件則申請人可提出請願，請求美國專利局所設定之期限自郵局之郵戳日起算：

(A) 申請人於收到審定書之二週內提出請願。

(B) 該審定書所定之回覆期限乃是繳交領證費之期限或者是1個月或30天之期限者。

(C) 請願書需包括(1)證據，證明收到審定書之日期(2)裝該審定書之信封以顯示郵戳日期(3)一聲明，說明上述延遲收到之事實。

MPEP 710.06 Situations When Reply Period Is Reset or Restarted

I.PETITIONS TO RESET A PERIOD FOR REPLY DUE TO LATE RECEIPT OF AN OFFICE ACTION

The Office will grant a petition to restart the previously set period for reply to an Office action to run from the date of receipt of the Office action at the correspondence address when the following criteria are met:

(A) the petition is filed within 2 weeks of the date of receipt of the Office action at the correspondence address;

(B) a substantial portion of the set reply period had elapsed on the date of receipt (e.g., at least 1 month of a 2- or 3-month reply period had elapsed); and

(C) the petition includes (1) evidence showing the date of receipt of the Office action at the correspondence address (e.g., a copy of the Office action having the date of receipt of the Office action at the correspondence address stamped thereon, a copy of the envelope (which contained the Office action) having the date of receipt of the Office action at the correspondence address stamped thereon, etc.), and (2) a statement setting forth the date of receipt of the Office action at the correspondence address and explaining how the evidence being presented establishes the date of receipt of the Office action at the correspondence address.

＊ ＊ ＊ ＊ ＊ ＊ ＊ ＊ ＊ ＊ ＊ ＊

II.PETITIONS TO RESET A PERIOD FOR REPLY DUE TO A POSTMARK DATE LATER THAN THE MAIL DATE PRINTED ON AN OFFICE ACTION

The Office will grant a petition to restart the previously set period for reply to an Office action to run from the postmark date shown on the Office mailing envelope which contained the Office action when the following criteria are met:

(A) the petition is filed within 2 weeks of the date of receipt of the Office action at the correspondence address;

(B) the reply period was for payment of the issue fee, or the reply period set was 1 month or 30 days; and

(C) the petition includes (1) evidence showing the date of receipt of the Office action at the correspondence address (e.g., copy of the Office action having the date of receipt of the Office action at the correspondence address stamped thereon, etc.), (2) a copy of the envelope which contained the Office action showing the postmark date, and (3) a statement setting forth the date of receipt of the Office action at the correspondence address and stating that the Office action was received in the postmarked envelope.

＊＊＊＊＊＊＊＊＊＊＊

§15.7 未收到審定書致使申請案放棄時之處置

如果申請人（代理人）確實未收到美國專利局發下之審定書而致審查委員聲明該申請案放棄（holding of abandonment）時，申請人可提出請願（petition，不必繳規費）請求美國專利局撤回放棄之聲明並重發審定書。此時申請人（代理人）需作一聲明確實未收到審定書並且附上應該收到日期左右之所有收文記錄及該申請案之檔案記錄，或者所有管制期限記錄之影本作為佐證。

MPEP 711.03(c) Petitions Relating to Abandonment

＊＊＊＊＊＊＊＊＊＊＊

A.Petition To Withdraw Holding of Abandonment Based on Failure To Receive Office Action

In Delgar v. Schulyer, 172 USPQ 513 (D.D.C. 1971), the court decided that the Office should mail a new Notice of Allowance in view of the evidence presented in support of the contention that the applicant's representative did not receive the original Notice of Allowance. Under the reasoning of Delgar, an allegation that an Office action was never received may be considered in a petition to withdraw the holding of abandonment. If adequately supported, the Office may grant the petition to withdraw the holding of abandonment and remail the Office action. That is, the reasoning of Delgar is applicable regardless of whether an application is held abandoned for failure to timely pay the issue fee (35 U.S.C. 151) or for failure to prosecute (35U.S.C. 133).

To minimize costs and burdens to practitioners and the Office, the Office has modified the showing required to establish nonreceipt of an Office action. The showing required to establish nonreceipt of an Office communication must include a statement from the practitioner **>describing the system used for recording an Office action received at the correspondence address of record with the USPTO. The statement should establish that the docketing system is sufficiently reliable. It is expected that the record would include, but not be limited to, the application number, attorney docket number, the mail date of the Office action and the due date for the response.

Practitioner must state that the Office action was not received at the correspondence address of record, and that a search of the practitioner's record(s), including any file jacket or the equivalent, and the application contents, indicates that the Office action was not received. A copy of the record(s) used by the practitioner where the non-received Office action would have been entered had it been received is required.

A copy of the practitioner's record(s) required to show non-receipt of the Office action should include the master docket for the firm. That is, if a three month period for reply was set in the nonreceived Office action, a copy of the master docket report showing all replies docketed for a date three months from the mail date of the nonreceived Office action must be submitted as documentary proof of nonreceipt of the Office action. If no such master docket exists, the practitioner should so state and provide other evidence such as, but not limited to, the following: the application file jacket; incoming mail log; calendar; reminder system; or the individual docket record for the application in question.<

§15.8　未對審定書作答辯致使申請案放棄時之處置

37CFR 1.137 規定了當申請人（代理人）因為非故意（unintentional）或不可避免（unavoidable）之延遲而未及時回答美國專利局發下之通知或審定書而致申請案放棄時，可提出請願（petition）附上規費，如有未答辯或未答覆之審定書或通知書需同時答辯或答覆及敘明可被專利局接受之聲明，請求申請案復活。

所謂非故意之延遲，例如申請人（代理人）因為疏忽未將答辯書及時送出或未繳延期規費等，但是由於申請人（代理人）刻意地去選擇不及時回復或答辯美國專利局之通知或審定之作為則不是 "非故意的延遲"，例如以下之情況：

(A) 申請人當時認為所請求之發明（claimed invention）對於所引證之先前技術並非可專利的，故不提出答辯。

(B) 申請人當時認為可核准的請求項範圍不夠廣，不值得花費用取得專利。

(C) 申請人當時認為此專利之價值不值得花費用去取得專利。

(D) 申請人當時認為不值得去繳維持費維持專利權。

(E) 申請人雖有興趣取得專利，但想延遲繳交所需之費用。

所謂不可避免之延遲，例如申請人（代理人）更改連絡地址，且已在該申請案中告知美國專利局已更改地址，但是美國專利局仍將審定書寄至舊的地址，致使申請人延遲收到審定書，致使專利申請案放棄。

37CFR 1.137

Revival of abandoned application, terminated or limited reexamination prosecution, or lapsed patent.

(a) Unavoidable. If the delay in reply by applicant or patent owner was unavoidable, a petition may be filed pursuant to this paragraph to revive an abandoned application, a reexamination prosecution terminated under §§ 1.550(d) or 1.957(b) or limited under § 1.957(c), or a lapsed patent. A grantable petition pursuant to this paragraph must be accompanied by:

(1) The reply required to the outstanding Office action or notice, unless previously filed;

(2) The petition fee as set forth in § 1.17(l);

(3) A showing to the satisfaction of the Director that the entire delay in filing the required reply from the due date for the reply until the filing of a grantable petition pursuant to this paragraph was unavoidable; and

(4) Any terminal disclaimer (and fee as set forth in § 1.20(d)) required pursuant to paragraph (d) of this section.

(b) Unintentional. If the delay in reply by applicant or patent owner was unintentional, a petition may be filed pursuant to this paragraph to revive an abandoned application, a reexamination prosecution terminated under §§ 1.550(d) or 1.957(b) or limited under § 1.957(c), or a lapsed patent. A grantable petition pursuant to this paragraph must be accompanied by:

(1) The reply required to the outstanding Office action or notice, unless previously filed;

(2) The petition fee as set forth in § 1.17(m);

(3) A statement that the entire delay in filing the required reply from the due date for the reply until the filing of a grantable petition pursuant to this paragraph was unintentional. The Director may require additional information where there is a question whether the delay was unintentional; and

(4) Any terminal disclaimer (and fee as set forth in § 1.20(d)) required pursuant to paragraph (d) of this section.

＊＊＊＊＊＊＊＊＊＊＊

MPEP711.03(c) Petitions Relating to Abandonment

＊＊＊＊＊＊＊＊＊＊＊

A delay resulting from a deliberately chosen course of action on the part of the applicant does not become an "unintentional" delay within the meaning of 37 CFR 1.137(b) because:

(A) the applicant does not consider the claims to be patentable over the references relied upon in an outstanding Office action;

(B) the applicant does not consider the allowed or patentable claims to be of sufficient breadth or scope to justify the financial expense of obtaining a patent;

(C) the applicant does not consider any patent to be of sufficient value to justify the financial expense of obtaining the patent;

(D) the applicant does not consider any patent to be of sufficient value to maintain an interest in obtaining the patent; or

(E) the applicant remains interested in eventually obtaining a patent, but simply seeks to defer patent fees and patent prosecution expenses.

＊＊＊＊＊＊＊＊＊＊＊

實例

審查委員發下一最初審定引證了先前技術 X 及 Y 以顯而易見核駁一專利申請案之所有請求項，此專利申請案之美國代理人經與該申請案之受讓人之公司的專利部內討論的結果，決定不提出答辯，且美國代理人依照 37CFR 1.138 提出一表明的放棄書（express abandonment）。美國專利局收到且確認此案放棄之後，該申請案之發明人告知美國代理人事實上本申請案並非顯而易知，故擬申請復活此申請案，並提出答辯，請問此申請案是否可復活？

無法復活，提出表明的放棄乃是故意的放棄，故只能重新提出申請。

§15.9　對非最終審定書答辯時需注意之事項

申請人對於美國專利發下之非最終審定要提出答辯的話必需以書面提出，答辯書必需清楚地指出審查委員之錯誤之處（supposed errors），而且需對審查委員之所有核駁理由（objections，形式之核駁）及（rejection，專利性之核駁）一一作答辯。答辯書可對說明書或請求項作修正，亦可不必修正但必需有說明欄（remarks）說明其請求項所定義之發明之專利性，不能只籠統的主張請求項界定了可予專利之發明，而不清楚地指出請求項之用語如何與先前技術之引證（prior art references）區別。如請求項有修正，則修正後之請求項需界定出在說明欄所說明之與引證之先前技術差異之可專利的新穎點（patentable novelty）。申請人同時需顯示這些修正如何避開了所引證之先前技術與形式核駁。

37CFR 1.111 Reply by applicant or patent owner to a non-final Office action

＊＊＊＊＊＊＊＊＊＊＊＊

(b) In order to be entitled to reconsideration or further examination, the applicant or patent owner must reply to the Office action. The reply by the applicant or patent owner must be reduced to a writing which distinctly and specifically points out the supposed errors in the examiner's action and must reply to every ground of objection and rejection in the prior Office action. The reply must present arguments pointing out the specific distinctions believed to render the claims, including any newly presented claims, patentable over any applied references. If the reply is with respect to an application, a request may be made that objections or requirements as to form not necessary to further consideration of the claims be held in abeyance until allowable subject matter is indicated. The applicant's or patent owner's reply must appear throughout to be a bona fide attempt to advance the application or the reexamination proceeding to final action. A general allegation that the claims define a patentable invention without specifically pointing out how the language of the claims patentably distinguishes them from the references does not comply with the requirements of this section.

(c) In amending in reply to a rejection of claims in an application or patent under reexamination, the applicant or patent owner must clearly point out the patentable novelty which he or she thinks the claims present in view of the state of the art disclosed by the references cited or the objections made. The applicant or patent owner must also show how the amendments avoid such references or objections.

§15.10 答辯書（修正書）之簽名

對於審定書中之答辯（修正）需由有權利答辯的人（authority to prosecute）簽名，例如發明人，受讓人之代表人，專利代理人（patent agent）或專利律師（patent attorney），如果答辯書（修正書）未簽名或不當簽名者則答辯及修正將不被接受。例如當專利申請案件未委任代理人但有二個發明人時，需由二發明人均簽名才會被接受。而如果原來有委任專利代理人（專利律師）但所委任之專利代理人(專利律師)已被辭任或資格已被撤銷，則其簽名之答辯（修正）書將不被接受。

再者，一專利案中雖然在申請時有委任專利代理人（專利律師），但在後續之程序中，申請人（發明人）在答辯書（修正）書上之簽字仍可被美國專利局接受（不必解除原代理人），而在下一程序中又委由原專利代理人（專利律師）簽字亦可。當答辯（修正）書未簽字或不當簽字時，美國專利局會發下通知書給予申請人一個月或 30 天之期限作補正（可依 37CFR 1.136(a)延期），如未在期限內補正則申請案將被視為未答辯而放棄。

MPEP714.01 Signatures to Amendments

An amendment must be signed by a person having authority to prosecute the application. An unsigned or improperly signed amendment will not be entered. See MPEP § 714.01(a).

To facilitate any telephone call that may become necessary, it is recommended that the complete telephone number with area code and extension be given, preferably near the signature.

An unsigned amendment or one not properly signed by a person having authority to prosecute the application is not entered. This applies, for instance, where the amendment is signed by only one of two applicants and the one signing has not been given a power of attorney by the other applicant.

If copies of papers which require an original signature as set forth in 37CFR 1.4(e) are filed, the signature must be applied after the copies are made. MPEP § 714.07.

When an unsigned or improperly signed amendment is received the amendment will be listed in the contents of the application file, but not entered. The examiner will notify applicant of the status of the application, advising him or her to furnish a duplicate amendment properly signed or to ratify the amendment already filed. In an application not under final rejection, applicant should be given a 1-month time period in which to ratify the previously filed amendment (37 CFR 1.135(c)).

Applicants may be advised of unsigned amendments by use of form paragraph 7.84.01.

The proposed reply filed on [1] has not been entered because it is unsigned. Since the above-mentioned reply appears to be bona fide, applicant is given a TIME PERIOD of ONE (1) MONTH or THIRTY (30) DAYS from the mailing date of this notice, whichever is longer, within which to supply the omission or correction in order to avoid abandonment. EXTENSIONS OF THIS TIME PERIOD MAY BE GRANTED UNDER 37 CFR 1.136(a).

Sometimes problems arising from unsigned or improperly signed amendments may be disposed of by calling in the local representative of the attorney or agent of record, since he or she may have the authority to sign the amendment.

An amendment signed by a person whose name is known to have been removed from the registers of attorneys and agents under the provisions of 37 CFR 10.11 is not entered. The file and unentered amendment are submitted to the Office of Enrollment and Discipline for appropriate action.

MPEP 714.01(c) Signed by Attorney or Agent Not of Record

See MPEP § 405. A registered attorney or agent acting in a representative capacity under 37 CFR 1.34, may sign amendments even though he or she does not have a power of attorney in the application. See MPEP § 402.

MPEP 714.01(d) Amendment Signed by Applicant but Not by Attorney or Agent

If an amendment signed by the applicant is received in an application in which there is a duly appointed attorney or agent, the amendment should be entered and acted upon. Attention should be called to 37 CFR 1.33(a) in patent applications and to 37 CFR 1.33(c) in reexamination proceedings. Two copies of the action should be prepared, one being sent to the attorney and the other directly to the applicant. The notation: "Copy to applicant" should appear on the original and on both copies.

§15.11 答辯不完全

申請人對於審定書之答辯必需針對審查委員之專利要件及程式要件之核駁均一一答辯。如果此不完全的答辯（non-responsive reply）是蓄意的（intentional），例如審查委員發下限制／選擇之審定書，但申請人認為審查委員之限制／選擇要求不合理，而不作任何的選擇／限制，則就被視為是不完整的答辯，而申請人如不主動在法定期間內提出補正，則該申請案將被放棄。

而如果申請人確實意欲提出答辯及讓其申請案進行至最終階段（a bona fide attempt to advance the application to final action）且答辯書實質上是完整的，僅有一些事項不小心省略了（inadvertently omitted）的話，實查委員可能作下列三項之一的處置：

(A) 將該答辯書（修正書）視為一完整之答辯，

(B) 通知申請人在原答辯期限內再作補充答辯，或

(C) 重新設定一個月 30 天之期限予申請人補充答辯。

MPEP714.03 Amendments Not Fully Responsive, Action To Be Taken

＊＊＊＊＊＊＊＊＊＊＊＊

An examiner may treat an amendment not fully responsive to a non-final Office action by:

(A) accepting the amendment as an adequate reply to the non-final Office action to avoid abandonment under 35 U.S.C. 133 and 37 CFR 1.135;

(B) notifying the applicant that the reply must be completed within the remaining period for reply to the non-final Office action (or within any extension pursuant to 37 CFR 1.136(a)) to avoid abandonment; or

(C) setting a new time period for applicant to complete the reply pursuant to 37 CFR 1.135(c).

The practice set forth in 37 CFR 1.135(c) does not apply where there has been a deliberate omission of some necessary part of a complete reply; rather, 37 CFR 1.135(c) is applicable only when the missing matter or lack of compliance is considered by the examiner as being "inadvertently omitted." For example, if an election of species has been required and applicant does not make an election because he or she believes the requirement to be improper, the amendment on its face is not a "bona fide attempt to advance the application to final action" (37 CFR 1.135(c)), and the examiner is without authority to postpone decision as to abandonment.

§15.12 對非最終審定書之補充答辯

　　申請人在對非最終審定書提出答辯之後，如果要再提出補充答辯需在仍對審定書可答辯之期限內為之。補充答辯並非申請人之權利，亦即可能不會被接受。但是如果補充答辯清楚地是限定於以下事項時，則此補充答辯可能會被接受：

(A) 刪除請求項；

(B) 採用審查委員之建議；

(C) 補充答辯可使申請案成為可以核准之狀態；

(D) 在第 1 次答辯之後對專利局之要求作回覆；

(E) 更正一些形式如打字之錯誤；及

(F) 將要訴願之議題簡化。

---

37 C.F.R§ 1.111 Reply by applicant or patent owner to a non-final Office action

＊＊＊＊＊＊＊＊＊＊＊

(2) Supplemental replies. (i) A reply that is supplemental to a reply that is in compliance with §1.111(b) will not be entered as a matter of right except as provided in paragraph (a)(2)(ii) of this section. The Office may enter a supplemental reply if the supplemental reply is clearly limited to:

(A) Cancellation of a claim(s);

(B) Adoption of the examiner suggestion(s);

(C) Placement of the application in condition for allowance;

(D) Reply to an Office requirement made after the first reply was filed;

(E) Correction of informalities (e.g., typographical errors); or

(F) Simplification of issues for appeal.

(ii) A supplemental reply will be entered if the supplemental reply is filed within the period during which action by the Office is suspended under §1.03(a) or (c).

＊＊＊＊＊＊＊＊＊＊＊

§15.13 與審查委員面談

在專利申請案審查過程中，申請人或其代理人可請求與審查委員面談，以加速審查程序或釐清爭議處。對於一新專利申請案只允許在審查委員發下第 1 次審定書後才可進行面談，在未發下審定書之前，無法與審查委員面談。但是對於延續申請案或取代申請案（substitute）則可在第 1 次審定書發下前就面談。此外，當審查委員發下最終審定書後，是否進行面談乃由審查委員自行決定，審查委員可能會允許面談，也可能不接受面談。此外，在訴願階段，抵觸程序階段或專利核准後，通常不會允許面談。

面談可為電話面談、視訊面談、電子郵件面談或面對面之面談方式進行。如果是以面對面之面談方式進行，則必需在美國專利局上班時間，而且必需在審查委員之辦公室或專利局內進行。如果是以電子郵件面談，則電子郵件之內容需印出置於專利局之卷宗內。面談完後，審查委員必需將面談內容作成面談摘要（interview summary），而申請人面談完後且需將面談之實質內容記錄在答辯書中仍需對審定書作完整之答辯。

37CFR 1.133 Interviews.

(a)(1) Interviews with examiners concerning applications and other matters pending before the Office must be conducted on Office premises and within Office hours, as the respective examiners may designate. Interviews will not be permitted at any other time or place without the authority of the Director.

(2) An interview for the discussion of the patentability of a pending application will not occur before the first Office action, unless the application is a continuing or substitute application or the examiner determines that such an interview would advance prosecution of the application.

(3) The examiner may require that an interview be scheduled in advance.

(b) In every instance where reconsideration is requested in view of an interview with an examiner, a complete written statement of the reasons presented at the interview as warranting favorable action must be filed by the applicant. An interview does not remove the necessity for reply to Office actions as specified in §§ 1.111 and 1.135.

§15.14 與審查委員面談後之程序

申請人之代理人在與審查委員面談之後，仍需對審定書作答辯，且在答辯書中需記錄面談之要點，包括下列各項：

(A) 簡述面談中所呈送之證據資料，

(B) 說明面談中所討論之請求項，

(C) 說明面談中所討論之先前技術，

(D) 說明所建議之請求項之修正，

(E) 面談中主要討論之內容，

(F) 面談中所討論的其他相關先前技術，

(G) 面談之結果，

(H) 如果是以電子郵件面談，則需附上電子郵件內容印出之紙本。

如果申請人在答辯書中所附之面談要不完整或不正確，審查委員將會給申請人一個月的期間作補正。

MPEP 713.04 Substance of Interview Must Be Made of Record

＊＊＊＊＊＊＊＊＊＊＊

Where an interview initiated by the applicant results in the allowance of the application, the applicant is advised to file a written record of the substance of the interview as soon as possible to prevent any possible delays in the issuance of a patent. Where an examiner initiated interview directly results in the allowance of the application, the examiner may check the appropriate box on the "Examiner Initiated Interview Summary" form, PTOL-413B, to indicate that the examiner will provide a written record of the substance of the interview with the Notice of Allowability.

It should be noted, however, that the Interview Summary form will not be considered a complete and proper recordation of the interview unless it includes, or is supplemented by the applicant, or the examiner to include, all of the applicable items required below concerning the substance of the interview.

The complete and proper recordation of the substance of any interview should include at least the following applicable items:

(A) a brief description of the nature of any exhibit shown or any demonstration conducted;

(B) identification of the claims discussed;

(C) identification of specific prior art discussed;

(D) identification of the principal proposed amendments of a substantive nature discussed, unless these are already described on the Interview Summary form completed by the examiner;

(E) the general thrust of the principal arguments of the applicant and the examiner should also be identified, even where the interview is initiated by the examiner. The identification of arguments need not be lengthy or elaborate. A verbatim or highly detailed description of the arguments is not required. The identification of the arguments is sufficient if the general nature or thrust of the principal arguments can be understood in the context of the application file. Of course, the applicant may desire to emphasize and fully describe those arguments which he or she feels were or might be persuasive to the examiner;

(F) a general indication of any other pertinent matters discussed;

(G) if appropriate, the general results or outcome of the interview; and

(H) in the case of an interview via electronic mail, a paper copy of the Internet e-mail contents MUST be made and placed in the patent application file as required by the Federal Records Act in the same manner as an Examiner Interview Summary Form, PTOL 413, is entered.

Examiners are expected to carefully review the applicant's record of the substance of an interview. If the record is not complete or accurate, the examiner may give the applicant a 1-month time period to complete the reply under 37 CFR 1.135(c) where the record of the substance of the interview is in a reply to a nonfinal Office action.

＊＊＊＊＊＊＊＊＊＊＊

§15.15 答辯時對說明書，圖式及請求項修正之格式

a.說明書本文之修正

　　說明書（不包括求項，電腦列表（computer listing）及序列表（sequence listing）必需依下列方式藉增加、刪除或取代一段落（paragraph）或者取代一節（section），或藉由一取代之說明書（substitute specification）來作修正，也可以加入整段／整節或刪除整段／整節。而取代之段落或節或取代之說明書之刪除部份以刪除線（劃在刪除部份之中央），增加部份以底線表示。亦可用雙括號來表示刪除部份。例如，當用刪除線不易表示時（例如 5 個字以下）可用雙括弧來表示刪除之部份。另外說明書修正時需同時附乾淨版本之說明書。

b.請求項之修正

　　如果對請求項作修正（增加，改變或刪除），需將所有請求項表列，且需在每一請求項之前面對其狀態作註記。例如（currently amended），（canceled），（original），（withdrawn），（previously presented），（new），（not entered）刪除的部份用刪除線，增加的部份用底線，刪除的部份亦可用雙括弧表示刪除部份（特別當只刪除 5 個字以下時）。當有新加入的請求項時，該些新的請求項之編號乃從原來的項數接下去編號，亦即當有請求項被刪除時，其他未刪除之請求項不重新編號。

c.圖式之修正

　　如對圖式作修正或補新的圖式，在該些圖式之上方需標註"new sheet"或"replacement sheet"。亦可同時附上經修正之圖式並註明"Annotated sheet"。如對圖作修正或有新的圖式時在答辯書之 "Remark" 欄需作說明有新的圖式時並在說明書之 "Brief Description of the Drawing"欄及說明書本文需作補充說明。

37CFR 1.121 Manner of making amendments in applications

＊＊＊＊＊＊＊＊＊＊＊＊

(b) Specification. Amendments to the specification, other than the claims, computer listings (§ 1.96) and sequence listings (§ 1.825), must be made by adding, deleting or replacing a paragraph, by replacing a section, or by a substitute specification, in the manner specified in this section.

(ii) The full text of any replacement paragraph with markings to show all the changes relative to the previous version of the paragraph. The text of any added subject matter must be shown by underlining the added text. The text of any deleted matter must be shown by strike-through except that double brackets placed before and after the deleted characters may be used to show deletion of five or fewer consecutive characters. The text of any deleted subject matter must be shown by being placed within double brackets if strikethrough cannot be easily perceived;

(c) Claims. Amendments to a claim must be made by rewriting the entire claim with all changes (e.g., additions and deletions) as indicated in this subsection, except when the claim is being canceled. Each amendment document that includes a change to an existing claim, cancellation of an existing claim or addition of a new claim, must include a complete listing of all claims ever presented, including the text of all pending and withdrawn claims, in the application. The claim listing, including the text of the claims, in the amendment document will serve to replace all prior versions of the claims, in the application. In the claim listing, the status of every claim must be indicated after its claim number by using one of the following identifiers in a parenthetical expression: (Original), (Currently amended), (Canceled), (Withdrawn), (Previously presented), (New), and (Not entered).

(2) When claim text with markings is required. All claims being currently amended in an amendment paper shall be presented in the claim listing, indicate a status of "currently amended," and be submitted with markings to indicate the changes that have been made relative to the immediate prior version of the claims. The text of any added subject matter must be shown by underlining the added text. The text of any deleted matter must be shown by strike-through except that double brackets placed before and after the deleted characters may be used to show deletion of five or fewer consecutive characters. The text of any deleted subject matter must be shown by being placed within double brackets if strike-through cannot be easily perceived. Only claims having the status of "currently amended," or "withdrawn" if also being amended, shall include markings. If a withdrawn claim is currently amended, its status in the claim listing may be identified as "withdrawn- currently amended."

(d) Drawings: One or more application drawings shall be amended in the following manner: Any changes to an application drawing must be in compliance with § 1.84 and must be submitted on a replacement sheet of drawings which shall be an attachment to the amendment document and, in the top margin, labeled "Replacement Sheet". Any replacement sheet of drawings shall include all of the figures appearing on the immediate prior version of the sheet, even if only one figure is amended. Any new sheet of drawings containing an additional figure must be labeled in the top margin as "New Sheet". All changes to the drawings shall be explained, in detail, in either the drawing amendment or remarks section of the amendment paper.

(1) A marked-up copy of any amended drawing figure, including annotations indicating the changes made, may be included. The marked-up copy must be clearly labeled as "Annotated Sheet" and must be presented in the amendment or remarks section that explains the change to the drawings.

(2) A marked-up copy of any amended drawing figure, including annotations indicating the changes made, must be provided when required by the examiner.

(e) Disclosure consistency. The disclosure must be amended, when required by the Office, to correct inaccuracies of description and definition, and to secure substantial correspondence between the claims, the remainder of the specification, and the drawings.

＊＊＊＊＊＊＊＊＊＊＊＊

37CFR 1.126 Numbering of claims

The original numbering of the claims must be preserved throughout the prosecution. When claims are canceled the remaining claims must not be renumbered. When claims are added, they must be numbered by the applicant consecutively beginning with the number next following the highest numbered claim previously presented (whether entered or not). When the application is ready for allowance, the examiner, if necessary, will renumber the claims consecutively in the order in which they appear or in such order as may have been requested by applicant.

§15.16 審查委員之修正書

當審查委員收到申請人對最終審定書之完整答辯書之後，如果發覺該申請案已可核准，但說明書本文有一些錯誤或請求項有錯誤或遺漏時可能會發下審查委員之修正書（Examiner's amendment）。申請人對於審查委員之修正書必需回覆是否同意。

　　如果申請人對最終審定書之答辯是在二個月之內提出，而審查委員之修正書是最終審定三個月之後才發下時，申請人對審查委員之修正書之回復可不必繳延期費。而如果申請人對於最終審定書之答辯未在二個月之內提出，則對於審查委員之修正書之回覆就必需在最終審定書之三個月之內，或在最終審定書之六個月內並繳延期費提出，否則專利申請案將被視為放棄。

MPEP 1302.04 Examiner's Amendments and Changes

＊＊＊＊＊＊＊＊＊＊＊

A formal examiner's amendment may be used to correct all other informalities in the body of the written portions of the specification as well as all errors and omissions in the claims, but such corrections must be made by a formal examiner's amendment, signed by the primary examiner, placed in the file and a copy sent to applicant.

＊＊＊＊＊＊＊＊＊＊＊

MPEP 706.07(f) Time for Reply to Final Rejection

＊＊＊＊＊＊＊＊＊＊＊

(F)Where a complete first reply to a final Office action has been filed within 2 months of the final Office action, an examiner's amendment to place the application in condition for allowance may be made without the payment of extension fees even if the examiner's amendment is made more than 3 months from the date of the final Office action. Note that an examiner's amendment may not be made more than 6 months from the date of the final Office action, as the application would be abandoned at that point by operation of law.

(G)Where a complete first reply to a final Office action has not been filed within 2 months of the final Office action, applicant's authorization to make an amendment to place the application in condition for allowance must be made either within the 3 month shortened statutory period or within an extended period for reply that has been petitioned and paid for by applicant pursuant to 37 CFR 1.136(a).

＊＊＊＊＊＊＊＊＊＊＊

<u>實例</u>

　　美國專利局發下一最終審定書，設定之答辯期限（SSP）為三個月，申請人於最終審定書之郵寄日起第 11 週提出一完整之答辯書，審查委員於收到答辯書的二個月

之後發下審查委員之修正書（examiner's amendment），請問申請人如果不對此審查委員作回覆，申請案是否會被放棄？

　　由於申請人之答辯是在最終審定書之二個月之後才提出，且審查委員之修正書發下時已超過自最終審定書起之三個月，故如果申請人不對審查委員之修正書作答覆，申請案會被放棄。申請人對審查委員修正書作答覆時仍需同時繳交延期費，申請案才不會被放棄。

§15.17 核准通知發下後之修正

　　依照 37CFR 1.312，在美國專利局發下核准通知之後，對申請案之修正並不是申請人之權利（matter of right），亦即，修正可能不被接受。此處所謂的修正包括(A)對說明書本文之修正(B)圖式之修正(C)請求項之修正(D)發明人之修正(E)呈送先前技術。但於宣誓書之補正則不被視為修正（但再發證之案件宣誓書之補正被視為修正）。

　　由於核准之後之修正並不是申請人之權利，而且此時審查委員（primary examiner）已不再管轄該案件，故通常修正需要有其主管（supervisory patent examiner）之允許才可修正，但是，如果僅是說明書、圖式、請求項之形式上之修正（打字錯誤，文法錯誤等）審查委員可以審查委員修正（examiner's amendment）方式處理並接受此修正。

　　任何上述核准後之修正必需在繳領證費之前提出。

MPEP 1302.04 Examiner's Amendments and Changes

37CFR 1.312

§1.312 Amendments after allowance.

　　No amendment may be made as a matter of right in an application after the mailing of the notice of allowance. Any amendment filed pursuant to this section must be filed before or with the payment of the issue fee, and may be entered on the recommendation of the primary examiner, approved by the Director, without withdrawing the application from issue.

MPEP 714.16 Amendment After Notice of Allowance, 37 CFR 1.312

＊＊＊＊＊＊＊＊＊＊＊＊

The amendment of an application by applicant after allowance falls within the guidelines of 37 CFR 1.312. Further, the amendment of an application broadly encompasses any change in the file record of the application. Accordingly, the following are examples of "amendments" by applicant after allowance which must comply with 37 CFR 1.312:

(A) an amendment to the specification,

(B) a change in the drawings,

(C) an amendment to the claims,

(D) a change in the inventorship,

(E) the submission of prior art, etc

Finally, it is pointed out that an amendment under 37 CFR 1.312 must be filed on or before the date the issue fee is paid.

The Commissioner has delegated the approval recommendations under 37 CFR 1.312 to the supervisory patent examiners.

With the exception of a supplemental oath or declaration submitted in a reissue, a supplemental oath or declaration is not treated as an amendment under 37 CFR 1.312. See MPEP § 603.01. A supplemental reissue oath or declaration is treated as an amendment under 37 CFR 1.312 because the correction of the patent which it provides is an amendment of the patent, even though no amendment is physically entered into the specification or claim(s). Thus, for a reissue oath or declaration submitted after allowance to be entered, the reissue applicant must comply with 37 CFR 1.312 in the manner set forth in this section.

After the Notice of Allowance has been mailed, the application is technically no longer under the jurisdiction of the primary examiner. He or she can, however, make examiner's amendments (see MPEP § 1302.04) and has authority to enter amendments submitted after Notice of Allowance of an application which embody merely the correction of formal matters in the specification or drawing, or formal matters in a claim without changing the scope thereof, or the cancellation of claims from the application, without forwarding to the supervisory patent examiner for approval.

Amendments other than those which merely embody the correction of formal matters without changing the scope of the claims require approval by the supervisory patent examiner. The Technology Center (TC) Director establishes TC policy with respect to the treatment of amendments directed to trivial informalities which seldom affect significantly the vital formal requirements of any patent, namely, (A) that its disclosure be adequately clear, and (B) that any invention present be defined with sufficient clarity to form an adequate basis for an enforceable contract.

§15.18 放棄專利申請案

申請人如果要放棄一專利申請案有二種方式，第一種方式就是當美國專利局發下通知（notice）或審定書（action）時不提出回覆或答辯，則期限到時該申請案就被放棄了。例如審查委員於 3 月 21 日發下最初審定書，其所定的答辯期限（SSP）為三個月，如果申請人於 6 月 21 日晚上 12 點前未提出答辯則該申請案自 6 月 22 日起被放棄，但若申請人於 9 月 21 日前提出答辯並繳適當之延期費則該申請案未放棄。

另一種方式是申請人（需有受讓人之同意）或其代理人在申請（審查）過程中主動提出表明或正式的放棄（express or formal abandonment）。如果提出表明的放棄則放棄函需申請人或其代理人簽名始有效。而放棄的日期為美國專利局之審查委員或職員確認其符合放棄規定之日期，或者是申請人（代理人）在放棄函中指定之日期。例如申請人提出一延續案，並指出在延續案提出之日放棄該申請案，則該申請案之放棄為延續案提出申請之日。

另外，在審查過程中或答辯時申請人將所有的請求項（claims）均刪除，但並未加入新的請求項，則此修正（答辯）書將不被視為是放棄該申請案而被視為答辯不完全（non-responsive），而不被接受（entered）。

如果申請人在專利申請案核准後提出表明的放棄書，需在繳領證費之前同時要依照 37CFR 1.313 提出請願（petition）並繳規費，而除非此表明的放棄書在發證前實際上被收到，否則此表明的放棄書將不被承認。如果已繳了領證費，則除了 37CFR 1.313(c)所列之 4 種情事(1)美國專利局核准專利有誤(2)專利申請案違反 37CFR 1.56（IDS，揭露義務）之規定(3)一項以上之請求項並無專利性(4)抵觸程序，以外均不能撤回發證程序。

申請人亦可提出表明的放棄以防止申請案在申請日（優先權日）之 18 個月時被核准前公告，但需於預定之核准前公告日之 4 週前提出。

MPEP 711 Abandonment of Patent Application

＊＊＊＊＊＊＊＊＊＊＊

An abandoned application, in accordance with 37CFR 1.135 and 1.138, is one which is removed from the Office docket of pending applications through:

(A) formal abandonment

(1)by the applicant (acquiesced in by the assignee if there is one),

(2)by the attorney or agent of record including an associate attorney or agent appointed by the principal attorney or agent and whose power is of record, or

(3)by a registered attorney or agent acting in a representative capacity under 37 CFR 1.34(a) when filing a continuing application; or

(B) failure of applicant to take appropriate action within a specified time at some stage in the prosecution of the application.

＊＊＊＊＊＊＊＊＊＊＊

MPEP 11.01 Express or Formal Abandonment

The applicant (acquiesced in by an assignee of record), or the attorney/agent of record, if any, can sign an express abandonment. It is imperative that the attorney or agent of record exercise every precaution in ascertaining that the abandonment of the application is in accordance with the desires and best interests of the applicant prior to signing a letter of express abandonment of a patent application. Moreover, special care should be taken to ensure that the appropriate application is correctly identified in the letter of abandonment.

A letter of abandonment properly signed becomes effective when an appropriate official of the Office takes action thereon. When so recognized, the date of abandonment may be the date of recognition or a later date if so specified in the letter itself. For example, where a continuing application is filed with a request to abandon the prior application as of the filing date accorded the continuing application, the date of the abandonment of the prior application will be in accordance with the request once it is recognized.

＊＊＊＊＊＊＊＊＊＊＊

Action in recognition of an express abandonment may take the form of an acknowledgement by the examiner or by the Publishing Division of the receipt of the express abandonment, indication that it is in compliance with 37 CFR 1.138.

＊＊＊＊＊＊＊＊＊＊＊

In view of the doctrine set forth in Ex parte Lasscell, 1884 C.D. 66, 29 O.G. 861 (Comm'r Pat. 1884), an amendment canceling all of the claims, even though said amendment is signed by the applicant himself/herself and the assignee, is not an express abandonment. The Office, however, will not enter any amendment that would cancel all of the claims in an application without presenting any new or substitute claims. See Exxon Corp. v. Phillips Petroleum Co., 265 F.3d 1249, 60 USPQ2d 1368 (Fed. Cir. 2001). Such an amendment is regarded as nonresponsive and is not a bona fide attempt to advance the application to final action. The practice set forth in 37 CFR 1.135(c) does not apply to such amendment. Applicant should be notified as explained in MPEP § 714.03 to § 714.05.

＊＊＊＊＊＊＊＊＊＊＊

37CFR 1.313 1.313 Withdrawal from issue.

＊＊＊＊＊＊＊＊＊＊＊

(b) Once the issue fee has been paid, the Office will not withdraw the application from issue at its own initiative for any reason except:

(1) A mistake on the part of the Office;

(2) A violation of § 1.56 or illegality in the application;

(3) Unpatentability of one or more claims; or

(4) For interference.

＊＊＊＊＊＊＊＊＊＊＊

第16章　重複專利之核駁與答辯

目　錄

§16.1　重複專利之意義

　　美國專利法 35 USC §101 規定"任何人發明或發現了任何新且有用之方法、機器、產品或組合物或其任何新且有用之改良，則依本法之條約及要件，可得到一專利。故一專利權人若擁有二個或以上之專利，而其中二專利的某一請求項（claim）所請求之發明與另一專利之任一請求項所請求之發明相同，而此二專利之專利權期限若非同時結束，則該專利權人對於該發明之專利權被不當地延長了。因而美國專利局發現有重複專利之情事就會發下重複專利之核駁（double patenting rejections）。通常有二種形式之重複專利的核駁，一種是法定上之重複專利核駁（statutory double patenting），就是二請求項所請求之發明完全一樣故依據 35 USC §101 只能有一專利核准。另一種是非法定之重複專利核駁，其是當二請求項所請求之發明雖非完全相同，但在專利性上是不能區別的時候就會發下此種顯而易見形式之重複專利之核駁以避免不當的延長專利權，此乃是美國專利局從判例上發展出之實務。另有一種非法定之重複專利核駁，其二請求項所請求之發明並非顯而易見，但事實上專利權人可能取得二個專利性上不能區別之專利。

---

35 USC §101

Inventions patentable.

　　Whoever invents or discovers any new and useful process, machine, manufacture, or composition of matter, or any new and useful improvement thereof, may obtain a patent therefor, subject to the conditions and requirements of this title.

MPEP 804 Definition of Double Patenting

＊＊＊＊＊＊＊＊＊＊＊

　　There are generally two types of double patenting rejections. One is the "same invention" type double patenting rejection based on 35 U.S.C. 101 which states in the singular that an inventor "may obtain a patent." The second is the "nonstatutory-type" double patenting rejection based on a judicially created doctrine grounded in public policy and which is primarily intended to prevent prolongation of the patent term by prohibiting claims in a second patent not patentably distinguishing from claims in a first patent. Nonstatutory double patenting includes rejections based on either a one-way determination of obviousness or a two-way determination of obviousness. Nonstatutory double patenting could include a rejection which is not the usual "obviousness-type" double patenting rejection. This type of double patenting rejection is rare and is limited to the particular facts of the case.

＊＊＊＊＊＊＊＊＊＊＊

實例

　　發明人 A 申請了第 1 個發明專利並於申請時就讓渡給 X 公司，在第 1 個發明專利發證六個月之後，發明人 A 又申請了第 2 個發明專利且在申請時將專利權讓渡給 Y 公司。發明人 A 之第 2 個申請案之 claim 1（獨立項）與第 1 個申請案核准之 claim 10 雖不相同但是在專利性上不可區分（patentably indistinguishable）。審查委員以顯而易見的重複專利用第 1 個申請案核駁第 2 個申請案之 claim 1，請問：(a) 審查委員用重複專利核駁是否恰當？(b) 申請人可否再不修改 claim 1 之情形下克服此重複專利之核駁？

　　此重複專利的核駁是適當的，當二申請案或一申請案與一專利有下列情形時，審查委員就會發下重複專利之核駁:

a)　相同之發明人（same inventive entity），即使受讓人不相同；

b)　相同之受讓人，即使發明人不同；

c)　只要發明人有一個相同（overlapping inventive entity），即使無相同之受讓人。

　　申請人必須修改 claims 才可克服此重複專利之核駁，因為第 1 個專利申請案與第 2 個專利申請案並非讓給同一受讓人，故不能以提出尾端棄權書之方式克服此核駁。

MPEP 804 Definition of Double Patenting

＊＊＊＊＊＊＊＊＊＊＊

Before consideration can be given to the issue of double patenting, there must be some common relationship of inventorship and/or ownership of two or more patents or applications. Since the doctrine of double patenting seeks to avoid unjustly extending patent rights at the expense of the public, the focus of any double patenting analysis necessarily is on the claims in the multiple patents or patent applications involved in the analysis.

＊＊＊＊＊＊＊＊＊＊＊

§16.2　重複專利發生之情況

有下列 4 種情形之一時，審查委員就會作出重複專利之核駁。

A.已發證之專利與一個或以上之專利申請案之間

如果一專利申請案與已發證之專利是由相同之發明人（inventive entity，指數個發明人完全相同或發明人之間有一個以上相同）所申請，及／或是由該專利之所有人所申請，而此已發證之專利與專利申請案有一 claim 相同或專利性上不能區分時。

B.二個或以上均審查中之申請案

如果一專利申請案與另一同在審查中之專利申請案是由相同之發明人（inventive entity，指數個發明人完全相同或發明人之間有一個以上相同）所申請，及／或是由相同之受讓人所申請，而此二審查中之專利申請案有一 claim 相同或專利性上不能區分時，審查委員就會發下暫時之重複專利核駁（provisional double patenting）。

C.一個或以上之專利申請案與一核准前公告之專利申請案

如果一專利申請案與一已核准前公告之專利申請案是由相同之發明人（數個發明人完全相同或發明人之間有一個以上相同時）所申請，及／或是相同之受讓人所申請時，而此專利申請案與已核准公告之專利申請案有一 claim 相同或專利性上不能區分時，審查委員亦會發下暫時之重複專利核駁。

D.再審查程序(re-examination proceedings)

重複專利之議題亦可能產生一專利之專利性之新問題，所以亦可再以重複專利提起再審查程序。

MPEP 804 Definition of Double Patenting

＊＊＊＊＊＊＊＊＊＊＊

INSTANCES WHERE DOUBLE PATENTING ISSUE CAN BE RAISED

A double patenting issue may arise between two or more pending applications, or between one or more pending applications and a patent. A double patenting issue may likewise arise in a reexamination proceeding between the patent claims being reexamined and the claims of one or more applications and/or patents. Double patenting does not relate to international applications which have not yet entered the national stage in the United States.

A.Between Issued Patent and One or More Applications

Double patenting may exist between an issued patent and an application filed by the same inventive entity, or by an inventive entity having a common inventor with the patent, and/or by the owner of the patent. Since the inventor/patent owner has already secured the issuance of a first patent, the examiner must determine whether the grant of a second patent would give rise to an unjustified extension of the rights granted in the first patent.

B.Between Copending Applications—Provisional Rejections

Occasionally, the examiner becomes aware of two copending applications filed by the same inventive entity, or by different inventive entities having a common inventor, and/or by a common assignee that would raise an issue of double patenting if one of the applications became a patent. Where this issue can be addressed without violating the confidential status of applications (35 U.S.C. 122), the courts have sanctioned the practice of making applicant aware of the potential double patenting problem if one of the applications became a patent by permitting the examiner to make a "provisional" rejection on the ground of double patenting. In re Mott, 539 F.2d 1291, 190 USPQ 536 (CCPA 1976); In re Wetterau,356 F.2d 556, 148 USPQ 499 (CCPA 1966). The merits of such a provisional rejection can be addressed by both the applicant and the examiner without waiting for the first patent to issue.

The "provisional" double patenting rejection should continue to be made by the examiner in each application as long as there are conflicting claims in more than one application unless that "provisional" double patenting rejection is the only rejection remaining in one of the applications. If the "provisional" double patenting rejection in one application is the only rejection remaining in that application, the examiner should then withdraw that rejection and permit the application to issue as a patent, thereby converting the "provisional" double patenting rejection in the other application(s) into a double patenting rejection at the time the one application issues as a patent.

If the "provisional" double patenting rejections in both applications are the only rejections remaining in those applications, the examiner should then withdraw that rejection in one of the applications (e.g., the application with the earlier filing date)and permit the application to issue as a patent. The examiner should maintain the double patenting rejection in the other application as a "provisional" double patenting rejection which will be converted into a double patenting rejection when the one application issues as a patent.

C.Between One or More Applications and a Published Application - Provisional Rejections

Double patenting may exist between a published patent application and an application filed by the same inventive entity, or by different inventive entities having a common inventor, and/or by a common assignee. Since the published application has not yet issued as a patent, the examiner is permitted to make a "provisional" rejection on the ground of double patenting. See the discussion regarding "provisional" double patenting rejections in subsection B. above.

D.Reexamination Proceedings

A double patenting issue may raise a substantial new question of patentability of a claim of a patent, and thus be addressed in a reexamination proceeding. In re Lonardo, 119 F.3d 960, 966, 43 USPQ2d 1262, 1266 (Fed. Cir. 1997) (In giving the Director authority under 35 U.S.C. 303(a) in determining the presence of a substantial new question of patentability, "Congress intended that the phrases 'patents and publications' and 'other patents or publications' in section 303(a) not be limited to prior art patents or printed publications." (emphasis added)).

實例 1

發明人 X 申請了一治療骨髓炎（osteomyelitis）之藥物，其核准之 claim 1 如下：

A medication for treating osteomyelitis containing an effective amount of an organic compound having a cyclopentadiene ring structure containing a metal ion held by coordination bonds, said metal ion being selected from the group consisting of osmium, iridium, platinum, and gold。在此專利發證之前，發明人 X 研究發現當所用之有機化合物上之 cyclopentadiene 環結構上之金屬離子為 iridium, platinum 及 gold 時效果較金屬離子為 osmium 時為佳，因而在上述專利發證前提出一 CIP 申請案，其 claim 1 為：

A medication for treating osteomyelitis containing an effective amount of an organic compound having a cyclopentadiene ring structure containing a metal ion held by coordination bonds, said metal ion being selected from the group consisting of iridium, platinum, and gold.

審查委員以本 CIP 申請案與發明人 X 之專利為法定上之重複專利（statutory double patenting）核駁，請問此核駁是否適當？

不適當。所謂法定上之重複專利，需二案之請求標的（subject matter）完全相同。

MPEP 804

* * * * * * * * * * * *

Statutory Double Patenting — 35 U.S.C. 101

In determining whether a statutory basis for a double patenting rejection exists, the question to be asked is: Is the same invention being claimed twice? 35 U.S.C. 101 prevents two patents from issuing on the same invention. "Same invention" means identical subject matter. Miller v. Eagle Mfg. Co., 151 U.S. 186 (1984); In re Vogel, 422 F.2d 438, 164 USPQ 619 (CCPA 1970); and In re Ockert, 245 F.2d 467, 114 USPQ 330 (CCPA 1957).

A reliable test for double patenting under 35 U.S.C. 101 is whether a claim in the application could be literally infringed without literally infringing a corresponding claim in the patent. In re Vogel, 422 F.2d 438, 164 USPQ 619 (CCPA 1970). Is there an embodiment of the invention that falls within the scope of one claim, but not the other? If there is such an embodiment, then identical subject matter is not defined by both claims and statutory double patenting would not exist. For example, the invention defined by a claim reciting a compound having a "halogen" substituent is not identical to or substantively the same as a claim reciting the same compound except having a "chlorine" substituent in place of the halogen because "halogen" is broader than "chlorine." On the other hand, claims may be differently worded and still define the same invention. Thus, a claim reciting a widget having a length of "36 inches" defines the same invention as a claim reciting the same widget having a length of "3 feet."

* * * * * * * * * * * *

實例 2

　　發明人 A 申請了一第 1 申請案請求防水組合物（Claim 1:一種防水組合物包括成份 X,Y,Z，說明書中揭露 X,Y,Z 但未揭露 T），此申請案並未讓渡給任何人。之後，發明人 A 又申請了一第二申請案請求一防水組合物（Claim 3:一種防水組合物，包括成份 X,Y,Z,T）。第 1 申請案在第 2 申請案提申之後核准，審查委員引證發明人 A 之第 1 申請案 Claim 1 及另一篇引證資料（揭露防水組合物包括成份 X,Y,T 並教示 Z 可加入防水組合物中）稱第 2 申請案之 Claim 3 由上述二引證是顯而易見的重複專利，請問此核駁是否恰當？

　　此顯而易見的重複專利核駁是適當的。因為第 2 申請案之申請是在第 1 申請案之後，故審查委員只要採用 one-way test 即可，（亦即，是否在第 2 申請案之 claim 3 所定義之發明是第 1 申請案之 claim 1 所定義的發明之顯而易見變形例（obvious variant）即可）。

MPEP 804

＊＊＊＊＊＊＊＊＊＊＊

One-Way Obviousness

If the application at issue is the later filed application or both are filed on the same day, only a one-way determination of obviousness is needed in resolving the issue of double patenting, i.e., whether the invention defined in a claim in the application is an obvious variation of the invention defined in a claim in the patent. See, e.g., In re Berg, 140 F.3d 1438, 46 USPQ2d 1226 (Fed. Cir. 1998) (the court applied a one-way test where both applications were filed the same day). If a claimed invention in the application is obvious over a claimed invention in the patent, there would be an unjustified timewise extension of the patent and an obvious-type double patenting rejection is proper. Unless a claimed invention in the application is obvious over a claimed invention in the patent, no double patenting rejection of the obvious-type should be made, but this does not necessarily preclude a rejection based on another type of nonstatutory double patenting (see MPEP § 804, paragraph II.B.2. below).

<u>實例 3</u>

發明人 A 申請了一第 1 申請案請求一防水組合物（Claim 1:一種防水組合物包括成份 X,Y,Z）。之後發明人又申請了一第 2 申請案亦請求一防水組合物（Claim 3:一種防水組合物包括成份 X,Y,Z,T）。第 2 申請案雖較第 1 申請案後申請但是先核准專利。審查委員在審查第 1 申請案時，認為第 1 申請案之 Claim 1 是第 2 申請案之 Claim3 是顯而易見的變形例（obvious variant）。請問審查委員如要以重複專利核駁第 1 申請案之 Claim 1，是否可採用 one-way test?

審查委員應採用 two-way test，亦即審查委員需同時證明 Claim 1 對 Claim 3 是顯而易見同時要證明 Claim 3 對 Claim 1 亦是顯而易見的，亦即審查委員需應用 Graham 分析二次方可。當申請人無法將二有衝突之 Claims 在一申請案中申請且審查委員在審查程序上有延遲時才需採用 two-way test。例如申請人為了早日取得專利讓範圍較小之 Claims（例如 Species）先核准，之後再申請延續案請求較大範圍的 Claims（如 genus），則審查委員只需採 one-way test 就可發下顯而易見之重複專利核駁。

MPEP 804

＊＊＊＊＊＊＊＊＊＊＊

Two-Way Obviousness

If the patent is the later filed application, the question of whether the timewise extension of the right to exclude granted by a patent is justified or unjustified must be addressed. A two-way test is to be applied only when the applicant could not have filed the claims in a single application and there is administrative delay. In re Berg, 46 USPQ2d 1226 (Fed. Cir. 1998) ("The two-way exception can only apply when the applicant could not avoid separate filings, and even then, only if the PTO controlled the rates of prosecution to cause the later filed species claims to issue before the claims for a genus in an earlier application

In Berg's case, the two applications could have been filed as one, so it is irrelevant to our disposition who actually controlled the respective rates of prosecution."). In the absence of administrative delay, a one-way test is appropriate. In re Goodman, 11 F.3d 1046, 29 USPQ2d 2010 (Fed. Cir. 1993) (applicant's voluntary decision to obtain early issuance of claims directed to a species and to pursue prosecution of previously rejected genus claims in a continuation is a considered election to postpone by the applicant and not administrative delay). Unless the record clearly shows administrative delay by the Office and that applicant could not have avoided filing separate applications, the examiner may use the one-way obviousness determination and shift the burden to applicant to show why a two-way obviousness determination is required.

When making a two-way obviousness determination where appropriate, it is necessary to apply the Graham obviousness analysis twice, once with the application claims as the claims in issue, and once with the patent claims as the claims in issue. Where a two-way obviousness determination is required, an obvious-type double patenting rejection is appropriate only where each analysis compels a conclusion that the invention defined in the claims in issue is an obvious variation of the invention defined in a claim in the other application/patent. If either analysis does not compel a conclusion of obviousness, no double patenting rejection of the obvious-type is made, but this does not necessarily preclude a nonstatutory double patenting rejection based on the fundamental reason to prevent unjustified timewise extension of the right to exclude granted by a patent. In re Schneller, 397 F.2d 350, 158 USPQ 210 (CCPA 1968).

§16.3　另一種型式之非法定重複專利

在 In re Schneller 之判例中同一發明人之二個先後申請之專利申請案所請求之發明是非顯而易見的，但審查委員仍發下重複專利之核駁。說明如下：

發明人 A 之第一申請案揭露了 A,B,C,X,Y 但 Claim 只請求包括 A,B,C,X 之次組合，但並未請求包括 A,B,C,X,Y 之組合。之後，發明人提出了第二申請案請求包括 A,B,C,X,Y 之組合。A,B,C,X,Y 之組合對於 A,B,C,X 之次組合並非是顯而易見的，但是從專利法上來說發明人 A 所請求之次組合及組合之範圍是重複的，故此二者具有重複專利之情事，會使得專利權人 A 的專利保護不當地延長。當審查委員遇到有上述之情況時需先與其主管商量，如主管同意發下重複專利之核駁需先得技術中心主管（technology center director）之同意。

MPEP 804

＊＊＊＊＊＊＊＊＊＊＊

Another Type of Nonstatutory Double Patenting Rejection

There are some unique circumstances where it has been recognized that another type of nonstatutory double patenting rejection is applicable even where the inventions claimed in two or more applications/patents are considered nonobvious over each other. These circumstances are illustrated by the facts before the court in In re Schneller, 397 F.2d 350, 158 USPQ 210 (CCPA 1968).

The decision in In re Schneller did not establish a rule of general application and thus is limited to the particular set of facts set forth in that decision. The court in Schneller cautioned "against the tendency to freeze into rules of general application what, at best, are statements applicable to particular fact situations." Schneller, 397 F.2d at 355, 158 USPQ at 215. Nonstatutory double patenting rejections based on Schneller will be rare. The Technology Center (TC) Director must approve any nonstatutory double patenting rejections based on Schneller. If an examiner determines that a double patenting rejection based on Schneller is appropriate in his or her application, the examiner should first consult with his or her supervisory patent examiner (SPE). If the SPE agrees with the examiner then approval of the TC Director must be obtained before such a nonstatutory double patenting rejection can be made.

§16.4　限制要求後之重複專利核駁之限制

依照 35 USC §121 之第 3 段之規定，審查委員不能用經限制要求後之獲得核准之專利案來當先前技術核駁其分割案，故亦不能稱母案與分割案為重複專利而核駁。但此禁止重複專利核駁並不適用下列之情況：

(A) 審查委員並未作限制要求，申請人主動提出分割案。

(B) 如果母案與分割案之請求項與當時審查委員要求作限制要求時之請求項不一樣時，例如申請人在分割案中加入其他請求項而所加入之請求項與原來之請求項在範圍上不一致。

(C) 審查委員發下之限制要求之原因是上位概念請求項（generic claim）或連接請求項（linking claim）不能核准，但之後上位概念請求項或連接請求項變得可以核准了故需撤回限制要求之情況時，可以發下重複專利之核駁。

(D) 限制要求是 PCT 國際申請案中由國際檢索機構或國際初步審查機構所作出的。

(E) 在母案專利發證之前，審查委員撤回限制要求時。

(F) 第二申請案（分割案）之請求項與母案請求完全相同之發明時。

MPEP 804.01 Prohibition of Double Patenting Rejections

The prohibition against holdings of double patenting applies to requirements for restriction between the related subjects treated in MPEP § 806.04 through § 806.05(i), namely, between combination and subcombination thereof, between

subcombinations disclosed as usable together, between process and apparatus for its practice, between process and product made by such process and between apparatus and claims in each application are filed as a result of such requirement.

The following are situations where the prohibition of double patenting rejections under 35 U.S.C. 121 does not apply:

(A)The applicant voluntarily files two or more applications without a restriction requirement by the examiner. In re Schneller, 397 F.2d 350, 158 USPQ 210 (CCPA 1968).

(B)The claims of the different applications or patents are not consonant with the restriction requirement made by the examiner, since the claims have been changed in material respects from the claims at the time the requirement was made. For example, the divisional application filed includes additional claims not consonant in scope to the original claims subject to restriction in the parent. Symbol Technologies, Inc. v. Opticon, Inc., 935 F.2d 1569, 19 USPQ2d 1241 (Fed. Cir. 1991) and Gerber Garment Technology, Inc. v. Lectra Systems, Inc., 916 F.2d 683, 16 USPQ2d 1436 (Fed. Cir. 1990). In order for consonance to exist, the line of demarcation between the independent and distinct inventions identified by the examiner in the requirement for restriction must be maintained. 916 F.2d at 688, 16 USPQ2d at 1440.

(C)The restriction requirement was written in a manner which made it clear to applicant that the requirement was made subject to the nonallowance of generic or other linking claims and such generic or linking claims are subsequently allowed. Therefore, if a generic or linking claim is subsequently allowed, the restriction requirement must be withdrawn.

(D)The requirement for restriction (holding of lack of unity of invention) was only made in an international application by the International Searching Authority or the International Preliminary Examining Authority.

(E)The requirement for restriction was withdrawn by the examiner before the patent issues. In re Ziegler, 443 F.2d 1211, 170 USPQ 129 (CCPA 1971).

(F)The claims of the second application are drawn to the "same invention" as the first application or patent. Studiengesellschaft Kohle mbH v. Northern Petrochemical Co., 784 F.2d 351, 228 USPQ 837 (Fed. Cir. 1986).

§16.5　克服重複專利之核駁

當審查委員當下重複專利之核駁時，可依下列方式克服重複專利之核駁：

1)　法定上之重複專利：

　　a)　主張被審查中之案件之該請求項與已核准之專利或另一申請案之請求項並非完全相同。

　　b)　刪除該審查中案件之請求項。

　　c)　加入限制條件至該審查中之案件之該請求項中。

2)　顯而易見之重複專利：

　　a)　主張被審查之案件之請求項與已核准之專利或另一申請案之請求項並非顯而易見。

　　b)　提出尾端棄權書（terminal disclaimer）：放棄從前一案件之申請日起超過 20 年部份之專利權以使二專利權期限同時結束，同時聲明其專利權只有當此二專利權之所有人相同時，專利權才是可執行的（enforceable），一旦兩專利權之所有人改變，專利將不可執行。

　　c)　如果兩案均是審查中之案件，則可另申請 CIP 將二案合併。

3)　非顯而易見之重複專利：

　　a)　提出尾端棄權書。

MPEP 804.02 Avoiding a Double Patenting Rejection

I.STATUTORY

A rejection based on the statutory type of double patenting can be avoided by amending the conflicting claims so that they are not coextensive in scope. Where the conflicting claims are in one or more pending applications and a patent, a rejection based on statutory type double patenting can also be avoided by canceling the conflicting claims in all the pending applications. Where the conflicting claims are in two or more pending applications, a provisional rejection based on statutory type double patenting can also be avoided by canceling the conflicting claims in all but one of the pending applications. A terminal disclaimer is not effective in overcoming a statutory double patenting rejection.

The use of a 37 CFR 1.131 affidavit in overcoming a statutory double patenting rejection is inappropriate.

II.NONSTATUTORY

A rejection based on a nonstatutory type of double patenting can be avoided by filing a terminal disclaimer in the application or proceeding in which the rejection is made. In re Vogel, 422 F.2d 438, 164 USPQ 619 (CCPA 1970); In re Knohl, 386 F.2d 476, 155 USPQ 586 (CCPA 1967); and In re Griswold, 365 F.2d 834, 150 USPQ 804 (CCPA 1966). The use of a terminal disclaimer in overcoming a nonstatutory double patenting rejection is in the public interest because it encourages the disclosure of additional developments, the earlier filing of applications, and the earlier expiration of patents whereby the inventions covered become freely available to the public. In re Jentoft, 392 F.2d 633, 157 USPQ 363 (CCPA 1968); In re Eckel, 393 F.2d 848, 157 USPQ 415 (CCPA 1968); and In re Braithwaite, 379 F.2d 594, 154 USPQ 29 (CCPA 1967).

35 U.S.C. 253 Disclaimer.

Whenever, without any deceptive intention, a claim of a patent is invalid the remaining claims shall not thereby be rendered invalid. A patentee, whether of the whole or any sectional interest therein, may, on payment of the fee required by law, make disclaimer of any complete claim, stating therein the extent of his interest in such patent. Such disclaimer shall be in writing and recorded in the Patent and Trademark Office, and it shall thereafter be considered as part of the original patent to the extent of the interest possessed by the disclaimant and by those claiming under him.

In like manner any patentee or applicant may disclaim or dedicate to the public the entire term, or any terminal part of the term, of the patent granted or to be granted.

§16.6　提出尾端棄權書需注意之事項

提出尾端棄權書雖可克服重複專利之核駁，但亦需注意以下事項：
1) 即使只有一請求項是重複專利，提出尾端棄權書時乃針對所有請求項，故所有核准的請求項其專利權期限均與另一申請案（或專利）同時結束。
2) 提出尾端棄權書並非承認該請求項與另一申請案或專利之請求項是顯而易見。

3)　提出尾端棄權書將不適用因訴願成功時專利權期限之調整。

4)　因為 FDA 及 USDA 之規定而造成藥品（農藥品）遲延上市而取得之專利延長部份不會因提出尾端棄權書而不能延長。

5)　如果經提出尾端棄權書後同時存在之兩專利之中有一專利因未繳維持費，或被法院判無效（invalid）或不可執行（unenforceable）或因再發證（reissue）或再審查（re-examination）而撤銷，則另一專利之專利權仍存在。

尾端棄權書可由美國代理人準備及提出，可不必發明人簽署及提出。

MPEP 804.02

A patentee or applicant may disclaim or dedicate to the public the entire term, or any terminal part of the term of a patent. 35 U.S.C. 253. The statute does not provide for a terminal disclaimer of only a specified claim or claims. The terminal disclaimer must operate with respect to all claims in the patent.

The filing of a terminal disclaimer to obviate a rejection based on nonstatutory double patenting is not an admission of the propriety of the rejection. Quad Environmental Technologies Corp. v. Union Sanitary District, 946 F.2d 870, 20 USPQ2d 1392 (Fed. Cir. 1991). The court indicated that the "filing of a terminal disclaimer simply serves the statutory function of removing the rejection of double patenting, and raises neither a presumption nor estoppel on the merits of the rejection."

A terminal disclaimer filed to obviate a double patenting rejection is effective only with respect to the application identified in the disclaimer, unless by its terms it extends to continuing applications. If an appropriate double patenting rejection of the nonstatutory type is made in two or more pending applications, an appropriate terminal disclaimer must be filed in each application.

Claims that differ from each other (aside from minor differences in language, punctuation, etc.), whether or not the difference is obvious, are not considered to be drawn to the same invention for double patenting purposes under 35 U.S.C. 101. In cases where the difference in claims is obvious, terminal disclaimers are effective to overcome double patenting rejections. However, such terminal disclaimers must include a provision that the patent shall be unenforceable if it ceases to be commonly owned with the other application or patent. Note 37 CFR 1.321(c). It should be emphasized that a terminal disclaimer cannot be used to overcome a rejection under 35 U.S.C. 102(e)/103(a).

37 CFR 1.321

§1.321 Statutory disclaimers, including terminal disclaimers.

(a) A patentee owning the whole or any sectional interest in a patent may disclaim any complete claim or claims in a patent. In like manner any patentee may disclaim or dedicate to the public the entire term, or any terminal part of the term, of the patent granted. Such disclaimer is binding upon the grantee and its successors or assigns. A notice of the disclaimer is published in the Official Gazette and attached to the printed copies of the specification. The disclaimer, to be recorded in the Patent and Trademark Office, must:

(1) Be signed by the patentee, or an attorney or agent of record;

(2) Identify the patent and complete claim or claims, or term being disclaimed. A disclaimer which is not a disclaimer of a complete claim or claims, or term will be refused recordation;

(3) State the present extent of patentee's ownership interest in the patent; and

(4) Be accompanied by the fee set forth in §1.20(d).

(b) An applicant or assignee may disclaim or dedicate to the public the entire term, or any terminal part of the term, of a patent to be granted. Such terminal disclaimer is binding upon the grantee and its successors or assigns. The terminal disclaimer, to be recorded in the Patent and Trademark Office, must:

(1) Be signed:

(i) By the applicant, or

(ii) If there is an assignee of record of an undivided part interest, by the applicant and such assignee, or

(iii) If there is an assignee of record of the entire interest, by such assignee, or

(iv) By an attorney or agent of record;

(2) Specify the portion of the term of the patent being disclaimed;

(3) State the present extent of applicant's or assignee's ownership interest in the patent to be granted; and

(4) Be accompanied by the fee set forth in § 1.20(d).

(c) A terminal disclaimer, when filed to obviate a judicially created double patenting rejection in a patent application or in a reexamination proceeding, must:

(1) Comply with the provisions of paragraphs (b)(2) through (b)(4) of this section;

(2) Be signed in accordance with paragraph (b)(1) of this section if filed in a patent application or in accordance with paragraph (a)(1) of this section if filed in a reexamination proceeding; and

(3) Include a provision that any patent granted on that application or any patent subject to the reexamination proceeding shall be enforceable only for and during such period that said patent is commonly owned with the application or patent which formed the basis for the rejection.

第 17 章　利用宣誓書克服核駁理由

目　錄

§17.1　Affidavit，Oath 及 Declaration

Affidavit，Oath 及 Declaration 均為宣誓書，宣誓書的目的是要提供在專利申請書，引證資料及其他文件上未明確顯示之事實證據, Affidavit 與 Oath 為同義字，宣誓人的 Affidavit 及 Oath 需經過公證(notarization)始有效。依照美國專利法 35USC §25 及 37CFR§1.168，宣誓人可使用 Declaration 來取代 Affidavit 及 Oath 而不必公證。但是如果宣誓人使用 Declaration 取代 Oath 或 Affidavit，則需在 Declaration 中包括下述之聲明：

a) 所有在此處基於本人所有之知識所作之聲明均為真實的。

b) 所有對於資訊之聲明相信均是真實的。

c) 宣誓人知道惡意虛偽之聲明，將依照美國法律 Title 18,section 1001 被處以罰金或監禁。

d) 宣誓人知道此惡意虛偽之聲明可使此專利申請案或所核准之專利無效。

---

35 U.S.C. 25 Declaration in lieu of oath.

(a) The Director may by rule prescribe that any document to be filed in the Patent and Trademark Office and which is required by any law, rule, or other regulation to be under oath may be subscribed to by a written declaration in such form as the Director may prescribe, such declaration to be in lieu of the oath otherwise required.

37CFR 1.68 Declaration in lieu of oath.

Any document to be filed in the Patent and Trademark Office and which is required by any law, rule, or other regulation to be under oath may be subscribed to by a written declaration. Such declaration may be used in lieu of the oath otherwise required, if, and only if, the declarant is on the same document, warned that willful false statements and the like are punishable by fine or imprisonment, or both (18 U.S.C. 1001) and may jeopardize the validity of the application or any patent issuing thereon. The declarant must set forth in the body of the declaration that all statements made of the declarant's own knowledge are true and that all statements made on information and belief are believed to be true.

§17.2 37CFR 1.131 宣誓書使用之目的

美國專利法基本上乃採先發明主義，亦即先發明者應取得專利權。然而在專利審查時專利審查委員並不知道專利申請案中各請求項（claim）所請求的發明的實際發明日，因而只能先推斷一專利申請案中各請求項之發明的發明日為有效申請日（如為新申請案則為在美國之申請日，如為延續案則為母案之申請日）來進行檢索先前技術（prior art），因而會發生審查委員引證用來核駁專利申請案之某請求項之先前技術之日期實際上是晚於該些被核駁的請求項之發明日。所以准許申請人藉37CFR§1.131 之宣誓書來宣誓其發明日早於所引證之先前技術之日期（Swearing behind a reference）。37CFR 1.131 之宣誓書可用來克服下列二種情況之核駁：

(A) 當審查委員引證之先前技術是 35USC§102(a)且非 35USC§102(b)之先前技術(亦即該先前技術之日期早於專利申請案之有效申請日且早於該專利申請案之有效申請日未超過一年)時可用來克服 35CFR§102(a) 及基於 35USC§102(a)事件之 35USC§103(a)之核駁。

(B) 當審查委員所引證之先前技術是 35USC§102(e)之先前技術（發證之美國專利或核准前公告之美國專利申請，其有效申請日在專利申請案之有效申請日之前，且並未請求（claim）相同之可專利的發明）時可用來克服 35USC§102(e)及基於 102(e)事件之 103(a)之核駁。

37CFR 1.131

§ 1.131 Affidavit or declaration of prior invention.

(a) When any claim of an application or a patent under reexamination is rejected, the inventor of the subject matter of the rejected claim, the owner of the patent under reexamination, or the party qualified under §§ 1.42, 1.43, or 1.47, may submit an appropriate oath or declaration to establish invention of the subject matter of the rejected claim prior to the effective date of the reference or activity on which the rejection is based. The effective date of a U.S. patent, U.S. patent application publication, or international application publication under PCT Article 21(2) is the earlier of its publication date or date that it is effective as a reference under 35 U.S.C. 102(e). Prior invention may not be established under this section in any country other than the United

States, a NAFTA country, or a WTO member country. Prior invention may not be established under this section before December 8, 1993, in a NAFTA country other than the United States, or before January 1, 1996, in a WTO member country other than a NAFTA country. Prior invention may not be established under this section if either:

(1) The rejection is based upon a U.S. patent or U.S. patent application publication of a pending or patented application to another or others which claims the same patentable invention as defined in § 41.203(a) of this title, in which case an applicant may suggest an interference pursuant to § 41.202(a) of this title; or

(2) The rejection is based upon a statutory bar.

MPEP 715

SITUATIONS WHERE 37 CFR 1.131 AFFIDAVITS OR DECLARATIONS CAN BE USED

Affidavits or declarations under 37 CFR 1.131 may be used, for example:

(A) To antedate a reference or activity that qualifies as prior art under 35 U.S.C. 102(a) and not under 35 U.S.C. 102(b), e.g., where the prior art date under 35 U.S.C. 102(a) of the patent, the publication or activity used to reject the claim(s) is less than 1 year prior to applicant's or patent owner's effective filing date.

(B) To antedate a reference that qualifies as prior art under 35 U.S.C. 102(e), where the reference has a prior art date under 35 U.S.C. 102(e) prior to applicant's effective filing date, and shows but does not claim the same patentable invention. See MPEP §715.05 for a discussion of "same patentable invention." See MPEP § 706.02(a) and § 2136 through § 2136.03 for an explanation of what references qualify as prior art under 35 U.S.C. 102(e).

§17.3　利用 37CFR 1.131 宣誓書證明發明在先

要利用 37CFR 1.131 之宣誓書證明發明在先時，首先需確定審查委員所引證之先前技術（prior art）或活動（activity）之日期。如果 35USC§102(a)之引證資料是外國之專利，則其先前技術日期是該外國專利取得專利權之日期，如果是外國刊物（包括核准前公告之公報)則其先前技術之日期為公告（發表）之日期，如果是美國之專利則先前技術之日期為發證之日期，如果是美國之刊物（包括核准前公告之公報）則其先前技術之日期為公告之日期，如果是先前之活動則先前技術之日期為在美國此活動首次被知道已發生之日期。如果是 35USC§102(e)之引證資料則先前技術之日期為美國專利或美國之核准前公告公報之有效申請日。

接著需確定被核駁的請求項所請求之發明所作的發明活動是否在美國，NAFTA 國家（需 1993 年 12 月 8 日以後之發明活動）或 WTO 會員國（需 1996 年 1 月 1 日）所進行者。如果是，才可適用 35CFR 1.131 之宣誓書，需注意的是台灣在 2002 年 1 月 1 日才加入 WTO，故發明活動如果在台灣進行需日期是在 2002 年 1 月 1 日以後才可適用 37 CFR 1.131 之宣誓書。

確定了先前技術之日期及發明活動是在美國、NAFTA 國家及 WTO 會員國進行之後，可以下列三種方式之一證明發明日是在先前技術日期之前：

(A) 該被核駁之請求項之發明付諸實施之日期早於先前技術之日期（如果是方法請求項則付諸實施之日期是方法，步驟成功地進行且達到所欲之目的之日期，如果是裝置或產品則是有形之實施例（模型）成功地作出，且可操作而達到所欲之目的之日期，如果是化合物或組合物則是將化合物/組合物製出，且其用途已確定之日期）。

(B) 該被核駁之請求項之發明構想日（conception）早於該先前技術之日期，且從該先前技術之日期之前直到付諸實施時有合理的勤勉（due diligence）去完成此發明。

(C) 該被核駁之請求項之發明的構想日早於該先前技術之日期，且從早於該先前技術之日期一直到專利提出申請時有合理的勤勉去完成此發明。

37CFR 1.131 之宣誓書可由下列人士簽署：

(A) 該專利申請案之所有發明人，

(B) 該請求項所請求之發明的發明人（可少於所有的發明人），

(C) 當所有或一部份之發明人無法或拒絕簽署時，由其法定代理人（監護人）或受讓人之代表人，

(D) 由受讓人或對該專利申請案有權利之人。

MPEP 715.07

THREE WAYS TO SHOW PRIOR INVENTION

The affidavit or declaration must state FACTS and produce such documentary evidence and exhibits in support thereof as are available to show conception and completion of invention in this country or in a NAFTA or WTO member country (MPEP § 715.07(c)), at least the conception being at a date prior to the effective date of the reference. Where there has not been reduction to practice prior to the date of the reference, the applicant or patent owner must also show diligence in the completion of his or her invention from a time just prior to the date of the reference continuously up to the date of an actual reduction to practice or up to the date of filing his or her application (filing constitutes a constructive reduction to practice, 37 CFR 1.131).

As discussed above, 37 CFR 1.131(b) provides three ways in which an applicant can establish prior invention of the claimed subject matter. The showing of facts must be sufficient to show:

(A) reduction to practice of the invention prior to the effective date of the reference; or

(B) conception of the invention prior to the effective date of the reference coupled with due diligence from prior to the reference date to a subsequent (actual) reduction to practice; or

(C) conception of the invention prior to the effective date of the reference coupled with due diligence from prior to the reference date to the filing date of the application (constructive reduction to practice).

MPEP 715.04

WHO MAY MAKE AFFIDAVIT OR DECLARATION

The following parties may make an affidavit or declaration under 37 CFR 1.131:

(A) All the inventors of the subject matter claimed.

(B) An affidavit or declaration by less than all named inventors of an application is accepted where it is shown that less than all named inventors of an application invented the subject matter of the claim or claims under rejection. For example, one of two joint inventors is accepted where it is shown that one of the joint inventors is the sole inventor of the claim or claims under rejection.

(C)A party qualified under 37 CFR 1.42 or 1.43, or 1.47 in situation where some or all of the inventors are not available or not capable of joining in the filing of the application.

(D) The assignee or other party in interest when it is not possible to produce the affidavit or declaration of the inventor. Ex parte Foster, 1903 C.D. 213, 105 O.G. 261 (Comm'r Pat. 1903).

§17.4 合理的勤勉

利用 37CFR 1.131 之宣誓書證明發明在先時，尚需考慮合理的勤勉。如果被核駁之請求項之發明之付諸實施日早於先前技術之日期就不必考慮勤勉。而如果被核駁之請求項所請求之發明之構想日晚於先前技術之日期時，也不可考慮勤勉之問題，只有當被核駁之請求項所請求發明之構想日早於先前技術之日期時才要考慮從先前技術日期之前至所請求之發明付諸實施（reduced to practice）或提出專利申請（constructive reduction to practice 推斷之付諸實施）之期間的勤勉。

勤勉包括二種：engineering-diligence（工程之勤勉）及 attorney- diligence（專利代理人（律師）之勤勉。亦即，發明人在該期間內是否有合理的勤勉及專利代理人（律師）受委託後是否積極的準備專利說明書，提出申請均需考慮。但是此二種勤勉並不是要求發明人（專利代理人）放下所有的其他工作專心在發明之付諸實施或提出專利申請上。只要發明人（專利代理人）有作發明（撰寫）的活動或未作發明（撰寫）的活動但可解釋其不活動之理由都有可能被認為是合理的勤勉。亦即勤勉是不分程度的，但是所作的活動必需是直接與該發明有關才可。例如發明人在該期間內不進行付諸實施之活動而去作商業的活動以圖取得專利產品上資金，將被視為不是合理的勤勉。又例如專利律師雖有其他未完成之案件，但其依照接案之先後順序承辦，一完成了先前委託之案件後立即進行了該專利申請案說明書之撰寫及申請手續將被為是合理的勤勉。

MPEP 715.07

715.07(a) Diligence

Where conception occurs prior to the date of the reference, but reduction to practice is afterward, it is not enough merely to allege that applicant or patent owner had been diligent. Ex parte Hunter, 1889 C.D. 218, 49 O.G. 733 (Comm'r Pat. 1889). Rather, applicant must show evidence of facts establishing diligence.

In determining the sufficiency of a 37 CFR 1.131 affidavit or declaration, diligence need not be considered unless conception of the invention prior to the effective date is clearly established, since diligence comes into question only after prior conception is established. Ex parte Kantor, 177 USPQ 455 (Bd. App. 1958).

What is meant by diligence is brought out in Christie v. Seybold, 1893 C.D. 515, 64 O.G. 1650 (6th Cir. 1893). In patent law, an inventor is either diligent at a given time or he is not diligent; there are no degrees of diligence. An applicant may be diligent within the meaning of the patent law when he or she is doing nothing, if his or her lack of activity is excused. Note, however, that the record must set forth an explanation or excuse for the inactivity; the USPTO or courts will not speculate on possible explanations for delay or inactivity. See In re Nelson, 420 F.2d 1079, 164 USPQ 458 (CCPA 1970). Diligence must be judged on the basis of the particular facts in each case. See MPEP § 2138.06 for a detailed discussion of the diligence requirement for proving prior invention.

Under 37 CFR 1.131, the critical period in which diligence must be shown begins just prior to the effective date of the reference or activity and ends with the date of a reduction to practice, either actual or constructive (i.e., filing a United States patent application). Note, therefore, that only diligence before reduction to practice is a material consideration. The "lapse of time between the completion or reduction to practice of an invention and the filing of an application thereon" is not relevant to an affidavit or declaration under 37 CFR 1.131. See Ex parte Merz, 75 USPQ 296 (Bd. App. 1947).

MPEP 2138.06

2138.06 "Reasonable Diligence" [R-1]

The diligence of 35 U.S.C. 102(g) relates to reasonable "attorney-diligence" and "engineering-diligence" (Keizer v. Bradley, 270 F.2d 396, 397, 123 USPQ 215, 216 (CCPA 1959)), which does not require that "an inventor or his attorney . drop all other work and concentrate on the particular invention involved......" Emery v. Ronden, 188 USPQ 264, 268 (Bd. Pat. Inter. 1974).

WORK RELIED UPON TO SHOW REASONABLE DILIGENCE MUST BE DIRECTLY RELATED TO THE REDUCTION TO PRACTICE

The work relied upon to show reasonable diligence must be directly related to the reduction to practice of the invention in issue. Naber v. Cricchi, 567 F.2d 382, 384, 196 USPQ 294, 296 (CCPA 1977), cert. denied, 439 U.S. 826 (1978). >See also Scott v. Koyama, 281 F.3d 1243, 1248-49, 61 USPQ2d 1856, 1859 (Fed. Cir. 2002) (Activities directed at building a plant to practice the claimed process of producing tetrafluoroethane on a large scale constituted efforts toward actual reduction to practice, and thus were evidence of diligence. The court distinguished cases where diligence was not found because inventors either discontinued development or failed to complete the invention while pursuing financing or other commercial activity.)

DILIGENCE REQUIRED IN PREPARING AND FILING PATENT APPLICATION

The diligence of attorney in preparing and filing patent application inures to the benefit of the inventor. Conception was established at least as early as the date a draft of a patent application was finished by a patent attorney on behalf of the inventor. Conception is less a matter of signature than it is one of disclosure. Attorney does not prepare a patent application on behalf of particular named persons, but on behalf of the true inventive entity. Six days to execute and file application is acceptable. Haskell v. Coleburne, 671 F.2d 1362, 213 USPQ 192, 195 (CCPA 1982). See also Bey v. Kollonitsch, 866 F.2d 1024, 231 USPQ 967 (Fed. Cir. 1986) (Reasonable diligence is all that is required of the attorney. Reasonable diligence is established if attorney worked reasonably hard on the application during the continuous critical period. If the attorney has a reasonable backlog of unrelated cases which he takes up in chronological order and carries out expeditiously, that is sufficient. Work on a related case(s) that contributed substantially to the ultimate preparation of an application can be credited as diligence.).

§17.5　37CFR 1.131 之宣誓書需顯示的涵蓋範圍

　　當使用 37CFR 1.131 之宣誓書證明所請求之發明在引證之先前技術之日期之前先發明時，需顯示多大的涵蓋範圍，可參考以下說明。

　　當審查委員以 35USC§102 僅引證一件先前技術核駁時，可分為以下幾種情況：

① 所請求之發明僅為一 species X，引證為 species X，則宣誓書只要證明 species X 之發明日在引證之 X 之前即可。

② 所請求之發明僅為一 species Y，引證資料為 species X，宣誓書如果能證明 species Y 之發明日在引證 X 之前，且證明 Y 為 X 之一顯而易見之變形例 (obvious variant)即可。

③ 所請求發明為一 genus 包括 species A,B,C,D,E…而所引證者為 species A，則宣誓書只要證明在 species A 之發明日在引證之 species A 之前即可，或者證明 species B 之發明日在引證 A 之前但 B 是 A 之顯而易見之變形例亦可。

④ 所請求發明為一 genus 包括 species A,B,C,D,E…而所引證者為多個 species A,B,C,D…則宣誓書能證明 species A,B,C,D 之發明日在引證之前即可。

⑤ 所請求發明為一 genus 包括 species A,B,C,D,E…而所引證者為多個 species A,B,C,D，但宣誓書只能證明 Species A 之發明日在引證之前，則此時需視此 genus 中之各 species 是否通常可被預測具有相同之物理/化學性質(功能)而定，例如 genus 為鹵素，引證為 F,Cl,Br,I，則宣誓書只要能證明 F 之發明日在引證之日期之前即可。

如果審查委員以 35 USC§103 引證二件或上之以先前技術以顯而易見核駁，申請人可以 37 CFR§1.131 之宣誓書證明被核駁之請求項之發明的發明日是在二件或以上之任何一件的先前技術日期之前來克服此顯而易見的核駁。但是，要注意的是要證明所請求的發明的發明日在任一件引證之先前技術日期之前，如果僅證明在任一件先前技術之日期之前已發明了該先前技術所揭露之技術，是不足以克服核駁理由的。

MPEP 715.03 Genus-Species, Practice Relative to Cases Where Predictability Is in Question

Where generic claims have been rejected on a reference or activity which discloses a species not antedated by the affidavit or declaration, the rejection will not ordinarily be withdrawn, subject to the rules set forth below, unless the applicant is able to establish that he or she was in possession of the generic invention prior to the effective date of the reference or activity. In other words, the affidavit or declaration under 37 CFR 1.131 must show as much as the minimum disclosure required by a patent specification to furnish support for a generic claim.

REFERENCE OR ACTIVITY DISCLOSES SPECIES

A.Species Claim

Where the claim under rejection recites a species and the reference or activity discloses the claimed species, the rejection can be overcome under 37 CFR 1.131 directly by showing prior completion of the claimed species or indirectly by a showing of prior completion of a different species coupled with a showing that the claimed species would have been an obvious modification of the species completed by applicant. See In re Spiller, 500 F.2d 1170, 182 USPQ 614 (CCPA 1974).

B.Genus Claim

The principle is well established that the disclosure of a species in a cited reference is sufficient to prevent a later applicant from obtaining a "generic claim." In re Gosteli, 872 F.2d 1008, 10 USPQ2d 1614 (Fed. Cir. 1989); In re Slayter, 276 F.2d 408, 125 USPQ 345 (CCPA 1960).

Where the only pertinent disclosure in the reference or activity is a single species of the claimed genus, the applicant can overcome the rejection directly under 37 CFR 1.131 by showing prior possession of the species disclosed in the reference or activity. On the other hand, a reference or activity which discloses several species of a claimed genus can be overcome directly under 37 CFR 1.131 only by a showing that the applicant completed, prior to the date of the reference or activity, all of the species shown in the reference. In re Stempel, 241 F.2d 755, 113 USPQ 77 (CCPA 1957).

Proof of prior completion of a species different from the species of the reference or activity will be sufficient to overcome a reference indirectly under 37 CFR 1.131 if the species shown in the reference or activity would have been obvious in view of the species shown to have been made by the applicant. In re Clarke, 356 F.2d 987, 148 USPQ 665 (CCPA 1966); In re Plumb, 470 F.2d 1403, 176 USPQ 323 (CCPA 1973); In re Hostettler, 356 F.2d 562, 148 USPQ 514 (CCPA 1966). Alternatively, if the applicant cannot show possession of the species of the reference or activity in this manner, the applicant may be able to antedate the reference or activity indirectly by, for example, showing prior completion of one or more species which put him or her in possession of the claimed genus prior to the reference's or activity's date. The test is whether the species completed by applicant prior to the reference date or the activity's date provided an adequate basis for inferring that the invention has generic applicability. In re Plumb, 470 F.2d 1403, 176 USPQ 323 (CCPA 1973); In re Rainer, 390 F.2d 771, 156 USPQ 334 (CCPA 1968); In re Clarke, 356 F.2d 987, 148 USPQ 665 (CCPA 1966); In re Shokal, 242 F.2d 771, 113 USPQ 283 (CCPA 1957).

It is not necessary for the affidavit evidence to show that the applicant viewed his or her invention as encompassing more than the species actually made. The test is whether the facts set out in the affidavit are such as would persuade one skilled in the art that the applicant possessed so much of the invention as is shown in the reference or activity. In re Schaub, 537 F.2d 509, 190 USPQ 324 (CCPA 1976).

C.Species Versus Embodiments

References or activities which disclose one or more embodiments of a single claimed invention, as opposed to species of a claimed genus, can be overcome by filing a 37 CFR 1.131 affidavit showing prior completion of a single embodiment of the invention, whether it is the same or a different embodiment from that disclosed in the reference or activity. See In re Fong, 288 F.2d 932, 129 USPQ 264 (CCPA 1961) (Where applicant discloses and claims a washing solution comprising a detergent and polyvinylpyrrolidone (PVP), with no criticality alleged as to the particular detergent used, the PVP being used as a soil-suspending agent to prevent the redeposition of the soil removed, the invention was viewed as the use of PVP as a soil-suspending agent in washing with a detergent. The disclosure in the reference of the use of PVP with two detergents, both of which differed from that shown in applicant's 37 CFR 1.131 affidavit, was considered a disclosure of different embodiments of a single invention, rather than species of a claimed genus); In re Defano, 392 F.2d 280, 157 USPQ 192 (CCPA 1968).

REFERENCE OR ACTIVITY DISCLOSES CLAIMED GENUS

In general, where the reference or activity discloses the claimed genus, a showing of completion of a single species within the genus is sufficient to antedate the reference or activity under 37 CFR 1.131. Ex parte Biesecker, 144 USPQ 129 (Bd. App. 1964).

In cases where predictability is in question, on the other hand, a showing of prior completion of one or a few species within the disclosed genus is generally not sufficient to overcome the reference or activity. In re Shokal, 242 F.2d 771, 113 USPQ 283 (CCPA 1957). The test is whether the species completed by applicant prior to the reference date or the date of the activity provided an adequate basis for inferring that the invention has generic applicability. In re Mantell, 454 F.2d 1398, 172 USPQ 530 (CCPA 1973); In re Rainer, 390 F.2d 771, 156 USPQ 334 (CCPA 1968); In re DeFano, 392 F.2d 280, 157 USPQ 192 (CCPA 1968); In re Clarke, 356 F.2d 987, 148 USPQ 665 (CCPA 1965). In the case of a small genus such as the halogens, which consists of four species, a reduction to practice of three, or perhaps even two, species might show possession of the generic invention, while in the case of a genus comprising hundreds of species, reduction to practice of a considerably larger number of species would be necessary. In re Shokal, supra.

It is not necessary for the affidavit evidence to show that the applicant viewed his or her invention as encompassing more than the species he or she actually made. The test is whether the facts set out in the affidavit are such as would persuade one skilled in the art that the applicant possessed so much of the invention as is shown in the reference. In re Schaub, 537 F. 509, 190 USPQ 324 (CCPA 1976).

MPEP 715(o)

SWEARING BEHIND ONE OF A PLURALITY OF COMBINED REFERENCES

Applicant may overcome a 35 U.S.C. 103 rejection based on a combination of references by showing completion of the invention by applicant prior to the effective date of any of the references; applicant need not antedate the reference with the earliest filing date. However, as discussed above, applicant's 37 CFR 1.131 affidavit must show possession of either the whole invention as claimed or something falling within the claim(s) prior to the effective date of the reference being antedated; it is not enough merely to show possession of what the reference happens to show if the reference does not teach the basic inventive concept.

Where a claim has been rejected under 35 U.S.C. 103 based on Reference A in view of Reference B, with the effective date of secondary Reference B,

§17.6　事實與書面證據

37CFR 1.131 宣誓書是由發明人以宣誓的方式證明先發明。而要證明先發明並非只由發明人作宣誓書主張即可，需附上令人滿意的事實證據以證明先發明。例如發明人可呈送以下證據或證物：

(A) 發明之草圖

(B) 發明之藍圖

(C) 發明之相片

(D) 發明記錄簿（發明提案書）

(E) 發明之模型

(F) 證人之聲明

(G) 抵觸程序中之證詞

(H) 揭露文件（申請人送至美國專利局用以證明發明日之發明揭露文件）

　　在宣誓書中宣誓人必需清楚地說明所附之事實證據與發明過程之關聯以證明在先前技術之日期之前已發明了所請求之發明。另外，在宣誓書中宣誓人可不必註明構想或附諸實施之確實日期，只主張早於一特定日期即可。但是對於勤勉之日期（期間）則需註明確實之日期。

MPEP 715.07 Facts and Documentary Evidence

GENERAL REQUIREMENTS

The essential thing to be shown under 37 CFR 1.131 is priority of invention and this may be done by any satisfactory evidence of the fact. FACTS, not conclusions, must be alleged. Evidence in the form of exhibits may accompany the affidavit or declaration. Each exhibit relied upon should be specifically referred to in the affidavit or declaration, in terms of what it is relied upon to show. For example, the allegations of fact might be supported by submitting as evidence one or more of the following:

(A) attached sketches;

(B) attached blueprints;

(C) attached photographs;

(D) attached reproductions of notebook entries;

(E) an accompanying model;

(F) attached supporting statements by witnesses, where verbal disclosures are the evidence relied upon. *Ex parte Ovshinsky*, 10 USPQ2d 1075 (Bd. Pat. App. & Inter. 1989);

(G) testimony given in an interference. Where interference testimony is used, the applicant must point out which parts of the testimony are being relied on; examiners cannot be expected to search the entire interference record for the evidence. *Ex parte Homan*, 1905 C.D. 288 (Comm'r Pat. 1905);

(H) Disclosure documents(MPEP§1706) may be used as documentary evidence of conception.

The affidavit or declaration and exhibits must clearly explain which facts or data applicant is relying on to show completion of his or her invention prior to the particular date. Vague and general statements in broad terms about what the exhibits describe along with a general assertion that the exhibits describe a reduction to practice "amounts essentially to mere pleading, unsupported by proof or a showing of facts" and, thus, does not satisfy the requirements of 37 CFR 1.131(b).

ESTABLISHMENT OF DATES

If the dates of the exhibits have been removed or blocked off, the matter of dates can be taken care of in the body of the oath or declaration.

When alleging that conception or a reduction to practice occurred prior to the effective date of the reference, the dates in the oath or declaration may be the actual dates or, if the applicant or patent owner does not desire to disclose his or her actual dates, he or she may merely allege that the acts referred to occurred prior to a specified date. However, the actual dates of acts relied on to establish diligence must be provided. See MPEP § 715.07(a) regarding the diligence requirement.

§17.7 不適合使用 37CFR 1.131 宣誓書之情況

在下述情況，不適合使用 37CFR 1.131 之宣誓書主張發明日早於引證之先前技術之日期：

(A) 當所引證之先前技術之公告日在申請案之有效申請日的一年之前，亦即 35USC§102(b)之事件的引證資料；

(B) 當引證之先前技術為美國專利或美國專利申請案且其請求相同、具專利性之發明時；

(C) 當所引證之先前技術為申請人外國之專利，從此外國專利之申請日至美國專利申請案之申請日已超過一年，且該外國專利申請案在美國專利申請案申請之前已取得專利，亦即為 35USC§102(d)事件的引證資料；

(D) 當美國專利申請案之有效申請日早於所引證之先前技術之日期時；

(E) 當所引證之先前技術為申請人之美國專利，且此美國專利請求相同之發明，亦即為重複專利之情形時；

(F) 當引證之先前技術為申請人之先前美國專利之揭露；

(G) 當引證之先前技術為申請人已承認為先前技術者；

(H) 當引證之先前技術是 35USC§102(f)事件者；

(I) 當引證之先前技術是 35USC§102(g)事件者；

(J) 當發明之標的為抵觸程序中輸掉的抵觸標的(lost count)者；

MPEP 715

SITUATIONS WHERE 37 CFR 1.131 AFFIDAVITS OR DECLARATIONS ARE INAPPROPRIATE

An affidavit or declaration under 37 CFR 1.131 is not appropriate in the following situations:

(A) Where the reference publication date is more than 1 year prior to applicant's or patent owner's effective filing date. Such a reference is a "statutory bar" under 35 U.S.C. 102(b) as referenced in 37 CFR 1.131(a)(2). A reference that only qualifies as prior art under 35 U.S.C. 102(a) or (e) is not a "statutory bar."

(B) Where the reference U.S. patent or U.S. patent application publication claims the same patentable invention. See MPEP § 715.05 for a discussion of "same patentable invention" and MPEP §2306. Where the reference patent and the application or patent under reexamination are commonly owned, and the inventions defined by the claims in the application or patent under reexamination and by the claims in the patent are not identical but are not patentably distinct, a terminal disclaimer and an affidavit or declaration under 37 CFR 1.130 may be used to overcome a rejection under 35 U.S.C. 103. See MPEP § 718.

(C) Where the reference is a foreign patent for the same invention to applicant or patent owner or his or her legal representatives or assigns issued prior to the filing date of the domestic application or patent on an application filed more than 12 months prior to the filing date of the domestic application. See 35 U.S.C. 102(d).

(D) Where the effective filing date of applicant's or patent owner's parent application or an International Convention proved filing date is prior to the effective date of the reference, an affidavit or declaration under 37 CFR 1.131 is unnecessary because the reference should not have been used. See MPEP § 201.11 to § 201.15.

(E) Where the reference is a prior U.S. patent to the same entity, claiming the same invention. The question involved is one of "double patenting."

(F) Where the reference is the disclosure of a prior U.S. patent to the same party, not copending. The question is one of dedication to the public. Note however, In re Gibbs, 437 F.2d 486, 168 USPQ 578 (CCPA 1971) which substantially did away with the doctrine of dedication.

(G) Where applicant has clearly admitted on the record that subject matter relied on in the reference is prior art. In this case, that subject matter may be used as a basis for rejecting his or her claims and may not be overcome by an affidavit or declaration under 37 CFR 1.131. In re Hellsund, 474 F.2d 1307, 177 USPQ 170 (CCPA 1973); In re Garfinkel, 437 F.2d 1000, 168 USPQ 659 (CCPA 1971); In re Blout, 333 F.2d 928, 142 USPQ 173 (CCPA 1964); In re Lopresti, 333 F.2d 932, 142 USPQ 177 (CCPA 1964).

(H) Where the subject matter relied upon is prior art under 35 U.S.C. 102(f).

(I) Where the subject matter relied on in the reference is prior art under 35 U.S.C. 102(g). 37 CFR 1.131 is designed to permit an applicant to overcome rejections based on references or activities which are not statutory bars, but which have dates prior to the effective filing date of the application but subsequent to the applicant's actual date of invention. However, when the subject matter relied on is also available under 35 U.S.C. 102(g), a 37 CFR 1.131 affidavit or declaration cannot be used to overcome it.

(J) Where the subject matter corresponding to a lost count in an interference is either prior art under 35 U.S.C. 102(g) or barred to applicant by the doctrine of interference estoppel.

§17.8　呈送 37CFR 1.131 宣誓書之時機

37CFR 1.131 之宣誓書必需及時送入美國專利局以被承認。如果 37CFR 1.131 之宣誓書在最終核駁之前或在未發下最終核駁之申請案提出訴願之前，送入將被視為是及時。如果是在最終核駁發下之後要送入 37CFR §1.131 之宣誓書則需同時呈送(1)為了克服在最終核駁之新的（第 1 次的）核駁理由之答辯書，或(2)顯示有充份理由為何此宣誓書未在最終核駁發下前呈送之聲明，才會被認為是及時。

如果審查委員認為所呈送之 37CFR §1.131 之宣誓書並不是及時的，申請人可以提出請願（petition）

MEP 715.09

Seasonable Presentation

Affidavits or declarations under 37 CFR 1.131 must be timely presented in order to be admitted. Affidavits and declarations submitted under 37 CFR 1.131 and other evidence traversing rejections are considered timely if submitted:

(A) prior to a final rejection;

(B) before appeal in an application not having a final rejection; or

(C) after final rejection **>,but before or on the same dare of filing an appeal, upon a showing of good and sufficient reasons why the affidavit or other evidence is necessary and was not earlier presented in compliance with 37CFR1.116(e);or

(D) after the prosecution is closed (e.g., after a final rejection, after appeal, or after allowance) if applicant files the affidavit or other evidence with a request for continued examination (RCE) under 37CFR 1.114 in a utility or plant application filed on or after June 8, 1995; or a continued prosecution application (CPA) under 37 CFR 1.53(d)in a design application.<

All admitted affidavits and declarations are acknowledged and commented upon by the examiner in his or her next succeeding action.

For affidavits or declarations under 37 CFR 1.131 filed after appeal, see37CFR*> 41.33(d)<and MPEP § *>1206and § 1211.03<.

Review of an examiner's refusal to enter an affidavit as untimely is by petition and not by appeal to the Board of Patent Appeals and Interferences. In reDerers,515F.2d 1152, 185 USPQ 644 (CCPA 1975);Ex parte Hale, 49USPQ 209(Bd.App. 1941). See MPEP § 715.08 regarding review of questions of propriety of 37CFR 1.131affidavits and declarations.

實例 1

一台灣之申請人於 2005 年 1 月 19 日申請了一台灣專利申請案並於 2005 年 5 月 24 日提出一美國專利申請案主張台灣專利申請案之優先權，美國專利局發下 Office Action 以一公告日為 2004 年 5 月 4 日之美國專利申請案用，35 USC §102(a) 核駁本案之 Claims。而台灣申請人之專利提案書於 2004 年 3 月 20 日提交其公司之專利審查部門審查（亦即發明人構思此發明之日期為 2004 年 3 月 20 日之前），請問申請人如何克服此 35 USC §102(a)之核駁？

由於此美國專利申請案之優先權日，2005 年 1 月 19 日仍晚於引證之 2004 年 5 月 4 日，但構思日為至少 2004 年 3 月 20 日之前，故如果發明人可證明其實際付諸實施之日在 2004 年 5 月 4 日之前且從 2004 年 3 月 20 日至 2005 年 5 月 4 日之間有持續之勤勉度（積極將構思付諸實施）則可以由發明人提出 37 CFR §1.131 之宣誓書證明其在 2004 年 3 月 20 日之前已發明，故發明日在該引證之前以克服 35 USC §102(a)之核駁。

另外，如果發明人未實際將此發明付諸實施，亦可證明從構思日（2004 年 3 月 20 日）至台灣專利提申日（2005 年 1 月 19 日）之間有勤勉度（積極準備說明書提出專利申請），但有可能審查委員認為從 2004 年 3 月 20 日構思至 2005 年 1 月 19 日才提出申請歷經 8 個月，為時太久沒有勤勉度。

另外，雖然此專利申請之發明活動乃在台灣進行，但台灣在 2002 年 1 月 1 日已加入 WTO，故仍符合美國專利法之規定。

實例 2

一專利申請案於 2007 年 2 月 4 日提出，其 claims 乃請求一具有多層構造之生物感知器。美國專利局引用一美國專利以 35 USC §102(a),(e)核駁本案之 claims，所引用之美國專利乃於 2006 年 10 月 1 日發證且其申請日為 2003 年 12 月 1 日，該引證之美國專利揭示了本案之多層生物感知器之構造及如何製造及使用，但該引證之美國專利並未請求多層生物感知器。被核駁之美國專利申請案之發明構思乃於 2006 年 9 月 20 日完成。請問本專利申請人應如何克服此 35 USC §102(a),(e)之核駁？

由於本專利申請案之發明日，2006 年 9 月 20 日晚於引證案之申請日，故申請人無法應用 37 CFR 1.131 之宣誓書主張發明日早於引證案，因而只能將原說明書中其他特徵加入 claim 中以與引證區別而克服 35 USC §102 之核駁。

實例 3

發明人 A 於 2006 年 6 月 15 日申請了一美國專利包括有二獨立項，claim 1 及 claim 2，均有關於一聚氨酯之製法。發明人 A 於 2005 年 1 月 15 日完成 claim 1 之製法並於 2005 年 5 月 12 日發表論文揭露 claim 1 之製法，並之後於 2006 年 1 月完成了 claim 2 之改良製法，美國專利局發下 Office Action 以該論文用 35 USC §102(b)核駁 Claim

1，並用一其他人之論文（日期為 2006 年 5 月 1 日）以 35 USC §102(a)核駁 claim 2。請問發明人 A 是否可以 37 CFR 1.131 之宣誓書來克服對 claim 1 及 claim 2 之核駁？

　　發明人可用 37 CFR 1.131 之宣誓書來克服 claim 2 之核駁，但是無法用 37 CFR 1.131 之宣誓書來克服 claim 1 之核駁，雖然為發明人自己發表之論文，但其發表日至本申請案之申請日已超過 1 年故違反 35 USC §102(b)之規定。

MPEP 2133.02

"Any invention described in a printed publication more than one year prior to the date of a patent application is prior art under Section 102(b), even if the printed publication was authored by the patent applicant." "Once an inventor has decided to lift the veil of secrecy from his [or her] work, he [or she] must choose between the protection of a federal patent, or the dedication of his [or her] idea to the public at large."

A rejection under 35 U.S.C. 102(b) cannot be overcome by affidavits and declarations under 37 CFR 1.131 (Rule 131 Declarations), foreign priority dates, or evidence that applicant himself invented the subject matter. Outside the 1-year grace period, applicant is barred from obtaining a patent containing any anticipated or obvious claims.

實例 4

　　台灣發明人 X 於 2004 年 1 月 10 日申請了一美國發明專利，此美國專利的 claim 乃是請求一由通式表示之化合物（genus），審查委員引證了發明人 Y 一美國專利其揭示了發明人 X 所請求之化合物之一（species），但未請求該化合物，發明人 Y 之專利於 2004 年 4 月 1 日發明，其申請日為 2003 年 12 月 10 日，發明人 Y 並未由發明人 X 得到任何資訊而產生其發明且發明人 Y 及發明人 X 並未有義務將其專利申請權讓予同一人，發明人 X 之發明活動乃於台灣境內完成，且發明於 2003 年 11 月 1 日完成，請問，發明人 X 可否使用 37 CFR 1.131 之宣誓書來克服此 35USC §102(e)核駁？

　　可以，該引證為美國專利且申請日在本案申請日之前而且揭示了本案化合物之一 species，故是為 35 USC §102(e)之核駁，此時可使用 37 CFR 1.131 之宣誓書證明，本案之發明日在引證之美國申請日之前以克服核駁。

實例 5

　　發明人 X 於 2005 年 2 月 11 日申請了一美國發明專利，其中 claim 1 乃有關一行動電話其包括了二特徵，一是天線構造，另一是可拆卸的吊帶環，審查委員引證了二先前技術以顯而易見核駁 claim 1，先前技術（一）乃是一 2004 年 11 月 2 日早期公告之美國專利申請案，其揭露了一具有相同天線構造之行動電話，但並未請求（claim）該行動電話，先前技術（二）則為一刊物，發行日為 2004 年 3 月 1 日其揭露了一具有相同構造之可拆卸的吊帶環之行動電話。發明人 X 提出 37CFR 1.131 之宣誓書主張其在至少 2004 年 5 月 1 日就已將所請求之行動電話付諸實施，並附上一刊物之論文作佐證，其顯示該論文在 2004 年 5 月 1 日被該刊物接受，並於 2004 年 6 月 1 日發表，該論文揭示了一行動電話具有該特殊構造之天線，但並未揭示可拆卸的吊帶環。請問此誓書是否可克服審查委員之顯而易見的核駁？

　　無法克服審查委顯而易見之核駁，發明人 X 之刊物雖可證明其在 2004 年 5 月 1 日就完成了發明，早於先前技術（一）之日期，但該刊物之發明並未揭示 claim 1 之另一特徵（可拆卸之吊帶環）故無法證明 claim 1 所請求之發明早於先前技術（一）。

實例 6

　　發明人 X 為一汽車製造公司之工程師於 2006 年 2 月 20 日構思了一發明，可改善車內使用之收音機之電路系統，發明人 X 於 2006 年 2 月 20 日與公司之專利律師討論其構想，專利律師 Y 於 2006 年 8 月 20 日完成了新穎性之檢索，發現此構思是新穎的。專利律師 Y 之後就著手於與該公司其他發明人討論他們的發明並作檢索及撰寫專利說明書，這些其他發明人的發明乃是有關於汽車全安性，且其發明均晚於發明人 X 之發明日。專利律師終於在 2007 年 11 月 1 日將發明人 X 之專利申請案提出申請。之後於審查過程中審查委員引證了一發明人 Z 之美國專利其申請日為 2006 年 4 月 21 日，發證日為 2007 年 12 月 1 日（35USC§102(e)之先前技術），該美國專利揭露了發明人 X 之專利申特徵，且未請求該電路系統，請問發明人可否藉由 37CFR§1.131 之宣誓書來克服 35USC§102(e)之核駁？請案的請求項之所有

　　無法克服 35USC§102(e)之核駁，發明人 X 之構思雖早於引證資料，但從發明人構思日，2006 年 2 月 20 日，至其專利申請日（推斷的付諸實施日），2007 年 11 月 1 日之間，專利律師並未有合理的勤勉，專利律師於作完新穎性檢索確認發明為新穎後，就應勤勉地準備專利說明並提出申請，專利師律 Y 卻先進行與發明人 X 之發明無關的案件的檢索、撰寫，且延遲一年多才提出申請，將被視為未合理的勤勉。

§17.9　37CFR 1.132 之宣誓書使用之目的

　　37 CFR§1.132 之宣誓書乃用來呈送答辯書中提到的證據。通常可藉由宣誓書之方式呈送證據以（A）移除一引證資料，例如當引證之資料是申請人發表者，或是引證資料是在敘述申請人的發明，或者引證資料中用相關內容實際上乃是衍生自申請人者；（B）克服顯而易見之核駁，克服無法據以實施之核駁（non-enabling），克服不具實用性，不能操作之核駁，或克服不符揭露性（敘述要件、據以實施要件，最佳模式要件）之核駁等等。

37CFR 1.132

Affidavits or declarations traversing rejections or objections.

When any claim of an application or a patent under reexamination is rejected or objected to, any evidence submitted to traverse the rejection or objection on a basis not otherwise provided for must be by way of an oath or declaration under this section.

§17.10 使用 37CFR 1.132 宣誓書移除先前技術

　　如果審查委員所引證之先前技術為一刊物，而申請人事實上是刊物的共同作者之一（co-author）時，申請人可提出 37CFR§1.132 之宣誓書證明所引證刊物乃在敘述申請人自己的工作，而其他作者乃是依照申請人之指示作實驗等。

　　如果審查委員所引證之專利，公告之專利申請案，或刊物中之標的是申請人自己的發明時，申請人亦可提出 37CFR§1.132 之宣誓書證明該專利之專利權人，申請人或作者乃是從申請人處得到相資訊。申請人於宣誓書中必需證明自己發明了在該專利、專利申請案或刊物中揭示的標的。

　　如果審查委員之核駁的理由是在申請人的專利申請案已被使用或知悉，申請人亦可提出一 37CFR§1.132 之宣誓書，證明上述被使用或知悉之活動乃是申請人自己實行者。

　　如果發明人與他人共同申請了一專利，其說明書揭露了標的 X 但並未請求該標的，之後發明人自己單獨申請了一專利申請案，請求該標的 X，則發明人與該他人之申請案就會成為發明人單獨申請的專利申請案之 35USC§102(a)、(e)或(f)之先前技術，此時如果發明人可提出 37CFR§1.132 之宣誓書很明確的（uncontradicted, unequivocal）證明自己單獨發明該標的 X，就可克服此先前技術之核駁。

MPEP 715.01(C)

CO-AUTHORSHIP

　　Where the applicant is one of the co-authors of a publication cited against his or her application, he or she may overcome the rejection by filing an affidavit or declaration under 37 CFR 1.131. Alternatively, the applicant may overcome the rejection by filing a specific affidavit or declaration under 37 CFR 1.132 establishing that the article is describing applicant's own work. An affidavit or declaration by applicant alone indicating that applicant is the sole inventor and that the others were merely working under his or her direction is sufficient to remove the publication as a reference under 35 U.S.C. 102(a). In re Katz, 687 F.2d 450, 215 USPQ 14 (CCPA 1982).

DERIVATION

　　When the unclaimed subject matter of a patent, application publication, or other publication is applicant's own invention, a rejection, which is not a statutory bar, on that patent or publication may be removed by submission of evidence establishing the fact that the patentee, applicant of the published application, or author derived his or her knowledge of the relevant subject matter from applicant. Moreover applicant must further show that he or she made the invention upon which the relevant disclosure in the patent, application publication, or other publication is based. In re Mathews, 408 F.2d 1393, 161 USPQ 276 (CCPA 1969); In re Facius, 408 F.2d 1396, 161 USPQ 294 (CCPA 1969).

MPEP 715.01(d) Activities Applied Against the Claims

Unless it is a statutory bar, a rejection based on an activity showing that the claimed invention was used or known prior to the filing date of the application may be overcome by an affidavit or declaration under 37 CFR 1.131 establishing a date of invention prior to the date of the activity. Alternatively, the applicant(s) may overcome the rejection by filing a specific affidavit or declaration under 37 CFR 1.132 showing that the activity was performed by the applicant(s).

MPEP 716.10

An uncontradicted "unequivocal statement" from the applicant regarding the subject matter disclosed in an article, patent, or published application will be accepted as establishing inventorship. In re DeBaun, 687 F.2d 459, 463, 214 USPQ 933, 936 (CCPA 1982). However, a statement by the applicants regarding their inventorship in view of an article, patent, or published application may not be sufficient where there is evidence to the contrary. Ex parte Kroger, 218 USPQ 370 (Bd. App. 1982) (a rejection under 35 U.S.C. 102(f) was affirmed notwithstanding declarations by the alleged actual inventors as to their inventorship in view of a nonapplicant author submitting a letter declaring the author's inventorship); In re Carreira, 532 F.2d 1356, 189 USPQ 461 (CCPA 1976) (disclaiming declarations from patentees were directed at the generic invention and not at the claimed species, hence no need to consider derivation of the subject matter).

實例 1

X 教授於 2008 年 5 月 31 日申請了一美國專利其為唯一之發明人，美國專利局引證了一篇論文，日期為 2008 年 1 月 31 日，此論文之作者為 X 教授及 Y，且所揭露之內容即為 X 之申請案之請求項所請求之發明，以 35 USC §102(a)核駁本案，請問此 102(a)之核駁是否適當？應如何克服？

該引證之論文的日期在 X 教授的專利申請案的申請日之前，且該論文之作者為 X 及 Y 與專利申請案 X 並不相同，故美國專利局以此論文作先前技術用 35 USC §102(a)核駁為恰當。

如果論文另一作者 Y 並非本申請案之發明人，只是依照 X 教授進行實驗，則 X 教授可依照 37 CFR 1.132 提出一宣誓書，指出該論文乃在敘述 X 自己的發明，且 X 為唯一之發明人，Y 只是依 X 之指示進行實驗，即可將此引證移除而克服 350 USC §102(a)之核駁。

實例 2

　　發明人 A 及 B 於 2006 年 11 月 24 日提出一美國專利申請案 X，審查委員引證了二件核准前公告之美國專利申請案 Y 及 Z，Y 之發明人為 A 及 C，其申請日在申請案 X 之前，但於 2007 年 5 月 3 日核准前公告，Z 之發明人為 A 及 D，申請日亦在申請案 X 之前，但於 2007 年 7 月 5 日核准前公告，專利申請案 Y 及 Z 中敘述到專利申請案 X 之 claims 1-10 之發明，但未請求該發明，審查委員認為申請案 X 之 claims 1-10 之發明之敘述在申請案 Y 及 Z 之中，故不具新穎性，請問此情形可否藉由 37CFR §1.132 之宣誓書主張專利申請案 Y 及 Z 中所敘述之發明乃是在敘述專利申請案之發明人 A 之發明？

　　如果專利申請案 X 之 claims 1-10 之發明人僅為 A，則由發明人 A 提出一 37CFR§1.132 之宣誓書主張專利申請案 Y 及 Z 中揭露之部份事實上為發明人 A 自己（一人）之發明，則此宣誓書應可被接受。

　　如果專利申請案 X 之 claims 1-10 之發明人為 A 及 B，則審查委員可能會質疑為何 B 未被列為專利申請案 Y 及 Z 之發明人。故通常要以 37CFR§1.132 主張引證之專利申請案或專利所揭露者為自己的發明或衍生自己或敘述自己之發明時，專利申請案之發明人（inventorship）應較引證案之發明人之人數為少（為其 subunit）。

§17.11 以 37CFR 1.132 宣誓書主張發明之不可預期效果

　　所請求之發明與先前技術有性質上之差異，而此性質上之差異確實是不可預期的，則可藉由 37CFR§1.132 之宣誓書來主張不是顯而易見的。但是性質上之差異怎樣才算是不可預期的，可分為下列數種情形：

(A) 大於所期待的結果（greater than expected results）

　　例如，將二成份相加時，原本預期性質會降低，但事實上將二成份相加時，所得到之性質是二成份之性質之相加之結果，此情形就是不可預期的。再例如將二成份相加時，原預期之性質乃是兩者相加之效果，但事實上得到之性質是大於兩者相加之效果（所謂的相乘效果 "synergisum"）且此相乘的效果不是可預期的話，亦可藉由 37CFR§1.132 之宣誓書來克服顯而易見之核駁。

(B) 較先前技術所具有的性質為優（Superior of a property shared with the prior art）

例如，一化合物的性質中某一性質較優（非預期地）。又例如，所請求的化合物在對抗厭氧菌之治療活性是非預期的優良時，均可藉由 37CFR§1.132 之宣誓書來克服顯而易見的核駁。

(C) 存在有不可預期的性質（presence of an unexpected property）

例如，所請求之化合物之化學結構與先前技術之化學結構很類似，但是所請求之化合物不可預期地具有抗發炎之性質（先前技術之化合物無此性質）時，可藉由 37CFR§1.132 之宣誓書來克服顯而易見的核駁。

(D) 不存在可預期的性質（Absence of an expected property）

例如，由先前技術的揭露可知所請求的化合物應該具有 β-and -renergic 阻斷之活性，但事實上所請求的化合物並不具有此活性時，可藉由 37CFR§1.132 之宣誓書來克服顯而易見的核駁。

申請人要藉由 37CFR§1.132 之宣誓書來主張不可預期之效果時，申請人需負舉證責任，證明具有不可預期之效果。申請人所提出之來證明不可預期效果之證據可為所請求之發明與最近的先前技術（closest prior art）在範圍為同等大小（commensurate in scope）之直接或間接的證據。所謂所請求的發明乃先前技術的範圍是同等大小的意思，是指所提出之不可預期的效果必需在所請求的發明的整個範圍均會發生才符合同等大小之條件。例如申請人所請求之發明乃是有關於使用某種離子交換樹脂在高溫下（elevated temperature）除銹的方法（claim 8 之溫度為大於 100℃），申請人提出在 110℃及 130℃使用先前技術之離子交換樹脂之比較測試結果以主張其發明之不可預期之效果。而所謂"高溫"涵蓋的範圍可最低至 60℃（在 60℃時先前技術之離子交換樹脂亦可達成所希望之效能）。因而原 claims 1-7 及 9-10 之核駁仍會被維持，只有 claim 8 會核准，因為 claims 1-7 及 9-10 之高溫範圍可為 60℃以上，而申請人所作之比較溫度範圍為 110℃及 130℃，範圍並非同等大小。然而，雖然申請人只針對一較窄的範圍作不可預期之效果之測試，而此技藝人士可合理地從比較窄的範圍延伸至整個範圍時，此較窄範圍之測試所呈現之不可預期效果亦可被接受來克服顯而易見之核駁。

使用 37CFR§1.132 之宣誓書主張不可預期之效果時，必需比較所請求之標的與最接近的先前技術，以克服顯而易見之核駁。實務上，申請人可比較所請求之發明與對所請求之發明相對上比審查委員所引證之先前技術更接近之先前技術。例如，申請人所請求之化合物為在 13-位置被氯取代之化合物（13-chloro substituted compound），審查委員用未被氯化之類似物（nonchlorinated analogs）以顯而易見核駁，則申請人可用 9-，12 及 14-位置被氯取代之化合物來作先前技術比較之。此外，當有二件同樣接近之先前技術時，如果只針對一件先前技術作比較將無法克服顯而易見之核駁，除非此二件先前技術揭露了實質上類似之內容。例如所請求之化合物與第 1 件引證資料之差異在於以三氟甲基取代氯自由基，而與第 2 件引證資料之差異在於不飽和之酯基取代飽和之酯基。申請人之 37CFR§1.132 之宣誓書如果只比較三氟甲基取代之化合物與氯自由基取代之化合物，仍不能克服顯而易見之核駁。

MPEP 716.02(a) Evidence Must Show Unexpected Results [R-2]

GREATER THAN EXPECTED RESULTS ARE EVIDENCE OF NONOBVIOUSNESS

"A greater than expected result is an evidentiary factor pertinent to the legal conclusion of obviousness ...... of the claims at issue." *In re Corkill*, 711 F.2d 1496, 226 USPQ 1005 (Fed. Cir. 1985). In *Corkhill*, the claimed combination showed an additive result when a diminished result would have been expected. This result was persuasive of nonobviousness even though the result was equal to that of one component alone. Evidence of a greater than expected result may also be shown by demonstrating an effect which is greater than the sum of each of the effects taken separately (i.e., demonstrating "synergism"). *Merck & Co. Inc. v. Biocraft Laboratories Inc.*, 874 F.2d 804, 10 USPQ2d 1843 (Fed. Cir.), *cert. denied*, 493 U.S. 975 (1989). However, a greater than additive effect is not necessarily sufficient to overcome a *prima facie* case of obviousness because such an effect can either be expected or unexpected.

Applicants must further show that the results were greater than those which would have been expected from the prior art to an unobvious extent, and that the results are of a significant, practical advantage. *Ex parte The NutraSweet Co.*, 19 USPQ2d 1586 (Bd. Pat. App. & Inter. 1991) (Evidence showing greater than additive sweetness resulting from the claimed mixture of saccharin and L-aspartyl-L-phenylalanine was not sufficient to outweigh the evidence of obviousness because the teachings of the prior art lead to a general expectation of greater than additive sweetening effects when using mixtures of synthetic sweeteners.).

SUPERIORITY OF A PROPERTY SHARED WITH THE PRIOR ART IS EVIDENCE OF NONOBVIOUSNESS

Evidence of unobvious or unexpected advantageous properties, such as superiority in a property the claimed compound shares with the prior art, can rebut *prima facie* obviousness. "Evidence that a compound is unexpectedly superior in one of a spectrum of common properties . . . can be enough to rebut a *prima facie* case of obviousness." No set number of examples of superiority is required. *In re Chupp*, 816 F.2d 643, 646, 2 USPQ2d 1437, 1439 (Fed. Cir. 1987) (Evidence showing that the claimed herbicidal compound was more effective than the closest prior art compound in controlling quackgrass and yellow nutsedge weeds in corn and soybean crops was sufficient to overcome the rejection under 35 U.S.C. 103, even though the specification indicated the claimed compound was an average performer on crops other than corn and soybean.). See also *Ex parte A*, 17 USPQ2d 1716 (Bd. Pat. App. & Inter. 1990) (unexpected superior therapeutic activity of claimed compound against anaerobic bacteria was sufficient to rebut *prima facie* obviousness even though there was no evidence that the compound was effective against all bacteria).

PRESENCE OF AN UNEXPECTED PROPERTY IS EVIDENCE OF NONOBVIOUSNESS

Presence of a property not possessed by the prior art is evidence of nonobviousness. *In re Papesch*, 315 F.2d 381, 137 USPQ 43 (CCPA 1963) (rejection of claims to compound structurally similar to the prior art compound was reversed because claimed compound unexpectedly possessed anti-inflammatory properties not possessed by the prior art compound); *Ex parte Thumm*, 132 USPQ 66 (Bd. App. 1961) (Appellant showed that the claimed range of ethylene diamine was effective for the purpose of producing "'regenerated cellulose consisting substantially entirely of skin'" whereas the prior art warned "this compound has 'practically no effect.' "). The submission of evidence that a new product possesses unexpected properties does not necessarily require a conclusion that the claimed invention is nonobvious. *In re Payne*, 606 F.2d 303, 203 USPQ 245 (CCPA 1979). See the discussion of latent properties and additional advantages in MPEP§2145.

ABSENCE OF AN EXPECTED PROPERTY IS EVIDENCE OF NONOBVIOUSNESS

Absence of property which a claimed invention would have been expected to possess based on the teachings of the prior art is evidence of unobviousness. *Ex parte Mead Johnson & Co.* 227 USPQ 78 (Bd. Pat. App. & Inter. 1985) (Based on prior art disclosures, claimed compounds would have been expected to possess beta-andrenergic blocking activity; the fact that claimed compounds did not possess such activity was an unexpected result sufficient to establish unobviousness within the meaning of 35 U.S.C. 103.).

MPEP 716.02(d) Unexpected Results Commensurate in Scope With Claimed Invention [R-1]

Whether the unexpected results are the result of unexpectedly improved results or a property not taught by the prior art, the "objective evidence of nonobviousness must be commensurate in scope with the claims which the evidence is offered to support." In other words, the showing of unexpected results must be reviewed to see if the results occur over the entire claimed range. In re Clemens, 622 F.2d 1029, 1036, 206 USPQ 289, 296 (CCPA 1980) (Claims were directed to a process for removing corrosion at "elevated temperatures" using a certain ion exchange resin (with the exception of claim 8 which recited a temperature in excess of 100°C). Appellant demonstrated unexpected results via comparative tests with the prior art ion exchange resin at 110°C and 130°C. The court affirmed the rejection of claims 1-7 and 9-10 because the term "elevated temperatures" encompassed temperatures as low as 60°C where the prior art ion exchange resin was known to perform well. The rejection of claim 8, directed to a temperature in excess of 100°C, was reversed.). See also In re Peterson, 315 F.3d 1325, 1329-31, 65 USPQ2d 1379, 1382-85 (Fed. Cir. 2003) (data showing improved alloy strength with the addition of 2% rhenium did not evidence unexpected results for the entire claimed range of about 1-3% rhenium); In re Grasselli, 713 F.2d 731, 741, 218 USPQ 769, 777 (Fed. Cir. 1983) (Claims were directed to certain catalysts containing an alkali metal. Evidence presented to rebut an obviousness rejection compared catalysts containing sodium with the prior art. The court held this evidence insufficient to rebut the prima facie case because experiments limited to sodium were not commensurate in scope with the claims.).

NONOBVIOUSNESS OF A GENUS OR CLAIMED RANGE MAY BE SUPPORTED BY DATA SHOWING UNEXPECTED RESULTS OF A SPECIES OR NARROWER RANGE UNDER CERTAIN CIRCUMSTANCES

The nonobviousness of a broader claimed range can be supported by evidence based on unexpected results from testing a narrower range if one of ordinary skill in the art would be able to determine a trend in the exemplified data which would allow the artisan to reasonably extend the probative value thereof.

716.02(e) Comparison With Closest Prior Art [R-2]

An affidavit or declaration under 37 CFR 1.132 must compare the claimed subject matter with the closest prior art to be effective to rebut a *prima facie* case of obviousness. *In re Burckel*, 592 F.2d 1175, 201 USPQ 67 (CCPA 1979). "A comparison of the *claimed* invention with the disclosure of each cited reference to determine the number of claim limitations in common with each reference, bearing in mind the relative importance of particular limitations, will usually yield the closest single prior art reference." *In re Merchant*, 575 F.2d 865, 868, 197 USPQ 785, 787 (CCPA 1978) (emphasis in original). Where the comparison is not identical with the reference disclosure, deviations therefrom should be explained, *In re Finley*, 174 F.2d 130, 81 USPQ 383 (CCPA 1949), and if not explained should be noted and evaluated, and if significant, explanation should be required. *In re Armstrong*, 280 F.2d 132, 126 USPQ 281 (CCPA 1960) (deviations from example were inconsequential).

THE CLAIMED INVENTION MAY BE COMPARED WITH PRIOR ART THAT IS CLOSER THAN THAT APPLIED BY THE EXAMINER

Applicants may compare the claimed invention with prior art that is more closely related to the invention than the prior art relied upon by the examiner. *In re Holladay*, 584 F.2d 384, 199 USPQ 516 (CCPA 1978); *Ex parte Humber*, 217 USPQ 265 (Bd. App. 1961) (Claims to a 13-chloro substituted compound were rejected as obvious over nonchlorinated analogs of the claimed compound. Evidence showing unexpected results for the claimed compound as compared with the 9-, 12-, and 14- chloro derivatives of the compound rebutted the *prima facie* case of obviousness because the compounds compared against were closer to the claimed invention than the prior art relied upon.).

COMPARISONS WHEN THERE ARE TWO EQUALLY CLOSE PRIOR ART REFERENCES

Showing unexpected results over one of two equally close prior art references will not rebut *prima facie* obviousness unless the teachings of the prior art references are sufficiently similar to each other that the testing of one showing unexpected results would provide the same information as to the other. *In re Johnson*, 747 F.2d 1456, 1461, 223 USPQ 1260, 1264 (Fed. Cir. 1984) (Claimed compounds differed from the prior art either by the presence of a trifluoromethyl group instead of a chloride radical, or by the presence of an unsaturated ester group instead of a saturated ester group. Although applicant compared the claimed invention with the prior art compound containing a chloride radical, the court found this evidence insufficient to rebut the *prima facie* case of obviousness because the evidence did not show relative effectiveness over all compounds of the closest prior art.

§17.12 以 37CFR 1.132 宣誓書主張商業上之成功

申請人可藉由 37CFR 1.132 之宣誓書主張商業上之成功，因而發明是非顯而易見的。但是要藉此主張非顯而易見性時，申請人需證明所請求之發明與商業成功之證據是有關聯的（nexus），所謂有關聯乃指商業成功之證據與所請求之發明之間具有事實與法律上足夠之關係而使得該證據具有在決定非顯而易見上是有證據價值者。在美國以外之商業成功的證據與在美國之商業成功證據一樣可被採納。

此外，商業成功的證據必需與所請求的發明的範圍同等大小，例如，所請求的發明是一種販賣機所使用由發泡熱型性塑膠所製之容器（container）。而所附商業成功之證據只是一種販賣機所使用之由發泡熱塑性塑膠所製成之杯子（cups）時，將被認為商業成功之證據與所請求之發明的範圍並非同等大小。

然而，當請求一範圍時，申請人並不需要顯示在該範圍以每一點均有商業上的成功。只要在該範圍內之一典型之點可達成商業上成功即可。商業成功必需直接衍生自所請求之發明，而不是由於強力的宣傳或廣告，所以申請人必需證明所請求之發明的特徵造成了商業上的成功，單純主張在商業上成功的產品是完全依照請求項（claim）之實例製出並不能證明商業成功與所請求之發明具有關聯。依照請求項請求發明所製出之產品之商業上的成功必須源自在專利說明書中有揭露之功能及優點。

在新式樣之專利申請案件中，要證明商業上的成功與所請求之新式樣有關聯，特別困難。因為必需證明商業上的成功是由於產品之外觀設計，而不是其商業（品牌）為消費大眾所認同。另外，產品之銷售總金額之增加並不一定代表商業上之成功，必需有市場占有率增加之證據才會被認為是商業上之成功。

MPEP716.03 Commercial Success

NEXUS BETWEEN CLAIMED INVENTION AND EVIDENCE OF COMMERCIAL SUCCESS REQUIRED

An applicant who is asserting commercial success to support its contention of nonobviousness bears the burden of proof of establishing a nexus between the claimed invention and evidence of commercial success.

The term "nexus" designates a factually and legally sufficient connection between the evidence of commercial success and the claimed invention so that the evidence is of probative value in the determination of nonobviousness. *Demaco Corp. v. F. Von Langsdorff Licensing Ltd.*, 851 F.2d 1387, 7 USPQ2d 1222 (Fed. Cir. 1988).

COMMERCIAL SUCCESS ABROAD IS RELEVANT

Commercial success abroad, as well as in the United States, is relevant in resolving the issue of nonobviousness. *Lindemann Maschinenfabrik GMBH v. American Hoist & Derrick Co.*, 730 F.2d 1452, 221 USPQ 481 (Fed. Cir. 1984).

716.03(a) Commercial Success Commensurate in Scope With Claimed Invention

EVIDENCE OF COMMERCIAL SUCCESS MUST BE COMMENSURATE IN SCOPE WITH THE CLAIMS

Objective evidence of nonobviousness including commercial success must be commensurate in scope with the claims. *In re Tiffin*, 448 F.2d 791, 171 USPQ 294 (CCPA 1971) (evidence showing commercial success of thermoplastic foam "cups" used in vending machines was not commensurate in scope with claims directed to thermoplastic foam "containers" broadly). In order to be commensurate *>in< scope with the claims, the commercial success must be due to claimed features, and not due to unclaimed features.

REQUIREMENTS WHEN CLAIMED INVENTION IS NOT COEXTENSIVE WITH COMMERCIAL PRODUCT OR PROCESS

If a particular range is claimed, applicant does not need to show commercial success at every point in the range. "Where, as here, the claims are directed to a combination of ranges and procedures not shown by the prior art, and where substantial commercial success is achieved at an apparently typical point within those ranges, and the affidavits definitely indicate that operation throughout the claimed ranges approximates that at the particular points involved in the commercial operation, we think the evidence as to commercial success is persuasive." *In re Hollingsworth*, 253 F.2d 238, 240, 117 USPQ 182, 184 (CCPA 1958). See also *Demaco Corp. v. F. Von Langsdorff Licensing Ltd.*, 851 F.2d 1387, 7 USPQ2d 1222 (Fed. Cir. 1988) (where the commercially successful product or process is not coextensive with the claimed invention, applicant must show a legally sufficient relationship between the claimed feature and the commercial product or process).

716.03(b) Commercial Success Derived From Claimed Invention

COMMERCIAL SUCCESS MUST BE DERIVED FROM THE CLAIMED INVENTION

In considering evidence of commercial success, care should be taken to determine that the commercial success alleged is directly derived from the invention claimed, in a marketplace where the consumer is free to choose on the basis of objective principles, and that such success is not the result of heavy promotion or advertising, shift in advertising, consumption by purchasers normally tied to applicant or assignee, or other business events extraneous to the merits of the claimed invention, etc. *In re Mageli*, 470 F.2d 1380, 176 USPQ 305 (CCPA 1973) (conclusory statements or opinions that increased sales were due to the merits of the invention are entitled to little weight); *In re Noznick*, 478 F.2d 1260, 178 USPQ 43 (CCPA 1973).

In *ex parte* proceedings before the Patent and Trademark Office, an applicant must show that the claimed features were responsible for the commercial success of an article if the evidence of nonobviousness is to be accorded substantial weight. See *In re Huang*, 100 F.3d 135, 140, 40 USPQ2d 1685, 1690 (Fed. Cir. 1996) (Inventor's opinion as to the purchaser's reason for buying the product is insufficient to demonstrate a nexus between the sales and the claimed invention.). Merely showing that there was commercial success of an article which embodied the invention is not sufficient.

COMMERCIAL SUCCESS MUST FLOW FROM THE FUNCTIONS AND ADVANTAGES DISCLOSED OR INHERENT IN THE SPECIFICATION DESCRIPTION

To be pertinent to the issue of nonobviousness, the commercial success of devices falling within the claims of the patent must flow from the functions and advantages disclosed or inherent in the description in the specification. Furthermore, the success of an embodiment within the claims may not be attributable to improvements or modifications made by others. *In re Vamco Machine & Tool, Inc.,* 752 F.2d 1564, 224 USPQ 617 (Fed. Cir. 1985).

IN DESIGN CASES, ESTABLISHMENT OF NEXUS IS ESPECIALLY DIFFICULT

Establishing a nexus between commercial success and the claimed invention is especially difficult in design cases. Evidence of commercial success must be clearly attributable to the design to be of probative value, and not to brand name recognition, improved performance, or some other factor. *Litton Systems, Inc. v. Whirlpool Corp.,* 728 F.2d 1423, 221 USPQ 97 (Fed. Cir. 1984) (showing of commercial success was not accompanied by evidence attributing commercial success of Litton microwave oven to the design thereof).

SALES FIGURES MUST BE ADEQUATELY DEFINED

Gross sales figures do not show commercial success absent evidence as to market share, *Cable Electric Products, Inc. v. Genmark, Inc.*, 770 F.2d 1015, 226 USPQ 881 (Fed. Cir. 1985), or as to the time period during which the product was sold, or as to what sales would normally be expected in the market, *Ex parte Standish*, 10 USPQ2d 1454 (Bd. Pat. App. & Inter. 1988).

§17.13 以 37CFR 1.132 宣誓書主張業界長期之需要

申請人要以 37CFR§1.132 之宣誓書主張所請求之發明為業界長期之需要（long-felt need）時，需能夠提出客觀的證據證明在業界一直存在的問題未解決。所提出之證據需注意下列數點。第一，此長期的需要必需是持續存在，被該技藝人士承認者。第二，此長期之需要必需在申請人之發明之前一直未被業界解決者。第三，申請人的發明必需事實上能滿足此長期的需要。另外，尚要考慮其他影響長期需要之因素，例如，該技藝人士不能解決此長期之需要可能不是缺乏技術能力，而是缺乏興趣，缺乏市場等。

MPEP 716.04 Long-Felt Need and Failure of Others

THE CLAIMED INVENTION MUST SATISFY A LONG-FELT NEED WHICH WAS RECOGNIZED, PERSISTENT, AND NOT SOLVED BY OTHERS

Establishing long-felt need requires objective evidence that an art recognized problem existed in the art for a long period of time without solution. The relevance of long-felt need and the failure of others to the issue of obviousness depends on several factors. First, the need must have been a persistent one that was recognized by those of ordinary skill in the art. *In re Gershon*, 372 F.2d 535, 539, 152 USPQ 602, 605 (CCPA 1967) ("Since the alleged problem in this case was first recognized by appellants, and others apparently have not yet become aware of its existence, it goes without saying that there could not possibly be any evidence of either a long felt need in the . . . art for a solution to a problem of dubious existence or failure of others skilled in the art who unsuccessfully attempted to solve a problem of which they were not aware."); *Orthopedic Equipment Co., Inc. v. All Orthopedic Appliances, Inc.,* 707 F.2d 1376, 217 USPQ 1281 (Fed. Cir. 1983) (Although the claimed invention achieved the desirable result of reducing inventories, there was no evidence of any prior unsuccessful attempts to do so.).

Second, the long-felt need must not have been satisfied by another before the invention by applicant. *Newell Companies v. Kenney Mfg. Co.*, 864 F.2d 757, 768, 9 USPQ2d 1417, 1426 (Fed. Cir. 1988) (Although at one time there was a long-felt need for a "do-it-yourself" window shade material which was adjustable without the use of tools, a prior art product fulfilled the need by using a scored plastic material which could be torn. "[O]nce another supplied the key element, there was no long-felt need or, indeed, a problem to be solved".)

Third, the invention must in fact satisfy the long-felt need. *In re Cavanagh*, 436 F.2d 491, 168 USPQ 466 (CCPA 1971).

LONG-FELT NEED IS MEASURED FROM THE DATE A PROBLEM IS IDENTIFIED AND EFFORTS ARE MADE TO SOLVE IT

Long-felt need is analyzed as of the date the problem is identified and articulated, and there is evidence of efforts to solve that problem, not as of the date of the most pertinent prior art references. *Texas Instruments Inc. v. Int'l Trade Comm'n*, 988 F.2d 1165, 1179, 26 USPQ2d 1018, 1029 (Fed. Cir. 1993).

OTHER FACTORS CONTRIBUTING TO THE PRESENCE OF A LONG-FELT NEED MUST BE CONSIDERED

The failure to solve a long-felt need may be due to factors such as lack of interest or lack of appreciation of an invention's potential or marketability rather than want of technical know-how. *Scully Signal Co. v. Electronics Corp. of America*, 570 F.2d 355, 196 USPQ 657 (1st. Cir. 1977).

See also *Environmental Designs, Ltd. v. Union Oil Co. of Cal.*, 713 F.2d 693, 698, 218 USPQ 865, 869 (Fed. Cir. 1983) (presence of legislative regulations for controlling sulfur dioxide emissions did not militate against existence of long-felt need to reduce the sulfur content in the air); *In re Tiffin*, 443 F.2d 344, 170 USPQ 88 (CCPA 1971) (fact that affidavit supporting contention of fulfillment of a long-felt need was sworn by a licensee adds to the weight to be accorded the affidavit, as long as there is a *bona fide* licensing agreement entered into at arm's length).

實例

申請人 A 提出一專利申請乃是有關於飲食用塑膠容器，如餐盤，杯子等，claim 1 乃是請求飲食用塑膠容器，claim 2 依附 claim 1，請求餐盤，claim 3 依附於 claim 1，請求杯子。審查委員引證一先前技術以顯而易見核駁 claims 1-3。申請人 A 提出 37CFR§1.132 之宣誓書，檢附證據證明在業界有對其所請求專利之杯子的長期需要，但業界的技術並無法解決此長期需要之問題，而其杯子實際上解決了此問題。申請人同時又提出證據證明在過去三年，其並未對該請求專利之杯子作任何之廣告及宣傳但其請求專利之杯子之銷售量大增，而且已授權在業界知名的塑膠杯製造商製造其杯子，故事實上請求專利之杯子在商業上極成功。然而，審查委員並不採納申請人所提之宣誓書，發下最終審定書，核駁 claims 1-3，請問審查委員之最終審定是否恰當？

審查委員之最終審定書不恰當，申請人已提供了足夠之證據證明所請求之杯子在商業上成功且解決了商業長期之需要問題，故 claim 3 應可核准，但由於所提出之證據只限於杯子與 claim 1 及 claim 2 之範圍非同等大小，故 claim 1 及 claim 2 仍不能核准。

§17.14 以 37CFR 1.132 之宣誓書主張發明被專家質疑

申請人可提出所請求的發明被專家質疑之宣誓書證明非顯而易見的有利證據。例如所請求的發明是一種將排放氣流中的硫化合物去除的方法，此方法先將硫化合物轉換成硫化氫，之後再處理硫化氫。但是，在此技術領域的專家長期以來認為先將硫化合物轉換成硫化氫，再加以處理並不能將排放氣流中之硫化合物完全除去，故懷疑此方法可達到完全去除硫化合物之目的。如果可提出宣誓書（通常由業界之專家宣誓）證明其在此之前對此發明存疑，亦可克服審查委員一顯而易見之核駁。

MPEP 716.05 Skepticism of Experts

"Expressions of disbelief by experts constitute strong evidence of nonobviousness." Environmental Designs, Ltd. v. Union Oil Co. of Cal., 713 F.2d 693, 698, 218 USPQ 865, 869 (Fed. Cir. 1983) (citing United States v. Adams, 383 U.S. 39, 52, 148 USPQ 479, 483-484 (1966)) (The patented process converted all the sulfur compounds in a certain effluent gas stream to hydrogen sulfide, and thereafter treated the resulting effluent for removal of hydrogen sulfide. Before learning of the patented process, chemical experts, aware of earlier failed efforts to reduce the sulfur content of effluent gas streams, were of the opinion that reducing sulfur compounds to hydrogen sulfide would not adequately solve the problem.).

"The skepticism of an expert, expressed before these inventors proved him wrong, is entitled to fair evidentiary weight, . . . as are the five to six years of research that preceded the claimed invention." In re Dow Chemical Co., 837 F.2d 469, 5 USPQ2d 1529 (Fed. Cir. 1988); Burlington Industries Inc. v. Quigg, 822 F.2d 1581, 3 USPQ2d 1436 (Fed. Cir. 1987) (testimony that the invention met with initial incredulity and skepticism of experts was sufficient to rebut the prima facie case of obviousness based on the prior art).

§17.15 以 37CFR 1.132 之宣誓書主張發明被仿冒之事實

　　申請人可以 37CFR 1.132 之宣誓書主張發明被仿冒，因而不是顯而易見。但是，僅憑被仿冒之事實主張所請求之發明是非顯而易見的不一定會被審查委員所接受。例如，仿冒之人可能不了解專利，或者認為專利權人並無能力去執行其專利權。如果仿冒者實質上花了很長的時間（例如 10 年）及費用想要設計出類似於申請人所請求而之產品或方法，但最終仍失敗因而才仿冒。如果是此種情形、附上證據 37CFR§1.132 之宣誓書主張非顯而易見就較易被接受。當仿冒品與所請求之產品並不相同，而且其他的製造商並未花很大的努力去發展自己的產品時，以 37CFR§1.132 之宣誓書主張所請求之發明被仿冒，因而是非顯而易見則不易被審查委員所接受。

MPEP 716.06 Copying [R-6]

　　Another form of secondary evidence which may be presented by applicants during prosecution of an application, but which is more often presented during litigation, is evidence that competitors in the marketplace are copying the invention instead of using the prior art. However, more than the mere fact of copying is necessary to make that action significant because copying may be attributable to other factors such as a lack of concern for patent property or contempt for the patentees ability to enforce the patent. *Cable Electric Products, Inc. v. Genmark, Inc.*, 770 F.2d 1015, 226 USPQ 881 (Fed. Cir. 1985). Evidence of copying was persuasive of nonobviousness when an alleged infringer tried for a substantial length of time to design a product or process similar to the claimed invention, but failed and then copied the claimed invention instead. *Dow Chem. Co. v. American Cyanamid Co.*, 837 F.2d 469,2 USPQ2d 1350 (Fed. Cir. 1987). Alleged copying is not persuasive of nonobviousness when the copy is not identical to the claimed product, and the other manufacturer had not expended great effort to develop its own solution. *Pentec, Inc. v. Graphic Controls Corp.*, 776 F.2d 309, 227 USPQ 766 (Fed. Cir. 1985). See also *Vandenberg v. Dairy Equipment Co.*, 740 F.2d 1560, 1568, 224 USPQ 195, 199 (Fed. Cir. 1984) (evidence of copying not found persuasive of nonobviousness) and *Panduit Corp. v. Dennison Manufacturing Co.*, 774 F.2d 1082, 1098-99, 227 USPQ 337, 348, 349 (Fed. Cir. 1985), *vacated on other grounds*, 475 U.S. 809, 229 USPQ 478 (1986), *on remand*, 810 F.2d 1561, 1 USPQ2d 1593 (Fed. Cir. 1987) (evidence of copying found persuasive of nonobviousness where admitted infringer failed to satisfactorily produce a solution after 10 years of effort and expense).

§17.16 以37CFR 1.132之宣誓書主張先前技術是不可操作的

申請人雖可以主張審查委員所引證之先前技術是不可操作的，但是卻需有占優勢的證據（preponderance of evidence），多量且有力之證據，才會被採信。特別是當審查委員所引證之先前技術如果是一專利文獻時尤然，因為每一專利都被假設是有效的，為可操作的。亦即，在此技藝人士使用一專利中所揭露的方法或作法被認為會達成該專利所揭露之產品或結果。因而，如果宣誓人只顯示該專利之方法可操作，但得不到該產品或結果，不是具有說服力的。因為得不到該產品或結果可能是操作之方法、條件不恰當或不願繼續作實驗所致。此外，如果申請人主張所引證之先前技術並不可操作的話，申請人所請求之發明必需要能與不能操作之證據之揭露有所區別，如果引證資料之專利中教示或建議了申請人所請求之發明，由該引證之專利的專利權人作宣誓說其並不希望所揭露的發明被如申請人所請求之發明般地被使用來主張引證之先前技術是不可操作的，也通常不易被接受。

但是，例如審查委員引證一文獻，文獻中揭示了申請人所請求之化合物，而事實上該文獻中所揭示之化合物明顯是打字之錯誤造成，該文獻之作者並非要揭示該化合物，則由該文獻之作者宣誓該文獻之化合物為打字錯誤造成則有說服力。

MPEP 716.07 Inoperability of References

Since every patent is presumed valid (35 U.S.C. 282), and since that presumption includes the presumption of operability *(Metropolitan Eng. Co. v. Coe*, 78 F.2d 199, 25 USPQ 216 (D.C.Cir. 1935), examiners should not express any opinion on the operability of a patent. Affidavits or declarations attacking the operability of a patent cited as a reference must rebut the presumption of operability by a preponderance of the evidence. *In re Sasse*, 629 F.2d 675, 207 USPQ 107 (CCPA 1980).

Further, since in a patent it is presumed that a process if used by one skilled in the art will produce the product or result described therein, such presumption is not overcome by a mere showing that it is possible to operate within the disclosure without obtaining the alleged product. *In re Weber*, 405 F.2d 1403, 160 USPQ 549 (CCPA 1969). It is to be presumed also that skilled workers would as a matter of course, if they do not immediately obtain desired results, make certain experiments and adaptations, within the skill of the competent worker.

The failures of experimenters who have no interest in succeeding should not be accorded great weight. *In re Michalek*, 162 F.2d 229, 74 USPQ 107 (CCPA 1947); *In re Reid*, 179 F.2d 998, 84 USPQ 478 (CCPA 1950).

Where the affidavit or declaration presented asserts inoperability in features of the reference which are not relied upon, the reference is still effective as to other features which are operative. *In re Shepherd*, 172 F.2d 560, 80 USPQ 495 (CCPA 1949).

Where the affidavit or declaration presented asserts that the reference relied upon is inoperative, the claims represented by applicant must distinguish from the alleged inoperative reference disclosure. *In re Crosby*, 157 F.2d 198, 71 USPQ 73 (CCPA 1946). See also *In re Epstein*, 32 F.3d 1559, 31 USPQ2d 1817 (Fed. Cir. 1994) (lack of diagrams, flow charts, and other details in the prior art references did not render them nonenabling in view of the fact that applicant's own specification failed to provide such detailed information, and that one skilled in the art would have known how to implement the features of the references).

If a patent teaches or suggests the claimed invention, an affidavit or declaration by patentee that he or she did not intend the disclosed invention to be used as claimed by applicant is immaterial. In re Pio, 217 F.2d 956, 104 USPQ 177 (CCPA 1954). Compare In re Yale, 434 F.2d 66, 168 USPQ 46 (CCPA 1970) (Correspondence from a co-author of a literature article confirming that the article misidentified a compound through a typographical error that would have been obvious to one of ordinary skill in the art was persuasive evidence that the erroneously typed compound was not put in the possession of the public.).

第 18 章　新事項

目　錄

§18.1　原始揭露

　　所謂原始揭露乃指申請時送入美國專利局之說明書本文、圖式、請求項（claims）。送入美國專利局之請求項（claims）本身是明確的、清楚的話亦構成原先揭露的一部份。而先前修正（preliminary amendment）如果是在申請之同時提出且在申請人（發明人）後補之宣誓書中有明確提到的話，亦為原始之揭露的一部份。

MPEP 608.01(l) Original Claims

　　In establishing a disclosure, applicant may rely not only on the description and drawing as filed but also on the original claims if their content justifies it.

　　Where subject matter not shown in the drawing or described in the description is claimed in the application as filed, and such original claim itself constitutes a clear disclosure of this subject matter, then the claim should be treated on its merits, and requirement made to amend the drawing and description to show this subject matter

MPEP 608.04(b) New Matter by Preliminary Amendment

　　An amendment is sometimes filed along with the filing of the application. Where a 37 CFR 1.53(b) application is filed without a signed oath or declaration and such application is accompanied by an amendment, that amendment is considered a part of the original disclosure. The subsequently filed oath or declaration must refer to both the application and the amendment. See MPEP §714.09.

§18.2　新事項

　　一專利申請案在提出申請後，申請人藉由修正將原始揭露中未揭露之內容加入，則此未揭露之內容即為新事項（new matter）。但是加入原始揭露所明示（explicit），隱含（implicit），固有（inherent），內含（intrinsic）之內容則不算是新事項。申請人不能藉由修正時將任何新事項導入新申請案或再發證申請案之原始揭露中。

由於原始揭露包括了說明書本文、圖式及請求項，因而包含在說明書本文，圖式及請求項三者中任一部份的資訊均可加入其他部份不被視為導入了新事項。

37CFR 1.121 Manner of making amendment in application

＊＊＊＊＊＊＊＊＊＊＊＊

(f) No new matter. No amendment may introduce new matter into the disclosure of an application.

＊＊＊＊＊＊＊＊＊＊＊

2163.06 Relationship of Written Description Requirement to New Matter

Lack of written description is an issue that generally arises with respect to the subject matter of a claim. If an applicant amends or attempts to amend the abstract, specification or drawings of an application, an issue of new matter will arise if the content of the amendment is not described in the application as filed. Stated another way, information contained in any one of the specification, claims or drawings of the application as filed may be added to any other part of the application without introducing new matter.

There are two statutory provisions that prohibit the introduction of new matter: 35 U.S.C. 132 - No amendment shall introduce new matter into the disclosure of the invention; and, similarly providing for a reissue application, 35 U.S.C 251- No new matter shall be introduced into the application for reissue.

§18.3　非新事項之修正類型

下列幾種類型之修正通常不會被審查委員認為是新事項：

I 重新敘述（定義）（rephrasing）

重新敘述說明書中的一段或者改變用語，通常不會被認為是新事項。將在申請時業界所承認之定義或字典上之定義加入說明書中通常不會被認為是新事項。但是當說明書中之用語（term）具有多種定義，而申請人要將一定義加入說明書中時，申請人要加入之定義必需從原始揭露中可以很清楚地表示，才不致被審查委員認為是新事項。

II 明顯之錯誤（obvious errors）

　　申請人修正原始揭露中的明顯錯誤通常不會造成新事項。但是美國專利申請案主張國外的優先權時，申請人有時並不能依賴所主張之國外優先權案件中之揭露來修正其說明書本文、圖式或請求項之錯誤，因為主張國外優先權僅僅表示美國專利申請案所請求之標的與該國外優先權案件中相同之標的可享有較早之申請日（優先權日）而已。並不表示美國專利申請案就可依照所主張之國外優先權之揭露作相對之修正。然而，對於 2004 年 9 月 21 日當天或以後申請之美國專利申請案，依照 37CFR 1.57(a)，如果明示地將該國外優先權文件作為參照併入（incorporate the foreign priority document by reference）時，可將其因疏忽而省略之部份（inadvertently omitted portion）加入原始揭露中而不被認為是新事項。此時，如果國外優先權文件為非英文時，需提出認證之英文翻譯本。

III 加入固有的功能、理論或優點（inherent function, theory or advantage）

　　例如，原始揭露中敘明到一裝置或產品或方法，其原本就可執行某種功能，具有某種性質，可達成某些優點。則將此裝置（產品、方法）之操作理論，可達到之功能，具有之性質，或具有之優點加入原始揭露中通常不會被認為是新事項。然而，通常需有外部之證據證實所加入之敘述（事項）是原來就有的（固有的），而且是由此技藝人士所承認的。

IV 加入參照併入（incorporation by reference）之內容

　　為了避免專利說明書過於冗長，申請人可不必重複其他文件所包含之資訊，而將這些其他文件之資料以參照併入的方式加入說明書之中。這種參照併入（incorporation by reference）的內容，可被視為申請時之專利說明書之一部份。因而將原來在說明書中參照併入之內容藉修正加入專利說明書中中通常不會被認為是新事項。一美國專利申請案如果主張國外優先權，亦可將國外優先權文件之內容作為參照併入。通常的作法是於專利說明書第 1 頁之最前面加入以下之敘述：

This application claims priority for Taiwan patent application no. XXXXXXXX filed XXXX, the content of which is incorporated by reference in its entirety.

V 將說明書之內容修正以使各部份一致

如前所述，申請時提出之說明書本文、圖式及請求項中之內容均為原始之揭露，故將說明書本文、圖式及請求項之內容作修正以使內容一致，例如將原請求項之內容加入說明書本文中，或將說明書本文之內容加入請求項之中，通常不會被認為是新事項。

VI 將已刪除之內容重新加入

將已刪除之說明書本文、圖式或已刪除之請求項（claims）重新藉由修正之方式加入說明書、請求項之中是被允許的，不會被認為是新事項。

VII 生物材料之託存

如果專利申請案是有關於生物材料，而於申請時之申請書（說明書）中有特別指明（specifically identified in the application）的話，在申請之後將該特別指明之生物材料託存，將不會被認為是新事項。但申請人必需呈送一聲明說明該託存之生物材料是在申請前之說明中所指明者。

MPEP 2163.07 Amendments to Application Which Are Supported in the Original Description

Amendments to an application which are supported in the original description are NOT new matter.

I.REPHRASING

Mere rephrasing of a passage does not constitute new matter. Accordingly, a rewording of a passage where the same meaning remains intact is permissible. In re Anderson, 471 F.2d 1237, 176 USPQ 331 (CCPA 1973). The mere inclusion of dictionary or art recognized definitions known at the time of filing an application would not be considered new matter. If there are multiple definitions for a term and a definition is added to the application, it must be clear from the application as filed that applicant intended a particular definition, in order to avoid an issue of new matter and/or lack of written description. See, e.g., Scarring Corp. v. Megan, Inc., 222 F.3d 1347, 1352-53, 55 USPQ2d 1650, 1654 (Fed. Cir. 2000). In Scarring, the original disclosure drawn to recombinant DNA molecules utilized the term "leukocyte interferon." Shortly after the filing date, a scientific committee abolished the term in favor of "IFN-(a)," since the latter term more specifically identified a particular polypeptide and since the committee found that leukocytes also produced other types of interferon. The court held that the subsequent amendment to the specification and claims substituting the term "IFN-(a)" for "leukocyte interferon" merely renamed the invention and did not constitute new matter. The claims were limited to cover only the interferon subtype coded for by the inventor's original deposits.

II.OBVIOUS ERRORS

An amendment to correct an obvious error does not constitute new matter where one skilled in the art would not only recognize the existence of error in the specification, but also the appropriate correction. In re Odd, 443 F.2d 1200, 170 USPQ 268 (CCPA 1971).

Where a foreign priority document under 35 U.S.C. 119 is of record in the U.S. application file, applicant may not rely on the disclosure of that document to support correction of an error in the pending U.S. application. Ex parte *>Bondiou<, 132 USPQ 356 (Bd. App. 1961). This prohibition applies regardless of the language of the foreign priority documents because a claim for priority is simply a claim for the benefit of an earlier filing date for subject matter that is common to two or more applications, and does not serve to incorporate the content of the priority document in the application in which the claim for priority is made. This prohibition does not apply where the U.S. application explicitly incorporates the foreign priority document by reference. For applications filed on or after September 21, 2004, where all or a portion of the specification or drawing(s) is inadvertently omitted from the U.S. application, a claim under 37 CFR 1.55 for priority of a prior-filed foreign application that is present on the filing date of the application is considered an incorporation by reference of the prior-filed foreign application as to the inadvertently omitted portion of the specification or drawing(s), subject to the conditions and requirements of 37 CFR 1.57(a). See 37 CFR 1.57(a) and MPEP §201.17.

Where a U.S. application as originally filed was in a non-English language and an English translation thereof was subsequently submitted pursuant to 37 CFR 1.52(d), if there is an error in the English translation, applicant may rely on the disclosure of the originally filed non-English language U.S. application to support correction of an error in the English translation document

2163.07(a) Inherent Function, Theory, or Advantage

By disclosing in a patent application a device that inherently performs a function or has a property, operates according to a theory or has an advantage, a patent application necessarily discloses that function, theory or advantage, even though it says nothing explicit concerning it. The application may later be amended to recite the function, theory or advantage without introducing prohibited new matter. In re Reynolds, 443 F.2d 384, 170 USPQ 94 (CCPA 1971); In re Smythe, 480 F. 2d 1376, 178 USPQ 279 (CCPA 1973). "To establish inherency, the extrinsic evidence 'must make clear that the missing descriptive matter is necessarily present in the thing described in the reference, and that it would be so recognized by persons of ordinary skill. Inherency, however, may not be established by probabilities or possibilities. The mere fact that a certain thing may result from a given set of circumstances is not sufficient.'" In re Robertson, 169 F.3d 743, 745, 49 USPQ2d 1949, 1950-51 (Fed. Cir. 1999) (citations omitted).

2163.07(b) Incorporation by Reference [R-3]

Instead of repeating some information contained in another document, an application may attempt to incorporate the content of another document or part thereof by reference to the document in the text of the specification. The information incorporated is as much a part of the application as filed as if the text was repeated in the application, and should be treated as part of the text of the application as filed. Replacing the identified material incorporated by reference

2163.06 Relationship of Written Description Requirement to New Matter

Lack of written description is an issue that generally arises with respect to the subject matter of a claim. If an applicant amends or attempts to amend the abstract, specification or drawings of an application, an issue of new matter will arise if the content of the amendment is not described in the application as filed. Stated another way, information contained in any one of the specification, claims or drawings of the application as filed may be added to any other part of the application without introducing new matter.

There are two statutory provisions that prohibit the introduction of new matter:35 U.S.C. 132 - No amendment shall introduce new matter into the disclosure of the invention; and, similarly providing for a reissue application,35 U.S.C. 251- No new matter shall be introduced into the application for reissue.

608.01(s) Restoration of Canceled Matter

　　Canceled text in the specification can be reinstated only by a subsequent amendment presenting the previously canceled matter as a new insertion.37 CFR 1.121(b)(4). A claim canceled by amendment (deleted in its entirety) may be reinstated only by a subsequent amendment presenting the claim as a new claim with a new claim number.37 CFR 1.121(c)(2).See MPEP§714.24

§1.804 Time of making an original deposit.

　　(a) Whenever a biological material is specifically identified in an application for patent as filed, an original deposit thereof may be made at any time before filing the application for patent or, subject to§1.809,during pendency of the application for patent.

　　(b) When the original deposit is made after the effective filing date of an application for patent, the applicant must promptly submit a statement from a person in a position to corroborate the fact, stating that the biological material which is deposited is a biological material specifically identified in the application as filed.

§18.4　新事項之修正類型

　　為了克服 35USC§112 第 1 段之核駁所加入之內容通常會被視為是新事項。亦即為了克服敘述要件（written description），據以實施要件（enablement）或最佳模式（best mode）要件之核駁所新加入之內容會是新事項，例如加入新的實施例，加入一發明之用途（utility），託存在說明書中未明確指明之生物材料，甚至對新式樣之圖式加上陰影線都可能被認為是新事項。

§18.5　美國專利局對新事項之處理

　　如果申請人於摘要或說明書本文中加入之內容，審查委員認為是新事項時，此修正加入之內容，審查委員通常會先准予進入審查（enter），但是在接著發下之審定書中審查委員會以 35USC§132 作形式核駁（objection），並要求此加入之內容刪除。為何先准予進入審查，再予以形式核駁乃是要給申請人答辯之機會，也許申請人認為其加入之內容只是重新敘述原有之內容而已，並非是新事項。

　　如果申請人於圖式中藉修正而加入之內容審查委員認為是新事項時，審查委員通常不予進入審查（enter denied），並在接著發下之審定書中依據 35USC§132(a)告知申請人修正不予進入審查。如果申請人於請求項（claims）藉修正加入之內容審查委員認為是新事項時，通常會准予進入審查（entered），之後在發下審定書中審查委員會以 35 USC§112 第 1 段作實質核駁（rejection）。而且審查委員仍會依據先前技術審查該新加入之內容標的是否可核准或核駁，因為申請人可能可以克服此新事項之核駁。

　　申請人對於新事項之形式核駁（objection）可提出請願（petition）來救濟，而對於新事項之實質核駁（rejection）則需提出訴願（appeal）來救濟，但如果說明書本文及請求項均有新事項之核駁時，申請人就需以訴願來尋求救濟。

---

35 U.S.C. 132 Notice of rejection; reexamination.

(a) Whenever, on examination, any claim for a patent is rejected, or any objection or requirement made, the Director shall notify the applicant thereof, stating the reasons for such rejection, or objection or requirement, together with such information and references as may be useful in judging of the propriety of continuing the prosecution of his application; and if after receiving such notice, the applicant persists in his claim for a patent, with or without amendment, the application shall be reexamined. No amendment shall introduce new matter into the disclosure of the invention.

* * * * * * * * * * *

MPEP 2163.07.

* * * * * * * * * * *

If new subject matter is added to the disclosure, whether it be in the abstract, the specification, or the drawings, the examiner should object to the introduction of new matter under 35 U.S.C. 132 or 251 as appropriate, and require applicant to cancel the new matter. If new matter is added to the claims, the examiner should reject the claims under 35 U.S.C. 112, first paragraph - written description requirement. In re Rasmussen, 650 F.2d 1212, 211 USPQ 323 (CCPA 1981). The examiner should still consider the subject matter added to the claim in making rejections based on prior art since the new matter rejection may be overcome by application.

In an instance in which the claims have not been amended, per se, but the specification has been amended to add new matter, a rejection of the claims under 35 U.S.C. 112, first paragraph should be made whenever any of the claim limitations are affected by the added material.

When and amendment is filed in reply to an objection or rejection based on 35 U.S.C. 112, first paragraph, a study of the entire application is often necessary to determine whether or not "new matter" is involved. Application should therefore specifically point out the support for any amendments made to the disclosure.

Ⅱ.REVIEW OF NEW MATTER OBJECTIONS AND/OR REJECTION

A rejection of claims is reviewable by the Board of Patent Appeals and Interferences, whereas an objection and requirement toe delete new matter is subject to supervisory review by petition under 37 CFR 1.181. If both the claims and specification contain new matter either directly or indirectly, and there has been both a rejection and objection by the examiner, the issue becomes appealable and should not be decided by petition.

實例 1

申請人之專利申請案之請求項乃是有關於感光層結構，其包括一支持層，一鹵化銀層及一粘合層將支持層與鹵化銀層粘合在一齊，其中該粘合層包括氰基丙烯酸酯粘合劑。審查委員審查時發現圖式中只顯示支持層與鹵化銀層呈接觸地粘合在一齊，而且說明書本文中並未有任何關於氰基丙烯酸酯粘合劑之敘述，因而以35USC§112 第 1 段核駁該請求項，請問申請人將該粘合層加入說明書本文及圖式中是否會被審查委員以新事項（new matter）作形式核駁（objection）？

不會被認為是新事項而作形式核駁，因為請求項之揭露亦為原先揭露之一部份，申請人只是將請求項中之敘述加入說明書本文及圖中以使其一致。

實例 2

申請人提出一專利申請乃是有關於一化學組合物，於 claim 1 中敘述之組合物包括有最多 6%之界面活性劑，claim 2 乃依附於 claim 1 且敘述該組合物包括 3-6%之界面活性劑。而說明書本文只提到該組合物包括最多 6%之界面活性劑。審查委員亦引證一先前技術其為一化學組合物包括 1-2%之該界面活性劑，因而核駁了 claim 1 及 claim 2。發明人在提出答辯之同時重新審閱說明書才發現說明書本文及 claim 1 及

claim 2 中界面活性劑之上限均應為 60%，但原來提申之說明書之記載 6%乃是誤繕，欲同時提出修正，請問此修正是否會被審查委員認為是新事項？

　　此修正將被審查委認為是新事項，因為在原來之說明書本文及請求項中均記載為 6%。

實例 3

　　一美國專利申請案乃有關於製造聚乙烯醇方法，此專利申請案有主張台灣之優先權，且在美國專利申請案中有敘明：This application claims priority for Taiwa n patent application no.XXXXXXXX, the content of cellulesis incorporated by reference in its entirety。但是在該美國專利申請案中之說明書本文及請求項均將聚乙烯醇翻譯成 polyvinyl ether（聚乙烯醚），提出申請後申請人發現此翻譯之錯誤提出修正。請問此修正是否會造成新事項？

　　雖然說明書本文及請求項均誤譯為 polyvinyl ether,但此專利申請案有主張台灣優先權，且該台灣優先權文件之內容乃有主張為參照併入（incorporation by reference）。故申請人可將原優先權文件作認證翻譯（certified translation），並由翻譯人宣誓乃為真實且正確之翻譯，則此修正將不會被視為是新事項。

第 19 章　延續案與 RCE

目　錄

§19.1　延續案

　　依照 35USC§120，一專利申請案其所請求之發明揭露在一先前向美國提出申請之先申請案中，且是由該先申請案的發明人之一所申請，且乃是在該先申請案獲得專利，放棄或法律程序結束之前提出申請，而且其包括了或修改後包括了主張該先申請案之特定資訊，則此後申請案所請求之發明與該先申請案有相同之效果。（所請求之發明具有與先申請案相同之有效申請日）。此後申請案就稱為先申請案（母案）之延續案（continuing application）。

　　如果此延續案未在其審查（pending）之期間內將主張優先權之資訊加入則不能擁有與先申請案相同之利益。

　　亦即，所謂延續案需符合下列 4 要件：

①　所請求之發明揭露在先申請案（母案）中；

②　發明人需至少有一人與先申請案相同；

③　需在先申請案審查（pending）之期間內提出；

④　需在申請時或審查中之期間提出優先權之申請。

---

35 U.S.C. 120 Benefit of earlier filing date in the United States.

　　An application for patent for an invention disclosed in the manner provided by the first paragraph of section 112 of this title in an application previously filed in the United States, or as provided by section 363 of this title, which is filed by an inventor or inventors named in the previously filed application shall have the same effect, as to such invention, as though filed on the date of the prior application, if filed before the patenting or abandonment of or termination of proceedings on the first application or on an application similarly entitled to the benefit of the filing date of the first application and if it contains or is amended to contain a specific reference to the earlier filed application. No application shall be entitled to the benefit of an earlier filed application under this section unless an amendment containing the specific reference to the earlier filed application is submitted at such time during the pendency of the application as required by the Director. The Director may consider the failure to submit such an amendment within that time period as a waiver of any benefit under this section. The Director may establish procedures, including the payment of a surcharge, to accept an unintentionally delayed submission of an amendment under this section.

§19.2　延續案之種類

延續案可分為下列四種：

(1)　一般延續案（Continuation Application, CA）

(2)　分割案（Divisional Application, DA）

(3)　部份延續案（Continuation In-Part, CIP）

(4)　繼續審查案（Continued Prosecution Applicaiion, CPA）

(1)一般延續案（CA）除了發明人需有一人與先申請案（母案）相同，且需在母案審查中之期間提出之外，一般延續案之揭露必需與母案完全相同，可包括下列數種情況：

　　(a)　母案包括請求項（claims)1～20 項，在審查過程中審查委員指出 claims 1～10 可核准，claims 11～20 核駁，則申請人可在母案發證前（或當天）提出一般延續案（CA）請求被核駁之 claims 11～20。此一般延續案之揭露與母案完全相同，但發明人僅為 claims 11~20 之發明人之一部份（與母案發明人至少一人相同）。

　　(b)　母案包括請求項（claims）1～20 項，母案一直審查到最終審定書發下審查委員仍核駁所有 claims，申請人如果不想提訴願，可於最終審定書發下日起之六個月內對原有之 claim 或放棄一部份 claims 提出一般延續案，此時一般延續案之揭露與母案完全相同。

(2)　分割案乃是母案提出申請後由於請求或揭露二種以上之標的，審查委員發下限制/選擇通知，認為是不同組之發明或者是專利性上不同之發明，因而申請人將未選擇之部份提出分割案。或者申請人自己認為請求了二個不同之發明或揭露了二個不同之發明而自願將其他的發明分割成另外的申請案。故分割案與母案乃請求不同之發明，分割案可包括下列數種情況：

　　(a)　母案之請求項包括 claims 1～10（裝置），claims 21～40（方法），審查委員發下限制／選擇審定書，申請人選擇了 claims 1~10（裝置），申請人可同時或在母案發證或放棄前對未選擇／限制之 claims 21~40 提出分割案。

(b) 母案之請求項包括 claims 1~10（裝置），claims 21～40（方法），審查委員未發下限制／選擇之審定書，申請人自己認為乃是不同之發明，故將 claims 1～10 或 claims 21～40 在母案未發證或放棄前提出分割案。

(c) 母案之請求項只包括 claims 1～10（製造方法），但說明書揭露了二個發明，製造方法及使用方法，claims 1～10 一直到發下最終審定書仍不能核准，申請人決定放棄 claim1～10 之發明，另撰寫使用方法之 claim 並於母案放棄前提出分割案。

(3) 部份延續案（CIP）必需在母案尚在審查中（pending）提出申請，其揭露之內容除了重複母案之重要部份或全部之外，尚增加了在母案中未揭露的部份。請求與母案重複部份內容之請求項（claims）之申請日與母案之申請日為同一天，請求母案未揭露部份之內容的請求項（claims）之申請日則為提出部份延續案之日期，可分為下列二種情況：

(a) 母案揭露了一發明包括 A+B 之組合及 A+B+C 之組合，但請求項（claims）乃請求 A+B 之組合。申請人於母案發證或放棄之前提出一部份延續案揭露 A+B+C 及 A+B+C+D，其包括二請求項，claim 1 請求 A+B+C，claim 2 請求 A+B+C+D，則部份延續案之 claim 1 的申請日與母案相同，claim 2 之申請日則為提出部份延續案之日期。

(b) 母案請求之裝置被審查委員認為說明書本文無足夠之支持致使此技藝人士無法據以實施（not enabling），申請人於說明書中補充資訊以證明可據以實施，但仍請求相同之裝置後提出部份延續案之申請，此部份延續案之申請日為提出部份延續案之日期。

(4) 繼續審理延續案（CPA）只適用於新式樣申請之延續案，實用專利申請案（utility）及植物專利申請案（plant）因自 2000 年 5 月 21 日以後已有 RCE 制度，故無法申請 CPA 案。

MPEP 201.07 Continuation Application [R-3]

A continuation is a second application for the same invention claimed in a prior nonprovisional application and filed before the original prior application becomes abandoned or patented. The continuation application may be filed under 37 CFR 1.53(b) (or 1.53(d) if the application is a design application). The applicant in the continuation application must include at least one inventor named in the prior nonprovisional application. The disclosure presented in the continuation must be the same as that of the original application; i.e., the continuation should not include anything which would constitute new matter if inserted in the original application. The continuation application must claim the benefit of the prior nonprovisional application under 35 U.S.C. 120 or 365(c). >For more information on claiming the benefit of a prior nonprovisional application, see MPEP § 201.11.<

An application claiming the benefits of a provisional application under 35 U.S.C. 119(e) should not be called a "continuation" of the provisional application since an application that claims benefit of a provisional application is a nonprovisional application of a provisional application, not a continuation, division, or continuation-in-part of the provisional application.

At any time before the patenting or abandonment of or termination of proceedings on his or her earlier nonprovisional application, an applicant may have recourse to filing a continuation in order to introduce into the application a new set of claims and to establish a right to further examination by the primary examiner. *>A continued prosecution< application >(CPA)< under 37 CFR 1.53(d) >(available only for design applications)<, however, must be filed prior to payment of the issue fee unless a petition under 37 CFR 1.313(c) is granted in the prior application. In addition, a continuation or divisional application may only be filed under 37 CFR 1.53(d) if the prior nonprovisional application is a design application that is complete as defined by 37 CFR 1.51(b).

MPEP 201.06 Divisional Application [R-2]

A later application for an independent or distinct invention, carved out of a pending application and disclosing and claiming only subject matter disclosed in the earlier or parent application, is known as a divisional application or "division." >A divisional application is often filed as a result of a restriction requirement made by the examiner.< The divisional application must claim the benefit of the prior nonprovisional application under 35 U.S.C. 121 or 365(c). *>See MPEP § 201.11 for the conditions for receiving the benefit of the filing date of the prior application. The divisional application should set forth at least the portion of the earlier disclosure that is germane to the invention as claimed in the divisional application.

Divisional applications of utility or plant applications must be filed under 37 CFR 1.53(b). Divisional applications of design applications< may be filed pursuant to 37 CFR 1.53(b) or 1.53(d). 37 CFR 1.60 and 1.62 have been deleted as of December 1, 1997.

>Effective July 14, 2003, continued prosecution application (CPA) practice set forth in 37 CFR 1.53(d) has been eliminated as to utility and plant applications.< An application claiming the benefits of a provisional application under 35 U.S.C. 119(e) should not be called a "division" of the provisional application since the application will have its patent term

calculated from its filing date, whereas an application filed under 35 U.S.C. 120, 121, or 365(c) will have its patent term calculated from the date on which the earliest application was filed, provided a specific reference is made to the earlier filed application(s). 35 U.S.C. 154(a)(2) and (a)(3).

MPEP 201.08 Continuation-in-Part Application [R-3]

A continuation-in-part is an application filed during the lifetime of an earlier nonprovisional application, repeating some substantial portion or all of the earlier nonprovisional application and adding matter not disclosed in the said earlier nonprovisional application. (In re Klein, 1930 C.D. 2, 393 O.G. 519 (Comm'r Pat. 1930)). The continuation-in-part application may only be filed under 37 CFR 1.53(b). The continuation-in-part application must claim the benefit of the prior nonprovisional application under 35 U.S.C. 120 or 365(c). >For more information on claiming the benefit of a prior nonprovisional application, see MPEP § 201.11.<

37CFR 1.53(d)

(d) Application filing requirements - Continued prosecution (nonprovisional) application.

(1) A continuation or divisional application (but not a continuation-in-part) of a prior nonprovisional application may be filed as a continued prosecution application under this paragraph, provided that:

(i) The application is for a design patent;

(ii) The prior nonprovisional application is a design application that is complete as defined by §1.51(b); and

(iii) The application under this paragraph is filed before the earliest of:

(A) Payment of the issue fee on the prior application, unless a petition under §1.313(c) is granted in the prior application;

(B) Abandonment of the prior application; or

(C) Termination of proceedings on the prior application.

§19.3　延續案需滿足 35 USC112 第 1 段之要件

延續案亦需滿足揭露要件，亦即一延續案所請求之發明（claimed invention）必需揭露在先申請之母案之中，而且在先申請之母案與延續案對於所請求之發明之揭露必需足夠滿足 35USC§112 第 1 段之要件。亦即，需足夠滿足①敘述要件（written description），②據以實施要件（enablement）及③最佳模式要件（best mode）三要件。

　　而所請求之發明是否滿足 35USC§112 第 1 段之要件乃是針對每一請求項獨立地審查。如一延續案之一請求項滿足上述三要件則其享有母案之有效申請日，而另一請求項不能完全滿足上述三要件，則其有效申請日為提出延續案之日期。

　　一般延續案（CA）之揭露必需與先申請案（母案）之揭露完全相同，分割案（DA）之揭露可與先申請案完全相同，亦可只包括分割案所請求之發明在先申請案中相對應之揭露部份，部份延續案（CIP）則可包括先申請案未揭露之部份（new matter，新事項）。

　　如果一延續案有一系列之共同審查中之先申請案，則延續案所請求之發明必需在所有一系列之先申請案中均有支持（揭露），且該些揭露均滿足 35USC§112 第 1 段之要件，才能享有較早之有效申請日。

MPEP 201.11

B.Claiming the Benefit of Nonprovisional Applications

The disclosure of a continuation application must be the same as the disclosure of the prior-filed application. See MPEP § 201.07. The disclosure of a divisional application must be the same as the disclosure of the prior-filed application, or include at least that portion of the disclosure of the prior-filed application that is germane to the invention claimed in the divisional application. See MPEP § 201.06. The disclosure of a continuation or divisional application cannot include anything which would constitute new matter if inserted in the prior-filed application. A continuation-in-part application may include matter not disclosed in the prior-filed application. See MPEP § 201.08. Only the claims of the continuation-in-part application that are disclosed in the manner provided by the first paragraph of 35 U.S.C. 112 in the prior-filed application are entitled to the benefit of the filing date of the prior-filed application. If there is a continuous chain of copending nonprovisional applications, each copending application must disclose the claimed invention of the later-filed application in the manner provided by the first paragraph of 35 U.S.C. 112, in order for the later-filed application to be entitled to the benefit of the earliest filing date.

<u>實例 1</u>

　　一專利申請案於 2004 年 1 月 11 日提出申請，此專利申請案包括了 5 項請求項（claims）均請求一塗覆裝置。2005 年 2 月 1 日審查委員發下最初審定書以 35USC§112 第 1 段核駁了該 5 項請求項，且形式核駁了說明書之揭露，認為此技藝人士無法據以實施。申請人之後於 2005 年 7 月 1 日提出一部份延續案，此部份延續案之揭露補充了相關之敘述以使此技藝人士可據以實施母案所請求之塗覆裝置，且新加入了一種塗覆組合物，此部份延續案包括原來之 Claims 1～5 請求與母案完全相同之塗覆裝置及 Claims 6～10 請求所揭露之塗覆組合物，請目此部份延續案 Claims 1～5 是否可享有母案之申請日之利益？

　　不行，只有當先申請（母案）及部份延續案之揭露均滿足 35USC§112 第 1 段之要件時，部份延續案（CIP）才能享有母案申請日之利益（排除由母案申請日至部份延續案申請日之期間的先行技術）。

<u>實例 2</u>

　　一美國專利申請案於 2002 年 2 月 1 日申請主張台灣專利申請之優先權。台灣專利申請案乃於 2001 年 3 月 2 日提出申請。該美國專利申請案揭露了一裝置包括元件 A，B，C 之組合，且 claim 1 乃請求此 A+B+C 之組合。該台灣專利申請於 2002 年 9 月 10 日核准前公告。申請人於 2003 年 10 月 10 日又提出一 CIP 專利申請，此 CIP 專利揭露並請求了一裝置其為元件 A+B+C+D 之組合。審查委員以申請人自已的台灣專利公報及另一先前技術以顯而易見核駁此 CIP 案之請求項，該另一先前技術教示了加入元件 D 可改良申請人之元件組合 A+B+C。請問申請人可否主張 CIP 案之請求項所請求之發明：A+B+C+D 之組合裝置之有效申請日為 2002 年 2 月 1 日而克服此核駁？

　　不行，克服此顯而易見之核駁。因為包括 A+B+C+D 組合之裝置在母案中並未揭露，故無法享用母案之申請日。

MPEP 2133.01 Rejections of Continuation-In-Part (CIP) Applications

　　When applicant files a continuation-in-part whose claims are not supported by the parent application, the effective filing date is the filing date of the child CIP. Any prior art disclosing the invention or an obvious variant thereof having a critical reference date more than 1 year prior to the filing date of the child will bar the issuance of a patent under 35 U.S.C. 102(b). Paperless Accounting v. Bay Area Rapid Transit System, 804 F.2d 659, 665, 231 USPQ 649, 653 (Fed. Cir. 1986).

實例 3

　　一發明人於 2001 年 2 月 13 日申請之美國專利申請案 A 之揭露包括了二發明 X 及 Y，但請求項只請求 X 部份。此專利申請案 A 於 2004 年 3 月 1 日發證，在 2004 年 2 月 11 日發明人又申請了一專利申請案 A 之部份延續案 B，其揭露了 X，Y 及 Z 三個發明但只請求 Y 部份，此部份延續案 B 於 2006 年 12 月 16 日發證。發明人在部份延續案 B 發證前發現發明 Z 未請求故於 2006 年 11 月 11 日提出一延續案 C，其揭露了發明 X，Y 及 Z，但只請求 Z，請問延續案 C 是否可主張專利申請案 A 之優先權？

　　此三案件之申請日程圖如下：

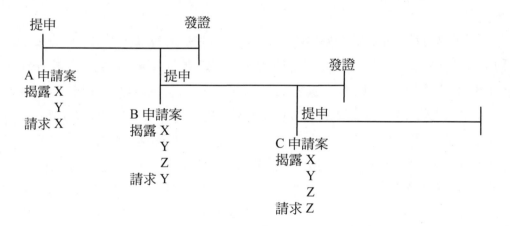

　　由於 C 申請案所請求之發明 Z 只在 B 申請案中有揭露，而在 A 申請案中未揭露，故無法主張 A 申請案之優先權只能主張 B 申請案之優先權。

實例 4

　　一專利申請案包括了二組請求項 A 與 B，均請求同一標的，A 之範圍較大，B 之範圍較窄，經審查之後，審查委員指出 B 請求項可核准，但 A 仍不能核准，請問申請人是否可申請一延續案（CA）繼續請求項 A 之審查？

　　申請人可將請求項 A 自專利申請案中刪除，而讓請求項 B 核准，並在專利申請案發證之前提出一延續案（CA）只請求之請求項 A 以繼續接受審查。但注意如果申請人未對延續案（CA）之請求項 A 作修改，審查委員可能發下的最初審定書（first office action）即為最終審定書（final）。

實例 5

　　A 與 B 為同一公司之研發工程師。A 於 2006 年 8 月 10 日申請了一美國專利揭露及請求一改良之底片。之後，B 看了 A 之發明，研發出一改良的底片、適可用於醫療用之 X 光技術上，由於 B 之發明乃為 A 之發明的改良，因而公司內部之專利律師在撰寫 B 之發明專利說明書時，就以參照併入（incorporated by reference）之方式將 A 之發明揭露敘述在 B 之專利說明書中，而只敘述 B 之發明的改良部份。請問對於 A 及 B 之專利說明書所揭露之相同標的，B 之專利申請案可否主張 A 之專利申請案之優先權的利益？

　　不行。此二專利申請案之發明人無一相同，故專利申請案 B 無法作為專利申請之延續案。

§19.4　延續案申請之時程

　　後申請案（延續案）如果要依照 35USC§120，121 或 365(c)主張先前申請案（母案）之優先權的利益，則後申請案必需與先前申請案是共同審查中（co-pending），或者後申請案必需與中間申請案（其同樣主張先前申請案之優先權)是共同審查中（co-pending）。所謂共同審查中（co-pending）乃指後申請案需在先申請案(A)取得專利，或(B)放棄，或(C)程序終止之前提出申請。

　　如果先申請案在某天發證，而後申請案在同一天或在某天之前提出申請，均被視為是共同審查中。通常先申請案繳了領證費之後的四週內會發下證書，故申請人最好於繳領證費之前提出延續案。而所謂先申請案放棄乃指申請人未提出答辯，自已提出正式的放棄書，未繳領證費，或申請人提出核准前不公告後未通知專利局，已提出國外申請所造成之放棄。而所謂程序之終止，例如，申請人上訴至聯邦巡迴法院(Court of Appeals for the Federal Circuit(CAFC)，而在 CAFC 發下決定書確定核駁所有的請求項，美國專利局收到 CAFC 之決定書之影本當日，即為法律程序終止之日。

MPEP 201.11

B.Claiming the Benefit of Nonprovisional Applications - Copendency

When a later-filed application is claiming the benefit of a prior-filed nonprovisional application under 35 U.S.C. 120, 121, or 365(c), the later-filed application must be copending with the prior application or with an intermediate nonprovisional application similarly entitled to the benefit of the filing date of the prior application. Copendency is defined in the clause which requires that the later-filed application must be filed before: (A) the patenting of the prior application; (B) the abandonment of the prior application; or (C) the termination of proceedings in the prior application.

If the prior application issues as a patent, it is sufficient for the later-filed application to be copending with it if the later-filed application is filed on the same date, or before the date that the patent issues on the prior application. Thus, the later-filed application may be filed under 37 CFR 1.53b) while the prior application is still pending before the examiner, or is in issue, or even between the time the issue fee is paid and the patent issues. Patents usually will be published within four weeks of payment of the issue fee. Applicants are encouraged to file any continuing applications no later than the date the issue fee is paid, to avoid issuance of the prior application before the continuing application is filed.

If the prior application is abandoned, the later-filed application must be filed before the abandonment in order for it to be copending with the prior application. The term "abandoned," refers to abandonment for failure to prosecute (MPEP § 711.02), express abandonment (MPEP § 711.01), abandonment for failure to pay the issue fee (37 CFR 1.316), and abandonment for failure to notify the Office of a foreign filing after filing a nonpublication request under 35 U.S.C. 122(b)(2)(B)(iii) (MPEP § 1124)<.

The expression "termination of proceedings" includes the situations when an application is abandoned or when a patent has been issued, and hence this expression is the broadest of the three.

After a decision by the Court of Appeals for the Federal Circuit in which the rejection of all claims is affirmed, the proceeding is terminated when the mandate is issued by the Court.

There are several other situations in which proceedings are terminated as is explained in MPEP § 711.02(c).

實例

　　一專利申請案之申請人在所有的請求項被最終核駁之後，決定提起訴願。訴願及抵觸委員會在 2006 年 9 月 2 日發下決定書仍維持原審查委員之核駁。申請人於 2006 年 11 月 13 日向聯邦巡迴法院提出上訴通知（notice of appeal）之後， 2006 年 11 月 30 日申請人覺得上訴之勝訴不大，想要申請一延續案將原請求項之範圍縮小。請問申請人之延續案是否可主張母案之優先權之利益？

　　不行。申請人對於訴願及抵觸委員之決定不服應該於決定書發下之二個月內，亦即 2006 年 11 月 2 日前，向聯邦巡迴法庭提出上訴。故本申請案已被放棄，故無法提出延續案之申請。

MPEP 201.11

　　Time for appeal or civil action.

　　(a)(1) The time for filing the notice of appeal to the U.S. Court of Appeals for the Federal Circuit (§1.302) or for commencing a civil action (§1.303) is two months from the date of the decision of the Board of Patent Appeals and Interferences. If a request for rehearing or reconsideration of the decision is filed within the time period provided under §41.52(a), § 41.79 (a), or §41.127 (d) of this title, the time for filing an appeal or commencing a civil action shall expire two months after action on the request. In contested cases before the Board of Patent Appeals and Interferences, the time for filing a cross-appeal or cross-action expires:

　　(i) Fourteen days after service of the notice of appeal or the summons and complaint; or

　　(ii) Two months after the date of decision of the Board of Patent Appeals and Interferences, whichever is later.

§19.5 延續案如何主張母案之優先權

一專利申請案要主張先前申請案之優先權時需在申請書（說明書）中作註明（specific reference）。此優先權主張可註明在專利說明書之發明名稱下之第 1 行且最好是自成一段落，或者註明在申請資料表（application data sheet）中。如果優先權主張只註明在申請資訊表中，則優先權主張會被印在之後核准之專利或核准前公告之公報的第 1 頁，而不會被印在說明書本文第 1 行之中。

此優先權主張註明方式如下：

This application is a continuation of prior application noXXXXXXXXX, filed xxx。

如果延續案主張數個前案之優先權，則可註明如下：

This application is a continuation of application no. C, filed……,which is a continuation of application no. B, filed……,which is a continuation of application no. A, filed……

一專利申請案其是實用（utility）或植物專利申請案於 2000 年 11 月 29 日當天或以後所提申者，如果要主張先前申請案之優先權時，此優先權之主張需在專利申請案仍審查中（during pendency）且在申請日起 4 個月或先前申請案之申請日起 16 個月中較晚的日期之前提出。而如果是國際申請案進入美國國家階段之申請案時，則需在進入國家階段之日期起 4 個月或先前申請案之申請日起 16 個月中較晚的日期之前提出。此主張期限不能延展，且如果未在此期間內提出將被視為放棄優先權之主張。

如果未在上述期間提出優先權之主張，則可提出一非故意的延遲請願（petition），此請願必需包括(A)優先權主張之聲明(B)規費（surcharge）(C)從主張優先權之到期日至提出優先權之主張之日期之期間內並無故意延遲之聲明。

另外，需注意的是，如果在申請時沒有同時主張優先權，之後要主張優先權時不能加入參照併入（incorporation by reference）之聲明。

MPEP 201.11

＊＊＊＊＊＊＊＊＊＊＊

III.REFERENCE TO PRIOR APPLICATION(S)

The third requirement of the statute is that the later-filed application must contain a specific reference to the prior application. This should appear as the first sentence(s) of the specification following the title preferably as a separate paragraph (37 CFR 1.78(a)) and/or in an application data sheet (37 CFR 1.76). If the specific reference is only contained in the application data sheet, then the benefit claim information will be included on the front page of any patent or patent application publication, but will not be included in the first sentence(s) of the specification. When a benefit claim is submitted after the filing of an application, the reference to the prior application cannot include an incorporation by reference statement of the prior application, unless an incorporation by reference statement of the prior application was presented upon filing of the application. See Dart Indus. v. Banner, 636 F.2d 684, 207 USPQ 273 (C.A.D.C. 1980).

A.Reference to Prior Nonprovisional Applications

Except for benefit claims to the prior application in a continued prosecution application (CPA), benefit claims under 35 U.S.C. 120, 121, and 365(c) must identify the prior application by application number, or by international application number and international filing date, and indicate the relationship between the applications. See 37 CFR 1.78(a)(2)(i). The relationship between the applications is whether the instant application is a continuation, divisional, or continuation-in-part of the prior nonprovisional application. An example of a proper benefit claim is "this application is a continuation of prior Application No. ---, filed ---."

V.The time period requirement under 37 CFR 1.78(a)(2) and (a)(5) is only applicable to utility or plant applications filed on or after November 29, 2000.

＊＊＊＊＊＊＊＊＊＊＊

If the application is a utility or plant application filed under 35 U.S.C. 111(a) on or after November 29, 2000, the benefit claim of the prior application under 35 U.S.C. 119(e), 120, 121, or 365(c) must be made during the pendency of the application and within the later of four months from the actual filing date of the later-filed application or sixteen months from the filing date of the prior-filed application. If the application is a nonprovisional application which entered the national stage from an international application after compliance with 35 U.S.C. 371, the benefit claim must be made within the later of: (1) four months from the date on which the national stage commenced under 37 U.S.C. 371(b) or (f); or (2) sixteen months from the filing date of the prior application. See 37 CFR 1.78(a)(2)(ii) and (a)(5)(ii). This time period is not extendable and a failure to submit the reference required by 35 U.S.C. 119(e) and/or 120, where applicable, within this time period is considered a waiver of any benefit of such prior application(s) under 35 U.S.C. 119(e), 120, 121 and 365(c).

If the reference required by 35 U.S.C. 120 and 37 CFR 1.78(a)(2) is not submitted within the required time period, a petition for an unintentionally delayed claim may be filed. The petition must be accompanied by: (A) the reference required by 35 U.S.C. 120 and 37 CFR 1.78(a)(2) to the prior application (unless previously submitted); (B) a surcharge under 37 CFR 1.17(t); and (C) a statement that the entire delay between the date the claim was due under 37 CFR 1.78(a)(2) and the date the claim was filed was unintentional. The Director may require additional information where there is a question whether the delay was unintentional. See 37 CFR 1.78(a)(3).

§19.6　申請延續案所需之文件

所有的非暫時申請案，包括延續案均需包括下列文件：

1. 說明書其符合 35USC§112 第 1 段規定之敘述要件，據以實施要件及最佳模式要件；

2. 至少一請求項（Claim）；

3. 圖式，如果需要；

4. 宣誓書，及

5. 申請規費。

要取得有效的申請日，必需於申請時提出說明書、請求項、圖式。宣誓書及申請規費可後補。

一般延續案（CA）及分割案（DA）之說明書、圖式乃沿用母案提申時之說明書及圖式（非申請後修正之版本）。在申請延續案時可在延續案之說明書或送美國專利局之書信（transmittal letter）上註明先前申請案（母案）為合併參照（incorporated by reference）。

而部份延續案（CIP）則需呈送新的說明書及圖式及請求項。

對於一般延續案及分割案如果符合下列條件可不必簽署新的宣誓書：

1. 先前申請案包括有符合規定之宣誓書；

2. 一般延續案或分割案是先前申請案之所有發明人或較少之發明人所申請的；

3. 一般延續案或分割案之說明書不包括先申請案之說明書中之新事項（new matter）；

4. 一般延續案或分割案有呈送先前申請案之宣誓書的影本，

其中記載有發明人的簽名。

部份延續案則需由發明人簽署新的宣誓書。

MPEP 201.06(c)

In order to be complete for filing date purposes, all applications filed under 37 CFR 1.53(b) must include a specification as prescribed by 35 U.S.C. 112 containing a description pursuant to 37 CFR 1.71 and at least one claim pursuant to 37 CFR 1.75, and any drawing required by 37 CFR 1.81(a). The statutory filing fee and an oath or declaration in compliance with 37 CFR 1.63 (and 37 CFR 1.175 (if a reissue) or 37 CFR 1.162 (if for a plant patent)) are also required by 37 CFR 1.51(b) for a complete application, but the filing fee and oath or declaration may be filed after the application filing date upon payment of the surcharge set forth in 37 CFR 1.16(f). See 37 CFR 1.53(f) and MPEP § 607.

Any application filed on or after December 1, 1997, which is identified by the applicant as an application filed under 37 CFR 1.60 will be processed as an application under 37 CFR 1.53(b) (using the copy of the specification, drawings and signed oath/declaration filed in the prior application supplied by the applicant). Any submission of an application including or relying on a copy of an oath or declaration that would have been proper under 37 CFR 1.60 will be a proper filing under 37 CFR 1.53(b).

37 CFR 1.63 Oath or Declartion.

＊＊＊＊＊＊＊＊＊＊＊

(d)1) A newly executed oath or declaration is not required under § 1.51(b)(2) and § 1.53(f) in a continuation or divisional application, provided that:

(i) The prior nonprovisional application contained an oath or declaration as prescribed by paragraphs (a) through (c) of this section;

(ii) The continuation or divisional application was filed by all or by fewer than all of the inventors named in the prior application;

(iii) The specification and drawings filed in the continuation or divisional application contain no matter that would have been new matter in the prior application; and

(iv) A copy of the executed oath or declaration filed in the prior application, showing the signature or an indication thereon that it was signed, is submitted for the continuation or divisional application.

實例

　　一美國專利申請案於 2004 年 2 月 1 日提出申請，並於 2005 年 8 月 25 日核准前公告。此專利申請案之說明書揭露了一裝置由 A 及 B 元件組合，此裝置亦可加上 C 元件為 A+B+C 之組合，請求項則請求 A+B 之組合。本案於 2006 年 11 月 2 日發下最終審定書，核駁所有的請求項，此時申請人已對原發明作了一些改良另加上了一元件 D，故想提出一部份延續案或獨立申請案，其中 claim 1 請求 A+B+C 之組合，claim 2 請求 A+B+C+D 之組合，說明書揭露了 A+B 之組合，A+B+C 之組合，亦揭露了 A+B+C+D 之組合。請問申請人應提出 CIP 申請案或是獨立申請案？

　　申請部份延續案或獨立申請案各有利弊。如果以 CIP 提出申請，則 claim 1 之有效申請日 2005 年 8 月 25 日，claim 2 之申請日為 CIP 提出申請之申請日，故審查委員要找到母案申請日之前的先前技術才能核駁 claim 1，而對於 Claim 2 審查委員要找到 CIP 申請日之前之先前技術才可核駁，不會被自己的核准前公告公報以 102(b) 核駁，但有可能被以母案的核准前公告公報作為 102(b) 之事件以 103(a) 核駁。

　　如果以獨立案提出申請，則審查委員可用母案的核准前公告公報核駁 claim 1，而找到 CIP 申請日之前之先前技術就可核駁 claim 1 及 claim 2。

　　提出部份延續案之缺點為其專利權期限乃自母案之申請日起算 20 年結束，而獨立申請案則從其申請日起算 20 年結束。

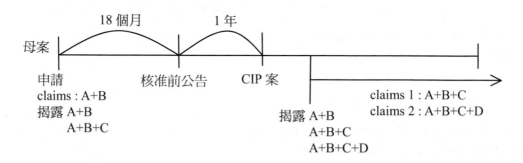

§19.7　RCE 提出之時機

　　一專利申請人要提出延續案（CA）通常是因為最終審定書（final action）發下，而申請人想再對請求項作修正，但修正卻不能進入審查，故提出延續案。但延續案（CA）為一新的案件，有新的案號，審查程序而又需經過形式審查，較麻煩。2000年 5 月 29 日 35 USC §132 及 37CFR 因應美國發明人保護法案（American Inventors Protection Act）作了修正，讓申請人可不必提出延續案而代以請求繼續審查（Request for Continued Examination，RCE）就可延續一案件之審查（並無新的申請案號），以簡化程序。

　　RCE 需在一案件之審查結束之後並在：

　　(1)繳領證費；

　　(2)申請案放棄；

　　(3)依照 35USC§141 對聯邦巡迴法庭（CAFC）提出上述通知或依照 35USC§145 或 146 開始民事訴訟；三者之中較早發生之事件之前提出申請。

　　所謂審查結束乃是指在訴願之中，發下最終審定書（final action），發下核准通知，或者是其他審查終結之審定書（如 Ex Parte Quayle OA）發下而言。亦即，如果是發下最終審定書（final）則需在放棄之前（最終審定書發下日期之六個月內）提出 RCE。如果是已發下核准通知，就需在繳領證費之前提出 RCE，如果已提出訴願則需在向 CAFC 提出上訴通知或開始民事訴訟之前提出 RCE。

---

35 U.S.C. 132 Notice of rejection; reexamination.

　　(b) The Director shall prescribe regulations to provide for the continued examination of applications for patent at the request of the applicant. The Director may establish appropriate fees for such continued examination and shall provide a 50 percent reduction in such fees for small entities that qualify for reduced fees under section 41(h)(1) of this title.

＊＊＊＊＊＊＊＊＊＊＊

37 CFR § 1.114 Request for continued examination.

(a) If prosecution in an application is closed, an applicant may request continued examination of the application by filing a submission and the fee set forth in § 1.17(e) prior to the earliest of:

(1) Payment of the issue fee, unless a petition under § 1.313 is granted;

(2) Abandonment of the application; or

(3) The filing of a notice of appeal to the U.S. Court of Appeals for the Federal Circuit under 35 U.S.C. 141, or the commencement of a civil action under 35 U.S.C. 145 or 146, unless the appeal or civil action is terminated.

(b) Prosecution in an application is closed as used in this section means that the application is under appeal, or that the last Office action is a final action (§ 1.113), a notice of allowance (§ 1.311), or an action that otherwise closes prosecution in the application.

§19.8　可利用 RCE 進行繼續審查之申請案

可利用 RCE 進行繼續審查之案件為 1995 年 6 月 8 日當天或以後申請之實用（utility）專利申請案，及植物（plant）專利申請案或 1995 年 6 月 8 日當天或以後申請之國際專利申請案進入美國階段者。不能利用 RCE 進行繼續審查的案件包括：

(A)　暫時申請案。

(B)　1995 年 6 月 8 日以前申請之發明專利申請案及植物專利申請案。

(C)　1995 年 6 月 8 日以後申請之國際專利申請案進入美國階段者。

(D)　新式樣專利申請案。

(E)　再審查申請案（patent under reexamination）

---

37 CFR 1.114 Request for continued examination

＊＊＊＊＊＊＊＊＊＊＊

(e) The provisions of this section do not apply to:

(1) A provisional application;

(2) An application for a utility or plant patent filed under 35 U.S.C. 111(a) before June 8, 1995;

(3) An international application filed under 35 U.S.C. 363 before June 8, 1995;

(4) An application for a design patent; or

(5) A patent under reexamination.

§19.9　RCE 之申請要件

要提出 RCE 需具備下列文件（費用）：

(1)　RCE 申請書。

(2)　新事證（submission）。

(3)　規費（不能後繳），但已繳過超項費之請求項不必再繳超項費。

所謂新事證（submission）包括下列數種：

Ｉ. 最終核駁審定書發下後提 RCE 之新事證：

如果申請人在最終核駁後提出之新事證必需符合 37CFR 1.111 之要件（完整的答辯），亦即需對審查委員之最終核駁的每一核駁（rejection）理由，形式核駁（objection）一一作答辯，則審查委員會撤回最終審定且此新事證會進入審查（entered）及被考慮。

通常申請人在提出 RCE 時可提出新的請求項修正（new claims amendment），或者將先前對最終審定書答辯時所提出但未進入審查之請求項修正本（previously-filed and not entered amendment）作為新事證。可送入答辯理由，證據或 IDS 資料作為新事證。需注意的是送入之新事證必需針對最終核駁審定書作完整的答辯（針對每一項核駁作答辯），如果不是完整的答辯, 則需是真正企圖（bona fide attempt）要對最終審定書作完整的答辯

因要答辯而將所有請求項刪除，作為新事證將不會被視為真正企圖要答辯，將所有請求項均刪除且未有新的請求項亦不被視為真正企圖要作完整答辯。在此情形

下至最終審定書發下起六個月時專利申請案將被放棄。但是下列二種情形雖非完整之答辯，但將被視為真正企圖要答辯：

(A) 如果申請 RCE 時所送之新事證為請求項及/或說明書之修正，但不符合 37CFR§1.121 之規定（修正之方式不符規定），將被視為真正企圖要答辯，審查委員會給 30 天或 1 個月之期間作補正。

(B) 申請人申請 RCE 時將原有進入審查之請求項全部刪除，但加入原揭露在說明書中但未請求之發明之請求項，則此修正就不能進入審查（not entered），但會被視為真正企圖要答辯，因而審查委員會給予 30 天或 1 個月時間作補正。

II. 核准通知或 Quayle Action 發下後提 RCE 之新事證

如果在發下 Ex Parte Quayle OA 之後要提出 RCE，則新事證必需包括對此 Ex Parte Quayle OA 之答辯（修正）。如果是收到核准通知後要提出 RCE，則新事證包括但不限定於 IDS 資料，請求項或說明書修正本，新的答辯理由或新的證據。如果核准通知書發下但尚未繳交領證費，則申請人附上新事證及繳交申請 RCE 之規費，提出 RCE 申請即可。但如果已繳交了領證費，仍要提出 RCE 則需同時提出請願（petition），但所繳交之領證費可能無法退還，如果申請 RCE 之後又核准，則此時原繳之領證費可抵後來要繳之領證費。

III. 提出訴願之後提出 RCE 之新事證

申請人在向專利訴願及抵觸委員會（BPAI）提出訴願之後，在 BPAI 尚未作決定之前提出 RCE，則不論所提之 RCE 是否適當，均將被視為撤回訴願並重開審查。此時所提之新事證必需包括對最終審定書之答辯書，亦即為包括訴願理由書（appeal brief）或訴願答辯書（reply brief）之答辯書。如果新事證符合 37CFR§1.111 而為一完整的答辯，則審查委員會撤回最終審定並重開審查。如果並非一完整的答辯（non-responsive），但是真正企圖要答辯時，審查委員會再給一個月期間作補充答辯。

如果申請人提出 RCE，但未繳 RCE 之規費或未提出新事證，或新事證必非真正企圖要提答辯，則此 RCE 將被視為不適當。此時所提出之訴願仍將被視為撤回。如果原來就有已核准之請求項（allowed claims）則此已核准之請求項會發證，但可核准之請求項（allowable claims）及被核駁之請求項（rejected claims）均會被刪除。如果沒有任何已核准之請求項則專利申請案將被放棄。

Ⅳ. 在訴願決定之後提 RCE 之新事證

　　申請人如果在訴願委員會（BPAI）作出訴願決定之後，且在向聯邦巡迴法庭（CAFC）或向聯邦地方法院提出上訴之前（通常期間為二個月），提出 RCE 的話，則最終審定就會被撤回且所提出之新事證（submission）會被考慮。在訴願決定之後提 RCE 需注意的是，訴願委員會所作之決定有 res judicata（ same legal issue has already been finally decided）的效果。亦即，訴願委員會對一申請案所作之決定為"law"，其可抗控制申請案及其他後續之相關申請案。所以當訴願委員會有新的核駁理由（new ground of rejection）時，申請人所提之答辯若未對請求項作修正或有新的事證通常會造成最終之核駁。所以，在此階段所提之新事證如果沒有對請求項具作修改或未有新的事實證據通常不能克服核駁理由。

　　如果申請人在此期間提出 RCE，但未附上 RCE 之規費，則審查委員會通知申請人所提出 RCE 不適當。而且如果過了二個月期間未補正，所提出之 RCE 無效。如果申請人在此二個月期間未提出 RCE 亦可提出上訴，則審查委員會發核准通知（如果有已核准之請求項）或發下放棄通知。

Ⅴ. 在向聯邦巡迴法庭（CAFC）或聯邦地方法院提出上訴之後提 RCE 之新事證

　　在申請人向 CAFC 或聯邦地方法院提出上訴之後，通常申請人就不能再提出 RCE。除非此上訴已經終止，且此申請案仍在審查中（pending）。如果在上訴期間申請人提出 RCE，審查委員會將此 RCE 送交審查委員主管（supervisory patent examiner）或專案審查委員（special program examiner）以決定所提之 RCE 是否適當。除非該申請案包括已核准之請求項或上訴法院很清楚地指出由美國專利局作進步一審查，否則此專利申請案將被放棄。

MPEP 706.07(h)

＊＊＊＊＊＊＊＊＊＊＊＊

Status of the Application	The Submission:
After Final	Must include a reply under 37 CFR 1.111 to the final rejection (e.g., an amendment filed with the RCE or a previously-filed after final amendment).
After Ex Parte Quayle action	Must include a reply to the Ex Parte Quayle action.
After allowance	Includes, but not limited to, an IDS, amendment, new arguments, or new evidence.
After appeal	Must include a reply under 37 CFR 1.111 to the final rejection (e.g., a statement that incorporates by reference the arguments in a previously filed appeal brief or reply brief).

V.　AFTER FINAL REJECTION

If an applicant timely files an RCE with the fee set forth in 37 CFR 1.17(e) and a submission that meets the reply requirements of 37 CFR 1.111, the Office will withdraw the finality of any Office action to which a reply is outstanding and the submission will be entered and considered. See 37 CFR 1.114(d). The submission meeting the reply requirements of 37 CFR 1.111 must be timely received to continue prosecution of an application. In other words, the mere request for, and payment of the fee for, continued examination will not operate to toll the running of any time period set in the previous Office action for reply to avoid abandonment of the application.

Any submission that is an amendment must comply with the manner of making amendments as set forth in 37 CFR 1.121. See MPEP §7140.03. The amendment must include markings showing the changes relative to the last entered amendment. Even though previously filed unentered amendments after final may satisfy the submission requirement under 37 CFR 1.114(c), applicants are encouraged to file an amendment at the time of filing the RCE that incorporates all of the desired changes, including changes presented in any previously filed unentered after final amendments, accompanied by instructions not to enter the unentered after final amendments. See subsection VI for treatment of not fully responsive submissions including noncompliant amendments.

If the RCE is proper, form paragraph 7.42.04 should be used to notify applicant that the finality of the previous Office action has been withdrawn.

VI.　NOT FULLY RESPONSIVE SUBMISSION

If reply to a final Office action is outstanding and the submission is not fully responsive to the final Office action, then it must be a bona fide attempt to provide a complete reply to the final Office action in order for the RCE to toll the period for reply.

If the submission is not a bona fide attempt to provide a complete reply, the RCE should be treated as an improper RCE. Thus, a "Notice of Improper Request for Continued Examination (RCE)," Form PTO-2051, should be prepared by the technical support personnel and mailed to the applicant indicating that the request was not accompanied by a submission complying with the requirements of 37 CFR 1.111 (see 37 CFR 1.114(c)). The RCE will not toll the period for reply and the application will be abandoned after the expiration of the statutory period for reply if no submission complying with 37 CFR 1.111 is filed. For example, if a reply to a final Office action is outstanding and the submission only includes an information disclosure statement (IDS), the submission will not be considered a bona fide attempt to provide a complete reply to the final Office action and the period for reply will not be tolled. Similarly, an amendment that would cancel all of the claims in an application and does not present any new or substitute claims is not a bona fide attempt to advance the application to final action. The Office will not enter such an amendment. See Exxon Corp. v. Phillips Petroleum Co., 265 F.3d 1249, 60 USPQ2d 1368 (Fed. Cir. 2001).

If the submission is a bona fide attempt to provide a complete reply, applicant should be informed that the submission is not fully responsive to the final Office action, along with the reasons why, and given a new shortened statutory period of one month or thirty days (whichever is longer) to complete the reply. See 37 CFR 1.135(c). Form paragraph 7.42. 08 set forth below should be used.

Situations where a submission is not a fully responsive submission, but is a bona fide attempt to provide a complete reply are:

(A) Non-compliant amendment - An RCE filed with a submission which is an amendment that is not in compliance with 37 CFR 1.121, but which is a bona fide attempt to provide a complete reply to the last Office action, should be treated as a proper RCE and a Notice of Noncompliant Amendment should be mailed to the applicant. Applicant is given a time period of one month or thirty days from the mailing date of the notice, whichever is longer, to provide an amendment complying with 37 CFR 1.121. See MPEP §714.03 for information on the amendment practice under 37 CFR 1.121.

(B) Presentation of claims for different invention - Applicants cannot file an RCE to obtain continued examination on the basis of claims that are independent and distinct from the claims previously claimed and examined as a matter of right (i.e., applicant cannot switch inventions). See 37 CFR 1.145. If an RCE is filed with an amendment canceling all claims drawn to the elected invention and presenting only claims drawn to a nonelected invention, the RCE should be treated as a proper RCE but the amendment should not be entered. The amendment is not fully responsive and applicant should be given a time period of one month or thirty days (whichever is longer) to submit a complete reply. See MPEP §821.03. Form paragraphs 8.04 or 8.26 should be used as appropriate.

VIII. FIRST ACTION FINAL AFTER FILING AN RCE

The action immediately subsequent to the filing of an RCE with a submission and fee under 37 CFR 1.114 may be made final only if the conditions set forth in MPEP § 706.07(b)** are met.

It would not be proper to make final a first Office action immediately after the filing of an RCE if the first Office action includes a new ground of rejection. See MPEP § 1207.03 for a discussion of what may constitute a new ground of rejection.

Form paragraph 7.42.09 should be used if it is appropriate to make the first action after the filing of the RCE final.

IX. AFTER ALLOWANCE OR QUAYLE ACTION

The phrase "withdraw the finality of any Office action" in 37 CFR 1.114(d)includes the withdrawal of the finality of a final rejection, as well as the closing of prosecution by an Office action under Ex parte Quayle, 25 USPQ 74, 453 O.G. 213 (Comm'r Pat. 1935), or notice of allowance under 35 U.S.C. 151 (or notice of allowability). Therefore, if an applicant files an RCE with the fee set forth in 37 CFR 1.17(e)and a submission in an application which has been allowed, prosecution will be reopened. If the issue fee has been paid, however, payment of the fee for an RCE and a submission without a petition under 37 CFR 1.313 to withdraw the application from issue will not avoid issuance of the application as a patent. If an RCE (with the fee and a submission) is filed in an allowed application prior to payment of the issue fee, a petition under 37 CFR 1.313 to withdraw the application from issue is not required.

If an RCE complying with the requirements of 37 CFR 1.114 is filed in an allowed application after the issue fee has been paid and a petition under 37 CFR 1.313 is also filed and granted, prosecution will be reopened. Applicant may not obtain a refund of the issue fee. If, however, the application is subsequently allowed, the Notice of Allowance will reflect an issue fee amount that is due that is the difference between the current issue fee amount and the issue fee that was previously paid.

X.AFTER APPEAL BUT BEFORE DECISION BY THE BOARD

If an applicant files an RCE under 37 CFR 1.114 after the filing of a Notice of Appeal to the Board of Patent Appeals and Interferences (Board), but prior to a decision on the appeal, it will be treated as a request to withdraw the appeal and to reopen prosecution of the application before the examiner, regardless of whether the RCE is proper or improper. See 37 CFR 1.114(d). The Office will withdraw the appeal upon the filing of an RCE. Applicants should advise the Board when an RCE under 37 CFR 1.114 is filed in an application containing an appeal awaiting decision. Otherwise, the Board of Patent Appeals and Interferences may refuse to vacate a decision rendered after the filing (but before the recognition by the Office) of an RCE under 37 CFR 1.114.

A. Proper RCE

If the RCE is accompanied by a fee (37 CFR 1.17(e)) and a submission that includes a reply which is responsive within the meaning of 37 CFR 1.111 to the last outstanding Office action, the Office will withdraw the finality of the last Office action and the submission will be entered and considered. If the submission is not fully responsive to the last outstanding Office action but is considered to be a bona fide attempt to provide a complete reply, applicant will be notified that the submission is not fully responsive, along with the reasons why, and will be given a new time period to complete the reply (using form paragraph 7.42.08). See 37 CFR 1.135 (c) and subsection VI.

If the RCE is proper, form paragraph 7.42.06 should be used to notify applicant that the appeal has been withdrawn and prosecution has been reopened.

B. Improper RCE

The appeal will be withdrawn even if the RCE is improper. If an RCE is filed in an application after appeal to the Board but the request does not include the fee required by 37 CFR 1.17 (e) or the submission required by 37 CFR 1.114, or both, the examiner should treat the request as an improper RCE and withdraw the appeal pursuant to 37 CFR 1.114(d). If the submission is not considered to be a bona fide attempt to provide a complete reply to the last outstanding Office action (e.g., an IDS only), the submission will be treated as an improper submission or no submission at all under 37 CFR 1.114 (c) (thus the request is an improper RCE). See subsection VI.

Upon withdrawal of the appeal, the application will be treated in accordance with MPEP § 1215.01 based on whether there are any allowed claims or not. The proceedings as to the rejected claims are considered terminated. Therefore, if no claim is allowed, the application is abandoned. Claims which are allowable except for their dependency from rejected claims will be treated as if they were rejected. See MPEP § 1215.01. If there is at least one allowed claim, the application should be passed to issue on the allowed claim(s). If there is at least one allowed claim but formal matters are outstanding, applicant should be given a shortened statutory period of one month or thirty days (whichever is longer) in which to correct the formal matters. Form paragraphs 7.42-7.42.14 should be used as appropriate.

XI. AFTER DECISION BY THE BOARD

A. Proper RCE After Board Decision

The filing of an RCE (accompanied by the fee and a submission) after a decision by the Board of Patent Appeals and Interferences, but before the filing of a Notice of Appeal to the Court of Appeals for the Federal Circuit (Federal Circuit) or the commencement of a civil action in federal district court, will also result in the finality of the rejection or action being withdrawn and the submission being considered. Generally, the time period for filing a notice of appeal to the Federal Circuit or for commencing a civil action is within two months of the Board's decision. See 37 CFR 1.304 and MPEP §1216. Thus, an RCE filed within this two month time period and before the filing of a notice of appeal to the Federal Circuit or the commencement of a civil action would be timely filed. In addition to the res judicata effect of a Board of Patent Appeals and Interferences decision in an application (see MPEP §706.06(w)), a Board decision in an application is the "law of the case," and is thus controlling in that application and any subsequent, related application. See MPEP §1214.01 (where a new ground of rejection is entered by the Board of Patent Appeals and Interferences pursuant to 37 CFR *>41.50(b)<, argument without either amendment of the claims so rejected or the submission of a showing of facts can only result in a final rejection of the claims, since the examiner is without authority to allow the claims unless amended or unless the rejection is overcome by a showing of facts not before the Board of Patent Appeals and Interferences). As such, a submission containing arguments without either amendment of the rejected claims or the submission of a showing of facts will not be effective to remove such rejection.

Form paragraph 7.42.07 should be used to notify applicant that the appeal has been withdrawn and prosecution has been reopened.

B. Improper RCE After Board Decision

If an RCE is filed after a decision by the Board of Patent Appeals and Interferences, but before the filing of a Notice of Appeal to the Federal Circuit or the commencement of a civil action in federal district court, and the RCE was not accompanied by the fee and/or the submission, the examiner should notify the applicant that the RCE is improper by using form paragraph 7.42.16 set forth below. If the time for seeking court review has passed without such review being sought, the examiner should include the form paragraph with the mailing of a Notice of Allowability or a Notice of Abandonment depending on the status of the claims. See

MPEP §1214.06. If the time for seeking court review remains, the examiner should include the form paragraph on a PTOL-90. No time period should be set. If a submission is filed with the RCE, but the fee is missing, the examiner should also include a statement as to whether or not the submission has been entered. In general, such a submission should not be entered. If, however, the submission is an amendment that obviously places the application in condition for allowance, it should be entered with the approval of the supervisory patent examiner. See MPEP §1214.07. Form paragraph 7.42.16 should not be used if the application is not a utility or plant application filed under 35 U.S.C. 111(a) on or after June 8, 1995, or an international application filed under 35 U.S.C. 363 on or after June 8, 1995. In that situation, a "Notice of Improper Request for Continued Examination (RCE)," Form PTO-2051, should be prepared and mailed by the technical support personnel to notify applicant that continued examination does not apply to the application. When the time for seeking court review has passed without such review being sought, the examiner must take up the application for consideration. See MPEP §1214.06 for guidance on the action to be taken.

XII. AFTER APPEAL TO THE FEDERAL CIRCUIT OR CIVIL ACTION

The procedure set forth in 37 CFR 1.114 is not available in an application after the filing of a Notice of Appeal to the Federal Circuit or the commencement of a civil action in federal district court, unless the appeal or civil action is terminated and the application is still pending. If an RCE is filed in an application that has undergone court review, the examiner should bring the application to the attention of the supervisory patent examiner or special program examiner in the TC to determine whether the RCE is proper. Unless an application contains allowed claims (or the court's mandate clearly indicates that further action is to be taken by the Office), the termination of an unsuccessful appeal or civil action results in abandonment of the application. See MPEP §1216.01.

實例

審查委員對一專利申請案於 2001 年 6 月 30 日發下核准通知書，專利申請人於 2001 年 8 月 20 日繳交了領證費，但於 2001 年 8 月 10 日發現有一篇論文與本案專利性有關。此篇論文審查委員在審查過程中並未考慮，申請人乃於 2001 年 9 月 2 日將此篇論文以 IDS 資料之方式送入美國專利局。請問審查委員是否會考慮此篇論文？

審查委員將不會考慮該篇論文，因為申請人已繳了領證費，如果要讓審查委員考慮該篇論文，申請人應將本案撤回領證，以該篇論文作為新事證（submission）申請 RCE。

第 20 章　訴願程序

目　錄

§20.1　提出訴願

　　一專利申請人（發明人）如果其專利申請案發下最終審定書或任一請求項（claim）被核駁二次，就可向專利訴願及抵觸程序委員會（BPAI, Board of Patent Appeals and Interference）提出訴願。當任一請求項被核駁二次時，不論請求項是否被最終核駁（final rejected），而且此被核駁二次並不限定要在一特定之申請案中發生。例如一請求項在母案中只被核駁了一次，申請人申請延續案，而在延續案中該請求項又被核駁，雖然不是最終核駁，申請人亦可針對被核駁之請求項提出訴願。

　　一專利之所有權人在再審查程序（re-examination proceeding）中任一請求項被最終核駁時可向 BPAI 提出訴願，在 Inter partes 程序中之第三者在審查委員作出對一專利之原有或修正或新的請求項有利之最終審定時，可向 BPAI 提出訴願。

---

35 U.S.C. 134 Appeal to the Board of Patent Appeals and Interferences.

(a) PATENT APPLICANT.- An applicant for a patent, any of whose claims has been twice rejected, may appeal from the decision of the primary examiner to the Board of Patent Appeals and Interferences, having once paid the fee for such appeal.

(b) PATENT OWNER.- A patent owner in any reexamination proceeding may appeal from the final rejection of any claim by the primary examiner to the Board of Patent Appeals and Interferences, having once paid the fee for such appeal.

(c) THIRD-PARTY.- A third-party requester in an inter partes proceeding may appeal to the Board of Patent Appeals and Interferences from the final decision of the primary examiner favorable to the patentability of any original or proposed amended or new claim of a patent, having once paid the fee for such appeal.

MPEP 1204 Notice of Appeal

＊＊＊＊＊＊＊＊＊＊＊

I.< APPEAL BY PATENT APPLICANT

Under 37 CFR *>41.31(a)(1)<, an applicant for a patent dissatisfied with the primary examiner's decision in the second ** rejection of his or her claims may appeal to the Board for review of the examiner's rejection by filing a notice of appeal and the required fee set forth in

37 CFR *>41.20(b)(1)< within the time period provided under 37 CFR 1.134 and 1.136. A notice of appeal may be filed after any of the claims has been twice rejected, regardless of whether the claim(s) has/have been finally rejected. The limitation of "twice ** rejected" does not have to be related to a particular application.>See Ex Parte Lemoine, 46 USPQ2d 1420, 1423 (Bd. Pat. App. & Inter. 1994) ("so long as the applicant has twice been denied a patent, an appeal may be filed").< For example, if any claim was rejected in a parent application, and the claim is again rejected in a continuing application, then applicant **>can choose< to file an appeal in the continuing application, even if the claim was rejected only once in the continuing application. >Applicant cannot file an appeal in a continuing application, or after filing a request for continued examination (RCE) under 37 CFR 1.114, until the application is under a rejection. Accordingly, applicant cannot file a notice of appeal with an RCE regardless of whether the application has been twice rejected prior to the filing of the RCE.<

＊＊＊＊＊＊＊＊＊＊＊＊

§20.2　專利訴願之流程

I.　正規之訴願流程如下圖所示：

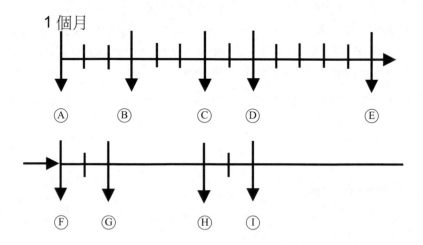

Ⓐ：最終審定書發下
Ⓑ：三個月答辯期限

Ⓒ：提出訴願聲明(notice of appeal)

Ⓓ：補訴願理由（appeal brief）

Ⓔ：最後補訴願理由

Ⓕ：審查委員之回應（Examiner's answer）（Ⓔ至Ⓕ二個月）

Ⓖ：提出訴願答覆聲明（reply brief）或請求口頭聽證（oral hearing）

Ⓗ：訴願決定書發下

Ⓘ：向聯邦巡迴法庭（CAFC）上訴

II. 美國專利局 2005 年 7 月 12 日起實施之請求訴願前審理之試行辦法，流程如下：

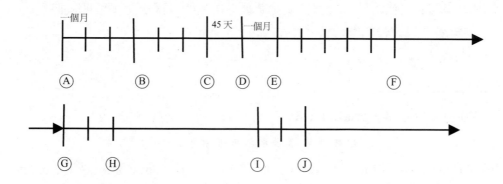

Ⓐ：最終審定書發下

Ⓑ：三個月答辯期限

Ⓒ：提出訴願聲明（notice of appeal）及訴願前審理請求（pre-appeal panel review request）

Ⓓ：發下訴願前審理之決定

Ⓔ：補訴願理由（appeal brief）

Ⓕ：最後補訴願理由

Ⓖ：審查委員回應（examiner's answer）發下

Ⓗ：提出訴願答覆聲明（reply brief）或請求口頭聽證（oral hearing）

Ⓘ：訴願決定書發下

Ⓙ：向聯邦巡迴法庭（CAFC）上訴

§20.3 提出訴願聲明之期限

訴願聲明（notice of appeal）必需在對最後一次審定書（last office action）之答辯期限，通常為三個月內提出。如果提出延期及繳延期費則亦可在最後一次審定書發下之六個月內提出。需注意的是，已對最終審定書（final office action）提出答辯及修正，但是一直到從最終審定書發下日期之六個月時審查委員仍未發下建議性審定書（advisory action）或核准通知時，如果不擬放棄專利申請案，則仍需在六個月期限屆滿前提出訴願聲明以免專利申請案被放棄。訴願聲明不必註明要訴願的請求項為哪些，亦不必由申請人或代理人簽名。但是提出訴願聲明之同時需繳交費用 USD 540。提出訴願聲明之後需於二個月內提出訴願理由。如果提出訴願聲明同時提出訴願理由，仍需繳交訴願聲明之規費（USD 540）。

MPEP 1204 Notice of Appeal [R-3]

＊＊＊＊＊＊＊＊＊＊＊＊

Although the rules **>do not< require that the notice of appeal identify the rejected claim(s) appealed, or be signed, applicants **>may< file notices of appeal which identify the appealed claims and are signed. ** It should be noted that the elimination of the requirement to sign a notice of appeal does not affect the requirements for other papers (such as an amendment under 37 CFR 1.116) submitted with the notice, or for other actions contained within the notice, e.g., an authorization to charge fees to a deposit account or to a credit card>, to be signed<. See MPEP § 509. Thus, failure to sign the notice of appeal may have unintended adverse consequences; for example, if an unsigned notice of appeal contains an (unsigned) authorization to charge the >notice of< appeal fee to a deposit account, the notice of appeal will be unacceptable because the >notice of< appeal fee is lacking.

The notice of appeal must be filed within the period for reply set in the last Office action, which is normally 3 months for applications. See MPEP § 714.13. *>For example, failure< to remove all grounds of rejection and otherwise place an application in condition for allowance or to file an appeal after final rejection will result in the application becoming abandoned, even if one or more claims have been allowed, except where claims suggested for interference have been copied. The notice of appeal and appropriate fee may be filed up to 6 months from the date of the **>Office action (e.g., a final rejection) from which the appeal was taken<, so long as an appropriate petition and fee for an extension of time >under 37 CFR 1.136(a)< is filed either prior to or with the notice of appeal.

>The use of a separate letter containing the notice of appeal is strongly recommended. Form PTO/SB/31 may be used for filing a notice of appeal. Appellant must file an appeal brief in compliance with 37 CFR 41.37 accompanied by the fee set forth in 37 CFR 41.20(b)(2) within two months from the date of filing the notice of appeal. See MPEP § 1205.<

* * * * * * * * * * * *

§20.4　提出訴願理由之期限

　　提出訴願理由（appeal brief）及訴願理由規費（USD 540）之期限為自訴願聲明日期（date of notice of appeal）起二個月內。所謂訴願聲明日期乃指美國專利局收到訴願聲明之日期，而非郵寄收據（Certificate of Mailing）上註明之日期。但是，如果申請人（訴願人）使用美國郵局（USPS）之郵局對收件之快速郵件（Express Mail Post Office to Addressee）送交訴願聲明，則補送訴願理由之日期即為快遞郵件之投件（date in）的日期。另外，需注意的是，申請人一旦提出訴願聲明，則原答覆最終審定書之日期就犧牲掉了（tolled）。例如，申請人在最終審定書發下之一個月內就提出訴願聲明，但是超過提出訴願聲明之二個月後才補訴願理由，雖然補訴願理由時期限仍在自最終審定書發下起之三個月之內，但已超過二個月才補訴願理由故仍需繳延期費。提出訴願理由之二個月期限可延期最多 5 個月，延期不必於期限前事先提出。但是進入再審查申請案之訴願程序，要延期補入訴願理由則需在期限前事先提出。

　　如果未在期限內提出訴願理由，雖訴願被視為不受理（dismissed），此時如果原申請案有已核准之請求項（allowed claims）則專利申請案不會被放棄，申請案會退回審查委員進行核准之請求項之核准程序，而如果申請案無已核准之請求項，則專利申請案自補訴願理由之到期日被放棄。需注意的是，依附至被核駁之請求項（rejected claim）之可核准的請求項（allowable claims）並不會核准，如果是再審查案件（re-examination）之訴願程序，未呈送訴願理由則美國專利局將逕自發下再審查證書（certificate of re-examination）。

MPEP 1205.01 Time for Filing Appeal Brief< [R-3]

37 CFR *> 41.37(a)< provides 2 months from the date of the notice of appeal for the appellant to file an appeal brief >and the appeal brief fee set forth in 37 CFR 41.20(b)(2)<. In an ex parte reexamination proceeding, the time period can be extended only under the provisions of 37 CFR 1.550(c). See also MPEP § 2274.

The usual period of time in which appellant must file his or her brief is 2 months from the date of appeal. The Office date of receipt of the notice of appeal (and not the date indicated on any Certificate of Mailing under 37 CFR 1.8) is the date from which this 2>-<month time period is measured

37 CFR 41.37(a) does not permit the brief to be filed within the time allowed for reply to the action from which the appeal was taken even if such time is later. Once appellant timely files a notice of appeal in compliance with 37 CFR 41.31, the time period for reply set forth in the last Office action is tolled and is no longer relevant for the time period for filing an appeal brief. For example, if appellant filed a notice of appeal within one month from the mailing of a final Office action which sets forth a 3-month shortened statutory period for reply, and then the appellant filed an appeal brief after 2 months from the filing date of the notice of appeal but within 3 months from the mailing of the final action, a petition for an extension of time for one month would be required.

✳ ✳ ✳ ✳ ✳ ✳ ✳ ✳ ✳ ✳ ✳

This 2-month time period for a patent application may be extended under 37 CFR 1.136(a), and if 37 CFR 1.136(a) is not available, under 37 CFR 1.136(b) for extraordinary circumstances.<

✳ ✳ ✳ ✳ ✳ ✳ ✳ ✳ ✳ ✳ ✳

Although failure to file the brief >and the required appeal brief fee< within the permissible time will result in dismissal of the appeal, if any claims stand allowed, the application does not become abandoned by the dismissal, but is returned to the examiner for action on the allowed claims. See MPEP § 1215.04. If there are no allowed claims, the application is abandoned as of the date the brief was due. Claims which have been objected to as dependent from a rejected claim do not stand allowed. In a reexamination proceeding failure to file the brief will result in the issuance of the certificate under 37 CFR 1.570 >or 1.997<.

✳ ✳ ✳ ✳ ✳ ✳ ✳ ✳ ✳ ✳ ✳

§20.5　訴願理由之內容

訴願理由必需針對審查委員之每一核駁理由逐一提出答辯以請求訴願委員會審理（review）。如果有核駁理由未在訴願理由中答辯，則訴願委員會在訴願決定書中會維持原來之核駁理由。

於 2004 年 9 月 13 日以後提出之訴願理由需依照 37CFR 41.37 之規定包括以下理由（內容），並同時繳交訴願理由規費（USD 540）。

Ⅰ. 真正的利害關係人（real party in interest）

此部份需聲明真正的利害關係人。例如，專利申請案已讓渡給一公司，而該公司又是其他公司之合資企業（joint ventures）時，需註明為"the real party in interest is XXXX corporation, the assignee of record, which is a subsidiary of a joint ventures. between YYYY corporation and ZZZZ corporation。註明真正的利害關係人可使訴願委員會釐清利益衝突之問題。如訴願理由中未註明真正之利益關係人，則訴願委員會給予一期間補正，如未補正，則訴願將視為不受理(dismissed)。

Ⅱ. 相關之訴願及抵觸程序（related appeals and interferences）

訴願人需註明所有相關之訴願程序及抵觸程序之案件，例如具有相同申請標的之共同擁有之案件，主張同一優先權之案件，且不限定於共同審查（copending）中之案件，需註明其申請案號，專利號碼，訴願編號，抵觸程序編號，並以附件方式附上影本一份。如果訴願人未註明任何相關案件，則視為無相關案件。

Ⅲ. 請求項之狀態（status of claims）

針對提出訴願之所有請求項需列出並註明其狀態為，例如：cancelled, allowed, rejected, withdrawn, objected to 等。

Ⅳ. 修改之狀態 (status of amendments)

需附上說明在最終審定書發下後之修正是否已進入審查 (entered)，或未進入審查 (denied entry) 以使訴願委員知道訴願中之請求項為哪些。

Ⅴ. 所請求之標的之概要 (summary of claimed subject matter)

需簡明地說明訴願之各獨立請求項所界定之標的，並指出說明書、圖式哪裡有這些標的之支持。另外，對於訴願之各獨立項或要分開爭論之依附項，各裝置功能用語在說明書，圖式中之對應的結構，材料或動作均需指明。

Ⅵ.要在訴願被審理之核駁理由 (grounds of rejection to be reviewed on appeal)

簡單說明各核駁理由，例如" Whether claims 1 and 2 are unpatentable under 35§103 over Smith in view of Jones"。

Ⅶ. 答辯理由 (argument.)

需針對每一核駁理由提出答辯，當二個以上之請求項被同一核駁理由核駁 (同樣之先前技術) 時，訴願人可對個別的請求項分別答辯 (argued separately)，也可就被同一核駁理由核駁之所有請求項一齊答辯 (argued as a group)，如果訴願人選擇一齊答辯，則訴願委員會將選擇一被核駁之請求項來審理審查委員之核駁理由是否合法，如果訴願人未選擇分別答辯，則視同放棄請求訴願委員會分別審理之權利。

Ⅷ. 請求項之附件 (claims appendix)

需附上訴願之請求項之乾淨版本 (不包括底線、括弧之版本) 一份，除非是再發證案之訴願理由。

IX. 證據之附件

X. 相關訴願，抵觸程序之附件

如果訴願人呈送之訴願理由不符合上述之規定，美國專利局會發下 non-compliant（不符合規定）之通知，訴願人若未在一個月（或 30 天）內提出修正，則訴願將被視為不受理（dismissed）。

MPEP 1205.02 Appeal Brief Content [R-3]

＊＊＊＊＊＊＊＊＊＊＊

An appellant's brief must be responsive to every ground of rejection stated by the examiner >that the appellant is presenting for review in the appeal. If a ground of rejection stated by the examiner is not addressed in the appellant's brief, that ground of rejection will be summarily sustained by the Board<.

＊＊＊＊＊＊＊＊＊＊＊

*>(i)< Real party in interest. A statement identifying >by name< the real party in interest *>even< if the party named in the caption of the brief is * the real party in interest. If appellant does not name *>the< real party in interest**>under this heading, the Office will notify appellant of the defect in the brief and give appellant a time period within which to file an amended brief. See 37 CFR 41.37(d). If the appellant fails to correct the defect in the real party in interest section of the brief within the time period set forth in the notice, the appeal will stand dismissed.<

The identification of the real party in interest **>allows< members of the Board to comply with ethics regulations associated with working in matters in which the member has a financial interest to avoid any potential conflict of interest. **>When an application is assigned to a subsidiary corporation, the real party in interest is both the assignee and either the parent corporation or corporations, in the case of joint ventures. One example of a statement identifying the real party in interest is: The real party in interest is XXXX corporation, the assignee of record, which is a subsidiary of a joint venture between YYYY corporation and ZZZZ corporation.<

*>(ii)< Related appeals and interferences. A statement identifying **>all prior and pending appeals, judicial proceedings or interferences known to the appellant which may be

related to, directly affect or be directly affected by or have a bearing on the Board's decision in the pending appeal. Appellant includes the appellant, the appellant's legal representative and the assignee. Such related proceedings must be identified by application number, patent number, appeal number (if available) or interference number (if available).< The statement is not limited to copending applications. **>The requirement to identify related proceedings requires appellant to identify every related proceeding (e.g., commonly owned applications having common subject matter, claim to a common priority application) which may be related to, directly affect or be directly affected by or have a bearing on the Board's decision in the pending appeal. Copies of any decisions rendered by a court or the Board in any proceeding identified under this paragraph must be included in an appendix as required by 37 CFR 41.37(c)(1)(x). If appellant does not identify any other items under this section, it will be presumed that there are none.<

*>(iii)< Status of Claims. A statement of the status of all the claims in the application, or patent under reexamination, i.e., for each claim in the case, appellant must state whether it is cancelled, allowed >or confirmed<, rejected, >withdrawn, objected to,< etc. Each claim on appeal must be identified.

*>(iv)< Status of Amendments. A statement of the status of any amendment filed subsequent to final rejection, i.e., whether or not the amendment has been acted upon by the examiner, and if so, whether it was entered, >or< denied entry**.

＊＊＊＊＊＊＊＊＊＊＊

**>(v) Summary of claimed subject matter. A concise explanation of the subject matter defined in each of the independent claims involved in the appeal, which must refer to the specification by page and line number, and to the drawing,

＊＊＊＊＊＊＊＊＊＊＊

For each independent claim involved in the appeal and for each dependent claim argued separately under the provisions of 37 CFR 41.37(c)(1)(vii), every means plus function and step plus function as permitted by 35 U.S.C. 112, sixth paragraph, must be identified and the structure, material, or acts described in the specification as corresponding to each claimed function must be set forth with reference to the specification by page and line number, and to the drawing, if any, by reference characters.

**>(vi) Grounds of rejection to be reviewed on appeal. A concise statement of each ground of rejection presented for review.< For example, the statement ** "Whether claims 1 and 2 are unpatentable" would not comply with **>the rule, while the statements< "Whether claims 1 and 2 are unpatentable under 35 U.S.C. 103 over Smith in view of Jones,"

＊＊＊＊＊＊＊＊＊＊＊

**>(vii)< Argument. The appellant's contentions with respect to each **>ground of rejection< presented ** and the basis for those contentions, including citations of authorities, statutes, and parts of the record relied on, should be presented in this section.

＊＊＊＊＊＊＊＊＊＊＊

>Each ground of rejection must be treated under a separate heading. For each ground of rejection applying to two or more claims, the claims may be argued separately or as a group. When multiple claims subject to the same ground of rejection are argued as a group by appellant, the Board may select a single claim from the group of claims that are argued together to decide the appeal with respect to the group of claims as to the ground of rejection on the basis of the selected claim alone. The failure of appellant to separately argue claims which appellant has grouped together constitutes a waiver of any argument that the Board must consider the patentability of any grouped claim separately.

＊＊＊＊＊＊＊＊＊＊＊

**>(viii) Claims appendix.< An appendix containing a copy of the claims involved in the appeal.

The copy of the claims ** should be a clean copy and should not include any markings such as brackets or underlining >except for claims in a reissue application<. See MPEP § 1454 for the presentation of the copy of the claims in a reissue application.

＊＊＊＊＊＊＊＊＊＊＊

**>(ix) Evidence appendix

＊＊＊＊＊＊＊＊＊＊＊

MPEP 1205.03 Non-Compliant Appeal Brief and Amended Brief< [R-3]

The question of whether a brief complies with the rule is a matter within the jurisdiction of the examiner **>and the Board. The examiner will review the brief to ensure that the required items of the brief are present. Both the Board and the examiner will review the brief for compliance with the content requirements of the brief (37 CFR 41.37(c)). 37 CFR 41.37(d)< provides that if a brief is filed which does not comply with all the requirements of paragraph (c), the appellant will be notified of the reasons for noncompliance. Appellant will be given ** 1 month or 30 days from the mailing of the notification of non-compliance, whichever is longer **>to file an amended brief.<

＊＊＊＊＊＊＊＊＊＊＊

§20.6 提出訴願後修正說明書、圖式、請求項及補充宣誓書及證據

在訴願階段時修正說明書、圖式、請求項或補充宣誓書、證據並不是訴願人之權利（not a matter of right）。亦即，修正及補充是否可進入審查（entry）乃依照 37 CFR§41.33 來辦理。

Ⅰ.在提出訴願聲明到補充訴願理由期間作下列之修正可被允許（可進入審查）：

(A) 刪除請求項。

(B) 符合先前之審定書中明白記載之格式上的修正。

(C) 將被核駁之請求項修正更適合在訴願時被考慮之形式的修正。

(D) 具有充份理由顯示此修正是必要，而且為何未在之前提出之理由之說明書及請求項之修正。

Ⅱ. 在補充訴願理由之時或之後只能作下列之修正：

(A) 刪除請求項，此刪除不能影響任何其他審查中之請求項之範圍。

(B) 將依附項改寫成獨立項。

Ⅲ. 補呈宣誓書及其他證據

在提出訴願聲明之後，補充訴願理由之前，訴願人如果補呈宣誓書及其他證據，如果審查委員認為(A)此宣誓書及證據可克服所有的核駁理由，(B)此宣誓書及證據是必要，且未在之前提出有充分之理由時就會允許補件，進入審查（entered）。

在訴願案件送入訴願委員會之後（管轄權已交至訴願委員會），則任何修正，補充宣誓書及證據均不會被考慮，除非在訴願委員會將訴願案件發回（remand）給審查委員重審時才會被考慮。

MPEP 1206 Amendments and Affidavits or Other Evidence< Filed With or After Appeal [R-3]

＊＊＊＊＊＊＊＊＊＊＊

I.AMENDMENTS

A new amendment must be submitted in a separate paper. Entry of a new amendment in an application on appeal is not a matter of right. The entry of an amendment (which may not include a new affidavit, declaration, exhibit or other evidence) submitted in an application on appeal is governed by 37 CFR 41.33, not 37 CFR 1.116.

Amendments filed after the filing of a notice of appeal, but prior to the date of filing a brief, may be admitted only to:

(A) cancel claims;

(B) comply with any requirement of form expressly set forth in a previous action;

(C) present rejected claims in better form for consideration on appeal; or

(D) amend the specification or claims upon a showing of good and sufficient reasons why the amendment is necessary and was not earlier presented. See 37 CFR 41.33(a).

＊＊＊＊＊＊＊＊＊＊＊

Amendments filed on or after the date of filing a brief pursuant to 37 CFR 41.37 may be admitted only to:

(A) cancel claims, where such cancellation does not affect the scope of any other pending claim in the proceeding; or

(B) rewrite dependent claims into independent form.

＊＊＊＊＊＊＊＊＊＊＊

II.AFFIDAVITS OR OTHER EVIDENCE

Affidavits or other evidence (e.g., declarations or exhibits) submitted after the date of filing a notice of appeal, but prior to the date of filing a brief pursuant to 37 CFR 41.37, may be admitted if the examiner determines that:

(A) the affidavits or other evidence overcomes all rejections under appeal; and

(B) a showing of good and sufficient reasons why the affidavit or other evidence is necessary and was not earlier presented has been made.

＊＊＊＊＊＊＊＊＊＊＊

An amendment>, affidavit or other evidence< received after jurisdiction has passed to the Board should not be considered by the examiner unless remanded >or returned< by the Board for such purpose. See MPEP § 1210 and § *> 1211.02<.

＊＊＊＊＊＊＊＊＊＊＊

§20.7　審查委員對訴願理由之回應

訴願人提出訴願理由之後,訴願委員會之審查委員會考慮相關議題,並作出下列之回應:

(A) 重開審查（Reopen prosecution）以發下新的核駁理由（需主管級審查委員同意）;

(B) 如果無新的核駁理由且原核駁理由可克服,撤回原核駁理由並核准申請案,或

(C) 由訴願會議小組召開訴願會議維持訴願之進行,並發下審查委員之回應（examiner's answer）（2004 年 9 月 13 日以後之訴願案件,審查委員的回應可包括新的核駁理由）。

訴願會議小組由三位審查委員組成,一位是員負責撰寫審查委員之回應的原審查委員,另一位是主管級審查委員（supervisory patent examiner,SPE）以及另一位審查委員,稱作評議員（conferee）。

如果負責撰寫審查委員之回應之審查委員認為不必進行訴願,要重開審查,則經 SPE 同意就會重開審查,發下新的審定書（Office Action）,如果其新的核駁理由

是(A)因為訴願人修正所造成，或(B)基於送入之 IDS 資料時，此重開審查之審定書
（OA）亦可為最終審定書（Final Office Action）。訴願人對於重開審查後發下之審
定書可作下列三項回覆：

(A)　依照 37CFR1.111 提出答辯，如果審定書並非最終審定書。

(B)　依照 37CFR§1.113 提出答辯，如果審定書為最終審定書。

(C)　再提出訴願聲明。

訴願會議小組維持訴願並發下審查委員之回應時必需先召開訴願會議，而且審
查委員之回應需於訴願人提出訴願理由之二個月內發下。審查委員之回應需對訴願
理由中之答辯理由作回答且需指出訴願人之請求項之錯誤，如果部份之核駁理由審
查委員要撤回，亦需於審查委員之回應中註明。如果訴願理由並未針對所有被核駁
之理由作答辯，則審查委員亦需註明為（grounds of rejection not on review）。再者，
如果審查委員之回應包括了對現有法律之新的解釋或適用，則需送至技術中心之審
查長（TC Director）請其考慮並批准，再送至專利局副局長作最後之核准並寄交訴
願人。

MPEP 1207 < Examiner's Answer [R-3]

* * * * * * * * * * * *

After an appeal brief under 37 CFR 41.37 has been filed and the examiner has considered the issues on appeal, the examiner may:

(A) reopen prosecution to enter a new ground of rejection with approval from the supervisory patent examiner (see MPEP § 1207.04);

(B) withdraw the final rejection and allow the application if the examiner determines that the rejections have been overcome and no new ground of rejection is appropriate; or

(C) maintain the appeal by conducting an appeal conference (MPEP § 1207.01) and draft an examiner's answer (MPEP § 1207.02). Any examiner's answer mailed on or after September 13, 2004 may include a new ground of rejection (MPEP § 1207.03).<

MPEP 207.04 < Reopening of Prosecution After Appeal [R-3]

The examiner may, with approval from the supervisory patent examiner, reopen prosecution to enter a new ground of rejection after appellant's brief or reply brief has been filed. The Office action containing a new ground of rejection may be made final if the new ground of rejection was (A) necessitated by amendment, or (B) based on information presented in an information disclosure statement under 37 CFR 1.97(c) where no statement under 37 CFR 1.97(e) was filed. See MPEP § 706.07(a). >Any after final amendment or affidavit or other evidence that was not entered before must be entered and considered on the merits.<

* * * * * * * * * * *

After reopening of prosecution, appellant must exercise one of the following options to avoid abandonment of the application:

(A) file a reply under 37 CFR 1.111, if the Office action is non-final;

(B) file a reply under 37 CFR 1.113, if the Office action is final; or

(C) **>initiate a new appeal by filing a new notice of appeal under 37 CFR 41.31<.

MPEP 1207.01 Appeal Conference< [R-3]

An appeal conference is mandatory in all cases in which an acceptable brief (MPEP § *>1205<) has been filed. However, if the examiner charged with the responsibility of preparing the examiner's answer reaches a conclusion that the appeal should not go forward and the supervisory patent examiner (SPE) approves, then no appeal conference is necessary. >In this case, the examiner may reopen prosecution and issue another Office action. See MPEP § 1207.04.<

The participants of the appeal conference should include (1) the examiner charged with preparation of the examiner's answer, (2) a supervisory patent examiner (SPE), and (3) another examiner, known as a conferee, having sufficient experience to be of assistance in the consideration of the merits of the issues on appeal.

* * * * * * * * * * *

MPEP1207.02 Contents of Examiner's Answer< [R-3]

The examiner should furnish the appellant with a written statement in answer to the appellant's brief within 2 months after the receipt of the brief by the examiner.

The answer should contain a response to the allegations or arguments in the brief and should call attention to any errors in appellant's copy of the claims. If any rejection is withdrawn, the withdrawal should be clearly stated in the examiner's answer under *>subheading "Grounds of Rejection Withdrawn"

* * * * * * * * * * *

If the brief in compliance with 37 CFR 41.37 fails to address all grounds of rejection advanced by the examiner, the examiner should identify each ground of rejection not addressed by the brief in the examiner's answer under a subheading "Grounds of Rejection Not on Review"

＊＊＊＊＊＊＊＊＊＊＊

If an examiner's answer is believed to contain a new interpretation or application of the existing patent law, the examiner's answer, application file, and an explanatory memorandum should be forwarded to the TC Director for consideration. See MPEP § 1003. If approved by the TC Director, the examiner's answer should be forwarded to the Office of the Deputy Commissioner for Patent Examination Policy for final approval.

＊＊＊＊＊＊＊＊＊＊＊

§20.8 審查委員之回應包含新的核駁理由

依照 37CFR41.39(a)(2)之規定，在 2004 年 9 月 13 日當天或之後之訴願案，審查委員對訴願之回應（answer）可包含新的核駁理由。但通常審查委員在回應中包含新的核駁理由情形很少見。例如，當訴願人在訴願理由中提出一新的答辯理由時，審查委員可能會新引證出一先前技術作出新的核駁理由。當然不只有當訴願人提出新的答辯理由時審查委員才可作出新的核駁理由。審查委員委員亦可自行決定是否要作出新的核駁理由。當審查委員作出新的核駁理由時可重開審查或在回應中記載新的核駁理由，但需符合下列二要件：

(A) 需技術中心（TC）之主管核准；

(B) 需在審查委員之回應之"Grounds of Rejection to be Reviewed on Appeal"欄及 "Grounds of Rejection" 中明確地指出。

需注意的是，如果審查委員的新的核駁理由（即使此新的核駁理由是基於原有的引證資料）是要用來核駁一已核准（allowed）或形式上核駁（objected to）之請求項的話，則經技術中心之主管同意後只能重開審查，不能在審查委員之回應中加入此新的核駁理由。另外，如果訴願人之訴願理由是在申請案審查時就提出，但審查委員當時未予以回應者，審查委員就不能在審查委員之回應中提出新的核駁理由，只能重開審查。

　　但是，如果審查委員用來核駁之法源（statutory basis）相同，而且所用之證據(引證)亦相同，雖然支持其核駁之用詞或理論（rational）並不相同，此核駁理由不能算是新的核駁理由。此外，如果訴願人來補充訴願理由時對請求項作出修正，而此修正是審查委員發下建議性審定書（advisory action）時告知訴願時可進入審查者(entered)，且會被個別的核駁理由核駁的話，則審查委員之核駁理由將不被視為是新的核駁理由。

MPEP 1207.03 < New Ground of Rejection in Examiner's Answer [R-3]

　　37 CFR **>41.39(a)(2) permits< the entry of a new ground of rejection in an examiner's answer >mailed on or after September 13, 2004. New grounds of rejection in an examiner's answer are envisioned to be rare, rather than a routine occurrence. For example, where appellant made a new argument for the first time in the appeal brief, the examiner may include a new ground of rejection in an examiner's answer to address the newly presented argument by adding a secondary reference from the prior art on the record. New grounds of rejection are not limited to only a rejection made in response to an argument presented for the first time in an appeal brief<. At the time of preparing the answer to an appeal brief, * the examiner may decide that he or she should apply a new ground of rejection against some or all of the appealed claims. In such an instance where a new ground of rejection is necessary, the examiner should **>either reopen prosecution or set forth the new ground of rejection in the answer<. The examiner must obtain supervisory approval in order to reopen prosecution after an appeal. See MPEP § 1002.02(d) *>and § 1207.04. A supplemental examiner's answer cannot include a new ground of rejection, except when a supplemental answer is written in response to a remand by the Board for further consideration of a rejection under 37 CFR 41.50(a). See MPEP § 1207.05.

I.REQUIREMENTS FOR A NEW GROUND OF REJECTION

　　Any new ground of rejection made by an examiner in an answer must be:

　　(A) approved by a Technology Center (TC) Director or designee; and

　　(B) prominently identified in the "Grounds of Rejection to be Reviewed on Appeal" section and the "Grounds of Rejection" section of the answer.

＊＊＊＊＊＊＊＊＊＊＊

II. SITUATIONS WHERE NEW GROUNDS OF REJECTION ARE NOT PERMISSIBLE

A new ground of rejection would not be permitted to reject a previously allowed or objected to claim even if the new ground of rejection would rely upon evidence already of record. In this instance, rather than making a new ground of rejection in an examiner's answer, if the basis for the new ground of rejection was approved by a supervisory patent examiner as currently set forth in MPEP § 1207.04, the examiner would reopen prosecution. In addition, if an appellant has clearly set forth an argument in a previous reply during prosecution of the application and the examiner has failed to address that argument, the examiner would not be permitted to add a new ground of rejection in the examiner's answer to respond to that argument but would be permitted to reopen prosecution, if appropriate.

III. SITUATIONS THAT ARE NOT CONSIDERED AS NEW GROUNDS OF REJECTION<

＊＊＊＊＊＊＊＊＊＊＊

Where the statutory basis for the rejection remains the same, and the evidence relied upon in support of the rejection remains the same, a change in the discussion of, or rationale in support of, the rejection does not necessarily constitute a new ground of rejection. Id. at 1303, 190 USPQ at 427 (reliance upon fewer references in affirming a rejection under 35 U.S.C. 103 does not constitute a new ground of rejection).

In addition, former< 37 CFR 1.193(a)(2) also *>provided< that if:

(A) an amendment under 37 CFR 1.116 >[or 41.33]< proposes to add or amend one or more claims;

(B) appellant was advised (through an advisory action) that the amendment would be entered for purposes of appeal; and

(C) the advisory action indicates which individual rejection(s) set forth in the action from which appeal has been taken would be used to reject the added or amended claims, then

(1) the appeal brief must address the rejection(s) of the added or amended claim(s) and

(2) the examiner's answer may include the rejection(s) of the added or amended claims. >Such rejection(s) made in the examiner's answer would not be considered as a new ground of rejection.<

實例

一專利申請案包括 claim 1（獨立項），claim 2（依附至 claim 1）及 claim 3（依附至 claim 1）。在最終審定時審查委員核駁 claims 1-3 之理由及引證如下：

請求項	引證	核駁理由
claim 1	A	35USC §102
claim 2	A+B	35USC §103
claim 3	A+C	35USC §103

在訴願人補訴願理由（appeal brief）時若訴願人分別對請求項作了以下之修正而審查委員在回應（Examiners Answer）作出下列對應之核駁。

	請求項之修正	核駁理由	引證資料
(1)	刪除 claim 2, claim 2 併入 claim 1	35 USC§103	A+B
(2)	刪除 claim 3, claim 3 併入 claim 1	35 USC§103	A+C
(3)	刪除 claim 2 及 claim 3,claim 2 及 claim3 併入 claim 1	35 USC§103	A+B+C

則上述是否可進入審查（entered）？審查委員之核駁理由是否是新的核駁理由？

上述(1)及(2)種之修正可進入審查，且審查委員之核駁理由並非新的核駁理由。而上述(3)之情形之修正不能進入審查（refused entry）因為是產生新的議題（new issue），審查委員在回應中之核駁理由為新的核駁理由。

§20.9　訴願人對新的核駁理由之回覆

如果審查委員之回應中包含有新的核駁理由時，訴願人可(A)請求重開審查，或(B)維持原訴願。

(A)請求重開審查

如果訴願人要請求重開審查，就需在審查委員之回應（examiner's answer）發下之二個月內提出答覆書（reply）針對審查委員提出之每一新的核駁理由作答覆，此答覆書亦可包括修正、證據或未受到新的核駁理由核駁之請求項之答辯。如果有最終核駁時未能進入審查之修正（或宣誓書或其他證據）亦可在此時提出。

一旦訴願人對審查委員之回應請求重開審查，則審查委員就必需重開審查。

(B)提出答覆聲明書（reply brief）請求維持原訴願

如果訴願人要維持原訴願，則需在審查委員之回應的二個月內提出答覆聲明書針對審查委員的新的核駁理由作答辯，答覆聲明書需包括下列項目（每一項目從另一頁開始）：

(1)　識別資料：記載訴願人姓名、申請案號、申請日、發明名稱、審查委員名稱及審查委員技術單位。

(2)　請求項之狀態資料。

(3)　在訴願時要被審理之核駁理由；及

(4)　答辯內容。

訴願人提出答覆聲明書後，審查委員可再提出補充的審查委員回應（supplement examiner's answer），此時訴願人亦可於二個月內再提出一次答覆聲明書。

(C)不請求重開審查亦不提出答覆聲明書

如果訴願人對於審查委員之新的核駁理由不請求重開審查，亦不提出答覆聲明書，則受此新的核駁理由核駁之請求項之訴願被視為不受理（sua sponte dismissal）。如果所有從訴願人請求項均被此新的核駁理由核駁，則整個訴願不受理。亦即，如果此時並無任何已核准之請求項（allowed claims）的話，則此申請案被放棄。

如果只有部份請求項被此新的核駁理由核駁，則此部份之請求項將被刪除，其餘的請求項維持訴願程序。

MPEP1207.03 < New Ground of Rejection in Examiner's Answer

＊＊＊＊＊＊＊＊＊＊＊

V.APPELLANT'S REPLY TO NEW GROUNDS OF REJECTION

＊＊＊＊＊＊＊＊＊＊＊

A.Request That Prosecution Be Reopened by Filing a Reply

If appellant requests that prosecution be reopened, the appellant must file a reply that addresses each new ground of rejection set forth in the examiner's answer in compliance with 37 CFR 1.111 within two months from the mailing of the examiner's answer. The reply may also include amendments, evidence, and/or arguments directed to claims not subject to the new ground of rejection or other rejections. If there is an after-final amendment (or affidavit or other evidence) that was not entered, appellant may include such amendment in the reply to the examiner's answer.

＊＊＊＊＊＊＊＊＊＊＊

Once appellant files a reply in compliance with 37 CFR 1.111 in response to an examiner's answer that contains a new ground of rejection, the examiner must reopen prosecution by entering and considering the reply.

＊＊＊＊＊＊＊＊＊＊＊

B.Request That the Appeal Be Maintained by Filing a Reply Brief

If appellant requests that the appeal be maintained, the appellant must file a reply brief that addresses each new ground of rejection set forth in the answer in compliance with 37 CFR 41.37(c)(1)(vii) within two months from the mailing of the answer. The reply brief should include the following items, with each item starting on a separate page, so as to follow the other requirements of a brief as set forth in 37 CFR 41.37(c):

(1) Identification page setting forth the appellant's name(s), the application number, the filing date of the application, the title of the invention, the name of the examiner, the art unit of the examiner and the title of the paper (i.e., Reply Brief);

(2) Status of claims page(s);

(3) Grounds of rejection to be reviewed on appeal page(s); and

(4) Argument page(s).

＊＊＊＊＊＊＊＊＊＊＊

The examiner may provide a supplemental examiner's answer (with TC Director or designee approval) to respond to any new issue raised in the reply brief. The supplemental examiner's answer responding to a reply brief cannot include any new grounds of rejection. See MPEP § 1207.05. In response to the supplemental examiner's answer, the appellant may file another reply brief under 37 CFR 41.41 within 2 months from the mailing of the supplemental examiner's answer.

＊＊＊＊＊＊＊＊＊＊＊

C.Failure To Reply to a New Ground of Rejection

If appellant fails to timely file a reply under 37 CFR 1.111 or a reply brief in response to an examiner's answer that contains a new ground of rejection, the appeal will be sua sponte dismissed as to the claims subject to the new ground of rejection. If all of the claims under appeal are subject to the new ground of rejection, the entire appeal will be dismissed. The examiner should follow the procedure set forth in MPEP §1215 to dismiss the appeal. For example, if there is no allowed claim in the application, the application would be abandoned when the two-month time expired.

If only some of the claims under appeal are subject to the new ground of rejection, the dismissal of the appeal as to those claims operates as an authorization to cancel those claims and the appeal continues as to the remaining claims.

§20.10 訴願人對審查委員之回應可採取之法律程序

當審查委員發下回應（examiner's answer）或補充回應之後，訴願人可在二個月之內提出答覆聲明書（reply brief）。通常當訴願人之答覆聲明書包含有新的議題時，審查委員會發下補充回應。一般說來，訴願人不必對審查委員之回應提出答覆聲明書，如果訴願人未提出答覆聲明書，則訴願案件就會移送至訴願委員會作決定。通常只有在以下情形時訴願人才需提出答覆之聲明書。

(A)　審查委員之回應包含有新的核駁理由，或

(B)　訴願委員會將訴願案件退回重審（remand）因而審查委員提出補充回應（此時通常包含有新的核駁理由）。

訴願人除了可提出答覆聲明書以回應審查委員之回應外，亦可在二個月之內提出口頭聽證（oral hearing）之請求並繳交規費。訴願人提出口頭聽證請求之後，訴願委員會就會通知訴願人口頭聽證之日期、時間及地點。如果訴願人未確認要參加口頭聽證或又取消口頭聽證或屆時未參加口頭聽證，案件就直接移至訴願委員會且不能退費。另外，訴願委員會如果覺得沒有口頭聽證之必要（例如訴願委員會擬將案件退回審查委員會重審）則會通知訴願人。訴願人如果希望口頭聽證在一特定日期、時間舉行亦可通知訴願委員會，在可配合之情形下訴願委員會亦會安排。

在口頭聽證時，訴願人或其代理人通常只有 20 分鐘可以作說明，如果訴願人覺得 20 分鐘不夠，可在口頭聽證之前先提出請求。在口頭聽證時訴願人只能針對先前進入審查及審查委員考慮過之證據及基於答覆聲明書（reply brief）之資訊作說明，如果有充份之理由，訴願人及／或審查委員可依據訴願委員會或聯邦法院之最近的相關決定提出新的論點。

口頭聽證時，如果審查委員想參加，或訴願委員會希望審查委員參加，則審查委員亦可在訴願人出席或未出席之狀況下參加口頭聽證，口頭聽證時，審查委員只有 15 分鐘作說明，且只能依據審查委員之回應補充回應中之資訊或證據作說明。

MPEP 1208 Reply Briefs and Examiner's Responses to Reply Brief

* * * * * * * * * * *

I.REPLY BRIEF

Under 37 CFR 41.41(a)(1) and 41.43(b), appellant may file a reply brief as a matter of right within 2 months from the mailing date of the examiner's answer or supplemental examiner's answer. Extensions of time to file the reply brief may be granted pursuant to 37 CFR 1.136(b) (for patent applications) or 1.550(c) (for ex parte reexamination proceedings). Extensions of time under 37 CFR 1.136(a) are not permitted. The examiner may provide a supplemental examiner's answer to respond to any reply brief that raises new issues. See MPEP § 1207.05. Normally, appellant is not required to file a reply brief to respond to an examiner's answer or a supplemental examiner's answer, and if appellant does not file a reply brief within the two month period of time, the application will be forwarded to the Board for decision on the appeal. In response to the following, however, appellant is required to file either a reply brief to maintain the appeal or a reply under 37 CFR 1.111 to reopen prosecution:

(A) An examiner's answer that contains a new ground of rejection pursuant to 37 CFR 41.39 (see MPEP § 1207.03); or

(B) A supplemental examiner's answer responding to a remand by the Board for further consideration of a rejection pursuant to 37 CFR 41.50(a) (see MPEP § 1207.05). Such a supplemental examiner's answer may contain a new ground of rejection (also see MPEP § 1207.03).

* * * * * * * * * * *

MPEP 1209 Oral Hearing [R-3]

＊＊＊＊＊＊＊＊＊＊＊

37 CFR *>41.47(b)< provides that an appellant who desires an oral hearing before the Board must request the hearing by filing, in a separate paper >captioned "REQUEST FOR ORAL HEARING,"< a written request therefor, accompanied by the appropriate fee set forth in 37 CFR *>41.20(b)(3)<, within 2 months after the date of the examiner's answer >or supplemental examiner's

＊＊＊＊＊＊＊＊＊＊

A notice of hearing, stating the date, the time, and the docket, is forwarded to the appellant in due course. If appellant fails to confirm >the hearing< within the time required in the notice of hearing >or the appellant waives the hearing<, the appeal will be removed from the hearing docket and assigned on brief in due course. No refund of the fee for requesting an oral hearing will be made. Similarly, after confirmation, if no appearance is made at the scheduled hearing, the appeal will be decided on brief. Since failure to notify the Board of waiver of hearing in advance of the assigned date results in a waste of the Board's resources, appellant should inform the Board of a change in plans at the earliest possible opportunity. If the Board determines that a hearing is not necessary (e.g., a remand to the examiner is necessary or it is clear that the rejection(s) cannot be sustained), appellant will be notified.

If appellant has any special request, such as for a particular date or day of the week, this will be taken into consideration in setting the hearing, if made known to the Board in advance, as long as such request does not unduly delay a decision in the case and does not place an undue administrative burden on the Board.

＊＊＊＊＊＊＊＊＊＊＊

Normally, 20 minutes are allowed for appellant to explain his or her position. If appellant believes that additional time will be necessary, a request for such time should be made well in advance and will be taken into consideration in assigning the hearing date. The final decision on whether additional time is to be granted rests within the discretion of the senior member of the panel hearing the case.

At the oral hearing, appellant may only rely on evidence that has been previously entered and considered by the primary examiner and present arguments that have been relied upon in the brief or reply brief. Upon a showing of good cause, appellant and/or the primary examiner may rely on a new argument based upon a recent relevant decision of either the Board or a Federal Court.

＊＊＊＊＊＊＊＊＊＊

PARTICIPATION BY EXAMINER

If the appellant has requested an oral hearing and the primary examiner wishes to appear and present an oral argument before the Board, a request to present oral argument must be **>set forth in a separate letter on a form PTOL-90 using form paragraph 12.163.<

In those appeals in which an oral hearing has been confirmed and either the primary examiner or the Board has indicated a desire for the examiner to participate in the oral argument, oral argument may be presented by the examiner whether or not appellant appears.

＊＊＊＊＊＊＊＊＊＊＊

At the hearing, after the appellant has made his or her presentation, the examiner will be allowed 15 minutes to reply as well as to present a statement which clearly sets forth his or her position with respect to the issues and rejections of record. >The primary examiner may only rely on argument and evidence relied upon in the examiner's answer or the supplemental examiner's answer.< Appellant may utilize any allotted time not used in the initial presentation for rebuttal.

§20.11 審查委員之補充回應

在下述情形下，審查委員可發下補充的回應：

Ⅰ. 如果訴願人提出之答覆聲明書（reply brief）包含有新的議題（new issue），則審查委員可提出補充回應。如果答覆聲明書無新的議題，則不能發出補充的回應。提出補充的回應可避免訴願委員會將訴願案件退回重審（remand）。

Ⅱ. 如果訴願委員會將訴願案件退回重審，則審查委員可提出補充回應，此補充回應可包含新的核駁理由。此時訴願人可(i)請求重開審查或(ii)提出答覆聲明書請求維持訴願。如果未請求重開審查亦未請求維持訴願，則訴願將被視為不受理（dismissed）。

MPEP 1207.05 Supplemental Examiner's Answer [R-3]

I. SUPPLEMENTAL EXAMINER'S ANSWER RESPONDING TO A REPLY BRIEF

In response to a reply brief filed in compliance with 37 CFR 41.41, the primary examiner may: (A) withdraw the final rejection and reopen prosecution (see MPEP § 1207.04); or (B) provide a supplemental examiner's answer responding to any new issue raised in the reply brief. The examiner cannot issue a supplemental examiner's answer if the reply brief raised no new issue.

See MPEP§1208 for more information on reply brief and examiner's response to reply brief. If the reply brief does raise new issues, providing a supplemental examiner's answer will avoid the need for the Board to remand the application or proceeding to the examiner to treat the new issues.

＊＊＊＊＊＊＊＊＊＊＊

II.　SUPPLEMENTAL EXAMINER'S AN-SWER RESPONDING TO A REMAND FOR FURTHER CONSIDERATION OF REJECTION

The examiner may provide a supplemental examiner's answer in response to a remand by the Board for further consideration of a rejection under 37 CFR 41.50(a). Appellant must respond to such supplemental examiner's answer and has the option to request that prosecution be reopened. A supplemental examiner's answer written in response to a remand by the Board for further consideration of a rejection pursuant to 37 CFR 41.50(a)(1) may set forth a new ground of rejection.

＊＊＊＊＊＊＊＊＊＊＊

§20.12 訴願委員會將案件發回重審

在下列二時點：(1)在審查委員發下回應（examiner's answer）或補充回應之二個月後，如果訴願人未提出答覆聲明書（reply brief）則，或(2)訴願人有提出答覆聲明書而審查委員書面告知訴願人答覆聲明書已進入審查之後，訴願案之檔案就從技術中心被移送至訴願委員會。

當訴願委員會認為有需要時，可將訴願案件發回重審，例如發回審查委員重新考慮引證資料之關聯性，IDS 資料，宣誓書，請求項之修正等，請審查委員重新作檢索，或準備審查委員補充回應以回答訴願人之答覆聲明書等。

MPEP 1210 Actions Subsequent to Examiner's Answer but Before Board's Decision [R-3]

＊＊＊＊＊＊＊＊＊＊＊

The application file and jurisdiction of the application are normally transferred from the Technology Centers to the Board at one of the following times:

(A) After 2 months from the examiner's answer >or supplemental examiner's answer<, plus mail room time, if no reply brief has been timely filed.

(B) After ** the examiner has notified the appellant by written communication that the reply brief has been entered and considered and that the application will be forwarded to the Board (for example, by mailing a PTOL-90 with form paragraph *>12.181<, as described in MPEP § *>1208<).

MPEP 1211 Remand by Board [R-3]

The Board has authority to remand a case to the examiner when it deems it necessary. For example, the Board may remand **>a case for further consideration of a rejection pursuant to 37 CFR 41.50(a)(1) such as< where the pertinence of the references is not clear, the Board may call upon the examiner for a further explanation. >See MPEP § 1211.01.< In the case of multiple rejections of a cumulative nature, the Board may also remand for selection of the preferred or best ground. The Board may also remand a case to the examiner for further search where it feels that the most pertinent art has not been cited, or to consider an amendment**. See MPEP * § 1211.02, * § 1211.03 >and § 1211.04<. Furthermore, the Board may remand an application to the examiner to prepare a supplemental examiner's answer in response to a reply brief **>which the examiner only acknowledged receipt and entry thereof (e.g., by using form paragraph 12.181 on form PTOL-90). See MPEP § 1207.05 for more information on supplemental examiner's answer<.

＊＊＊＊＊＊＊＊＊＊＊

§20.13 訴願委員會決定之內容及後續程序

訴願委員會乃由美國專利商標局長、副局長、專利處處長（Commissioner for Patents）、商標處處長（Commissioner for Trademarks）及專利法行政官（Administrative Patent Judges, APJ）所組成。通常一訴願案件由三位專利行政法官（AJP）來判決。訴願決定可全部或部份確認審查委員之決定，或駁回審查委員之決定，有時亦可能包括新的核駁理由，可能指出提起訴願之請求項應如何修正才可克服審查委員之核駁理由。訴願決定書通常在審查委員回應之後的 6 個月～1 年之內會發下，但有時會等相關之民事訴訟或向聯邦巡迴法庭之上訴有一結論之後才會發下。

如果訴願決定全部或部份確認審查委員對提起訴願之請求項之核駁，則訴願人可在二個月內向聯邦巡迴法庭上訴，或向哥倫比亞地方法院提出民事訴訟或提出 RCE（提出 RCE 之最後機會）。

　　如果在二個月之內訴願人未提出任何後續程序，則此專利申請案被視為非審查中（如果是 ex pate 之再審案件則發下再審查證書）。

　　需注意的是，在訴願前被認為可核准之請求項（allowable claims，其依附至被核駁之請求項）將被視為是核駁。例如，在訴願前審查中之請求項為 claim 1 及 claim 2，而 claim 1 被核駁，claim 2 依附至 claim 1 但是為可核准的，如果訴願委員會確認了 claim 1 之核駁，則 claim 2 一起被核駁，故申請案被放棄。而如果訴願前審查中之請求項為 claim 1 及 claim 2（依附至 claim 2）而 claim 1 及 claim 2 均被核駁。如果訴願委員會確認了 claim 1 之核駁，但駁回了 claim 2 之核駁，則審查委員會①以審查委員之修正將 claim 2 修正為獨立項，並核准發證並刪除所有確認核駁之請求項，或②依訴願人一個月之期間將 claim 2（依附項）修改為獨立項。另外，如果原本就是核准之請求項（allowed claims），訴願人不必作任何回覆，審查委員會對已核准之請求項發證並刪除其他請求項。

　　如果訴願決定駁回審查委員之決定，則審查委員會重開審查或者核准提起訴願之請求項。

　　如果訴願決定書中包括了明確的聲明請求項如何修正就可克服特定的核駁理由的話，訴願人就可依對請求項作修正。審查委員需確認訴願人之修正符合訴願決定書中之聲明。但是，訴願決定書之此聲明並不表示訴願人依此修正，請求項就是可核准的。審查委員仍可以新的核駁理由來核駁此修正的請求項，但此新的核駁理由需由技術中心的審查長批准。如果訴願委員會自行對提起訴願之請求項作出新的核駁理由（亦可對已核准之請求項作出新的核駁理由）的話，則訴願人可在二個月內(A)呈送請求項之修正本及/或新證據請求審查委員重開審查，或(B)請求訴願委員會再議（rehearing）。

---

35 U.S.C. 6 Board of Patent Appeals and Interferences.

(a) ESTABLISHMENT AND COMPOSITION.- There shall be in the United States Patent and Trademark Office a Board of Patent Appeals and Interferences. The Director, the Deputy Commissioner, the Commissioner for Patents, the Commissioner for Trademarks, and the administrative patent judges shall constitute the Board. The administrative patent judges shall be persons of competent legal knowledge and scientific ability who are appointed by the Director.

＊＊＊＊＊＊＊＊＊＊＊＊

MPEP 1213 Decision by Board [R-3]

After consideration of the record including appellant's *>briefs< and the examiner's *>answers<, the Board writes its decision, affirming the examiner in whole or in part, or reversing the examiner's decision, sometimes also setting forth a new ground of rejection.

＊ ＊ ＊ ＊ ＊ ＊ ＊ ＊ ＊ ＊ ＊

On occasion, the Board has refused to consider an appeal until after the conclusion of a pending civil action or appeal to the Court of Appeals for the Federal Circuit involving issues identical with and/or similar to those presented in the later appeal.

＊ ＊ ＊ ＊ ＊ ＊ ＊ ＊ ＊ ＊ ＊

MPEP 1213.01 Statement **>by Board of How an Appealed Claim May Be Amended To Overcome a Specific Rejection< [R-3]

＊ ＊ ＊ ＊ ＊ ＊ ＊ ＊ ＊ ＊ ＊

If the Board's decision includes an explicit statement **>how a claim on appeal may be amended to overcome a specific rejection<, appellant may amend the claim in conformity with the statement **. The examiner should make certain that the amendment does in fact conform to the statement in the Board's decision.

＊ ＊ ＊ ＊ ＊ ＊ ＊ ＊ ＊ ＊ ＊

An explicit statement by the Board on how a claim on appeal may be amended to overcome a specific rejection is not a statement that a claim so-amended is allowable. The examiner may reject a claim so-amended, provided that the rejection constitutes a new ground of rejection. Any new ground of rejection made by an examiner following the Board's decision must be approved by a Technology Center Director and must be prominently identified as such in the action setting forth the new ground of rejection.

MPEP 1213.02 New Grounds of Rejection by Board [R-3]

＊ ＊ ＊ ＊ ＊ ＊ ＊ ＊ ＊ ＊ ＊

Under 37 CFR *>41.50(b)<, the Board may, in its decision, make a new rejection of one or more of any of the claims pending in the case, including claims which have been allowed by the examiner.

MPEP 1214.01 Procedure Following New Ground of Rejection by Board [R-3]

When the Board makes a new rejection under 37 CFR *>41.50(b)<, the appellant, as to each claim so rejected, has the option of:

(A) >reopening prosecution before the examiner by< submitting an appropriate amendment and/or **>new evidence (37 CFR 41.50(b)(1))<; or

(B) requesting rehearing **>before the Board (37 CFR 41.50(b)(2))<.

****

MPEP 1214.06 Examiner Sustained in Whole or in Part [R-3]

****

I.NO CLAIMS STAND ALLOWED

The proceedings in an application or ex parte reexamination proceeding are terminated as of the date of the expiration of the time for filing court action. The application is no longer considered as pending. It is to be stamped abandoned and sent to abandoned files. In an ex parte reexamination proceeding, a reexamination certificate should be issued under 37 CFR 1.570.

Claims indicated as allowable prior to appeal except for their dependency from rejected claims will be treated as if they were rejected. The following examples illustrate the appropriate approach to be taken by the examiner in various situations:

(A) If claims 1-2 are pending, and the Board affirms a rejection of claim 1 and claim 2 was objected to prior to appeal as being allowable except for its dependency from claim 1, the examiner should hold the application abandoned.

(B) If the Board or court affirms a rejection against an independent claim and reverses all rejections against a claim dependent thereon, ** after expiration of the period for further appeal, >the examiner< should proceed in one of two ways:

(1) Convert the dependent claim into independent form by examiner's amendment, cancel all claims in which the rejection was affirmed, and issue the application; or

(2) Set a 1-month time limit in which appellant may rewrite the dependent claim(s) in independent form. Extensions of time under 37 CFR 1.136(a) will not be permitted. If no timely reply is received, the examiner will cancel all rejected and objected to claims and issue the application with the allowed claims only.

****

II.CLAIMS STAND ALLOWED

The appellant is not required to file a reply. The examiner issues the application or ex parte reexamination certificate on the claims which stand allowed. *>For paper files, a< red-ink line should be drawn through the refused claims and the notion "Board Decision" written in the margin in red ink.

If the Board affirms a rejection of claim 1, claim 2 was objected to prior to appeal as being allowable except for its dependency from claim 1 and independent claim 3 is allowed, the examiner should cancel claims 1 and 2 and issue the application or ex parte reexamination certificate with claim 3 only.

>If the Board affirms a rejection against independent claim 1, reverses all rejections against dependent claim 2 and claim 3 is allowed, after expiration of the period for further appeal, the examiner should either:

(A) Convert dependent claim 2 into independent form by examiner's amendment, cancel claim 1 in which the rejection was affirmed, and issue the application with claims 2 and 3; or

(B) Set a 1-month time limit in which appellant may rewrite dependent claim 2 in independent form. Extensions of time under 37 CFR 1.136(a) will not be permitted. If no timely reply is received, the examiner will cancel claims 1 and 2 and issue the application with allowed claim 3 only. The following form paragraph may be used where appropriate:<

****

§20.14 撤回訴願之程序

訴願人在提出訴願理由之後，訴願委員會發下決定之前，訴願人可隨時撤回訴願，如果撤回訴願時並無已核准之請求項（allowed claims），則申請案被視為自撤回訴願之日起放棄。

在訴願委員會作出決定之前，如果訴願人撤回訴願之目的是希望重新審查，則可依照 35CFR§1.114 附上新事證（submission），答辯理由及規費提出 RCE。提出 RCE 就被視為是撤回訴願，不管 RCE 提出時是否有附上新事證及規費。因而，如果要提出 RCE 時未附上新事證及規費，而且該申請案無已核准之請求項，則該申請案將被放棄。

需注意的是，撤回訴願時如果是可核准之請求項（allowable claims），其只有依附至被核駁之請求項之問題，會被視為是被核駁。例如下列之例子：

(A) claim 1 已核准，claim 2、claim 3 被核駁，如果撤回訴願時則審查委員將刪除 claim 2、3，核准 claim 1 並發證。

(B) claim 1-3 被核駁，如果撤回訴願, 申請案被放棄。

(C) claim 1 被核駁，claim 2 依附至 claim 1，但是為可核准的，如果撤回訴願，則申請案被放棄。

(D) claim 1 被核駁，claim 2 依附至 claim 1，但是為可核准的，claim 3（獨立項）為已核准，如果撤回訴願，則 claim 1，claim 2 被刪除，claim 3 核准發證。

如果訴願人在提出訴願理由之後，訴願委員會發下決定之前撤回部份請求項，則審查委員會刪除這些撤回的請求項，只針對剩餘的請求項進行訴願程序。

MPEP 1215.01 Withdrawal of Appeal [R-3]

＊＊＊＊＊＊＊＊＊＊＊

Where, after an appeal has been filed and before decision by the Board, an applicant withdraws the appeal after the period for reply to the final rejection has expired, the application is to be considered abandoned as of the date on which the appeal was withdrawn unless there are allowed claims in the case.

＊＊＊＊＊＊＊＊＊＊＊

Prior to a decision by the Board, if an applicant wishes to withdraw an application from appeal and to reopen * prosecution of the application, applicant can file a request for continued examination (RCE) under 37 CFR 1.114, accompanied by a submission (i.e., a reply responsive within the meaning of 37 CFR 1.111 to the last outstanding Office action) and the RCE fee set forth under 37 CFR 1.17(e).

＊＊＊＊＊＊＊＊＊＊＊

The filing of an RCE will be treated as a withdrawal of the appeal by the applicant, regardless of whether the RCE includes the appropriate fee or a submission. Therefore, when an RCE is filed without the appropriate fee or a submission in an application that has no allowed claims, the application will be considered abandoned. To avoid abandonment, the RCE should be filed in compliance with 37 CFR 1.114. See MPEP § 706.07(h), paragraphs I-II.

＊＊＊＊＊＊＊＊＊＊＊

*>Upon the withdrawal of an appeal, an application< having no allowed claims will be abandoned. Claims which are allowable except for their dependency from rejected claims will be treated as if they were rejected. The following examples illustrate the appropriate approach to be taken by the examiner in various situations:

(A) Claim 1 is allowed; claims 2 and 3 are rejected. The examiner should cancel claims 2 and 3 and issue the application with claim 1 only.

(B) Claims 1 - 3 are rejected. The examiner should hold the application abandoned.

(C) Claim 1 is rejected and claim 2 is objected to as being allowable except for its dependency from claim 1. The examiner should hold the application abandoned.

(D) Claim 1 is rejected and claim 2 is objected to as being allowable except for its dependency from claim 1; independent claim 3 is allowed. The examiner should cancel claims 1 and 2 and issue the application with claim 3 only.

In an ex parte reexamination proceeding, an ex parte reexamination certificate should be issued under 37 CFR 1.570.

＊＊＊＊＊＊＊＊＊＊＊

MPEP 1215.03 Partial Withdrawal [R-3]

A withdrawal of the appeal as to some of the claims on appeal operates as an authorization to cancel those claims from the application or reexamination proceeding and the appeal continues as to the remaining claims. The withdrawn claims will be canceled from an application by direction of the examiner at the **>time of the withdrawal of the appeal as to those claims.

＊＊＊＊＊＊＊＊＊＊＊

§20.15 訴願不受理

如果訴願人在提出訴願聲明之後未提出訴願理由，則該訴願被視為不受理（dismissed）。此時如果訴願案件無任何已核准之請求項（allowed claims），則案件被放棄。另外，需注意的是，依附至被核駁的請求項，但可核准（allowable）之請求項亦被視為被核駁。以下例子說明各種情況：

(A) claim 1 已核准，claim 2、claim 3 被核駁，如果不提訴願理由，則審查委員會刪除 claim 2、3 而只核准發證 claim 1。

(B) claim 1-3 均核駁，如果不提訴願理由，則申請案被放棄。

(C) claim 1 被核駁，claim 2 依附至 claim 1 但可核准，如不提出訴願理由，申請案被放棄。

(D) claim 1 被核駁，claim 2 依附至 claim 1 但可核准，claim 3 已核准，如果不提訴願理由，審查委員將刪除 claim 1,Claim 2，核准 claim 3。另外，如果訴願人未對審查委員之回應中的新核駁理由提出答覆聲明書（reply brief）則訴願亦視為未受理（sua sponte dismissed）。

MPEP 1215.04 Dismissal of Appeal [R-3]

If no brief is filed within the time prescribed by 37 CFR *>41.37<, the appeal stands dismissed by operation of the rule.

＊＊＊＊＊＊＊＊＊＊＊＊

Applications having no allowed claims will be abandoned. Claims which are allowable except for their dependency from rejected claims will be treated as if they were rejected. The following examples illustrate the appropriate approach to be taken by the examiner in various situations:

(A) Claim 1 is allowed; claims 2 and 3 are rejected. The examiner should cancel claims 2 and 3 and issue the application with claim 1 only.

(B) Claims 1 - 3 are rejected. The examiner should hold the application abandoned.

(C) Claim 1 is rejected and claim 2 is objected to as being allowable except for its dependency from claim 1. The examiner should hold the application abandoned.

(D) Claim 1 is rejected and claim 2 is objected to as being allowable except for its dependency from claim 1; independent claim 3 is allowed. The examiner should cancel claims 1 and 2 and issue the application with claim 3 only.

＊＊＊＊＊＊＊＊＊＊＊＊

MPEP1215.02 Claims Standing Allowed

＊＊＊＊＊＊＊＊＊＊＊＊

If appellant fails to a new ground of rejection made in an examiner's answer by either filing a reply brief or a reply under 37 CFR 1.111 within 2 months from the mailing of the examiner's answer, the appeal is sua sponte dismissed as to the claims subject to the new ground of rejection. See MPEP §1207.03

＊＊＊＊＊＊＊＊＊＊＊＊

第 21 章　核准、領證及專利權之維護

目　錄

§21.1　Ex parte Quayle Action

Ex parte Quayle Action 為專利性審查已經結束，審查委員認為已經沒有 35 USC 101、102、103、112 等法定核駁（rejections）問題，但有形式缺陷需修正（例如圖式、說明等之誤繕修正、新式樣之圖面不對稱之修正、或者刪除不審的請求項）（objections），為了加速發證而發出的核駁通知。一般答辯期限為兩個月，可付費展延。

此類形式修正要求類似最終審定書（final OA）後之修正，亦即：

> ➤　修改後不可產生專利性上的新議題（new issue）

> ➤　不可要求進一步的檢索

> ➤　不可產生 35 USC §112, ¶1 缺乏支持的問題，亦即修正不可產生新事項（new matter）

修正核可之後，審查委員將發出核准通知。

MPEP 714.14 Amendments After Allowance of All Claims

Under the decision in Ex parte Quayle, 25 USPQ 74, 1935 C.D. 11; 453 O.G. 213 (Comm'r Pat. 1935), after all claims in an application have been allowed the prosecution of the application on the merits is closed even though there may be outstanding formal objections which preclude fully closing the prosecution.

Amendments touching the merits are treated in a manner similar to amendments after final rejection, though the prosecution may be continued as to the formal matters. See MPEP § 714.12 and § 714.13.

See MPEP § 714.20 for amendments entered in part.

See MPEP § 607 for additional fee requirements.

See MPEP § 714 for non-compliant amendments.

Use form paragraph 7.51 to issue an Ex parte Quayle action.

¶ 7.51 Quayle Action

This application is in condition for allowance except for the following formal matters: [1].

Prosecution on the merits is closed in accordance with the practice under Ex parte Quayle, 25 USPQ 74, 453 O.G. 213 (Comm'r Pat. 1935).

A shortened statutory period for reply to this action is set to expire TWO MONTHS from the mailing date of this letter.

§21.2　核准通知的發出與內容

當案件處於可核准的狀態，審查委員即可發出 PTOL-37 之核准通知（Notice of Allowability），明確記載可核准的請求項與案件審查狀態。並可附載說明審查委員核准之理由及／或其將修正之事項及／或先前面詢之記錄。

領證通知（Notice of Allowance）內容包含領證期限、費用、繳費單、PTA 延長天數等等。領證期限為三個月，原則上不可延期。逾期則該申請案視為放棄。若需延期，需先提出請願（petition）並繳納請願費及載明延期理由。最多僅能延期一個月。

MPEP 1302.03 Notice of Allowability

A Notice of Allowability form PTOL-37 is used whenever an application has been placed in condition for allowance. The date of any communication and/or interview which resulted in the allowance should be included in the notice.

In all instances, both before and after final rejection, in which an application is placed in condition for allowance, applicant should be notified promptly of allowability of the claims by a Notice of Allowability PTOL-37. ** Prompt notice to applicant is important because it may avoid an unnecessary appeal and act as a safeguard against a holding of abandonment.

MPEP 1303 Notice of Allowance [R-5]

37 CFR 1.311 Notice of Allowance.

(a) If, on examination, it appears that the applicant is entitled to a patent under the law, a notice of allowance will be sent to the applicant at the correspondence address indicated in § 1.33. The notice of allowance shall specify a sum constituting the issue fee which must be paid within three months from the date of mailing of the notice of allowance to avoid abandonment of the application. The sum specified in the notice of allowance may also include the publication fee, in which case the issue fee and publication fee (§ 1.211(e)) must both be paid within three months from the date of mailing of the notice of allowance to avoid abandonment of the application. This three-month period is not extendable.

(b) An authorization to charge the issue fee or other post-allowance fees set forth in § 1.18 to a deposit account may be filed in an individual application only after mailing of the notice of allowance. The submission of either of the following after the mailing of a notice of allowance will operate as a request to charge the correct issue fee or any publication fee due to any deposit account identified in a previously filed authorization to charge such fees:

(1) An incorrect issue fee or publication fee; or

(2) A fee transmittal form (or letter) for payment of issue fee or publication fee.

A Notice of Allowance is prepared and mailed, and the mailing date appearing thereon is recorded on the paper or image file wrapper table of contents.

If an application is subject to publication under 37 CFR 1.211, the Notice of Allowance will require both the issue fee and the publication fee. See 37 CFR 1.211(e). The Notice of Allowance and Issue Fee Due form (PTOL-85) has been revised and the revised form is entitled "Notice of Allowance and Fee(s) Due." Revision of the form was necessary to include the amount of any required publication fee, as provided in 37 CFR 1.211(e) and 1.311, and to more clearly communicate the amount of any patent term extension or adjustment earned under 35 U.S.C. 154(b). As revised, the PTOL-85 form is three pages long, with all three pages being mailed to the applicant and a duplicate being retained in the application file. The first two pages of the revised form include an indication that the publication fee is due, if the application was subject to publication and the publication fee has not already been paid. Part B of the revised form (PTOL-85B) *>must< be returned to the Office with the payment of the issue fee. ** Applicants are reminded to transmit an extra copy of the PTOL-85B when payment of the issue fee is by way of authorization to debit a Deposit Account. See MPEP § 509.01.

There are different versions of page three of the revised PTOL-85 form, depending upon the filing date of the application >and the application type:

(A) For applications filed before June 8, 1995, page three will state that "This application was filed prior to June 8, 1995, thus no Patent Term Extension or Adjustment applies." Utility and plant applications filed before June 8, 1995 are eligible for a 17 year term and thus are not eligible for patent term extension or adjustment under 35 U.S.C. 154(b).

(B) For applications filed on or after June 8, 1995 and before May 29, 2000, page three will state that "The Patent Term Extension is _ day(s). Any patent to issue from the above identified application will include an indication of the _ extension on the front page. If a Continued Prosecution Application (CPA) was filed in the above identified application, the filing date that determines Patent Term Extension is the filing date of the most recent CPA." Utility and plant applications filed on or after June 8, 1995 and before May 29, 2000 may be eligible for patent term extension. See 35 U.S.C. 154(b), effective June 8, 1995, and 37 CFR 1.701.

(C) For applications filed on or after May 29, 2000, page three will state that "The Patent Term Adjustment to date is _ day(s). If the issue fee is paid on a date that is three months after the mailing date of this notice, and the patent issues on the Tuesday before the date that is 28 weeks (six and a half months) after the mailing date of this notice, the Patent Term Adjustment will be_day(s). If a Continued Prosecution Application (CPA) was filed in the above-identified application, the filing date that determines Patent Term Adjustment is the filing date of the most recent CPA." Utility and plant applications filed on or after May 29, 2000 may be eligible for patent term adjustment. See 35 U.S.C. 154(b), effective May 29, 2000, and 37 CFR 1.702 - 1.705, especially 37 CFR 1.705(a).

(D) For reissue applications, page three will state that "A reissue patent is for 'the unexpired part of the term of the original patent.' See 35 U.S.C. 251. Accordingly, the above-identified reissue application is not eligible for Patent Term Extension or Adjustment under 35 U.S.C. 154(b)."

(E) For design applications, page three will state that "Design patents have a term measured from the issue date of the patent and the term remains the same length regardless of the time that the application for the design patent was pending. Since the above-identified application is an application for a design patent, the patent is not eligible for Patent Term Extension or Adjustment under 35 U.S.C. 154(b).

For more information about eighteen month publication, publication fees, and patent term adjustment, visit the USPTO Internet web site at www.uspto.gov.

§21.3　核准理由之內容與效力

　　審查委員認為說明核准理由會對整體審查程序更為完整時，可於核准通知中記載核准理由。對於審查委員在核准通知中提到的核准理由（Reason for Allowance），申請人若有不同意之處，可以於繳交領證費時一併提出意見（statement for commenting）。但申請人提出的意見本身，對於專利性或專利文字的解讀可能構成禁反言的約束。再者，審查委員收到申請人對於核准理由的意見後，沒有義務要再做回應。但是申請人意見將存卷記錄。

MPEP 1302.14 Reasons for Allowance

37 CFR 1.104 Nature of examination.

　　(e) Reasons for allowance. If the examiner believes that the record of the prosecution as a whole does not make clear his or her reasons for allowing a claim or claims, the examiner may set forth such reasoning. The reasons shall be incorporated into an Office action rejecting other claims of the application or patent under reexamination or be the subject of a separate communication to the applicant or patent owner. The applicant or patent owner may file a statement commenting on the reasons for allowance within such time as may be specified by the examiner. Failure by the examiner to respond to any statement commenting on reasons for allowance does not give rise to any implication.

I. < REASONS FOR ALLOWANCE

One of the primary purposes of 37 CFR 1.104(e) is to improve the quality and reliability of issued patents by providing a complete file history which should clearly reflect, as much as is reasonably possible, the reasons why the application was allowed. Such information facilitates evaluation of the scope and strength of a patent by the patentee and the public and may help avoid or simplify litigation of a patent.

The practice of stating the reasons for allowance is not new, and the rule merely formalizes the examiner's existing authority to do so and provides applicants or patent owners an opportunity to comment upon any such statement of the examiner.

It should be noted that the setting forth of reasons for allowance is not mandatory on the examiner's part. However, in meeting the need for the application file history to speak for itself, it is incumbent upon the examiner in exercising his or her responsibility to the public, to see that the file history is as complete as is reasonably possible.

When an application is finally acted upon and allowed, the examiner is expected to determine, at the same time, whether the reasons why the application is being allowed are evident from the record.

Prior to allowance, the examiner may also specify allowable subject matter and provide reasons for indicating such allowable subject matter in an Office communication.

In determining whether reasons for allowance should be recorded, the primary consideration lies in the first sentence of 37 CFR 1.104(e) which states:

If the examiner believes that the record of the prosecution as a whole does not make clear his or her reasons for allowing a claim or claims, the examiner may set forth such reasoning. (Emphasis added).

In most cases, the examiner's actions and the applicant's replies make evident the reasons for allowance, satisfying the "record as a whole" proviso of the rule. This is particularly true when applicant fully complies with 37 CFR 1.111 (b) and (c) and 37 CFR 1.133(b). Thus, where the examiner's actions clearly point out the reasons for rejection and the applicant's reply explicitly presents reasons why claims are patentable over the reference, the reasons for allowance are in all probability evident from the record and no statement should be necessary.

Conversely, where the record is not explicit as to reasons, but allowance is in order, then a logical extension of 37 CFR 1.111 and 1.133 would dictate that the examiner should make reasons of record and such reasons should be specific.

Where specific reasons are recorded by the examiner, care must be taken to ensure that statements of reasons for allowance (or indication of allowable subject matter) are accurate, precise, and do not place unwarranted interpretations, whether broad or narrow, upon the claims. The examiner should keep in mind the possible misinterpretations of his or her statement that may be made and its possible effects. Each statement should include at least (1) the major difference in the claims not found in the prior art of record, and (2) the reasons why that difference is considered to define patentably over the prior art if either of these reasons for allowance is not clear in the record. The statement is not intended to necessarily state all the reasons for allowance or all the details why claims are allowed and should not be written to specifically or impliedly state that all the reasons for allowance are set forth. Where the examiner has a large number of reasons for allowing a claim, it may suffice to state only the major or important reasons, being careful to so couch the statement. For example, a statement might start: "The primary reason for the allowance of the claims is the inclusion of the limitation in all the claims which is not found in the prior art references," with further amplification as necessary.

Stock paragraphs with meaningless or uninformative statements of the reasons for the allowance should not be used. It is improper to use a statement of reasons for allowance to attempt to narrow a claim by providing a special definition to a claim limitation which is argued by applicant, but not supported by a special definition in the description in cases where the ordinary meaning of the term in the prior art demonstrates that the claim remains unpatentable for the reasons of record, and where such claim narrowing is only tangential to patentability. Cf. Festo Corp. v. Shoketsu Kinzoku Kogyo Kabushiki Co., 535 U.S. 722, 741, 62 USPQ2d 1705, 1714 (2002). The statement of reasons for allowance by the examiner is intended to provide information equivalent to that contained in a file in which the examiner's Office actions and the applicant's replies make evident the examiner's reasons for allowing claims.

Examiners are urged to carefully carry out their responsibilities to see that the application file contains a complete and accurate picture of the Office's consideration of the patentability of the application.

Under the rule, the examiner must make a judgment of the individual record to determine whether or not reasons for allowance should be set out in that record. These guidelines, then, are intended to aid the examiner in making that judgment. They comprise illustrative examples as to applicability and appropriate content. They are not intended to be exhaustive.

§21.4 核准通知中的審查委員修正之內容與效力

審查委員在核准通知中需一併載明 USPTO 發證時將進行的修正，包含：

➢ 一般明顯的形式修正：例如文字或圖式之誤繕

➢ 之前審查委員與申請人／代理人面詢的達成的修正共識

若要求圖式修正，則需申請人準備與領證費一起主動提交，審查委員修正之效力僅直接修正說明書文字，USPTO 並不會主動修正圖式，申請人需自行依審查委員要求遞交圖式修正頁。

MPEP 1302.04 Examiner's Amendments and Changes

Except by formal examiner's amendment duly signed or as hereinafter provided, no corrections, erasures, or interlineations may be made in the body of written portions of the specification or any other paper filed in the application for patent. (See 37 CFR 1.121.)

If the application file is a paper file, an informal examiner's amendment may be used for the correction of the following obvious errors and omissions only in the body of the written portions of the specification and may only be made with pen by the examiner of the application who will then initial in the margin and assume full responsibility for the change:

(A) Misspelled words.

(B) Disagreement of a noun with its verb.

(C) Inconsistent "case" of a pronoun.

(D) Disagreement between a reference character as used in the description and on the drawing. The character may be corrected in the description but only when the examiner is certain of the propriety of the change.

(E) Correction of reversed figure numbers. Garrett v. Cox, 233 F.2d 343, 345, 110 USPQ 52, 54 (CCPA 1956).

(F) Other obvious minor grammatical errors such as misplaced or omitted commas, improper parentheses, quotation marks, etc.

(G) Obvious informalities in the application, other than the ones noted above, or of purely grammatical nature.

Informal examiner's amendments are not permitted if the application is an Image File Wrapper (IFW) application. Any amendment of an IFW application must be by way of a formal examiner's amendment or be an amendment made by the applicant.

For continuing applications filed under 37 CFR 1.53(b), where a reference to the parent application has been inadvertently omitted by the applicant, an examiner should not add a reference to the prior application without the approval of the applicant and a formal examiner's amendment since applicant may decide to delete the priority claim in the application filed under 37 CFR 1.53(b). Furthermore, a petition under 37 CFR 1.78 to accept an unintentionally delayed benefit claim may be required if the application is a utility or plant application filed on or after November 29, 2000. See MPEP § 201.11.

When correcting originally filed papers in applications with a paper application file wrapper, clean red ink must be used (not blue or black ink).

A formal examiner's amendment may be used to correct all other informalities in the body of the written portions of the specification as well as all errors and omissions in the claims. The formal examiner's amendment must be signed by the primary examiner, placed in the file and a copy sent to applicant. The changes specified in the amendment are entered by the technical support staff in the regular way. A formal examiner's amendment should include form paragraph 13.02 and form paragraph 13.02.01. Form paragraph 13.02.02 should be used if an extension of time is required.

¶ 13.02 Formal Examiner's Amendment

An examiner's amendment to the record appears below. Should the changes and/or additions be unacceptable to applicant, an amendment may be filed as provided by 37 CFR 1.312. To ensure consideration of such an amendment, it MUST be submitted no later than the payment of the issue fee.

§21.5　收到核准通知後請求修正

收到核准通知後，申請人可提出 37 CFR 1.312 核准後修正請求，此請求必須在繳領證費之前或同時。核准後的申請人提交的修正是否可被接受（entered），必須經由審查委員確認。一般形式問題需由 Primary examiner 批示，實質修正（例如更動請求項範圍）必須由 supervisory examiner 批示。申請人必須陳述：修正的理由、為何不需要進一步的審查、為何請求項可以核准、為何沒有在之前呈送該修正等等。

申請人在核准前就準備修正，但在發出核准通知後才送達 USPTO，由於該修正是在核准通知發下後才送入，因此其地位仍然比照 1.312 核准後的修正。是否被接受要等審查委員確認。

37 CFR § 1.312 Amendments after allowance.

No amendment may be made as a matter of right in an application after the mailing of the notice of allowance. Any amendment filed pursuant to this section must be filed before or with the payment of the issue fee, and may be entered on the recommendation of the primary examiner, approved by the Director, without withdrawing the application from issue.

MPEP 714.15 Amendment Received in Technology Center After Mailing of Notice of Allowance [R-3]

Where an amendment, even though prepared by applicant prior to allowance, does not reach the Office until after the notice of allowance has been mailed, such amendment has the status of one filed under 37 CFR 1.312. Its entry is a matter of grace. For discussion of amendments filed under 37 CFR 1.312, see MPEP 714.16 to MPEP 714.16(e).

If * the amendment is filed in the Office prior to the mailing * of the notice of allowance, but is received by the examiner after the mailing of the notice of allowance, it **>may also not be approved for entry. If the amendment is a supplemental reply filed when action is not suspended, such an amendment will not be approved for entry because supplemental replies are not entered as matter of right. See 37 CFR 1.111(a)(2) and MPEP § 714.03(a). If the amendment is a preliminary amendment, such an amendment may be disapproved under 37 CFR 1.115(b). See MPEP § 714.01(e). If the amendment is approved for entry, the examiner may enter the amendment and provide a supplemental notice of allowance, or withdraw the application from issue and provide an Office action.

The application will not be withdrawn from issue for the entry of an amendment that would reopen the prosecution if the Office action next preceding the notice of allowance closed the application to further amendment, i.e., by indicating the patentability of all of the claims, or by allowing some and finally rejecting the remainder.

After an applicant has been notified that the claims are all allowable, further prosecution of the merits of the application is a matter of grace and not of right. Ex parte Quayle, 25 USPQ 74, 1935 C.D. 11, 453 O.G. 213 (Comm'r Pat. 1935).

714.16 (a)-(e) Amendment After Notice of Allowance, 37 CFR 1.312 [R-3]

收到核准通知後，若請求項範圍在審查期間有做出修改，因而導致對應核准範圍的發明人（inventorship）改變，此時遞交發明人補充宣誓書（supplemental oath / declaration）以增減發明人，不算 1.312 的修正。核准領證時，可一併呈送發明人的補充宣誓書（supplemental oath / declaration）增減發明人。

MPEP 603.01 Supplemental Oath or Declaration Filed After Allowance [R-7]

Since the decision in Cutter Co. v. Metropolitan Electric Mfg. Co., 275 F. 158 (2d Cir. 1921), many supplemental oaths and declarations covering the claims in the application have been filed after the applications were allowed. Such oaths and declarations may be filed as a matter of right and when received they will be placed in the file by the Office of **>Data Management<, but their receipt will not be acknowledged to the party filing them. They should not be filed or considered as amendments under 37 CFR 1.312, since they make no change in the wording of the papers on file. See MPEP § 714.16.

§21.6　撤回核准

收到核准通知之後，撤回核准分為兩個階段：

◆ 繳交領證費之前
  ◆ 審查委員主動撤回：審查委員提出新的引證案或理由
  ◆ 申請人主動撤回：提出請願與費用
◆ 繳交領證費之後：
  ◆ 審查委員主動撤回（可辦理退領證費）
    ● USPTO 的錯誤
    ● 不正行為／不法行為（inequitable conduct/illegality）
    ● 請求項不可專利（unpatentability of claim(s)）
    ● 抵觸程序（interference）
  ◆ 申請人主動撤回：提出請願與費用，並說明理由
    ● 不可專利性（需有明確的說明）

● 提出 RCE（e.g.,提交 IDS 並要求審查委員考慮）
● 表明放棄（express abandonment）

MPEP 1308 Withdrawal From Issue [R-5]

37 CFR 1.313 Withdrawal from issue.

(a) Applications may be withdrawn from issue for further action at the initiative of the Office or upon petition by the applicant. To request that the Office withdraw an application from issue, applicant must file a petition under this section including the fee set forth in § 1.17(h) and a showing of good and sufficient reasons why withdrawal of the application from issue is necessary. A petition under this section is not required if a request for continued examination under § 1.114 is filed prior to payment of the issue fee. If the Office withdraws the application from issue, the Office will issue a new notice of allowance if the Office again allows the application.

(b) Once the issue fee has been paid, the Office will not withdraw the application from issue at its own initiative for any reason except:

(1) A mistake on the part of the Office;

(2) A violation of § 1.56 or illegality in the application;

(3) Unpatentability of one or more claims; or

(4) For interference.

(c) Once the issue fee has been paid, the application will not be withdrawn from issue upon petition by the applicant for any reason except:

(1) Unpatentability of one of more claims, which petition must be accompanied by an unequivocal statement that one or more claims are unpatentable, an amendment to such claim or claims, and an explanation as to how the amendment causes such claim or claims to be patentable;

(2) Consideration of a request for continued examination in compliance with § 1.114; or

(3) Express abandonment of the application. Such express abandonment may be in favor of a continuing application.

(d) A petition under this section will not be effective to withdraw the application from issue unless it is actually received and granted by the appropriate officials before the date of issue. Withdrawal of an application from issue after payment of the issue fee may not be effective to avoid publication of application information.

§21.7　繳納領證費的注意事項

　　繳交金額原則上參照核准通知所檢附的繳費單，但實際需繳交的金額視繳費當時的 USPTO 實際費率為準。若繳費當時，申請人已經從小實體變成大實體，則需繳納大實體的費用。若從大實體變為小實體，亦可聲明轉為小實體，繳納小實體費用。領證費三個月內必須繳納，原則上不可展延。若申請案已經被公開或將被公開，繳費單上會一併列出公開費。若未載明，則專利將以申請人名義(也就是發明人)發證。若要以受讓人(assignee)名義，例如公司，作為專利權人(patentee)發證，則需於繳費時領證繳費單上註明。(有無註明並不影響讓渡效力，只影響專利公告首頁上的資料)

MPEP 1306 Issue Fee [R-5]

The issue fee and any required publication fee are due 3 months from the date of the Notice of Allowance. The amount of the issue fee and any required publication fee are shown on the Notice of Allowance. The Notice of Allowance will also reflect any issue fee previously paid in the application. The issue fee due does not reflect a credit for any previously paid issue fee in the application. If an issue fee has previously been paid in the application as reflected in the Notice of Allowance, the return of Part B (Fee(s) Transmittal form) will be considered a request to reapply the previously paid issue fee toward the issue fee that is now due. For example, if the application was allowed and the issue fee paid, but applicant withdrew the application from issue and filed a Request for Continued Examination (RCE) and the application was later allowed, the Notice of Allowance will reflect an issue fee amount that is due ** and the issue fee that was previously paid. Had applicant filed a Continued Prosecution Application (CPA) instead of an RCE, the Notice of Allowance< would not reflect any issue fee paid before the CPA was filed because the issue fee was paid in a prior application. Note that because the amount of the fees(s) due is determined by the fees set forth in 37 CFR 1.18 which are in effect as of the date of submission of payment of the fees(s), the amount due at the time the fee(s) are paid may differ from the amount indicated on the Notice of Allowance. Accordingly, applicants are encouraged, at the time of submitting payment of the fees(s), to determine whether the amount of the issue fee due or any required publication fee has changed to avoid the patent lapsing for failure to pay the balance of the issue fee due (37 CFR 1.317) or becoming abandoned for failure to pay the publication fee. The amounts due under 35 U.S.C.

41(a) (i.e., the issue fee, but not the publication fee) are reduced by 50 per centum for small entities.

Applicants and their attorneys or agents are urged to use the Fee(s) Transmittal form (PTOL-85B) provided with the Notice of Allowance when submitting their payments. Unless otherwise directed, all post allowance correspondence should be addressed "Mail Stop Issue Fee."

Where it is clear that an applicant actually intends to pay the issue fee and required publication fee, but the proper fee payment is not made, for example, an incorrect issue fee amount is supplied, or a PTOL-85B Fee(s) Transmittal form is filed without payment of the issue fee, a general authorization to pay fees or a specific authorization to pay the issue fee, submitted prior to the mailing of a notice of allowance, will be allowed to act as payment of the correct issue fee. 37 CFR 1.311(b). In addition, where the deposit account information is added to the Fee(s) Transmittal form (PTOL-85B), but the check box authorizing that the deposit account be charged the issue fee is not checked, the deposit account will still be charged the required issue fee and any required publication fee.

Technology Center personnel should forward all post allowance correspondence to the Office of Initial Patent Examination (OIPE). The papers received by the OIPE will be scanned and matched with the appropriate application and the entire application will be forwarded to the appropriate Technology Center for processing.

The payment of the issue fee due may be simplified by using a U.S. Patent and Trademark Office Deposit Account or a credit card payment with form PTO-2038 for such a fee. See MPEP § 509. However, any such payment must be specifically authorized by reference to the "issue fee" or "fees due under 37 CFR 1.18."

The fee(s) due will be accepted from the applicant, assignee, or a registered attorney or agent, either of record or under 37 CFR 1.34*.

The Director has no authority to extend the time for paying the issue fee. Intentional failure to pay the issue fee within the 3 months permitted by 35 U.S.C. 151 does not amount to unavoidable or unintentional delay in making payment.

§21.8　請求延遲或加速發證

申請人可以提出請願請求延遲公告發證，若理由良好被接受，可以最多延遲一個月發證。若要加速發證，則沒有正式的程序。發證程序由 USPTO 的 Final Data Capture（FDC）單位處理，此單位負責將案件資料完成電子化並準備公告發證。FDC 平均的處理速度為 5 週。

MPEP 1306.01 Deferring Issuance of a Patent [R-5]

37 CFR 1.314 Issuance of patent.

If applicant timely pays the issue fee, the Office will issue the patent in regular course unless the application is withdrawn from issue (§ 1.313) or the Office defers issuance of the patent. To request that the Office defer issuance of a patent, applicant must file a petition under this section including the fee set forth in § 1.17(h) and a showing of good and sufficient reasons why it is necessary to defer issuance of the patent.

There is a public policy that the patent will issue in regular course once the issue fee is timely paid. 37 CFR 1.314. It has been the policy of the U.S. Patent and Trademark Office to defer issuance of a patent, upon request, for a period of up to 1 month only, in the absence of extraordinary circumstances or requirement of the regulations (e.g., 37 CFR 1.177) which would dictate a longer period. Situations like negotiation of licenses, time for filing in foreign countries, collection of data for filing a continuation-in-part application, or a desire for simultaneous issuance of related applications are not considered to amount to extraordinary circumstances.

A petition to defer issuance of a patent is not appropriate until the issue fee is paid. Issuance of a patent cannot be deferred after an allowed application receives a patent number and issue date unless the application is withdrawn from issue under 37 CFR 1.313(b) or (c). The petition to defer is considered at the time the petition is correlated with the application file before the appropriate deciding official (MPEP § 1002.02(b)). In order to facilitate consideration of a petition for deferment of issue, the petition should be firmly attached to the **>Fee(s)< Transmittal form (PTOL-85B) and clearly labeled as a Petition to Defer Issue; Attention: Office of Petitions.

§21.9 發證通知

原則上 FDC 處理完案件基本資料後，會先發出發證通知（Issue Notification）給申請人，上面會記載預計發證公告日與證書號。發證通知一般會在發證前 2-3 週發出。另外，美國專利固定於每週二發證公告。

§21.10 美國專利維持費的繳納

有別於多數國家專利年費為逐年繳納，美國實用專利核准之後，乃每隔 3.5、4、4 年繳納維持費。美國實用專利維持費乃由發證公告日（issue date）起算，每 3.5、7.5、11.5 年各繳納一次，之後不論剩餘期間還有多長都不用再繳，因此最多三次。另外，若專利權人為小實體，維持費規費有減半優惠。美國新式樣與植物專利核准後不用繳納年費。

維持費並非隨時可繳，也不可預繳，預設於維持費繳納期限前的前六個月開放繳納，期限後的六個月結束滯納。

例：發證日為 2006.6.1，需在 2009.6.1-2009.12.1 之間繳納第 3.5 年次的年費。

● 維持費逾期：超過繳納期限的 6 個月期間，加繳納滯納金（USD 130，小實體亦可減半）

● 例：2009.12.2-2010.5.31 之間繳納維持費及滯納金

原則上專利權人或其授權人均可繳費。但由於 USPTO 並不檢核繳費人的身份，因此實務上任何人均可繳納維持費，無需透過原美國代理人。許多公司委託專業年費公司代為繳納。也有許多美國事務所不提供繳納維持費的服務。

若他人先誤繳了維持費之後，無論第二次的繳納是否確為專利權人所繳，USPTO 將不會再接受第二次的繳納。因此，必須要求第一繳納人辦理退費後，才能再為繳納。原則上如為他人誤繳，退費後 USPTO 會通知該專利案的年費聯繫人補繳維持費，因此不生逾限繳納之問題。

---

37 CFR § 1.362 Time for payment of maintenance fees.

(a) Maintenance fees as set forth in §§ 1.20(e) through (g) are required to be paid in all patents based on applications filed on or after December 12, 1980, except as noted in paragraph (b) of this section, to maintain a patent in force beyond 4, 8 and 12 years after the date of grant.

(b) Maintenance fees are not required for any plant patents or for any design patents. Maintenance fees are not required for a reissue patent if the patent being reissued did not require maintenance fees.

(c) The application filing dates for purposes of payment of maintenance fees are as follows:

(1) For an application not claiming benefit of an earlier application, the actual United States filing date of the application.

(2) For an application claiming benefit of an earlier foreign application under 35 U.S.C. 119, the United States filing date of the application.

(3) For a continuing (continuation, division, continuation-in-part) application claiming the benefit of a prior patent application under 35 U.S.C. 120, the actual United States filing date of the continuing application.

(4) For a reissue application, including a continuing reissue application claiming the benefit of a reissue application under 35 U.S.C. 120, the United States filing date of the original non-reissue application on which the patent reissued is based.

(5) For an international application which has entered the United States as a Designated Office under 35 U.S.C. 371, the international filing date granted under Article 11(1) of the Patent Cooperation Treaty which is considered to be the United States filing date under 35 U.S.C. 363.

(d) Maintenance fees may be paid in patents without surcharge during the periods extending respectively from:

(1) 3 years through 3 years and 6 months after grant for the first maintenance fee,

(2) 7 years through 7 years and 6 months after grant for the second maintenance fee, and

(3) 11 years through 11 years and 6 months after grant for the third maintenance fee.

(e) Maintenance fees may be paid with the surcharge set forth in § 1.20(h) during the respective grace periods after:

(1) 3 years and 6 months and through the day of the 4th anniversary of the grant for the first maintenance fee.

(2) 7 years and 6 months and through the day of the 8th anniversary of the grant for the second maintenance fee, and

(3) 11 years and 6 months and through the day of the 12th anniversary of the grant for the third maintenance fee.

(f) If the last day for paying a maintenance fee without surcharge set forth in paragraph (d) of this section, or the last day for paying a maintenance fee with surcharge set forth in paragraph (e) of this section, falls on a Saturday, Sunday, or a federal holiday within the District of Columbia, the maintenance fee and any necessary surcharge may be paid under paragraph (d) or paragraph (e) respectively on the next succeeding day which is not a Saturday, Sunday, or Federal holiday.

(g) Unless the maintenance fee and any applicable surcharge is paid within the time periods set forth in paragraphs (d), (e) or (f) of this section, the patent will expire as of the end of the grace period set forth in paragraph (e) of this section. A patent which expires for the failure to pay the maintenance fee will expire at the end of the same date (anniversary date) the patent was granted in the 4th, 8th, or 12th year after grant.

(h) The periods specified in §§1.362(d) and (e) with respect to a reissue application, including a continuing reissue application thereof, are counted from the date of grant of the original non-reissue application on which the reissued patent is based.

§21.11 逾期未繳納維持費的效力與補救方式

超過滯納期限仍未繳交維持費：自發證日第 4 年、第 8 年或第 12 年起視為專利權過期。如果要辦理恢復，必須提出請願（petition）及足額維持費與請求恢復費用：證明不可避免的延誤（unavoidable delay），或是非蓄意之延誤（unintentional delay）。一般兩年之內的非蓄意之延誤多半會被接受，但超過兩年，將被從嚴審核。

---

37 CFR § 1.378 Acceptance of delayed payment of maintenance fee in expired patent to reinstate patent.

(a) The Director may accept the payment of any maintenance fee due on a patent after expiration of the patent if, upon petition, the delay in payment of the maintenance fee is shown to the satisfaction of the Director to have been unavoidable (paragraph (b) of this section) or unintentional (paragraph (c) of this section) and if the surcharge required by §1.20(i) is paid as a condition of accepting payment of the maintenance fee. If the Director accepts payment of the maintenance fee upon petition, the patent shall be considered as not having expired, but will be subject to the conditions set forth in 35 U.S.C. 41(c)(2).

(b) Any petition to accept an unavoidably delayed payment of a maintenance fee filed under paragraph (a) of this section must include:

(1) The required maintenance fee set forth in §1.20 (e) through (g);

(2) The surcharge set forth in §1.20(i)(1); and

(3) A showing that the delay was unavoidable since reasonable care was taken to ensure that the maintenance fee would be paid timely and that the petition was filed promptly after the patentee was notified of, or otherwise became aware of, the expiration of the patent. The showing must enumerate the steps taken to ensure timely payment of the maintenance fee, the date and the manner in which patentee became aware of the expiration of the patent, and the steps taken to file the petition promptly.

(c) Any petition to accept an unintentionally delayed payment of a maintenance fee filed under paragraph (a) of this section must be filed within twenty-four months after the six-month grace period provided in §1.362(e) and must include:

(1) The required maintenance fee set forth in §1.20 (e) through (g);

(2) The surcharge set forth in §1.20(i)(2); and

(3) A statement that the delay in payment of the maintenance fee was unintentional.

(d) Any petition under this section must be signed by an attorney or agent registered to practice before the Patent and Trademark Office, or by the patentee, the assignee, or other party in interest.

(e) Reconsideration of a decision refusing to accept a maintenance fee upon petition filed pursuant to paragraph (a) of this section may be obtained by filing a petition for reconsideration within two months of, or such other time as set in the decision refusing to accept the delayed payment of the maintenance fee. Any such petition for reconsideration must be accompanied by the petition fee set forth in §1.17(f). After the decision on the petition for reconsideration, no further reconsideration or review of the matter will be undertaken by the Director. If the delayed payment of the maintenance fee is not accepted, the maintenance fee and the surcharge set forth in §1.20(i) will be refunded following the decision on the petition for reconsideration, or after the expiration of the time for filing such a petition for reconsideration, if none is filed. Any petition fee under this section will not be refunded unless the refusal to accept and record the maintenance fee is determined to result from an error by the Patent and Trademark Office.

第 22 章　專利說明書之修正與棄權聲明書

目　錄

§22.1　美國專利發証公告之後的修正方式及其修正限制

美國專利發証公告之後，可以主動提出修正，也可能因為其他情況導致修正，但各有限制。主要有以下幾種修正樣態：

- 請求證書修正（request of certificate of correction）：更正申請人或 USPTO 之錯誤或修正發明人
- 提出棄權聲明書（disclaimer），主動放棄一個或多個請求項
- 再發證申請案（reissue application）（詳見第 23 章）
- 請求再審查（*Ex parte/Inter partes* reexamination of patent）（詳見第 24 章）
- 抵觸程序（interference proceedings）（詳見第 25 章）

上述修正都只能更正基於善良意圖之錯誤（correct good faith errors）。不能導入新事項（new matter）。美國所有種類的專利均可循上述方式請求修正。

MPEP 1400.01 Introduction [R-2]

A patent may be corrected or amended in four ways, namely:

(A) by reissue,

(B) by the issuance of a certificate of correction which becomes a part of the patent,

(C) by disclaimer, and

(D) by reexamination.

The first three ways are discussed in this chapter while the fourth way (reexamination) is discussed in MPEP Chapter 2200 for ex parte reexamination and MPEP Chapter 2600 for inter partes reexamination.

§22.2　請求證書修正

請求證書修正（request of certificate of correction）僅能修正不影響權利範圍之形式上（formality）錯誤或申請資料的錯誤。不論源於 USPTO 的錯誤或申請人的錯誤，均可請求修正之。

§22.3 請求證書修正,源於 USPTO 之錯誤

若為 USPTO 之錯誤,下列人士可請求證書修正(request of certificate of correction):

- 專利權人
- 專利權之受讓人
- USPTO 自行發現
- 任意第三人:若為第三人向 USPTO 提出請求,USPTO 沒有義務要考慮,且需先通知專利權人或其登記在案之代理人

若肇因於 USPTO 疏失之錯誤,專利權人不需繳納規費,即可提出修正之請求。一般可以修正非實質問題的任何項目。例如:內容補回漏列、錯列之請求項或局部說明書、圖式、優先權日號等等,任何肇因於 USPTO 疏失的錯誤均可申請更正。另外,若專利權人欲行使權利或提起侵權訴訟前,應審慎確認專利內容是否正確,若有錯誤,應先提出修正。

MPEP 1480 Certificates of Correction - Office Mistake [R-3]

35 U.S.C. 254 Certificate of correction of Patent and Trademark Office mistake.

Whenever a mistake in a patent, incurred through the fault of the Patent and Trademark Office, is clearly disclosed by the records of the Office, the Director may issue a certificate of correction stating the fact and nature of such mistake, under seal, without charge, to be recorded in the records of patents. A printed copy thereof shall be attached to each printed copy of the patent, and such certificate shall be considered as part of the original patent. Every such patent, together with such certificate, shall have the same effect and operation in law on the trial of actions for causes thereafter arising as if the same had been originally issued in such corrected form. The Director may issue a corrected patent without charge in lieu of and with like effect as a certificate of correction.

37 CFR 1.322 Certificate of correction of Office mistake.

(a)

(1) The Director may issue a certificate of correction pursuant to 35 U.S.C. 254 to correct a mistake in a patent, incurred through the fault of the Office, which mistake is clearly disclosed in the records of the Office:

(i) At the request of the patentee or the patentee's assignee;

(ii) Acting sua sponte for mistakes that the Office discovers; or

(iii) Acting on information about a mistake supplied by a third party.

(2)

(i) There is no obligation on the Office to act on or respond to a submission of information or request to issue a certificate of correction by a third party under paragraph (a)(1)(iii) of this section.

(ii) Papers submitted by a third party under this section will not be made of record in the file that they relate to nor be retained by the Office.

(3) If the request relates to a patent involved in an interference, the request must comply with the requirements of this section and be accompanied by a motion under § 41.121(a)(2) or § 41.121(a)(3) of this title.

(4) The Office will not issue a certificate of correction under this section without first notifying the patentee (including any assignee of record) at the correspondence address of record as specified in § 1.33(a) and affording the patentee or an assignee an opportunity to be heard.

(b) If the nature of the mistake on the part of the Office is such that a certificate of correction is deemed inappropriate in form, the Director may issue a corrected patent in lieu thereof as a more appropriate form for certificate of correction, without expense to the patentee.

Mistakes incurred through the fault of the Office may be the subject of Certificates of Correction under 37 CFR 1.322. The Office, however, has discretion under 35 U.S.C. 254 to decline to issue a Certificate of Correction even though an Office mistake exists. If Office mistakes are of such a nature that the meaning intended is obvious from the context, the Office may decline to issue a certificate and merely place the correspondence in the patented file, where it serves to call attention to the matter in case any question as to it subsequently arises. Such is the case, even where a correction is requested by the patentee or patentee's assignee.

In order to expedite all proper requests, a Certificate of Correction should be requested only for errors of consequence. Instead of a request for a Certificate of Correction, letters making errors of record should be utilized whenever possible. Thus, where errors are of a minor typographical nature, or are readily apparent to one skilled in the art, a letter making the error(s) of record can be submitted in lieu of a request for a Certificate of Correction. There is no fee for the submission of such a letter.

It is strongly advised that the text of the correction requested be submitted on a Certificate of Correction form, PTO/SB/44 (also referred to as PTO 1050). Submission of this form in duplicate is not necessary. The location of the error in the printed patent should be identified on form PTO/SB/44 by column and line number or claim and line number. See MPEP § 1485 for a discussion of the preparation and submission of a request for a Certificate of Correction.

I. THIRD PARTY INFORMATION ON MISTAKES IN PATENT

Third parties do not have standing to demand that the Office issue, or refuse to issue, a Certificate of Correction. See Hallmark Cards, Inc. v. Lehman, 959 F. Supp. 539, 543-44, 42 USPQ2d 1134, 1138 (D.D.C. 1997). 37 CFR 1.322(a)(2) makes it clear that third parties do not have standing to demand that the Office act on, respond to, issue, or refuse to issue a Certificate of Correction. The Office is, however, cognizant of the need for the public to have correct information about published patents and may therefore accept information about mistakes in patents from third parties. 37 CFR 1.322(a)(1)(iii). Where appropriate, the Office may issue certificates of correction based on information supplied by third parties, whether or not such information is accompanied by a specific request for issuance of a Certificate of Correction.

While third parties are permitted to submit information about mistakes in patents which information will be reviewed, the Office need not act on that information nor deny any accompanying request for issuance of a Certificate of Correction. Accordingly, a fee for submission of the information by a third party has not been imposed. The Office may, however, choose to issue a Certificate of Correction on its own initiative based on the information supplied by a third party, if it desires to do so. Regardless of whether the third party information is acted upon, the information will not be made of record in the file that it relates to, nor be retained by the Office. 37 CFR 1.322(a)(2)(ii).

When such third party information (about mistakes in patents) is received by the Office, the Office will not correspond with third parties about the information they submitted either (1) to inform the third parties of whether it intends to issue a Certificate of Correction, or (2) to issue a denial of any request for issuance of a Certificate of Correction that may accompany the information. The Office will confirm to the party submitting such information that the Office has in fact received the information if a stamped, self-addressed post card has been submitted. See MPEP § 503.

II. PUBLICATION IN THE OFFICIAL GAZETTE

Each issue of the Official Gazette (patents section) numerically lists all United States patents having Certificates of Correction. The list appears under the heading "Certificates of Correction for the week of (date)."

§22.4　請求證書修正，源於申請人之錯誤

專利權人或專利權受讓人可以繳交規費（無小個體減半優惠）並提出修正請求。

USPTO 在檢核肇因於申請人之錯誤而請求證書修正時，必須確認該請求內容符合以下兩種條件：

I.　該錯誤本質上必須為 1）筆誤（of a clerical nature）、2）印刷的（of a typographical nature）或者 3）一些非重要性的錯誤（a mistake of minor character）。以及

II.　該更正本身並無新事項（new matter）或需要實質審查。

如果不能同時符合上述兩種條件，則需考慮藉由申請再發證（reissue）案更正。

---

35 U.S.C. 255 Certificate of correction of applicant's mistake.

Whenever a mistake of a clerical or typographical nature, or of minor character, which was not the fault of the Patent and Trademark Office, appears in a patent and a showing has been made that such mistake occurred in good faith, the Director may, upon payment of the required fee, issue a certificate of correction, if the correction does not involve such changes in the patent as would constitute new matter or would require reexamination. Such patent, together with the certificate, shall have the same effect and operation in law on the trial of actions for causes thereafter arising as if the same had been originally issued in such corrected form.

---

37 CFR 1.323 Certificate of correction of applicant's mistake.

The Office may issue a certificate of correction under the conditions specified in 35 U.S.C. 255 at the request of the patentee or the patentee's assignee, upon payment of the fee set forth in § 1.20(a). If the request relates to a patent involved in an interference, the request must comply with the requirements of this section and be accompanied by a motion under § 41.121(a)(2) or §　41.121(a)(3) of this title.

37 CFR 1.323 relates to the issuance of Certificates of Correction for the correction of errors which were not the fault of the Office. Mistakes in a patent which are not correctable by Certificate of Correction may be correctable via filing a reissue application (see MPEP §

1401 - § 1460). See Novo Industries, L.P. v. Micro Molds Corporation, 350 F.3d 1348, 69 USPQ2d 1128 (Fed. Cir. 2003) (The Federal Circuit stated that when Congress in 1952 defined USPTO authority to make corrections with prospective effect, it did not deny correction authority to the district courts. A court, however, can correct only if "(1) the correction is not subject to reasonable debate based on consideration of the claim language and the specification and (2) the prosecution history does not suggest a different interpretation......").

In re Arnott, 19 USPQ2d 1049, 1052 (Comm'r Pat. 1991) specifies the criteria of 35 U.S.C. 255 (for a Certificate of Correction) as follows:

Two separate statutory requirements must be met before a Certificate of Correction for an applicant's mistake may issue. The first statutory requirement concerns the nature, i.e., type, of the mistake for which a correction is sought. The mistake must be:

(1) of a clerical nature,

(2) of a typographical nature, or

(3) a mistake of minor character.

The second statutory requirement concerns the nature of the proposed correction. The correction must not involve changes which would:

(1) constitute new matter or

(2) require reexamination.

If the above criteria are not satisfied, then a Certificate of Correction for an applicant's mistake will not issue, and reissue must be employed as the vehicle to "correct" the patent. Usually, any mistake affecting claim scope must be corrected by reissue.

A mistake is not considered to be of the "minor" character required for the issuance of a Certificate of Correction if the requested change would materially affect the scope or meaning of the patent. See also MPEP § 1412.04 as to correction of inventorship via certificate of correction or reissue.

The fee for providing a correction of applicant's mistake, other than inventorship, is set forth in 37 CFR 1.20(a). The fee for correction of inventorship in a patent is set forth in 37 CFR 1.20(b).

§22.5 請求證書修正以更正專利權受讓人

美國專利公告版上的受讓人（Assignee）與地址是以領證繳費單（Fees Transmittal Form portion (PTOL-85B)）上所填的資料為準。若未填寫，只會以申請人名義(亦即發明人)公告。任何在繳納領證費之後才要求列出受讓人或者更正受讓人者，必須在專利發證前已提交其讓渡文件、提出修正請求與繳納規費（37 CFR 1.20(a)請求更正費與 37 CFR 1.17(i)處理費，均無小實體減半優惠）。

MPEP 1481.01 Correction of Assignees' Names [R-3]

The Fee(s) Transmittal Form portion (PTOL-85B) of the Notice of Allowance provides a space (item 3) for assignment data which should be completed in order to comply with 37 CFR 3.81. Unless an assignee's name and address are identified in the appropriate space for specifying the assignee, (i.e., item 3 of the Fee(s) Transmittal Form PTOL-85B), the patent will issue to the applicant. Assignment data printed on the patent will be based solely on the information so supplied.

Any request for the issuance of an application in the name of the assignee submitted after the date of payment of the issue fee, and any request for a patent to be corrected to state the name of the assignee must:

(A) state that the assignment was submitted for recordation as set forth in 37 CFR 3.11 before issuance of the patent;

(B) include a request for a certificate of correction under 37 CFR 1.323 along with the fee set forth in 37 CFR 1.20(a); and

(C) include the processing fee set forth in 37 CFR 1.17(i).

§22.6 請求證書修正以更正發明人資料

如果是發明人的名字或資料誤繕，可以檢附相關宣誓書或證明資料申請更正。而在沒有欺騙意圖的前提下，可以藉由此程序申請增加、減少發明人的數目或順序。

增減發明人詳細的處理的方式可參考第 2 章內容。原則上必須要所有發明人的對於增／減發明人之同意書以及專利權受讓人之同意增／減發明人之同意書。新增發明人之宣誓書與讓渡書（如果有專利權受讓人）。另外必須繳納§ 1.20(b)處理費（無小實體減半優惠）。

---

35 U.S.C. 256 Correction of named inventor.

Whenever through error a person is named in an issued patent as the inventor, or through error an inventor is not named in an issued patent and such error arose without any deceptive intention on his part, the Director may, on application of all the parties and assignees, with proof of the facts and such other requirements as may be imposed, issue a certificate correcting such error.

The error of omitting inventors or naming persons who are not inventors shall not invalidate the patent in which such error occurred if it can be corrected as provided in this section. The court before which such matter is called in question may order correction of the patent on notice and hearing of all parties concerned and the Director shall issue a certificate accordingly.

In requesting the Office to effectuate a court order correcting inventorship in a patent pursuant to 35 U.S.C. 256, a copy of the court order and a Certificate of Correction under 37 CFR 1.323 should be submitted to the Certificates of Corrections Branch.

37 CFR 1.324 Correction of inventorship in patent, pursuant to 35 U.S.C. 256.

(a) Whenever through error a person is named in an issued patent as the inventor, or through error an inventor is not named in an issued patent and such error arose without any deceptive intention on his or her part, the Director, pursuant to 35 U.S.C. 256, may, on application of all the parties and assignees, or on order of a court before which such matter is called in question, issue a certificate naming only the actual inventor or inventors. A petition to correct inventorship of a patent involved in an interference must comply with the requirements of this section and must be accompanied by a motion under § 41.121(a)(2) or § 41.121(a)(3) of this title.

(b) Any request to correct inventorship of a patent pursuant to paragraph (a) of this section must be accompanied by:

(1) Where one or more persons are being added, a statement from each person who is being added as an inventor that the inventorship error occurred without any deceptive intention on his or her part;

(2) A statement from the current named inventors who have not submitted a statement under paragraph (b)(1) of this section either agreeing to the change of inventorship or stating that they have no disagreement in regard to the requested change;

(3) A statement from all assignees of the parties submitting a statement under paragraphs (b)(1) and (b)(2) of this section agreeing to the change of inventorship in the patent, which statement must comply with the requirements of § 3.73(b) of this chapter; and

(4) The fee set forth in § 1.20(b).

(c) For correction of inventorship in an application, see §§ 1.48 and 1.497.

(d) In a contested case before the Board of Patent Appeals and Interferences under part 41, subpart D, of this title, a request for correction of a patent must be in the form of a motion under § 41.121(a)(2) or § 41.121(a)(3) of this title.

The petition to correct inventorship under 37 CFR 1.324 must include the statements and fee required by 37 CFR 1.324(b).

Under 37 CFR 1.324(b)(1), a statement is required from each person who is being added as an inventor that the inventorship error occurred without any deceptive intention on their part. In order to satisfy this, a statement such as the following is sufficient:

"The inventorship error of failing to include John Smith as an inventor of the patent occurred without any deceptive intention on the part of John Smith."

Nothing more is required. The examiner will determine only whether the statement contains the required language; the examiner will not make any comment as to whether or not it appears that there was in fact deceptive intention (see MPEP § 2022.05).

Under 37 CFR 1.324(b)(2), all current inventors who did not submit a statement under 37 CFR 1.324(b)(1) must submit a statement either agreeing to the change of inventorship, or stating that they have no disagreement with regard to the requested change. "Current inventors" include the inventor(s) being retained as such and the inventor(s) to be deleted. These current inventors need not make a statement as to whether the inventorship error occurred without deceptive intention.

If an inventor is not available, or refuses, to submit a statement, the assignee of the patent may wish to consider filing a reissue application to correct inventorship, because the inventor's statement is not required for a non-broadening reissue application to correct inventorship. See MPEP § 1412.04.

Under 37 CFR 1.324(b)(3), a statement is required from the assignee(s) of the patent agreeing to the change of inventorship in the patent. The assignee statement agreeing to the change of inventorship must be accompanied by a proper statement under 37 CFR 3.73(b) establishing ownership, unless a proper 37 CFR 3.73(b) statement is already in the file. See MPEP § 324 as to the requirements of a statement under 37 CFR 3.73(b).

While a request under 37 CFR 1.48 is appropriate to correct inventorship in a nonprovisional application, a petition under 37 CFR 1.324 is the appropriate vehicle to correct inventorship in a patent. If a request under 37 CFR 1.48(a), (b), or (c) is inadvertently filed in a patent, the request may be treated as a petition under 37 CFR 1.324, and if it is grantable, form paragraph 10.14 set forth below should be used.

Similarly, if a request under 37 CFR 1.48(a), (b), or (c) is filed in a pending application but not acted upon until after the application becomes a patent, the request may be treated as a petition under 37 CFR 1.324, and if it is grantable, form paragraph 10.14 set forth below should be used.

The statutory basis for correction of inventorship in a patent under 37 CFR 1.324 is 35 U.S.C. 256. It is important to recognize that 35 U.S.C. 256 is stricter than 35 U.S.C. 116, the statutory basis for corrections of inventorship in applications under 37 CFR 1.48. 35 U.S.C. 256 requires "on application of all the parties and assignees," while 35 U.S.C. 116 does not have the same requirement. Under 35 U.S.C. 116 and 37 CFR 1.48, waiver requests under 37 CFR 1.183 may be submitted (see, e.g., MPEP § 201.03, under the heading "Statement of Lack of Deceptive Intention"). This is not possible under 35 U.S.C. 256 and 37 CFR 1.324. In correction of inventorship in a nonprovisional application under 37 CFR 1.48(a), the requirement for a statement by each originally named inventor may be waived pursuant to 37 CFR 1.183; however, correction of inventorship in a patent under 37 CFR 1.324 requires petition of all the parties, i.e., originally named inventors and assignees, in accordance with statute (35 U.S.C. 256) and thus the requirement cannot be waived. Correction of inventorship requests under 37 CFR 1.324 should be directed to the Supervisory Patent Examiner whose unit handles the subject matter of the patent. Form paragraphs 10.13 through 10.18 may be used.

§22.7　請求證書修正以更正優先權主張

　　請求證書修正程序原則上只能更正優先權資料中的一些明顯的疏誤，例如優先日、號或國別等之誤繕或誤植或者增加主張仍審查中（pending）的美國前案優先權。但是無法用於新增國內外優先權主張或者更改國內外優先權主張，該等更改必須以申請再發證案（reissue）進行。

MPEP 1481.03 Correction of 35 U.S.C. 119 and 35 U.S.C. 120 Benefits [R-7]

＊＊＊＊＊＊＊＊＊＊＊＊

Under certain conditions specified below, a Certificate of Correction can be used, with respect to 35 U.S.C. 120 and 119(e) priority, to correct:

(A) the failure to make reference to a prior copending application pursuant to 37 CFR 1.78(a)(2) and (a)(4); or

(B) an incorrect reference to a prior copending application pursuant to 37 CFR 1.78(a)(2) and (a)(4).

For all situations other than where priority is based upon 35 U.S.C. 365(c), the conditions are as follows:

(A) for 35 U.S.C. 120 priority, all requirements set forth in 37 CFR 1.78(a)(1) must have been met in the application which became the patent to be corrected;

(B) for 35 U.S.C. 119(e) priority, all requirements set forth in 37 CFR 1.78(a)(3) must have been met in the application which became the patent to be corrected; and

(C) it must be clear from the record of the patent and the parent application(s) that priority is appropriate. See MPEP § 201.11 for requirements under 35 U.S.C. 119(e) and 120.

Where 35 U.S.C. 120 and 365(c) priority based on an international application is to be asserted or corrected in a patent via a Certificate of Correction, the following conditions must be satisfied:

(A) all requirements set forth in 37 CFR 1.78(a)(1) must have been met in the application which became the patent to be corrected;

(B) it must be clear from the record of the patent and the parent application(s) that priority is appropriate (see MPEP § 201.11); and

(C) the patentee must submit with the request for the certificate copies of documentation showing designation of states and any other information needed to make it clear from the record that the 35 U.S.C. 120 priority is appropriate. See MPEP § 201.13(b) as to the requirements for 35 U.S.C. 120 priority based on an international application.

If all the above-stated conditions are satisfied, a Certificate of Correction can be used to amend the patent to make reference to a prior copending application, or to correct an incorrect reference to the prior copending application. Note In re Schuurs, 218 USPQ 443 (Comm'r Pat. 1983) which suggests that a Certificate of Correction is an appropriate remedy for correcting, in a patent, reference to a prior copending application. Also, note In re Lambrech, 202 USPQ 620 (Comm'r Pat. 1976), citing In re Van Esdonk, 187 USPQ 671 (Comm'r Pat. 1975).

If any of the above-stated conditions is not satisfied, the filing of a reissue application (see MPEP § 1401 - § 1460) would be appropriate to pursue the desired correction of the patent.

＊＊＊＊＊＊＊＊＊＊＊

§22.8　提交請求證書修正後的程序

申請人提出請求時無須繳回原證書。待 USPTO 審核批准通過後，會發出證書修正通知（Certificate of Correction）並公告。所修正之內容會附載於在新公告的專利說明書之最後一頁，並會羅列更正項目。或者查詢官方公告網頁：http://www.uspto.gov/news/og/index.jsp

如果 USPTO 沒有批准全部的修正請求，只有局部批准或者不准，將會先通知請求人並載明原因。

MPEP 1485 Handling of Request for Certificates of Correction [R-7]

＊＊＊＊＊＊＊＊＊＊＊

Requests for Certificates of Correction will be forwarded to the Certificate of Correction Branch of the Office of Patent Publication, where they will be listed in a permanent record book.

If the patent is involved in an interference, a Certificate of Correction under 37 CFR 1.324 will not be issued unless a corresponding motion under 37 CFR 41.121(a)(2) or 41.121(a)(3) has been granted by the administrative patent judge. Otherwise, determination as to whether an error has been made, the responsibility for the error, if any, and whether the error is of such a nature as to justify the issuance of a Certificate of Correction will be made by the Certificate of Correction Branch. If a report is necessary in making such determination, the case will be forwarded to the appropriate group with a request that the report be furnished. If no certificate is to issue, the party making the request is so notified and the request, report, if any, and copy of the communication to the person making the request are placed in the file wrapper (for a paper file) or entered into the file history (for an IFW file), and entered into the "Contents" for the file by the Certificate of Correction Branch. The case is then returned to the patented files. If a certificate is to issue, it will be prepared and forwarded to the person making the request by the Office of Patent Publication. In that case, the request, the report, if any, and a copy of the letter transmitting the Certificate of Correction to the person making the request will be placed in the file wrapper (for a paper file) or entered into the file history (for an IFW file), and entered into the "Contents" for the file.

Applicants, or their attorneys or agents, are urged to submit the text of the correction on a special Certificate of Correction form, PTO/SB/44 (also referred to as Form PTO-1050), which can serve as the camera copy for use in direct offset printing of the Certificate of Correction.

Where only a part of a request can be approved, or where the Office discovers and includes additional corrections, the appropriate alterations are made on the form PTO/SB/44 by the Office. The patentee is notified of the changes on the Notification of Approval-in-part form PTOL-404. The certificate is issued approximately 6 weeks thereafter.

Form PTO/SB/44 should be used exclusively regardless of the length or complexity of the subject matter. Intricate chemical formulas or page of specification or drawings may be reproduced and mounted on a blank copy of PTO/SB/44. Failure to use the form has frequently delayed issuance because the text must be retyped by the Office onto a PTO/SB/44.

The exact page and line number where the errors occur in the application file should be identified on the request. However, on form PTO/SB/44, only the column and line number in the printed patent should be used.

The patent grant should be retained by the patentee. The Office does not attach the Certificate of Correction to patentee's copy of the patent. The patent grant will be returned to the patentee if submitted.

Below is a sample form illustrating a variety of corrections and the suggested manner of setting out the format. Particular attention is directed to:

(A) Identification of the exact point of error by reference to column and line number of the printed patent for changes in the specification or to claim number and line where a claim is involved.

(B) Conservation of space on the form by typing single space, beginning two lines down from the printed message.

(C) Starting the correction to each separate column as a sentence, and using semicolons to separate corrections within the same column, where possible.

(D) Leaving a two-inch space blank at bottom of the last sheet for the signature of the attesting officer.

(E) Using quotation marks to enclose the exact subject matter to be deleted or corrected; using double hyphens (-- --) to enclose subject matter to be added, except for formulas.

(F) Where a formula is involved, setting out only that portion thereof which is to be corrected or, if necessary, pasting a photocopy onto form PTO/SB/44.

＊＊＊＊＊＊＊＊＊＊＊

ELECTRONIC PUBLICATION OF CERTIFICATES OF CORRECTION WITH LATER LISTING IN THE OFFICIAL GAZETTE

Effective August 2001, the U.S. Patent and Trademark Office (USPTO) publishes on the USPTO web site at http://www.uspto.gov/web/patents/certofcorrect a listing by patent number of the patents for which certificates of correction are being issued.

The USPTO is now automating the publication process for certificates of correction. This new process will result in certificates of correction being published quicker electronically on the USPTO's web site as compared to their paper publication and the listing of the certificates of correction in the Official Gazette. Under the newly automated process, each issue of certificates of correction will be electronically published on the USPTO web site at http://www. uspto.gov/web/patents/certofcorrect, and will also subsequently be listed in the Official Gazette (and in the Official Gazette Notices posted at http://www.uspto.gov/web/offices/com/sol/og) approximately three weeks thereafter. The listing of certificates of correction in the Official Gazette will include the certificate's date of issuance.

On the date on which the listing of certificates of correction is electronically published on the USPTO web site: (A) the certificate of correction will be entered into the file wrapper of a paper-file patent, or entered into the file history of an IFW-file patent and will be available to the public; (B) a printed copy of the certificate of correction will be mailed to the patentee or the patent's assignee; and (C) an image of the printed certificate of correction will be added to the image of the patent on the patent database at **>http://www.uspto.gov/patft<. Dissemination of all other paper copies of the certificate of correction will occur shortly thereafter.

The date on which the USPTO makes the certificate of correction available to the public (e.g., by adding the certificate of correction to the file wrapper/file history) will be regarded as the date of issuance of the certificate of correction, not the date of the certificate of correction appearing in the Official Gazette. (For IFW processing, see IFW Manual.) Certificates of correction published in the above-described manner will provide the public with prompt notice and access, and this is consistent with the legislative intent behind the American Inventors Protection Act of 1999. See 35 U.S.C. 10(a) (authorizing the USPTO to publish in electronic form).

The listing of certificates of correction can be electronically accessed on the day of issuance at http://www.uspto.gov/web/patents/certofcorrect. The electronic image of the printed certificate of correction can be accessed on the patent database at http://www. uspto.gov/patft and the listing of the certificates of correction, as published in the Official Gazette three weeks later, will be electronically accessible at http://www.uspto.gov/web/ offices/com/sol/og.

§22.9　證書修正之效力

　　一般來說，請求證書修正並沒有法定上對積極勤勉（diligence）的要求，因此並無限定申請人何時需要提出修正。而批准後的證書修正亦可能在專利過期後發出，發出後係溯及既往。但證書修正程序不能用於已經因無效等實質理由被刪除的請求項。

§22.10 棄權聲明書

　　棄權聲明書（disclaimer）係由專利權人主動聲明放棄專利權的中全部或部份請求項之全部或部份權利之聲明。此種聲明必須書面提出，並經 USPTO 登記方生效。

---

35 U.S.C. 253 Disclaimer.

Whenever, without any deceptive intention, a claim of a patent is invalid the remaining claims shall not thereby be rendered invalid. A patentee, whether of the whole or any sectional interest therein, may, on payment of the fee required by law, make disclaimer of any complete claim, stating therein the extent of his interest in such patent. Such disclaimer shall be in writing, and recorded in the Patent and Trademark Office; and it shall thereafter be considered as part of the original patent to the extent of the interest possessed by the disclaimant and by those claiming under him.

In like manner any patentee or applicant may disclaim or dedicate to the public the entire term, or any terminal part of the term, of the patent granted or to be granted.

§22.11 棄權聲明書的種類

棄權聲明書（disclaimer）可分成法定棄權書（statutory disclaimer）與尾端棄權書（terminal disclaimer）。

法定棄權書（statutory disclaimer）係專利權人主動聲明將專利中的一個或數個請求項放棄。或者放棄專利效期。例如當專利訴訟時，判定特定請求項無效時，專利權人必須向 USPTO 提交棄權聲明，以公示大眾。

尾端棄權書（terminal disclaimer）則是當專利公告後，專利權人發現與自己擁有的其他專利有重複（double patenting）的問題，可向 USPTO 提出專利權效期之尾端棄權書（terminal disclaimer），簽署之後，除專利權效期一起結束外，同時該等專利權不可分開行使或分開讓與他人。

§22.12 提出棄權聲明書之程序

專利權人或其登記在案之代理人必須簽署書面聲明提交。聲明中必須包含：哪些請求項或者專利權期效要放棄（請求項必須整項放棄，不可部份放棄）、目前的專利權歸屬情況、以及繳納§ 1.20(d)規費（小實體減半）。

對於一般尾端棄權書，專利權人可以聲明放棄全部或尾端專利效期。亦需簽署書面、指明其放棄之效期、目前專利權歸屬情況與 1.20(d)規費（小實體減半）。

對於為了避免重複專利而提出的尾端棄權書，專利權人亦需聲明其尾端專利效期。亦需簽署書面、指明其放棄之效期、目前專利權歸屬情況與 1.20(d)規費（小實體減半）。

上述棄權聲明書經 USPTO 檢核批准後，均將公告大眾。

---

37 CFR 1.321 Statutory disclaimers, including terminal disclaimers.

(a) A patentee owning the whole or any sectional interest in a patent may disclaim any complete claim or claims in a patent. In like manner any patentee may disclaim or dedicate to the public the entire term, or any terminal part of the term, of the patent granted. Such disclaimer is binding upon the grantee and its successors or assigns. A notice of the disclaimer is published in the Official Gazette and attached to the printed copies of the specification. The disclaimer, to be recorded in the Patent and Trademark Office, must:

(1) be signed by the patentee, or an attorney or agent of record;

(2) identify the patent and complete claim or claims, or term being disclaimed. A disclaimer which is not a disclaimer of a complete claim or claims, or term, will be refused recordation;

(3) state the present extent of patentee's ownership interest in the patent; and

(4) be accompanied by the fee set forth in § 1.20(d).

(b) An applicant or assignee may disclaim or dedicate to the public the entire term, or any terminal part of the term, of a patent to be granted. Such terminal disclaimer is binding upon the grantee and its successors or assigns. The terminal disclaimer, to be recorded in the Patent and Trademark Office, must:

(1) be signed:

(i) by the applicant, or

(ii) if there is an assignee of record of an undivided part interest, by the applicant and such assignee, or

(iii) if there is an assignee of record of the entire interest, by such assignee, or

(iv) by an attorney or agent of record;

(2) specify the portion of the term of the patent being disclaimed;

(3) state the present extent of applicant's or assignee's ownership interest in the patent to be granted; and

(4) be accompanied by the fee set forth in § 1.20(d).

(c) A terminal disclaimer, when filed to obviate judicially created double patenting in a patent application or in a reexamination proceeding except as provided for in paragraph (d) of this section, must:

(1) Comply with the provisions of paragraphs (b)(2) through (b)(4) of this section;

(2) Be signed in accordance with paragraph (b)(1) of this section if filed in a patent application or in accordance with paragraph (a)(1) of this section if filed in a reexamination proceeding; and

(3) Include a provision that any patent granted on that application or any patent subject to the reexamination proceeding shall be enforceable only for and during such period that said

patent is commonly owned with the application or patent which formed the basis for the judicially created double patenting.

(d) A terminal disclaimer, when filed in a patent application or in a reexamination proceeding to obviate double patenting based upon a patent or application that is not commonly owned but was disqualified under 35 U.S.C. 103(c) as resulting from activities undertaken within the scope of a joint research agreement, must:

(1) Comply with the provisions of paragraphs (b)(2) through (b)(4) of this section;

(2) Be signed in accordance with paragraph (b)(1) of this section if filed in a patent application or be signed in accordance with paragraph (a)(1) of this section if filed in a reexamination proceeding;

(3) Include a provision waiving the right to separately enforce any patent granted on that application or any patent subject to the reexamination proceeding and the patent or any patent granted on the application which formed the basis for the double patenting, and that any patent granted on that application or any patent subject to the reexamination proceeding shall be enforceable only for and during such period that said patent and the patent, or any patent granted on the application, which formed the basis for the double patenting are not separately enforced.

＊＊＊＊＊＊＊＊＊＊＊

第 23 章　再發證程序

目　錄

§23.1　提出再發證申請

依據 35USC§251，當一美國專利由於無欺騙意圖之錯誤，此錯誤是因為說明書或圖式之缺陷或專利權人請求太多或太少之權利而造成，且使得該專利被認為是全部或部份不能操作或無效時，專利權人可提出再發證之申請。美國專利局長將在專利權人放棄原有專利及繳交規費之情況下，對於揭露在原專利中之發明再發證，且依據新的修正的申請書給予原專利未期滿之期間之專利權。

通常專利權人因為下列理由而提出再發證之申請：

(A)　請求項之範圍太窄或太廣；

(B)　說明書之揭露包含有錯誤；

(C)　申請人未主張國外優先權或不正確地主張國外優先權；及

(D)　申請人未參照併入或不正確地參照併入先前共同審查中之專利申請案。

除了上述 4 種情況下，專利權人可提出再發證申請以起始與他人之抵觸程序（interference proceeding）。再者，一延續案（continuation application）之美國專利之專利權人不能藉再發證程序將先前主張之母案之優先權撤銷以使專利權期限延長，因為再發證後之專利權期限只是專利權未期滿之部份。

---

35 U.S.C. 251 Reissue of defective patents.

Whenever any patent is, through error without any deceptive intention, deemed wholly or partly inoperative or invalid, by reason of a defective specification or drawing, or by reason of the patentee claiming more or less than he had a right to claim in the patent, the Director shall, on the surrender of such patent and the payment of the fee required by law, reissue the patent for the invention disclosed in the original patent, and in accordance with a new and amended application, for the unexpired part of the term of the original patent. No new matter shall be introduced into the application for reissue.

＊＊＊＊＊＊＊＊＊＊＊

MPEP 1402 Grounds for Filing [R-7]

These corrections may be made via a certificate of correction; see MPEP § 1481.

The most common bases for filing a reissue application are:

(A)　the claims are too narrow or too broad;

(B)　the disclosure contains inaccuracies;

(C)　applicant failed to or incorrectly claimed foreign priority; and

(D)　applicant failed to make reference to or incorrectly made reference to prior copending applications.

＊＊＊＊＊＊＊＊＊＊＊

MPEP 1405　Reissue and Patent Term [R-2]

＊＊＊＊＊＊＊＊＊＊＊

The maximum term of the original patent is fixed at the time the patent is granted. While the term may be subsequently shortened, e.g., through the filing of a terminal disclaimer, it cannot be extended through the filing of a reissue. Accordingly, a deletion in a reissue application of an earlier-obtained benefit claim under 35 U.S.C.120 will not operate to lengthen the term of the patent to be reissued.

＊＊＊＊＊＊＊＊＊＊＊

§23.2　再發證申請需提出之文件

提出再發證之申請需附上下列文件：

(1)　宣誓書，由發明人簽署，需於宣誓書中聲明錯誤是無欺騙的意圖及其相信原來之專利是完全或部份不能操作或無效的，且需註明理由為①說明書及圖之缺陷，或②請求項範圍太窄或太廣，或③其他理由。特別是當要擴大請求項之範圍時需於宣誓書中敘明理由。另外，如果再發證並不是要擴大請求項之範圍時，宣誓書可由完全權利之受讓人（assignee of the entire interest）簽署。

(2)　所有受讓人之同意書（assent by assignee for filing a reissue application）。

(3)　37 CFR 3.73(b)之建立所有權之文件，由受讓人簽署並於申請再發證時提出以支持其同意書。

(4) 專利說明書及圖式，專利說明書之格式為將原發證之專利以每頁兩欄（亦即直接影印即可）之方式呈現。圖式需呈送修正後之版本（乾淨之版本），如對說明書及請求項有修正可以(A)將原專利之要修改部份切開，並加入修正之部份方式或(B)以先前修正之方式送入。說明書或請求項之修正不能有任何的新事項（new matter）。

(5) Information Disclosure Statement（IDS）資料亦可同時提出。

(6) 再發證申請規費（小實體有優惠，且原專利已繳之超項費不需繳）。

(7) 主張國外優先權，國外優先權之主張需重新聲明，但不必再補呈優先權文件。

MPEP 1410.01 Reissue Applicant, Oath or Declaration, and Consent of all Assignees [R-7]

＊＊＊＊＊＊＊＊＊＊＊

The reissue oath must be signed and sworn to by all the inventors, or declaration made by all the inventors, except as otherwise provided in 37 CFR 1.42 , 1.43, and 1.47 (see MPEP § 409). Pursuant to 37 CFR 1.172, where the reissue application does not seek to enlarge the scope of any of the claims of the original patent, the reissue oath may be made and sworn to, or declaration made, by the assignee of the entire interest.

＊＊＊＊＊＊＊＊＊＊＊

The reissue oath or declaration must be accompanied by *>a< written consent of all assignees

＊＊＊＊＊＊＊＊＊＊＊

The assignee that consents to the filing of the reissue application (as discussed above) must also establish that it is the assignee, i.e., the owner, of the patent. See 37 CFR 1.172. Accordingly, a 37 CFR 3.73(b) paper establishing the ownership of the assignee should be submitted at the time of filing the reissue application, in order to support the consent of the assignee.

＊＊＊＊＊＊＊＊＊＊＊

For reissue applications filed on or after November 7, 2000, 37 CFR 1.173(a)(1) requires that the application specification, including the claims, must be furnished in the form of a copy of the printed patent in double column format (so that the patent can be simply copied without cutting). Applicants are required to submit a clean copy of each drawing sheet of the printed patent at the time the reissue application is filed (37 CFR 1.173(a)(2)). Any changes to the drawings must be made in accordance with 37 CFR 1.173(b)(3). Thus, a full copy of the printed patent (including the front page) is used to provide the abstract, drawings, specification, and claims of the patent for the reissue application. Each page of the patent must appear on only one side of each individual page of the specification of the reissue application; a two-sided copy of the patent is not proper. It should be noted that a re-typed specification is not acceptable in a reissue application; the full copy of the printed patent must be used. If, however, the changes to be made to the patent are so extensive/numerous that reading and understanding the specification is extremely difficult and error-prone, a clean copy of the specification may be submitted if accompanied by a grantable petition under 37 CFR 1.183 for waiver of 37 CFR 1.125(d) and 37 CFR 1.173(a)(1).

Pursuant to 37 CFR 1.173(b), amendments may be made at the time of filing of a reissue application. The amendment may be made either by:

(A) physically incorporating the changes within the specification by cutting the column of the printed patent and inserting the added material and rejoining the remainder of the column and then joining the resulting modified column to the other column of the printed patent. Markings pursuant to 37 CFR 1.173(d) must be used to show the changes. The columnar structure of the printed patent must be preserved, and the physically modified page must comply with 37 CFR 1.52(a)(1). As to compliance with 37 CFR 1.52(a)(1)(iv), the "written either by a typewriter or machine printer in permanent dark ink or its equivalent" requirement is deemed to be satisfied where a caret and line are drawn from a position within the text to a newly added phrase, clause, sentence, etc. typed legibly in the margin; or

(B) providing a preliminary amendment (a separate amendment paper) directing that specified changes be made to the copy of the printed patent.

＊＊＊＊＊＊＊＊＊＊＊

New matter, that is, matter not present in the patent sought to be reissued, is excluded from a reissue application in accordance with 35 U.S.C. 251.

＊＊＊＊＊＊＊＊＊＊＊

§23.3　再發證申請公開給大眾審閱

　　依照 37CFR 1.11(b)，在 1977 年 3 月 1 日以後申請之再發證申請案均會公開給大眾作檢視，而且大眾均可藉由付費取得其影本。再發證申請案之申請（除了 CPA 外）均會在美國專利局之公報上正式宣告（announcement），此宣告給予相關人士有機會提供與此再發證申請案之專利性有關之資訊給審查委員作參考。此正式宣告包括申請日，再發證申請案號，原專利號，發明名稱，專利分類及次分類，發明人名稱，所有權人名稱，代理人名稱及審查此案之技術中心。當再發證案是為了修改發明人時亦會宣告再發證案要修改之發明人的名字。

MPEP 1430 Reissue Files Open to the Public and, Notice of Filing Reissue Announced in, Official Gazette [R-7]

＊＊＊＊＊＊＊＊＊＊＊

　　Under 37 CFR 1.11(b) all reissue applications filed after March 1, 1977, are open to inspection by the general public, and copies may be furnished upon paying the fee therefor. The filing of reissue applications (except for continued prosecution applications (CPA's) filed under 37 CFR 1.53(d)) will be announced in the Official Gazette. The announcement gives interested members of the public an opportunity to submit to the examiner information pertinent to the patentability of the reissue application. The announcement includes the filing date, reissue application and original patent numbers, title, class and subclass, name of the *>inventor(s)<, name of the owner of record, name of the attorney or agent of record, and the Technology Center (TC) to which the reissue application is initially assigned. **>Where a reissue application seeks to change the inventorship of a patent, the names of the inventors of record of the patent file are set forth in the announcement, not the filing receipt, which sets forth the names of the inventors that the reissue application is seeking to make of record upon reissue of the patent.<

＊＊＊＊＊＊＊＊＊＊＊

§23.4　再發證申請案的請求項可請求之內容

一再發證申請案之請求項必需請求在原來之專利中揭露之發明。而在考慮是否是在原來之專利中揭露之發明時，不只考慮原來之請求項，而是考慮完整的揭露書。

符合下列兩種條件時，再發證申請案之請求項將被視為符合 35 §251 是在請求原來專利中揭露之發明：

(A) 再發證申請案之請求項乃敘述在原來之說明書（包括圖式之中）。而且符合 35USC112 第 1 段之據以實施要件；以及

(B) 在原來之說明書中並未有任何部份顯示申請人不想請求在再發證申請案中所請求之發明。

另外，如果在申請過程中有發下限制／選擇要求，而申請人作了限制／選擇而讓所限制／選擇之發明發下證書而未去申請分割案以請求未選擇之發明的話，則不能再以再發證申請案將未選擇之發明列在再發證申請案之請求項之中。

* * * * * * * * * *

MPEP 1412.01 Reissue Claims Must Be for Same General Invention [R-7]

The reissue claims must be for the same invention as that disclosed as being the invention in the original patent, as required by 35 U.S.C. 251. **The entire disclosure, not just the claim(s), is considered in determining what the patentee objectively intended as his or her invention.

* * * * * * * * * *

Claims presented in a reissue application are considered to satisfy the requirement of 35 U.S.C. 251 that the claims be "for the invention disclosed in the original patent" where:

(A) the claims presented in the reissue application are described in the original patent specification and enabled by the original patent specification such that 35 U.S.C. 112 first paragraph is satisfied; and

(B) nothing in the original patent specification indicates an intent not to claim the subject matter of the claims presented in the reissue application.

* * * * * * * * * *

FAILURE TO TIMELY FILE A DIVISIONAL APPLICATION PRIOR TO ISSUANCE OF ORIGINAL PATENT

Where a restriction >(or an election of species)< requirement was made in an application and applicant permitted the elected invention to issue as a patent without * filing * a divisional application on the non-elected invention(s), the non-elected invention(s) cannot be recovered by filing a reissue application. A reissue applicant's failure to timely file a divisional application covering the non-elected invention(s) in response to a restriction >(or an election of species)< requirement is not considered to be error causing a patent granted on the elected claims to be partially inoperative by reason of claiming less than the applicant had a right to claim. Accordingly,

＊＊＊＊＊＊＊＊＊＊＊＊

實例

一專利說明書中揭露了二組合物 X 及 Y，在說明書本文中敘述到組合物 X 不適合作為成型之用，因為其不能夠在短時間內乾燥，專利申請案後來核准，請求項只包括了 Y 之組合物。在專利發證之後專利權人發現組合物 X 雖然不能快速乾燥，但具有其他優良的性質，事實上亦可用作成型之用。專利權人是否可藉由再發證申請案請求組合物 X？

雖然組合物 X 揭露在原來之說明書之中，但是原來之說明書中已明白顯示申請人並不想請求組合物 X 之發明，故無法藉由再發證申請案請求組合物 X 之發明。

§23.5　再獲得法則

所謂再獲得法則就是專利權人不能藉由再發證申請案去再獲得在申請時為了得到專利所放棄的標的之專利權。因為在專利申請時為了避開先前技術，申請人將請求項之範圍縮小，如果再藉由再發證申請獲得與原放棄（刪除）之請求項之範圍相同或更大之範圍則顯然不合理。

美國專利局採以下三步驟判斷一再發證申請案是否違反再獲得法則：

(A) 是否再發證申請案之任一請求項有擴大權利之範圍，所謂擴大請求項之範圍，乃指一專利請求項任一（或部分）限制條件在再發證申請案之請求項中不再存在（刪除）。

(B) 是否再發證之請求項之任何擴大之部份（any broadening aspect）與原來放棄之標的有關。亦即審查委員需對再發證申請案之請求項中各被刪除或擴大之限制條件均檢視是否與原放棄之標的有關。此部份審查委員會檢視是否在審查過程中申請人對該部份有作修正，是否有提出爭論（argument）或聲明，如有對擴大的部份作修正、爭論或聲明則視為與原來放棄之標的有關。

(C) 是否再發證申請案之請求項在其他部份實質上縮小了範圍，而使得請求項並未擴大範圍，因此可避免再獲得法則之適用。

美國專利局之審查委員採用下列之流程來判定是否有再獲得法則之適用。

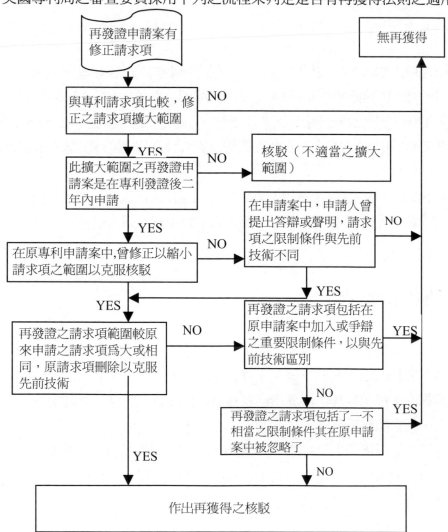

MPEP 1412.02 Recapture of Canceled Subject Matter [R-7]

A reissue will not be granted to "recapture" claimed subject matter which was surrendered in an application to obtain the original patent

＊＊＊＊＊＊＊＊＊＊＊＊

A. The First Step - Was There Broadening?

In every reissue application, the examiner must first review each claim for the presence of broadening, as compared with the scope of the claims of the patent to be reissued. A reissue claim is broadened where some limitation of the patent claims is no longer required in the reissue claim; see MPEP § 1412.03 for guidance as to the nature of a "broadening claim." If the reissue claim is not broadened in any respect as compared to the patent claims, the analysis ends; there is no recapture.

B. The Second Step - Does Any Broadening Aspect of the Reissued Claim Relate to Surrendered Subject Matter?

Where a claim in a reissue application is broadened in some respect as compared to the patent claims, the examiner must next determine whether the broadening aspect(s) of that reissue claim relate(s) to subject matter that applicant previously surrendered during the prosecution of the original application (which became the patent to be reissued). Each limitation of the patent claims, which is omitted or broadened in the reissue claim, must be reviewed for this determination. This involves two sub-steps:

＊＊＊＊＊＊＊＊＊＊＊

C. The Third Step - Were the reissued claims materially narrowed in other respects **>, so that the claims may not have been enlarged, and hence< avoid the recapture rule?

＊＊＊＊＊＊＊＊＊＊＊

Reissue Recapture - Determining its presence or absence

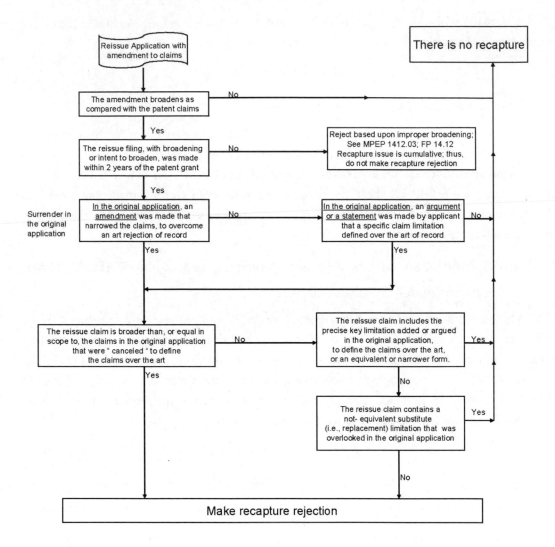

實例

　　發明人 A 於 2005 年 1 月 1 日提出一專利申請，是有關於一種將家禽之羽毛去除之裝置。審查委員對於請求項 1 引證了一先前技術核駁，發明人 A 之代理人針對此先前技術爭論其請求項 1 中之一限制條件並未揭示在該先前技術中。最後，請求項 1 於 2008 年 3 月 2 日取得專利，而且最後請求項 1 之該限制條件並未作任何之修改。2010 年 2 月 1 日發明人 A 發現所獲得的專利之請求項 1 範圍太窄，當時與審查委員爭論之限制條件可不必包括在請求項 1 之中，因而於 2010 年 3 月 1 日提出一再發證申請案將請求項 1 之該限制條件刪除。請問審查委員對此再發證之申請案之請求項 1，是否會發下再獲得之核駁？

　　雖然請求項 1 在原專利申請過程中並未因先前技術而作修正，但是發明人 A 之代理人在答辯過程中針對該限制條件作了爭論（argument），故於再發證以申請案中將之刪除，違反再獲得法則。

§23.6　藉由再發證申請擴大請求項範圍之限制

　　依照 35USC§251，如果專利權人要藉由再發證申請來擴大請求之範圍，需於專利發證之日起二年間為之。

Ⅰ. 請求項範圍擴大之意義

　　所謂請求項範圍擴大乃指一請求項其範圍較原專利之各個及每一個請求項之範圍為大。如果在提出再發證申後情之前有提出請求項之棄權聲明書（disclaimer），則被棄權之請求項（disclaimed claims）不是原專利之一部份。此外，一請求項雖然某一部份範圍較小，但只要至少一部份範圍變大，即是擴大範圍之請求項。

　　而如果再發證申請案之請求項包括了原專利之請求項未涵蓋之標的，則此請求項擴大了請求項之範圍。例如，再發證申請案中新修正之請求項或新增加之請求項包括了任何產品或製法其不會侵害原專利之請求項之權利者，則此請求項擴大了原專利之範圍。另外，於再發證申請案中加入組合請求項（在原專利中只有次組合請求項）是否擴大了請求項之範圍，則視所加入的組合請求項是否包括了原專利之次組合請求項之所有限制條件改定，如果組合請求項包括了次組合請求項之所有限制條件，則侵害組合請求項之權利必然就侵害了次組合請求項之權利，故此種情形必無擴大請求項範圍之情事。

II. 擴大依附項之範圍並不是請求項範圍之擴大

例如，claim 2 依附至 claim 1，claim 1 為中間產物之製法，claim 2 加上一步驟將中間產物轉換成最終產物，如果將 claim 2 之範圍擴大，並非請求項範圍之擴大，因為依附項被解譯成包括了獨立項之所有限制條件。

III. 於再發證申請案中加入新領域之發明通常是請求項範圍之擴大

例如，原專利只有產品 A 之請求項，現在於再發證申請案中加入了產品 A 製法，則此再發證申請案擴大了請求項之範圍。但是，如果原專利之請求項包括產品 A 之請求項，而於再發證之請求項中加入下列二請求項：

(1) 使用產品 A 以製造產品 B 之方法；
(2) 使用產品 A 以實行一方法 B

則此再發證申請案並未擴大請求項之範圍，因為新增之請求項包括了原專利之請求項之所有限制條件。

IV. 擴大範圍之請求項提出之時間點

擴大請求項範圍之再發證申請案必需在原專利發證之日起二年內提出。亦即，在提出再發證申請案時至少要有一請求項是擴大範圍的，如果有另外的擴大範圍之請求項亦可在二年之後提出。

MPEP 1412.03 Broadening Reissue Claims [R-7]

35 U.S.C. 251 prescribes a 2-year limit for filing applications for broadening reissues:

No reissue patent shall be granted enlarging the scope of the original patent unless applied for within two years from the grant of the original patent.

I.　MEANING OF "BROADENED REISSUE CLAIM"

A broadened reissue claim is a claim which enlarges the scope of the claims of the patent, i.e., a claim which is greater in scope than each and every claim of the original patent. If a disclaimer is filed in the patent prior to the filing of a reissue application, the disclaimed claims are not part of the "original patent"

＊＊＊＊＊＊＊＊＊＊＊

A claim in the reissue which includes subject matter not covered by the patent claims enlarges the scope of the patent claims. For example, if any amended or newly added claim in the reissue contains within its scope any conceivable product or process which would not have infringed the patent, then that reissue claim would be broader than the patent claims

＊＊＊＊＊＊＊＊＊＊＊

The addition of combination claims in a reissue application where only subcombination claims were present in the original patent could be a broadening of the invention. The question which must be resolved in this case is whether the combination claims added in the reissue would be for "the invention as claimed" in the original patent. See Ex parte Wikdahl, 10 USPQ2d at 1549. The newly added combination claims should be analyzed to determine whether they contain every limitation of the subcombination of any claim of the original patent. If the combination claims (added in the reissue) contain every limitation of the subcombination (which was claimed in the original application), then infringement of the combination must also result in infringement of the subcombination. Accordingly, the patent owner could not, if a reissue patent issues with the combination claims, sue any new party for infringement who could not have been sued for infringement of the original patent. Therefore, broadening does not exist.

＊＊＊＊＊＊＊＊＊＊＊

II.　SCOPE OF DEPENDENT CLAIM ENLARGED - NOT BROADENING

＊＊＊＊＊＊＊＊＊＊＊

III. NEW CATEGORY OF INVENTION ADDED IN REISSUE - GENERALLY IS BROADENING

The addition of process claims as a new category of invention to be claimed in the patent (i.e., where there were no method claims present in the original patent) is generally considered as being a broadening of the invention. See Ex parte Wikdahl, 10 USPQ2d 1546 (Bd. Pat. App. & Inter. 1989). A situation may arise, however, where the reissue application adds a limitation (or limitations) to process A of making the product A claimed in the original patent claims. For example:

(1) a process of using the product A (made by the process of the original patent) to make a product B, disclosed but not claimed in the original patent; or

(2) a process of using the product A to carry out a process B disclosed but not claimed in the original patent.

Although this amendment of the claims adds a method of making product B or adds a method of using product A, this is not broadening (i.e., this is not an enlargement of the scope of the original patent) because the "newly claimed invention" contains all the limitations of the original patent claim(s).

＊＊＊＊＊＊＊＊＊＊＊

實例 1

發明人 A 之美國專利在 2008 年 9 月 2 日發證。之後發明人發現取得專利之請求項之範圍太小，遂於 2010 年 8 月 1 日提出擴大範圍之再發證申請案。接著在 2010 年 12 月 13 日發明人 A 覺得需再加入一些請求項以得到更佳之保護，故再以先前修正之方式加入了 claim 20～40，其中 claims 20～30 之範圍較 2010 年 8 月 1 日提出之再發證申請案之範圍為大，claim 31～35 之範圍較 2010 年 8 月 1 日提出之再發證申請範圍為小，但是大於 2008 年 9 月 2 日發證之專利範圍，claims 36～40 則較 2008 年 9 月 2 日發證之專利的範圍為小。此加入之 claims 20～40 均由原專利申請案之揭露中有支持。請問審查委員會對此加入之 claims 20～40 作哪些專利性之審查？

claims 20～40 均會被審查。雖然 claims 20～40 加入時已超過原專利發證起之二年期限。但是其擴大範圍之再發證專利申請案是在原專利發證之二年之內提出，故 claims 20～40 均會被審查。

實例 2

申請人將其已核准專利之 claim 1 修正如下（底線表加入，〔〕表刪除）：

1. A <u>fluorescent</u> light fixture comprising:

a semicylindrical enclosure;

a <u>fluorescent</u> light lockably connected in the enclosure; a pair of 〔bayonet〕 sockets on opposite ends of said enclosure which engage the <u>fluorescent</u> light, and

a longitudinal adjustment screw〔means〕positioned between each〔bayonet〕socket and one of said ends of said enclosure to secure <u>fluorescent</u> light to said pair of〔bayonet〕sockets.

請問此 claim 經上述修正後，是否範圍變大？

範圍變大。雖然加入 fluorescent，但是刪除了〔bayonet〕故有可能原來不侵權之產品，會變成侵權，故範圍變大。

§23.7　藉由再發證申請修正發明人之錯誤

依照 Ex parte Scudder 之判例，法院判定當 35USC §256 及 37CFR 1.324 不適合來修正已發證之專利之發明人的錯誤時，35 USC§251 亦准許以再發證申請案來修正發明人（inventorship）之錯誤。

依照 35USC §256 當一發明之專利，發明人有錯誤，而此錯誤並非發明人有欺騙之意圖時，應以 certificate of corrction 作修正。而依照 37CFR 1.324 要對此發明人之修正時需所有發明人同意此發明人之修正。但是，當無法取得所有發明人之同意時，所有權利之受讓人則可依照 37CFR 1.172 之規定提出再發證申請案來修正發明人之錯誤，因為修正發明人之錯誤並未擴大原來專利之請求項之範圍。

MPEP 1412.04 Correction of Inventorship [R-7]

The correction of misjoinder of inventors has been held to be a ground for reissue. See Ex parte Scudder, 169 USPQ 814, 815 (Bd. App. 1971) wherein the Board held that 35 U.S.C.

251 authorizes reissue applications to correct misjoinder of inventors where 35 U.S.C. 256 is inadequate. See also A.F. Stoddard & Co. v. Dann, 564 F.2d 556, 567 n.16, 195 USPQ 97, 106 n.16 (D.C. Cir. 1977) wherein correction of inventorship from sole inventor A to sole inventor B was permitted in a reissue application. The court noted that reissue by itself is a vehicle for correcting inventorship in a patent.

＊＊＊＊＊＊＊＊＊＊＊

II.REISSUE AS A VEHICLE FOR CORRECTING INVENTORSHIP

Where the provisions of 35 U.S.C. 256 and 37 CFR 1.324 do not apply, a reissue application is the appropriate vehicle to correct inventorship. The failure to name the correct inventive entity is an error in the patent which is correctable under 35 U.S.C. 251. The reissue oath or declaration pursuant to 37 CFR 1.175 must state that the applicant believes the original patent to be wholly or partly inoperative or invalid through error of a person being incorrectly named in an issued patent as the inventor, or through error of an inventor incorrectly not named in an issued patent, and that such error arose without any deceptive intention on the part of the applicant. The reissue oath or declaration must, as stated in 37 CFR 1.175, also comply with 37 CFR 1.63.

The correction of inventorship does not enlarge the scope of the patent claims. Where a reissue application does not seek to enlarge the scope of the claims of the original patent, the reissue oath may be made and sworn to, or the declaration made, by the assignee of the entire interest under 37 CFR 1.172. An assignee of part interest may not file a reissue application to correct inventorship where the other co-owner did not join in the reissue application and has not consented to the reissue proceeding. See Baker Hughes Inc. v. Kirk, 921 F. Supp. 801, 809, 38 USPQ2d 1885, 1892 (D.D.C. 1995). See 35 U.S.C. 251, third paragraph. Thus, the signatures of the inventors are not needed on the reissue oath or declaration where the assignee of the entire interest signs the reissue oath/declaration. Accordingly, an assignee of the entire interest can add or delete the name of an inventor by reissue (e.g., correct inventorship from inventor A to inventors A and B) without the original inventor's consent.

＊＊＊＊＊＊＊＊＊＊＊

37 CFR § 1.172 Applicants, assignees

(a) A reissue oath must be signed and sworn to or declaration made by the inventor or inventors except as otherwise provided (see §1.42,1.43,1.47),and must be accompanied by the written consent of all assignees, if any, owning an undivided interest in the patent, but a reissue oath may be made and sworn to or declaration made by the assignee of the entire interest if the application does not seek to enlarge the scope of the claims of the original patent.

All assignees consenting to the reissue must establish their ownership interest in the patent by filing in the reissue application a submission in accordance with the provisions of § 3.73(b) of this chapter.

§23.8　再發證申請案之審查方式及程序

　　依照 37CFR 1.176 再發證申請案之審查與一般之非再發證，非暫時申請案之審查方式是相同的。亦即，審查委員如認為適當，可對請求項作出核駁，不論再發證之申請案之請求項與原專利請求項是否相同。亦即，原來已核准之請求項可能被審查委員以原申請案之引證或理由核駁。也就是說，再發證申請案之請求項並非假定是有效的。另外，審查委員對於再發證申請案作先前技術檢索之基準日為原申請案之申請日。所有再發證申請案均被視為"特別的"案件會優先審查，但是，再發證申請案在公報上發表（announced）之後的二個月之內通常不會作審查，亦即會先延遲二個月，以讓公眾可對再發證申請案以發證前異議（protest）之方式提出相關之資訊給美國專利局作審查。

　　再發證申請案之審查委員只允許對原專利之請求項與新增的請求項之間作出限制要求，而不會對原專利之請求項作出任何限制要求。當審查委員認為原專利之請求項與新增之請求項需作出限制要求之後，原來專利的請求項就視為是推斷地被選擇，因此審查委員會發下審定書指出①限制要求（新增的請求項被推斷地未選擇因而視為撤回）②只針對原專利之請求項審查專利要件，③通知申請人如果原請求項可核准，則原未選擇之請求項需作分割。

　　如果申請人對原專利的請求項提出棄權聲明（disclaimer），則新增的請求項會被審查。但是申請人對原請求項之棄權聲明必需在審查委員發下限制要求前提出，否則審查委員就主動只審查原專利之請求項。

　　再發證申請案與一般非暫時申請案一樣可主動提出分割案、延續案，亦可提出 RCE。

MPEP 1440 Examination of Reissue Application [R-3]

＊＊＊＊＊＊＊＊＊＊＊＊

Reissue applications with related litigation will be acted on by the examiner before any other special applications, and will be acted on immediately by the examiner, subject only to a 2-month delay after publication for examining reissue applications; see MPEP § 1441.

The original patent file wrapper /file history should always be obtained and reviewed when examining a reissue application thereof.

1441 Two-Month Delay Period [R-7]

37 CFR 1.176 provides that reissue applications will be acted on by the examiner in advance of other applications, i.e., "special." Generally, a reissue application will not be acted on sooner than 2 months after announcement of the filing of the reissue has appeared in the Official Gazette. The 2-month delay is provided in order that members of the public may have time to review the reissue application and submit pertinent information to the Office before the examiner's action. The pertinent information is submitted in the form of a protest under 37 CFR 1.291(a).

* * * * * * * * * * *

MPEP 1445 Reissue Application Examined in Same Manner as Original Application

As stated in 37 CFR 1.176, a reissue application, including all the claims therein, is subject to "be examined in the same manner as a non-reissue, non-provisional application." Accordingly, the claims in a reissue application are subject to any and all rejections which the examiner deems appropriate. It does not matter whether the claims are identical to those of the patent or changed from those in the patent. It also does not matter that a rejection was not made in the prosecution of the patent, or could have been made, or was in fact made and dropped during prosecution of the patent; the prior action in the prosecution of the patent does not prevent that rejection from being made in the reissue application. Claims in a reissue application enjoy no "presumption of validity." In re Doyle, 482 F.2d 1385, 1392, 179 USPQ 227, 232-233 (CCPA 1973); In re Sneed, 710 F.2d 1544, 1550 n.4, 218 USPQ 385, 389 n.4 (Fed. Cir. 1983). Likewise, the fact that during prosecution of the patent the examiner considered, may have considered, or should have considered information such as, for example, a specific prior art document, does not have any bearing on, or prevent, its use as prior art during prosecution of the reissue application.

MPEP 1450 Restriction and Election of Species Made in Reissue Application [R-7]

* * * * * * * * * * *

37 CFR 1.176(b) permits the examiner to require restriction in a reissue application between claims newly added in a reissue application and the original patent claims, where the

added claims are directed to an invention which is separate and distinct from the invention(s) defined by the original patent claims.

＊＊＊＊＊＊＊＊＊＊＊

Where a restriction requirement is made by the examiner, the original patent claims will be held to be constructively elected (except for the limited situation where a disclaimer is filed as discussed in the next paragraph). Thus, the examiner will issue an Office action in the reissue application (1) providing notification of the restriction requirement, (2) holding the added claims to be constructively non-elected and withdrawn from consideration, (3) treating the original patent claims on the merits, and (4) informing applicant that if the original claims are found allowable, and a divisional application has been filed for the non-elected claims, further action in the application will be suspended, pending resolution of the divisional application.

If a disclaimer of all the original patent claims is filed in the reissue application containing newly added claims that are separate and distinct from the original patent claims, only the newly added claims will be present for examination. In this situation, the examiner's Office action will treat the newly added claims in the reissue application on the merits. The disclaimer of all the original patent claims must be filed in the reissue application **>before< the issuance of the examiner's Office action containing the restriction requirement, in order for the newly added claims to be treated on the merits. Once the examiner has issued the Office action providing notification of the restriction requirement and treating the patent claims on the merits, it is too late to obtain an examination on the added claims in the reissue application by filing a disclaimer of all the original patent claims. If reissue applicant wishes to have the newly added claims be treated on the merits, a divisional reissue application must be filed to obtain examination of the added claims.

＊＊＊＊＊＊＊＊＊＊＊

MPEP 1451 Divisional Reissue Applications; Continuation Reissue Applications Where the Parent is Pending [R-7]

＊＊＊＊＊＊＊＊＊＊＊

I.DIVISIONAL REISSUE APPLICATIONS

37 CFR 1.176(b) permits the examiner to require restriction in a reissue application between the original claims of the patent and any newly added claims which are directed to a separate and distinct invention(s). See also MPEP § 1450. As a result of such a restriction requirement, divisional >reissue< applications may be filed for each of the inventions identified in the restriction requirement.

In addition, applicant may initiate a division of the claims by filing more than one reissue application in accordance with 37 CFR 1.177. The multiple reissue applications which are filed may contain different groups of claims from among the original patent claims, or some of the reissue applications may contain newly added groups (not present in the original patent). There is no requirement that the claims of the multiple reissue applications be independent and distinct from one another; if they are not independent and distinct from one another, the examiner must apply the appropriate double patenting rejections

＊＊＊＊＊＊＊＊＊＊＊

II.CONTINUATION REISSUE APPLI-CATIONS

Accordingly, prosecution of a continuation >reissue application< of a >parent< reissue application will be permitted (despite the existence of the pending parent reissue application) where the continuation >reissue application< complies with the rules for reissue.

＊＊＊＊＊＊＊＊＊＊＊

MPEP1452 Request for Continued Examination of Reissue Application [R-7]

A request for continued examination (RCE) under 37 CFR 1.114 is available for a reissue application. Effective May 29, 2000, an applicant in a reissue application may file a request for continued examination of the reissue application, if the reissue application was filed on or after June 8, 1995. This applies even where the application, which resulted in the original patent, was filed **>before< June 8, 1995.

＊＊＊＊＊＊＊＊＊＊＊

實例

發明人 A 之專利於 2008 年 6 月 1 日發下，該專利包括了一物品。發明人 A 之專利之原說明書中揭露了該物品及製造物品之方法。但製造物品之方法在申請時並未請求。發明人 A 於 2010 年 5 月 1 日提出擴大範圍之再發證申請案將物品 A 之製法請求項加入，請問審查委員將如何處置此再發證申請案？

審查委員如認為該製法與製品乃屬不同之發明，將發下限制要求之審定書作限制要求，並將原物品作為推斷的被選擇的發明，並告知將只針對物品作審查，並告知申請人如果物品仍可核准，則製法部份需提出分割之再發證申請案。

§23.9　再發證申請案之審查與相關訴訟之關係

所有再發證之申請案都被視為是"特別狀態"（special status）之案件，會以特別快的速度處理及審查。而涉入訴訟之再發證案件又被視為更特別之案件，會比其他再發證案件以更快的速度處理。因此，審查委員在審查再發證申請案之前，會先查明再發證申請案是否有涉入訴訟。如果發現有涉入訴訟，而申請人並未指明，則審查委員在審定書（Office Action）中要求申請人提供相關之資訊。有涉及訴訟之再發證申請案，審查委員通常只會給予一個月之答辯期限，如要延期，則需依 37CFR 1.136(b)顯示有明確正當理由（clear justification）。

審查委員在審查一再發證申請案時，如果發現有正在進行中之相關訴訟時，通常會先暫時中止（suspend）審查，除非有下列四種情事發生：

(A)　實際上該訴訟已擱置（stayed）；

(B)　該訴訟已中止；

(C)　再發證申請案與該訴訟並無重要的重複議題；或

(D)　申請人希望再發證申請案被儘速審查。

如果實際上訴訟已擱置（stayed），則審查委員會以比其他再發證申請案更快之速度去審查。

MPEP 1442 Special Status [R-7]

All reissue applications are taken up "special," and remain "special" even *>if< applicant does not respond promptly.

All reissue applications, except those under suspension because of litigation, will be taken up for action ahead of other "special" applications; this means that all issues not deferred will be treated and responded to immediately. Furthermore, reissue applications involved in litigation will be taken up for action in advance of other reissue applications.

MPEP 1442.01 Litigation-Related Reissues [R-7]

During initial review, the examiner should determine whether the patent for which the reissue has been filed is involved in litigation, and if so, the status of that litigation. If the examiner becomes aware of litigation involving the patent sought to be reissued during examination of the reissue application, and applicant has not made the details regarding that litigation of record in the reissue application, the examiner, in the next Office action, will inquire regarding the specific details of the litigation.

* * * * * * * * * * *

Applicants will normally be given 1 month to reply to Office actions in all reissue applications *>that< are being examined during litigation, or after litigation had been stayed, dismissed, etc., to allow for consideration of the reissue by the Office. This 1-month period may be extended only upon a showing of clear justification **>under< 37 CFR 1.136(b). The Office action will inform applicant that the provisions of 37CFR 1.136(a) are not available. Of course, up to 3 months may be >initially< set for reply if the examiner>, consultating with his/her supervisor,< determines such a period is clearly justified.

MPEP 1442.02 Concurrent Litigation [R-7]

**>To< avoid **>duplicating< effort, action in reissue applications in which there is an indication of concurrent litigation will be suspended *>sua sponte< unless and until it is evident to the examiner, or the applicant indicates, that any one of the following applies:

(A) a stay of the litigation is in effect;

(B) the litigation has been terminated;

(C) there are no significant overlapping issues between the application and the litigation; or

(D) it is applicant's desire that the application be examined at that time.

* * * * * * * * * * *

MPEP 1442.03 Litigation Stayed [R-7]

All reissue applications, except those under suspension because of litigation, will be taken up for action ahead of other "special" applications; this means that all issues not deferred will be treated and responded to immediately. Furthermore, reissue applications involved in "stayed litigation" will be taken up for action in advance of other reissue applications. Great emphasis is placed on the expedited processing of such reissue applications. The courts are especially interested in expedited processing in the Office where litigation is stayed.

In reissue applications with "stayed litigation," the Office will entertain petitions under 37 CFR 1.182, which are accompanied by the fee under 37 CFR 1.17(f), to not apply the 2-month delay period stated in MPEP § 1441. Such petitions are decided by the Office of Patent Legal Administration.

＊＊＊＊＊＊＊＊＊＊＊＊

§23.10 再發證之效果

當再發證申請案發證時，原有之專利之放棄（surrender）就生效。再發證之專利被視為是原有之專利以修正之方式核准，故在再發證之專利的效果與一般之專利之效果相同，具有排除他人製造、銷售、要約銷售，使用及輸入至美國該專利所保護之發明之權利。但是其專利權不及於再發證專利申請案發證之前之先使用者之製造、銷售、要約銷售、使用及輸入至美國該發明之行為。亦即先使用者具有先使用權（intervening rights）。此先使用權亦包括在再發證申請案發證前已完成實質之準備的先使用者。

---

35 U.S.C. 252 Effect of reissue.

The surrender of the original patent shall take effect upon the issue of the reissued patent, and every reissued patent shall have the same effect and operation in law, on the trial of actions for causes thereafter arising, as if the same had been originally granted in such amended form, but in so far as the claims of the original and reissued patents are substantially identical, such surrender shall not affect any action then pending nor abate any cause of action then existing, and the reissued patent, to the extent that its claims are substantially identical with the original patent, shall constitute a continuation thereof and have effect continuously from the date of the original patent.

A reissued patent shall not abridge or affect the right of any person or that person's successors in business who, prior to the grant of a reissue, made, purchased, offered to sell, or used within the United States, or imported into the United States, anything patented by the reissued patent, to continue the use of, to offer to sell, or to sell to others to be used, offered for sale, or sold, the specific thing so made, purchased, offered for sale, used, or imported unless the making, using, offering for sale, or selling of such thing infringes a valid claim of the reissued patent which was in the original patent. The court before which such matter is in question may provide for the continued manufacture, use, offer for sale, or sale of the thing made, purchased, offered for sale, used, or imported as specified, or for the manufacture, use, offer for sale, or sale in the United States of which substantial preparation was made before the grant of the reissue, and the court may also provide for the continued practice of any process patented by the reissue that is practiced, or for the practice of which substantial preparation was made, before the grant of the reissue, to the extent and under such terms as the court deems equitable for the protection of investments made or business commenced before the grant of the reissue.

MPEP 1460 Effect of Reissue [R-2]

＊＊＊＊＊＊＊＊＊＊＊＊

The effect of the reissue of a patent is stated in 35 U.S.C. 252. With respect to the Office treatment of the reissued patent, the reissued patent will be viewed as if the original patent had been originally granted in the amended form provided by the reissue. >With respect to intervening rights resulting from the reissue of an original patent, the second paragraph of 35 U.S.C.252 provides for two separate and distinct defenses to patent infringement under the doctrine of intervening rights:

"Absolute" intervening rights are available for a party that "prior to the grant of a reissue, made, purchased, offered to sell, or used within the United States, or imported into the United States, anything patented by the reissued patent," and "equitable" intervening rights may be provided where "substantial preparation was made before the grant of the reissue." See BIC Leisure Prods., Inc., v. Windsurfing Int'l, Inc., 1 F.3d 1214, 1220, 27 USPQ2d 1671, 1676 (Fed. Cir. 1993).<

§23.11 再發證專利之維持費繳交

提出再發證申請並不改變原專利之維持費之繳交日（繳費期限）。亦即，如果原專利未繳維持費而導致專利過期，則提出再發證之申請不會再發證。審查委員在要對一再發證申請案進行審查之前會先確認原專利是否已繳交了維持費。而在再發證申請案之審查及答辯過程中，亦會再去檢查是否原專利需繳的維持費均已繳交。而如果原專利已不用再繳交維持費了（例如已繳完了最後一次（第 11.5 年）之維持費），則再發證申請案發證後亦不用繳交維持費。而一旦再發證申請案獲准專利（發證時），原專利就無到期之維持費，而是再發證之專利有到期之維持費。亦即，繳維持費需對再發證之專利繳交，而非對原專利繳交，而其繳費之期限為原專利之期限（日程）。另外，當一個以上之再發證申請案對原來的單一專利發證時，此多個再發證專利一不會改變原專利之繳維持費之日程。在此情況下，只要將維持費繳交至最後發證之再發證專利即可，而不必對所有的再發證專利均繳維持費。

MPEP 1415.01 Maintenance Fees on the Original Patent [R-7]

The filing of a reissue application does not alter the schedule of payments of maintenance fees on the original patent. If maintenance fees have not been paid on the original patent as required by 35 U.S.C. 41(b) and 37 CFR 1.20, and the patent has expired, no reissue patent can be granted. 35 U.S.C. 251, first paragraph, only authorizes the granting of a reissue patent for the unexpired term of the original patent. Once a patent has expired, the Director of the USPTO no longer has the authority under 35 U.S.C. 251 to reissue the patent. See In re Morgan, 990 F.2d 1230, 26USPQ2d 1392 (Fed. Cir. 1993).

The examiner should determine whether all required maintenance fees have been paid **>before< conducting an examination of a reissue application. In addition, during the process of preparing the reissue application for issue, the examiner should again determine whether all * maintenance fees >required to date< have been paid **.

* * * * * * * * * * *

PAYMENT OF MAINTENANCE FEES WHERE THE PATENT HAS BEEN REISSUED

Pursuant to 37 CFR 1.362(b), maintenance fees are not required for a reissue patent if the original patent that was reissued did not require maintenance fees.

Where the original patent that was reissued did require maintenance fees, the schedule of payments of maintenance fees on the original patent will continue for the reissue patent. 37 CFR 1.362(h). Once an original patent reissues, maintenance fees are no longer due in the original patent, but rather the maintenance fees are due in the reissue patent. This is because upon the issuance of the reissue patent, the original patent is surrendered and ceases to exist.

In some instances, more than one reissue *>patent< will be granted to replace a single original patent. The issuance of more than one reissue patent does not alter the schedule of payments of maintenance fees on the original patent. The existence of multiple reissue patents for one original patent can arise where multiple divisional reissue applications are filed for the same patent, and the multiple applications issue as reissue patents (all to replace the same original patent). In addition, a divisional application or continuation application of an existing reissue application may be filed, and both may then issue as reissue patents. In such instances, 35 U.S.C. 41 does not provide for the charging of more than one maintenance fee for the multiple reissues. Thus, **>only one maintenance fee is required for all the multiple reissue patents that replaced the single original patent

第 24 章　再審查程序

目　錄

§24.1　再審查程序之特點

　　依照美國專利法 35USC§302，任何人在專利可執行期間之任何時間可提出先前技術請求美國專利局對一專利之任何請求項作再審查（re-examination）。此再審查程序乃始於 1981 年 7 月 1 日，為一種 Ex Parte Reexamination 程序，之後，美國專利局又於 1999 年導入 Inter Partes Reexamination 程序。當有人提出先前技術請求美國專利局對一專利之某些請求項作再審查時，美國專利局會先決定是否有專利性之實質新問題（substantial new question of patentability）。如果判定有專利性之實質新問題，就會發下進行再審查之指令（order），進行再審查。Ex Parte 再審查申請案之審查程序與一般專利申請案之程序類似，但有一些特點，說明如下：

(A) 任何人（包括專利權人）在專利可執行期間（指專利權期滿加上六年之期間）之任何時間點均可提出再審查；

(B) 提出之先前技術限於符合 35 USC 102 及 103 之專利及印製之刊物；

(C) 要有專利性之實質新問題時才會進行再審查程序；

(D) 決定是否進行再審查程序會在提出再審查申請之三個月內完成，且隨後之審查程序列為特別優先；

(E) 再審查程序通常會一直進行（即使專利權人不作答辯）到完成，並發下再審查證書為止；

(G) 所有再審查案件之檔案及審查過程均開放大眾審閱；

(F) 請求項之範圍不能藉由修正再擴大。

---

35 U.S.C. 302 Request for reexamination.

Any person at any time may file a request for reexamination by the Office of any claim of a patent on the basis of any prior art cited under the provisions of section 301 of this title. The request must be in writing and must be accompanied by payment of a reexamination fee established by the Director pursuant to the provisions of section 41 of this title. The request must set forth the pertinency and manner of applying cited prior art to every claim for which reexamination is requested. Unless the requesting person is the owner of the patent, the Director promptly will send a copy of the request to the owner of record of the patent.

MPEP 2209 Ex Parte Reexamination [R-7]

Procedures for reexamination of issued patents began on July 1, 1981, the date when the reexamination provisions of Public Law 96-517 came into effect.

The reexamination statute and rules permit any person to file a request for an ex parte reexamination containing certain elements and the fee required under 37 CFR 1.20(c)(1). The Office initially determines if "a substantial new question of patentability" (35 U.S.C. 303(a)) is presented. If such a new question has been presented, reexamination will be ordered.

* * * * * * * * * * * *

The basic characteristics of ex parte reexamination are as follows:

(A) Anyone can request reexamination at any time during the period of enforceability of the patent;

(B) Prior art considered during reexamination is limited to prior art patents or printed publications applied under the appropriate parts of 35 U.S.C.102 and 103;

(C) A substantial new question of patentability must be present for reexamination to be ordered;

(D) If ordered, the actual reexamination proceeding is ex parte in nature;

(E) Decision on the request must be made no later than 3 months from its filing, and the remainder of proceedings must proceed with "special dispatch" within the Office;

(F) If ordered, a reexamination proceeding will normally be conducted to its conclusion and the issuance of a reexamination certificate;

(G) The scope of a claim cannot be enlarged by amendment;

(H) All reexamination and patent files are open to the public, but see paragraph (I) below;

(I) The reexamination file is scanned into IFW to provide an electronic format copy of the file. All public access to and copying of the reexamination file may be made from the electronic format copy available through PAIR. Any remaining paper files are not available to the public.

* * * * * * * * * * * *

§24.2　再審查申請案之種類

　　再審查程序（Reexamination）分為 Ex Parte 再審查與 Inter Partes 再審查兩種。Ex Parte 之再審查程序可由專利權人或第三者（可匿名）提出，而其程序在第 1 階段（美國專利局判定是否有專利性之實質新問題）之前之所有文件均會一起通知第三者（如果由第三者提出），但是在之後的實際審查程序，則第三者不會被通知，第三者亦不能對審定書作答辯。而 Inter Partes 再審查程序只能由第三者（不能匿名）提出，而且第三者在再審查之第二階段亦可參加相關程序，例如審定書（Office Action）發下，專利權人提出答辯後，第三者亦可對專利權人之答辯提出意見，亦可提出訴願等。另外，申請 Ex Parte 再審查之規費為 USD 2520，而申請 Inter Partes 再審查之規費則為 USD 8800。

§24.3　提出 Ex Parte 再審查申請案之要件

　　任何人，在一專利可執行之期間內之任何時間，可依據專利或刊物之先前技術對該專利之任一請求項提出 Ex Parte 再審查申請。一旦提出了 Ex Parte 再審查申請及繳交了規費（USD 2520），則此再審查申請會進行至完成，而不會放棄、撤回。所謂專利可執行期間乃為專利權終止之期間加上六年。而實用專利之專利權終止期間乃由維持年費是否繳交，是否有棄權書（disclaimer）而縮短專利權期限，及專利權期限之調整及延長來決定。

　　所謂任何人，亦包括法人，政府組織，專利權人本身亦可請求 Ex Parte 再審查申請。美國專利局局長亦可提出審查申請案，但此種情形很少見，通常是有和公共政策相關時才會提。一般會提出 Ex Parte 再審查申請者包括專利權人，被授權人，律師，抵觸程序申請人，ITC 程序之答辯人等。Ex Parte 再審查申請案之申請人亦可以匿名方式提出。

　　在 Ex Parte 再審查程序中，如果請求再審查者並非專利權人，則美國專利局之人員，特別是專利審查委員不會回答請求再審查者（requester）或與之討論任何先前技術及證據及是否一請求項會核准等問題。

如果請求再審查者（requester）委託一專利代理人或專利律師提出再審查，則需附上委任書。如果專利代理人或專利律師受另一人（法人）委託以專利代理人或專利律師作為請求人提出再審查，則此專利代理人或專利律師即為第三者之請求者。除非特別註明，則美國專利局所有關於再審查案件之連絡書信均會寄至再審查請求人之代表人（代理人處）。

ExParte 再審查之申請需附上以下文件：

(1) 指出根據專利及刊物之先前技術之各專利性之實質新問題之聲明；

(2) 將請求再審查之請求項列出，並詳細說明與先前技術之關係及請求項與先前技術之區別；

(3) 作為先前技術之專利及刊物之影本，以及相關之英譯本（如果先前技術並非英文者）；

(4) 請求再審查之專利之整份影本，包括首頁、圖式、說明書及請求項，以及棄權書，證書之修正本（certificate of correction）或再審查證書之影本；

(5) 證明書，證明已將一份再審查請求書之影本遞交專利權人，並需指出被送交者之姓名及住址。

MPEP 2210 Request for Ex Parte Reexamination [R-7]

＊＊＊＊＊＊＊＊＊＊＊

Any person, at any time during the period of enforceability of a patent, may file a request for ex parte reexamination by the U.S. Patent and Trademark Office of any claim of the patent based on prior art patents or printed publications.

＊＊＊＊＊＊＊＊＊＊＊

After the request for reexamination, including the entire fee for requesting reexamination, is received in the Office, no abandonment, withdrawal, or striking of the request is possible, regardless of who requests the same.

＊＊＊＊＊＊＊＊＊＊＊

MPEP 2211 Time for Requesting *>Ex Parte Reexamination< [R-2]

The period of enforceability is determined by adding 6 years to the date on which the patent expires. The patent expiration date for a utility patent, for example, is determined by taking into account the term of the patent, whether maintenance fees have been paid for the patent, * whether any disclaimer was filed as to the patent to shorten its term>, any patent term extensions or adjustments for delays within the Office under 35 U.S.C. 154 (see MPEP § 2710, et seq.), and any patent term extensions available under 35 U.S.C. 156 for premarket regulatory review (see MPEP § 2750 et. seq.)<.

＊＊＊＊＊＊＊＊＊＊＊＊

MPEP 2212 Persons Who May File a Request >for Ex Parte Reexamination< [R-2]

＊＊＊＊＊＊＊＊＊＊＊＊

35 U.S.C. 302 and 37 CFR 1.510(a)both indicate that "any person" may file a request for reexamination of a patent. Accordingly, there are no persons who are excluded from being able to seek reexamination. Corporations and/or governmental entities are included within the scope of the term "any person." The patent owner can ask for reexamination which will be limited to an ex parte consideration of prior >art< patents or printed publications. If the patent owner wishes to have a wider consideration of issues by the Office, including matters such as prior public use or >on< sale, the patent owner may file a reissue application. It is also possible for the *>Director of the Office< to initiate reexamination on the *>Director's< own initiative under37 CFR 1.520.Reexamination will be initiated by the *>Director's< on a very limited basis, such as where a general public policy question is at issue and there is no interest by "any other person." Some of the persons likely to use reexamination are patentees, licensees, potential licensees, attorneys without identification of their real client in interest, infringers, potential exporters, patent litigants, interference applicants, and International Trade Commission respondents. The name of the person who files the request will not be maintained in confidence.

MPEP 2212.01 Inquiries from Persons Other Than the Patent Owner [R-7]

＊＊＊＊＊＊＊＊＊＊＊＊

Employees of the Office, particularly patent examiners who conducted a concluded reexamination proceeding, should not discuss or answer inquiries from any person outside the Office as to whether a certain reference or other particular evidence was considered during the proceeding and whether a claim would have been allowed over that reference or other evidence had it been considered during the proceeding.

＊＊＊＊＊＊＊＊＊＊＊＊

MPEP 2213 Representative of Requester [R-7]

* * * * * * * * * * * *

Where an attorney or agent files a request for an identified client (the requester), he or she may act under either a power of attorney from the client, or act in a representative capacity under 37 CFR 1.34*, see 37 CFR 1.510(f).

* * * * * * * * * * *

>If an attorney or agent files a request for reexamination for another entity (e.g., a corporation) that wishes to remain anonymous, then that attorney or agent is the third party requester.<

If any question of authority to act is raised, proof of authority may be required by the Office.

All correspondence for a requester that is not the patent owner **>is< addressed to the representative of the requester, unless a specific indication is made to forward correspondence to another address.

* * * * * * * * * * *

37 CFR 1.510 Request for ex parte reexamination.

(a) Any person may, at any time during the period of enforceability of a patent, file a request for an ex parte reexamination by the Office of any claim of the patent on the basis of prior art patents or printed publications cited under § 1.501. The request must be accompanied by the fee for requesting reexamination set in § 1.20(c)(1).

(b) Any request for reexamination must include the following parts:

(1) A statement pointing out each substantial new question of patentability based on prior patents and printed publications.

(2) An identification of every claim for which reexamination is requested, and a detailed explanation of the pertinency and manner of applying the cited prior art to every claim for which reexamination is requested. If appropriate the party requesting reexamination may also point out how claims distinguish over cited prior art.

(3) A copy of every patent or printed publication relied upon or referred to in paragraph (b)(1) and (2) of this section accompanied by an English language translation of all the necessary and pertinent parts of any non-English language patent or printed publication.

(4) A copy of the entire patent including the front face, drawings, and specification/claims (in double column format) for which reexamination is requested, and a copy of any disclaimer, certificate of correction, or reexamination certificate issued in the patent. All copies must have each page plainly written on only one side of a sheet of paper.

(5) A certification that a copy of the request filed by a person other than the patent owner has been served in its entirety on the patent owner at the address as provided for in § 1.33(c).

The name and address of the party served must be indicated. If service was not possible, a duplicate copy must be supplied to the Office.

§24.4　專利性之實質新問題

依照 35USC§304，美國專利局必需決定是否有專利性之實質性新問題，如果有，才會發下再審查之指令（order）進行再審查，而再審查請求者需提出專利性之實質性新問題之聲明。亦即，再審查請求者必需指出所提出之專利之問題實質上如何與先前審查時不同。亦即，需提出所引用之先前技術新的，非重覆的技術上之教示，其是在先前之審查過程中未被考慮，未被討論者。

此專利性之實質性問題可以是依據先前美國專利局所考慮之先前技術，但是新的以不同的方式被解讀分析者。

另外，專利及刊物之先前技術以外之先前技術，例如公開使用及銷售等均不被當作是專利性之實質新問題之先前技術。在決定是否有專利性之實質新問題時亦會考慮宣誓書、書面證據等解釋先前技術之相關日期之內容。

＊＊＊＊＊＊＊＊＊＊＊

MPEP 2216 Substantial New Question of Patentability [R-7]

＊＊＊＊＊＊＊＊＊＊＊

The request *>must< point out how any questions of patentability raised are substantially different from those raised in the previous examination of the patent before the Office. **

>It is not sufficient that a request for reexamination merely proposes one or more rejections of a patent claim or claims as a basis for reexamination. It must first be demonstrated that a patent or printed publication that is relied upon in a proposed rejection presents a new, non-cumulative technological teaching that was not previously considered and discussed on the record during the prosecution of the application that resulted in the patent for which reexamination is requested, and during the prosecution of any other prior proceeding involving the patent for which reexamination is requested. See also MPEP § 2242.

The legal standard for ordering ex parte reexamination, as set forth in 35 U.S.C.303(a), requires a substantial new question of patentability. The substantial new question of patentability may be based on art previously considered by the Office if the reference is presented in a new light or a different way that escaped review during earlier examination.

* * * * * * * * * * *

After the enactment of the Patent and Trademark Office Authorization Act of 2002 ("the 2002 Act"), a substantial new question of patentability can be raised by patents and printed publications "previously cited by or to the Office or considered by the Office" ("old art"). The 2002 Act did not negate the statutory requirement for a substantial new question of patentability that requires raising new questions about pre-existing technology. In the implementation of the 2002 Act, MPEP §2242, subsection II.A. was revised. The revision permits raising a substantial new question of patentability based solely on old art, but only if the old art is "presented/viewed in a new light, or in a different way, as compared with its use in the earlier concluded examination(s), in view of a material new argument or interpretation presented in the request." Thus, a request may properly raise an substantial new question of patentability by raising a material new analysis of previously considered reference(s) under the rationales authorized by KSR. <

Questions relating to grounds of rejection other than those based on prior art patents or printed publications should not be included in the request and will not be considered by the examiner if included. Examples of such questions that will not be considered are public use, on sale, and *>conduct by parties<.

Affidavits or declarations or other written evidence which explain the contents or pertinent dates of prior art patents or printed publications in more detail may be considered in reexamination. See MPEP § 2258.

§24.5 Ex Parte 再審查申請案之審查程序

當 Ex Prate 再審查申請案所有的文件齊備，付與申請日及案號（90／×××/×××）之後約 4-5 週，此 Ex Parte 再審查之申請案會公告在專利公報上，開放給大眾檢閱。

而審查委員必需在 Ex Prate 再審查申請日之三個月內決定是否有專利性之實質新問題，以決定是否發下進行再審查之指令（order）。而審查委員在決定是否有專利性之實質新問題之前必需請求科技資訊中心（STIC，Science and Technology Information Center）檢索是否再審查申請案有涉入訴訟。如果有涉入訴訟法則會註明，並請中央再審查單位（CRU, Central Reexamination Unit）注意，而主管級審查委員（SPE）就會審閱（review）審查委員作之決定及審定。

當審查委員（Primary Examiner）初步決定要發下再審查之指令，亦即，認為有專利性之實質新問題時，審查委員會先起草一初步之理由並通知 CRU 之 SPE，其欲發下再審查之指令。此時 SPE 就會召集另二位評論員（conferee），組成三人審查小組召開會議，如果會議結果決定進行再審查，則會發下再審查之指令，如會議結果不進行再審查，則審查委員將重新評估是否要發下指令進行再審查。如果審查委員初步決定要否定再審查之申請，亦即無專利性之實質新問題，不進行再審查，同樣會通知 CRU 之 SPE，由 SPE 召集兩位評論員組成小組，開會決定是否要否定再審查申請案。

再審查申請案，確定被否定之後，再審查申請案之管轄權就回到 CRU，如果請求者在一個月內提出請願（petition），則再審查申請案就會送到 CRU，由 CRU 之審查長作決定是否要核准此請願，如果核准請願則會將再審查申請案發給另一審查委員重新決定是否要發下進行再審查申請之指令，如果審查長不核准請願，則為最後之決定，不能上訴。

在審查委員發下進行再審查序之指令（request is granted）之兩個月之內，專利權人可提出聲明（statement），指出為何請求項是具有專利性的及任何縮小請求項範圍之修正。如果請求者不是專利權人時必需同時將此聲明及修正送交請求者。請求者可在收到聲明書及修正之二個月內提出回覆（不能延期），而回覆不限於聲明書中提到之議題，可再提供另外的專利及刊物作為先前技術。而如果專利權人未作聲明（修正）時，請求者就不能提出回覆。

上述期限過後，進入實際的審查程序之後，所有的程序就會是 Ex Prate，亦即第三者之請求者被排除在審查程序之外，不再能表達意見。審查程序為一特別加速程序。審查乃由決定是否發下再審查指令之同一審查委員進行之。而審查委員引用之先前技術通常（主要）只是在提出 Ex Prate 再審查請求中提到之專利及刊物，但是有些情形下審查委員亦會考慮以下之先前技術：

(A) 由另一個再審查請求者引證之先前技術；

(B) 專利權人之聲明或第三請求者之回覆中提到之先前技術；

(C) 專利權人之 IDS 中引證之先前技術；

(D) 審查委員在檢索中發現之先前技術；

(E) 在先前審查時卷宗記錄中的先前技術；

(F) 在再審查指令發下前依照 37CFR1.501 引證之先前技術。

Ex Prate 再審查申請案之審定書（Office Action）的答辯期限（縮短的法定期限 SSP）通常只有二個月，而如果再審查申請是因法院之命令或有相關的訴訟被擱置時，則 SSP 會設定只有一個月。或者，如果(A)有相關的訴訟與此再審查申請案同時進行中，且(B)再審查申請案已進行了一年以上時，美國專利局亦可能將對審定書之答辯期限設定為一個月或 30 天（視哪一日期為長）。而對審定書答辯之延期並非依照 37 CFR 1.136(a)(b) 之條款，而是依照 37 CFR 1.550(c)之條款。亦即，要請求延期答辯需 (A)到期前以書面提出，(B)必需有充份的理由，(C)必需繳交延期規費。一般以申請人在國外，來不及回覆等理由通常不會被接受，而期限到期就會造成再審查申請案被放棄。

再審查申請案之最終審定書發下後之審查流程與一般之專利申請案相同，亦即專利權人可提出訴願聲明（notice of appeal）及訴願理由，如最終不服訴願決定可向聯邦巡迴法庭上訴。

Ex Parte 再審查申請案之審查流程圖如下所示：

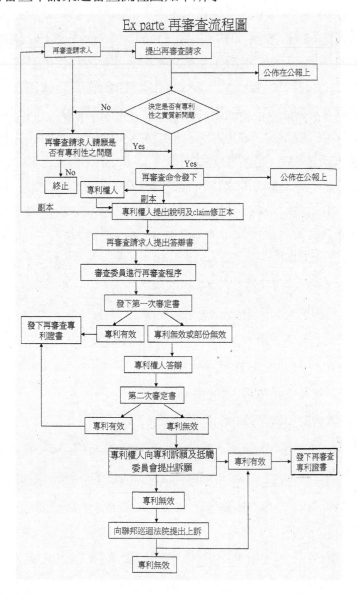

Ex parte 再審查流程圖

MPEP 2229 Notice of Request for Ex Parte Reexamination in Official Gazette [R-7]

Notice of filing of all complete ex parte reexamination requests will be published in the Official Gazette, approximately 4 - 5 weeks after filing.

＊＊＊＊＊＊＊＊＊＊＊＊

MPEP 2232 Public Access [R-7]

Reexamination files are open to inspection by the general public by way of the Public PAIR via the USPTO Internet site.

＊＊＊＊＊＊＊＊＊＊＊

MPEP 2240 Decision on Request [R-7]

Before making a determination on the request for reexamination, the examiner must request a litigation * search by the Scientific and Technical Information Center (STIC) to check if the patent has been, or is, involved in litigation.

＊＊＊＊＊＊＊＊＊＊＊

If the patent is or was involved in litigation, and a paper referring to the court proceeding has been filed, reference to the paper by number should be made in the "Litigation Review" box on the reexamination IFW file jacket form as, for example, "litigation; see paper filed 7-14-2005. If a litigation records search is already noted on the file, the examiner need not repeat or update it.

If litigation has concluded or is taking place in the patent on which a request for reexamination has been filed, the request must be promptly brought to the attention of the Central Reexamination Unit (CRU) **>Supervisory Patent Examiner (SPE)<, who should review the decision on the request and any examiner's action to ensure that it conforms to the current Office litigation policy and guidelines. See MPEP § 2286.

35 U.S.C. 303 requires that within 3 months following the filing of a request for reexamination, the Director of the USPTO will determine whether or not the request raises a "substantial new question of patentability" affecting any claim of the patent of which reexamination is desired.

＊＊＊＊＊＊＊＊＊＊＊＊

MPEP 2246 Decision Ordering Reexamination [R-7]

* * * * * * * * * * * *

If a request for reexamination is granted, the examiner's decision granting the request will conclude that a substantial new question of patentability has been raised by (A) identifying all claims and issues, (B) identifying the patents and/or printed publications relied on, and (C) providing a brief statement of the rationale supporting each new question.

* * * * * * * * * * * *

I.PANEL REVIEW CONFERENCE

After an examiner has determined that the reexamination proceeding is ready for granting reexamination, the examiner will formulate a draft preliminary order granting reexamination. The examiner will then inform his/her **>Central Reexamination Unit (CRU) Supervisory Patent Examiner (SPE)< of his/her intent to issue an order granting reexamination. The *>CRU SPE< will convene a panel review conference, and the conference members will review the matter. See MPEP § 2271.01 for the make-up of the panel. If the conference confirms the examiner's preliminary decision to grant reexamination, the proposed order granting reexamination shall be issued and signed by the examiner, with the two other conferees initialing the action (as "conferee") to indicate their presence in the conference. If the conference does not confirm the examiner's preliminary decision, the examiner will reevaluate and issue an appropriate communication.

* * * * * * * * * * * *

MPEP 2247 Decision on Request for Reexam-ination, Request Denied [R-7]

The request for reexamination will be denied if a substantial new question of patentability is not found based on patents or printed publications.

If the examiner concludes that no substantial new question of patentability has been raised, the examiner should prepare a decision denying the reexamination request. Form paragraph 22.02 should be used as the introductory paragraph in a decision denying reexamination.

* * * * * * * * * * * *

PANEL REVIEW CONFERENCE

After an examiner has determined that the reexamination proceeding is ready for denying reexamination, the examiner will formulate a draft preliminary order denying reexamination.

The examiner will then inform his/her **>CRU Supervisory Patent Examiner (SPE)< of his/her intent to issue an order denying reexamination. The *>CRU SPE< will convene a panel review conference, and the conference members will review the matter. See MPEP § 2271.01 for the make-up of the panel. If the conference confirms the examiner's preliminary decision to deny reexamination, the proposed order denying reexamination shall be issued and signed by the examiner, with the two other conferees initialing the action (as "conferee") to indicate their presence in the conference. If the conference does not confirm the examiner's preliminary decision, the examiner will reevaluate and issue an appropriate communication.

****

MPEP 2248 Petition From Denial of Request [R-7]

****

PROCESSING OF PETITION UNDER 37 CFR 1.515(c)

After a request for reexamination has been denied, jurisdiction over the reexamination proceeding is retained by the Central Reexamination Unit (CRU), to await the possibility of a petition seeking review of the examiner's determination refusing reexamination. If a petition seeking review of the examiner's determination refusing reexamination is not filed within one (1) month of the examiner's determination, the CRU will then process the reexamination file as a concluded reexamination file. See MPEP § 2247 and § 2294.

If a petition seeking review of the examiner's determination refusing reexamination is filed, it is forwarded (together with the reexamination file) to the Office of the CRU Director for decision. Where a petition is filed, the CRU Director will review the examiner's determination that a substantial new question of patentability has not been raised. The CRU Director's review will be de novo. Each decision by the CRU Director will conclude with the paragraph:

This decision is final and nonappealable. See 35 U.S.C. 303(c) and 37 CFR 1.515(c). No further communication on this matter will be acknowledged or considered.

If the petition is granted, the decision of the CRU Director should include a sentence setting a 2-month period for filing a statement under 37 CFR 1.530; the reexamination file will then be returned to the **>CRU Supervisory Patent Examiner (SPE)< of the art unit that will handle the reexamination for consideration of reassignment to another examiner.

****

MPEP 2249 Patent Owner's Statement [R-7]

****

If reexamination is ordered, the decision will set a period of not less than 2 months within which period the patent owner may file a statement and any narrowing amendments to the patent claims. If necessary, an extension of time beyond the 2 months may be requested under 37 CFR 1.550(c) by the patent owner. Such request is decided by the Technology Center (TC) or Central Reexamination Unit (CRU) Director.

Any statement filed must clearly point out why the patent claims are believed to be patentable, considering the cited prior art patents or printed publications alone or in any reasonable combination.

A copy of the statement must be served by the patent owner on the requester, unless the request was filed by the patent owner.

* * * * * * * * * * * *

MPEP 2251 Reply by Third Party Requester

* * * * * * * * * * *

If the patent owner files a statement in a timely manner, the third party requester is given a period of 2 months from the date of service to reply. Since the statute, 35 U.S.C. 304, provides this time period, there will be no extensions of time granted.

The reply need not be limited to the issues raised in the statement. The reply may include additional prior art patents and printed publications and may raise any issue appropriate for reexamination.

If no statement is filed by the patent owner, no reply is permitted from the third party requester.

* * * * * * * * * * * *

MPEP 2254 Conduct of Ex Parte Reexamination Proceedings [R-7]

* * * * * * * * * * *

Once ex parte reexamination is ordered pursuant to 35 U.S.C. 304 and the times for submitting any responses to the order have expired, no further active participation by a third party reexamination requester is allowed, and no third party submissions will be acknowledged or considered unless they are in accordance with 37 CFR 1.510. The reexamination proceedings will be ex parte, even if ordered based on a request filed by a third party, because this was the intention of the legislation. Ex parte proceedings preclude the introduction of arguments and issues by the third party requester which are not within the intent of 35 U.S.C. 305 ("reexamination will be conducted according to the procedures established for initial examination under the provisions of sections 132 and 133 of this title").

* * * * * * * * * * *

The examination will be conducted in accordance with 37 CFR 1.104, 1.105, 1.110-1.113, and 1.116 (35 U.S.C. 132 and 133) and will result in the issuance of a reexamination certificate under 37 CFR 1.570. The proceeding shall be conducted with special dispatch within the Office pursuant to 35 U.S.C. 305, last sentence. A full search will not routinely be made by the examiner. The third party reexamination requester will be sent copies of Office actions and the patent owner must serve responses on the requester. Citations submitted in the patent file prior to issuance of an order for reexamination will be considered by the examiner during the reexamination. Reexamination will proceed even if the copy of the order sent to the patent owner is returned undelivered. The notice under 37 CFR 1.11(c) is constructive notice to the patent owner and lack of response from the patent owner will not delay reexamination. See MPEP §2230.

* * * * * * * * * * *

MPEP 2255 Who Reexamines [R-5]

* * * * * * * * * * *

The examination will ordinarily be conducted by the same patent examiner ** who made the decision on whether the reexamination request should be granted. See MPEP § 2236.

* * * * * * * * * * *

MPEP 2256 Prior Art Patents and Printed Publications Reviewed by Examiner in Reexamination [R-7]

Typically, the primary source of prior art will be the patents and printed publications cited in the request for ex parte reexamination.

Subject to the discussion provided below in this section, the examiner must also consider patents and printed publications:

(A) cited by another reexamination requester under 37CFR 1.510 or 37CFR 1.915;

(B) cited in a patent owner's statement under 37 CFR 1.530 or a requester's reply under 37 CFR 1.535 if they comply with 37 CFR 1.98;

(C) cited by the patent owner under a duty of disclosure (37 CFR 1.555)in compliance with 37 CFR 1.98;

(D) discovered by the examiner in searching;

(E) of record in the patent file from earlier examination; and

(F) of record in the patent file from any 37 CFR 1.501 submission prior to date of an order if it complies with 37 CFR 1.98.

＊＊＊＊＊＊＊＊＊＊＊

MPEP 2263 Time for Response [R-5]

A shortened statutory period of 2 months will be set for response to Office actions in reexaminations, except *>as follows. Where< the reexamination results from a court order or litigation is stayed for purposes of reexamination, ** the shortened statutory period will be set at 1 month. >In addition, if (A) there is litigation concurrent with an ex parte reexamination proceeding and (B) the reexamination proceeding has been pending for more than one year, the Director or Deputy Director of the Office of Patent Legal Administration (OPLA), Director of the Central Reexamination Unit (CRU), Director of the Technology Center (TC) in which the reexamination is being conducted, or a Senior Legal Advisor of the OPLA, may approve Office actions in such reexamination proceeding setting a one-month or thirty days, whichever is longer, shortened statutory period for response rather than the two months usually set in reexamination proceedings.

＊＊＊＊＊＊＊＊＊＊＊

MPEP 2265 Extension of Time [R-7]

＊＊＊＊＊＊＊＊＊＊＊

An extension of time in an ex parte reexamination proceeding is requested pursuant to 37 CFR 1.550(c). Accordingly, a request for an extension (A) must be filed on or before the day on which action by the patent owner is due and (B) must set forth sufficient reason for the extension, and (C) must be accompanied by the petition fee set forth in 37 CFR 1.17(g). Requests for an extension of time in an ex parte reexamination proceeding will be considered only after the decision to grant or deny reexamination is mailed. Any request filed before that decision will be denied.

＊＊＊＊＊＊＊＊＊＊＊

MPEP 2273 Appeal in Ex Parte Reexamination [R-7]

＊＊＊＊＊＊＊＊＊＊＊

A patent owner who is dissatisfied with the primary examiner's decision to reject claims in an ex parte reexamination proceeding may appeal to the Board of Patent Appeals and Interferences for review of the examiner's rejection by filing a notice of appeal within the required time. A third party requester may not appeal, and may not participate in the patent owner's appeal.

＊＊＊＊＊＊＊＊＊＊＊

In an ex parte reexamination filed on or after November 29, 1999, the patent owner may appeal to the Board only after the final rejection of the claims. This is based on the current version of 35 U.S.C. 134.

＊＊＊＊＊＊＊＊＊＊＊

MPEP 2279 Appeal to Courts [R-3]

A patent owner >who is< not satisfied with the decision of the Board of Patent Appeals and Interferences may seek judicial review.

＊＊＊＊＊＊＊＊＊＊＊

In an ex parte reexamination filed on or after November 29, 1999, the patent owner may appeal the decision of the Board of Patent Appeals and Interferences only to the United States Court of Appeals for the Federal Circuit pursuant to 35 U.S.C. 141.

＊＊＊＊＊＊＊＊＊＊＊

§24.6　Ex Parte 再審查申請案與其他法律程序之關係

依照 37CFR 1.565，在 Ex Prate 再審查申請案之審查程序中，專利權人必需告知及提供資料給美國專利局任何之前或正在進行中之與該專利關的法律程序，例如：抵觸程序、再發證案件、Ex Prate 再審查申請案，Inter Prates 再審查申請案，訴訟案以及這些相關法律程序之結果。

當審查委員發下指令進行一專利之 Ex Prate 再審查申請案之審查程序時，如在之前有另一對此專利之 Ex Prate 再審查申請案時，通常美國專利局會將此二再審查申請案合併審查。但有時亦可能將之前的再審查申請案暫時中止審查，以等專利權人對後來之再審查申請案中作聲明（statement）及第三請求者作回覆後再進行合併審查。合併審查通常是後案併入前案由前一 Ex Prate 再審查申請案之審查委員來審查。

當 Ex Prate 再審查申請案在被審查之期間涉入一抵觸程序（interference proceeding）時，美國專利局之一般政策是不將再審查申請案之審查程序延緩或擱置（stayed），因為再審查申請案是以特別加速審查之方式進行。

　　如果有一專利同時請求 Ex Prate 再審查程序及再發證程序時，美國專利局不會分別處置以防止不一致或互相衝突之請求項（claims）被分別導入兩個不同之程序之中。因此，當有對一專利同時有 Ex Prate 再審查申請案及再發證申請案時，美國專利局會(A)將二程序合併處置或(B)擱置其中一程序。

　　如果 Ex Prate 再審查申請案涉入一專利訴訟，則美國專利局會依以下之情形作處置：

Ⅰ：如果請求者在請求 Ex Prate 再審查申請時指出：(A)是因法院之命令或訴訟雙方之同意提出 Ex Prate 再審查申請案，或(B)訴訟已被擱置以等待再審查請求之提出時，審查委員會在提出再審查請求之 6 週內決定是否發下進行再審查之指令，而且之後之所有程序均會以加速方式處理，例如對審定書答辯期限為一個月，而非二個月。

Ⅱ：當審查委員正在決定是否有專利性之實質新問題時，法院已對專利之請求項作出有效之判決時，並不表示沒有專利性之實質新問題，故審查委員仍可逕自決定是否要發下進行再審查程序之指令。但是當審查委員在決定是否有專利性之實質新問題時，法院已發下最終之判決，判定專利之請求項為無效或不具執行力時，則審查委員就不會再提出專利性之實質新問題，亦即不會發下進行再審查程序之指令。

Ⅲ：如果審查委員已發下進行再審查程序之指令，但法院對專利請求項之有效性尚未判決時，則再審查申請案之審查程序就會一直進行到知道法院判決的結果為止。

Ⅳ：如果審查委員已發下進行再審查程序之指令，而在審查期間，法院發下最終判決專利之請求項為有效，則此判決並不具拘束力，再審查申請案之審查仍會繼續下去，但如果法院之最終判決是請求項無效，則有拘束力，再審查申請案會被撤回。

MPEP 2283 Multiple Copending Ex Parte Reexamination Proceedings [R-7]

＊＊＊＊＊＊＊＊＊＊＊＊

I.PROCEEDINGS MERGED

**>

Where a second request for reexamination is filed and reexamination is ordered, and a first reexamination proceeding is pending, 37 CFR 1.565(c) provides that the proceedings will usually be merged.

* * * * * * * * * * * *

II.WHEN PROCEEDING IS SUSPENDED

It may also be desirable in certain situations to suspend a proceeding for a short and specified period of time. For example, a suspension of a first reexamination proceeding may be issued to allow time for the patent owner's statement and the requester's reply in a second proceeding prior to merging.

* * * * * * * * * * *

III.MERGER OF REEXAMINATIONS

The second request (i.e., Request 2) should be processed as quickly as possible and assigned to the same examiner to whom the first request (i.e., Request 1) is assigned

* * * * * * * * * * * *

MPEP 2284 Copending Ex Parte Reexamination and Interference Proceedings [R-5]

* * * * * * * * * * *

A patent being reexamined in an ex parte reexamination proceeding may be involved in an interference proceeding with at least one application, where the patent and the application are claiming the same patentable invention, and at least one of the application's claims to that invention are patentable to the applicant. See MPEP Chapter 2300.

The general policy of the Office is that a reexamination proceeding will not be delayed, or stayed, because of an interference or the possibility of an interference. The reason for this policy is the requirement of 35 U.S.C. 305 that all reexamination proceedings be conducted with "special dispatch" within the Office. In general, the Office will follow the practice of making the required and necessary decisions in the reexamination proceeding and, at the same time, going forward with the interference to the extent desirable.

* * * * * * * * * * *

MPEP 2285 Copending Ex Parte Reexamination and Reissue Proceedings [R-7]

＊＊＊＊＊＊＊＊＊＊＊＊

The general policy of the Office is that a reissue application examination and an ex parte reexamination proceeding will not be conducted separately at the same time as to a particular patent. The reason for this policy is to permit timely resolution of both proceedings to the extent possible and to prevent inconsistent, and possibly conflicting, amendments from being introduced into the two proceedings on behalf of the patent owner. Accordingly, if both a reissue application and an ex parte reexamination proceeding are pending concurrently on a patent, a decision will normally be made (A) to merge the two proceedings or (B) to stay one of the two proceedings.

＊＊＊＊＊＊＊＊＊＊＊＊

MPEP 2286 Ex Parte Reeexamination and Litigation Proceedings [R-7]

I. COURT->ORDERED/< SANCTIONED REEXAMINATION PROCEEDING, LITIGATION STAYED FOR REEXAMINATION, OR EXTENDED PENDENCY OF REEXAMINATION PROCEEDING CONCURRENT WITH LITIGATION

*>Where a< request for ex parte reexamination * indicates (A) that it is filed as a result of >an order by a court or< an agreement by parties to litigation which agreement is sanctioned by a court, or (B) that litigation is stayed for the filing of a reexamination request >, the request< will be taken up by the examiner for decision 6 weeks after the request was filed >, and all aspects of the proceeding will be expedited to the extent possible<. See MPEP § 2241. If reexamination is ordered, the examination following the statement by the patent owner under 37 CFR 1.530 and the reply by the requester under 37 CFR 1.535 will be expedited to the extent possible. Office actions in these reexamination proceedings will normally set a 1-month shortened statutory period for response rather than the 2 months usually set in reexamination proceedings.

＊＊＊＊＊＊＊＊＊＊＊＊

II. FEDERAL COURT DECISION KNOWN TO EXAMINER AT THE TIME THE DETERMINATION ON THE REQUEST FOR REEXAMINATION IS MADE

If a Federal Court decision on the merits of a patent is known to the examiner at the time the determination on the request for ex parte reexamination is made, the following guidelines will be followed by the examiner, whether or not the person who filed the request was a party to the litigation. When the initial question as to whether the prior art raises a substantial new question of patentability as to a patent claim is under consideration, the existence of a final court decision of claim validity in view of the same or different prior art does not necessarily mean that no new question is present.

Thus, while the Office may accord deference to factual findings made by the court, the determination of whether a substantial new question of patentability exists will be made independently of the court's decision on validity as it is not controlling on the Office. A non-final holding of claim invalidity or unenforceability will not be controlling on the question of whether a substantial new question of patentability is present. A final holding of claim invalidity or unenforceability (after all appeals), however, is controlling on the Office. In such cases, a substantial new question of patentability would not be present as to the claims held invalid or unenforceable.

* * * * * * * * * * *

III. REEXAMINATION WITH CONCURRENT LITIGATION BUT ORDERED PRIOR TO FEDERAL COURT DECISION

In view of the statutory mandate to make the determination on the request within 3 months, the determination on the request based on the record before the examiner will be made without awaiting a decision by the Federal Court. It is not realistic to attempt to determine what issues will be treated by the Federal Court prior to the court decision. Accordingly, the determination on the request will be made without considering the issues allegedly before the court. If an ex parte reexamination is ordered, the reexamination will continue until the Office becomes aware that a court decision has issued.

* * * * * * * * * * *

IV. FEDERAL COURT DECISION ISSUES AFTER EX PARTE REEXAMINATION ORDERED

The issuance of a final Federal Court decision upholding validity during an ex parte reexamination also will have no binding effect on the examination of the reexamination. This is because the court states in Ethicon v. Quigg, 849 F.2d 1422, 1428, 7 USPQ2d 1152, 1157 (Fed. Cir. 1988) that the Office is not bound by a court's holding of patent validity and should continue the reexamination.

* * * * * * * * * * *

On the other hand, a final Federal Court holding of invalidity or unenforceability (after all appeals), is binding on the Office. Upon the issuance of a final holding of invalidity or unenforceability, the claims being examined which are held invalid or unenforceable will be withdrawn from consideration in the reexamination.

* * * * * * * * * * *

§24.7 Ex Parte 再審查證書之發行內容及先使用權

一旦審查委員下發再審查之指令進行再審查程序，再審查之程序就不能停止。當再審查程序結束時，美國專利局就會發下再審查證書，即使申請人對審定書不答辯，美國專利局仍會發下再審查證書。再審查證書會記載再審查程序之結果及內容，通常包括以下項目：

(A) 刪除不能給予專利之請求項；

(B) 確認可予專利之請求項；

(C) 加入可予專利之修正的請求項或新的請求項；

(D) 對再審查程序中認可之說明書之修正；

(E) 加入專利權人之任何法定棄權書或尾端棄權書；

(F) 指出在其他程序被判定為無效之未修改的請求項；

(G) 指出未作再審查之請求項；

(H) 將再審查證書寄給專利權人及副本給第三請求者；

(I) 指出可予專利且依附至修正之請求項之請求項。

Ex Prate 再審查證書之格式與一般專利證書之首頁的格式大致相同。但在最上端印上"REEXAMINATION CERTIFICATE"再審查證書之專利號碼為原來之專利號碼之前（2001年1月2日前發證之再審查案）或之後（2001年1月2日以後發證之再審查案）加上二碼，例如2001年1月2日前發下再審查證書為B1，而2001年1月2日以後發下者為C1。（號碼1表示第1次的再審查證書，如有第2次則為C2）。

另外，在再審查證書發下之前之先使用者與再發證之情形一樣，具有先使用權。

MPEP 2288 Issuance of Ex Parte Reexamination Certificate [R-7]

＊＊＊＊＊＊＊＊＊＊＊

Since abandonment is not possible in a reexamination proceeding, a reexamination certificate will be issued >and published< at the conclusion of the proceeding in each patent in which a reexamination proceeding has been ordered under 37 CFR 1.525 except where the reexamination has been concluded by vacating the reexamination proceeding or by the grant of a reissue patent on the same patent in which case.

The reexamination certificate will set forth the results of the proceeding and the content of the patent following the reexamination proceeding. The certificate will:

(A) cancel any patent claims determined to be unpatentable;

(B) confirm any patent claims determined to be patentable;

(C) incorporate into the patent any amended or new claims determined to be patentable;

(D) make any changes in the description approved during reexamination;

(E) include any statutory disclaimer or terminal disclaimer filed by the patent owner;

(F) identify unamended claims which were held invalid on final holding by another forum on any grounds;

(G) identify any patent claims not reexamined;

(H) be mailed on the day it is dated to the patent owner at the address provided for in 37 CFR 1.33(c) and a copy will be mailed to the third party requester; and

(I) identify patent claims, dependent on amended claims, determined to be patentable.

＊＊＊＊＊＊＊＊＊＊＊

MPEP 2290 Format of Ex Parte Reexamination Certificate [R-5]

＊＊＊＊＊＊＊＊＊＊＊

The ex parte reexamination certificate is formatted much the same as the title page of current U.S. patents.

The ex parte reexamination certificate number will always be the patent number of the original patent followed by a two-character "kind code" suffix. The first letter of the "kind code" suffix is "B" for reexamination certificates published prior to January 2, 2001, and "C" for reexamination certificates published on or after January 2, 2001. The second letter of the "kind code" suffix is the number of the reexamination proceeding of that patent, and thus shows how many times that patent has been reexamined.

＊＊＊＊＊＊＊＊＊＊＊

MPEP 2293 Intervening Rights

＊＊＊＊＊＊＊＊＊＊＊

The situation of intervening rights resulting from reexamination proceedings parallels the intervening rights situation resulting from reissue proceedings, and the rights detailed in 35 U.S.C. 252 apply equally in reexamination and reissue situations.

§24.8　Inter Partes 再審查申請案之審查程序

Inter Prates 再審查申請案之審查程序如下之流程圖所示：

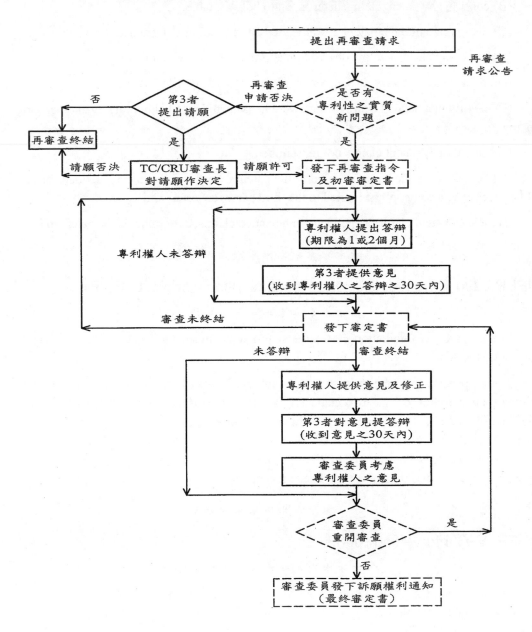

Inter Partes 再審查程序（訴願前）

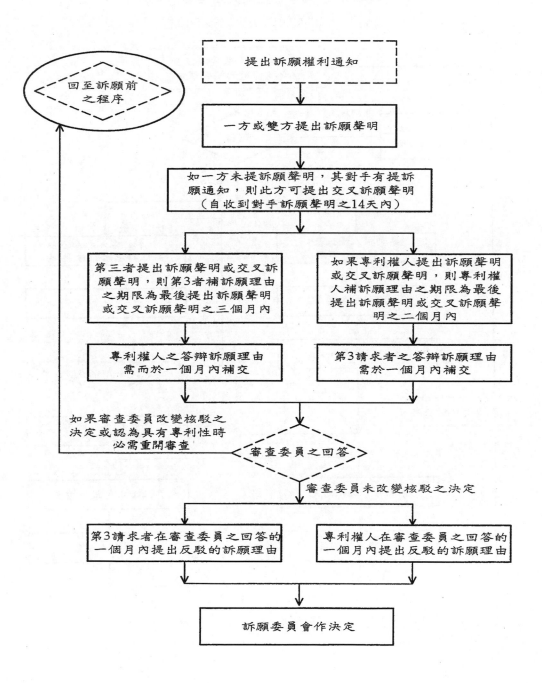

2002 年 11 月 2 日之前開始之 Inter Partes
再審查申請案訴願決定之後的流程

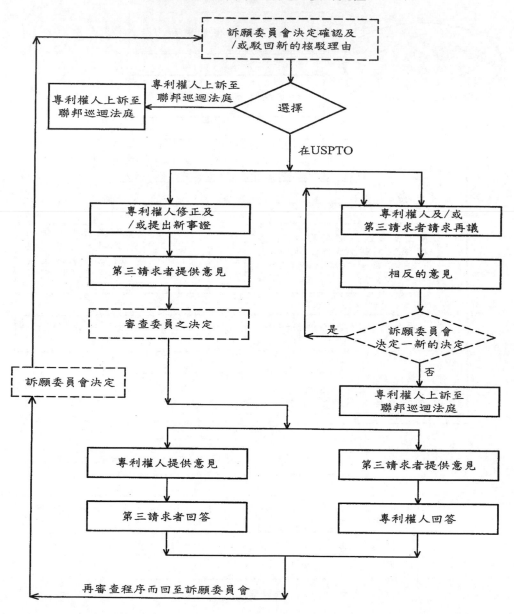

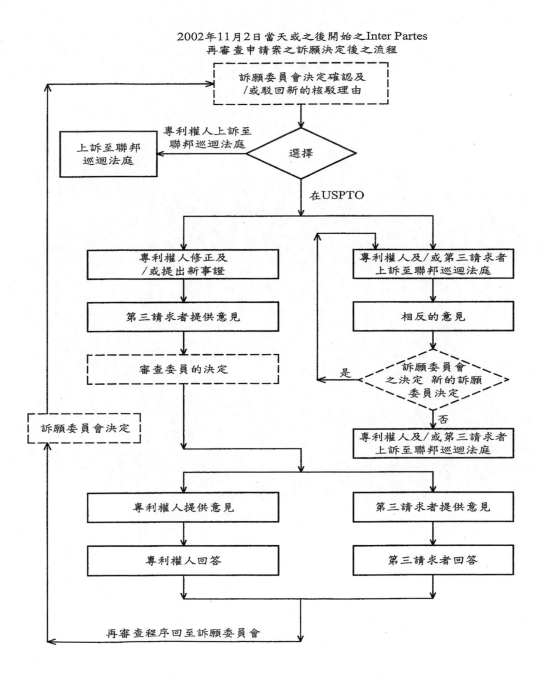

2002年11月2日當天或之後開始之Inter Partes
再審查申請案之訴願決定後之流程

第 25 章　抵觸程序

目　錄

§25.1　抵觸程序

依據 35USC§102(g)(1)："在符合第 351 條或第 291 條進行之抵觸程序（interference）期間，在專利申請人之發明日之前，其他的發明人依本法第 104 條確立該發明已被該其他發明人發明，且未放棄，不發表或隱藏"則此專利申請人不得取得專利。亦即，當有二方或以上請求相同之可予專利之發明時（例如一專利申請案與另一專利申請案（專利）之請求項請求之相同之發明），就宣告抵觸程序以協助美國專利局長決定哪一方先發明了共同請求之發明，亦即哪一方具有發明之優先權。通常的程序乃由審查委員建議抵觸程序或由申請人於申請案中加入或修正請求項之以起始抵觸程序，再交由專利訴願及抵觸委員會（BPAI）之行政專利法官（Administrative Patent Judge）宣告抵觸程序之後，由 BPAI 決定發明優先權及專利性。

在審查委員建議了抵觸程序，審查委員就不去審查該專利申請案一直到抵觸程序終止為止。

在抵觸程序終止之後，專利申請案就會回到原審查委員手中，並作適當之處理。例如，如有可核准之請求項則會發下核准通知。如果申請人在抵觸程序中輸掉，則該請求項就不會再被審查，而未涉入抵觸程序之請求項則會繼續接受審查。

---

35 U.S.C. 102 Conditions for patentability; novelty and loss of right to patent.

A person shall be entitled to a patent unless –

(g)(1) during the course of an interference conducted under section 135 or section 291, another inventor involved therein establishes, to the extent permitted in section 104, that before such person's invention thereof the invention was made by such other inventor and not abandoned, suppressed, or concealed, or

＊＊＊＊＊＊＊＊＊＊＊

35 U.S.C. 135 Interferences.

(a) Whenever an application is made for a patent which, in the opinion of the Director, would interfere with any pending application, or with any unexpired patent, an interference may be declared and the Director shall give notice of such declaration to the applicants, or applicant and patentee, as the case may be. The Board of Patent Appeals and Interferences shall determine questions of priority of the inventions and may determine questions of patentability. Any final decision, if adverse to the claim of an applicant, shall constitute the final refusal by the Patent and Trademark Office of the claims involved, and the Director may issue a patent to the applicant who is adjudged the prior inventor. A final judgment adverse to a patentee from which no appeal or other review has been or can be taken or had shall constitute cancellation of the claims involved in the patent, and notice of such cancellation shall be endorsed on copies of the patent distributed after such cancellation by the Patent and Trademark Office.

＊＊＊＊＊＊＊＊＊＊＊＊

MPEP 2301 Introduction [R-4]

An interference is a contest under 35 U.S.C. 135(a) between an application and either another application or a patent. An interference is declared to assist the Director of the United States Patent and Trademark Office in determining priority, that is, which party first invented the commonly claimed invention within the meaning of 35 U.S.C. 102(g)(1). See MPEP § 2301.03. Once an interference has been suggested under 37 CFR 41.202, the examiner refers the suggested interference to the Board of Patent Appeals and Interferences (Board). An administrative patent judge declares the interference, which is then administered at the Board. A panel of Board members enters final judgment on questions of priority and patentability arising in an interference.

Once the interference is declared, the examiner generally will not see the application again until the interference has been terminated. Occasionally, however, the Board may refer a matter to the examiner or may consult with the examiner on an issue. Given the very tight deadlines in an interference, any action on a consultation or referral from the Board must occur with special dispatch.

The application returns to the examiner after the interference has been terminated. Depending on the nature of the judgment in the case, the examiner may need to take further action in the application. For instance, if there are remaining allowable claims, the application may need to be passed to issue. The Board may have entered a recommendation for further action by the examiner in the case. If the applicant has lost an issue in the interference, the applicant may be barred from taking action in the application or any subsequent application that would be inconsistent with that loss.

＊＊＊＊＊＊＊＊＊＊＊

§25.2　抵觸程序之標的

　　當二個或以上之發明人各有一相同的發明之可予專利的請求項時，就稱為一發明人之請求項與另一發明人請求項抵觸。所謂相同的發明乃指專利性上不可區分之發明（patentably indistinct inventions）。亦即，當一發明人所請求之發明與另一發明人所請求之發明可區分時，就沒有抵觸之情事。請求項使用相同之語句去定義並不一定保證是相同的發明。每一請求項必需依據申請書之內容去解讀。使用裝置加功能用語（means-plus function）之請求項會依照說明書中對應結構而有不同之範圍。

　　當抵觸程序被宣告時，抵觸之標的就稱之為"count"（抵觸標的）而每一方均會有至少一請求項對應於"count"，但亦可能一請求項對應於一個以上之"count"。在決定一請求項是否對應於一"count"時，是把"count"之標的當作先前技術，如果請求項被"count"預見（anticipated）或是顯而易見的，則該請求項對應於該"count"。

MPEP 2301.03 Interfering Subject Matter [R-4]

＊＊＊＊＊＊＊＊＊＊＊

A claim of one inventor can be said to interfere with the claim of another inventor if they each have a patentable claim to the same invention.

＊＊＊＊＊＊＊＊＊＊＊

If the claimed invention of either party is patentably distinct from the claimed invention of the other party, then there is no interference-in-fact.

Identical language in claims does not guarantee that they are drawn to the same invention. Every claim must be construed in light of the application in which it appears. 37 CFR 41.200(b). Claims reciting means-plus-function limitations, in particular, might have different scopes depending on the corresponding structure described in the written description.

When an interference is declared, there is a description of the interfering subject matter, which is called a "count." Claim correspondence identifies claims that would no longer be allowable or patentable to a party if it loses the priority determination for the count. To determine whether a claim corresponds to a count, the subject matter of the count is assumed to be prior art to the party. If the count would have anticipated or supported an obviousness determination against the claim, then the claim corresponds to the count. 37 CFR 41.207(b)(2). Every count must have at least one corresponding claim for each party, but it is possible for a claim to correspond to more than one count.

＊＊＊＊＊＊＊＊＊＊＊＊

實例 1

　　一美國專利之一項請求項乃為一通式代表之化合物，其中 R 為 alkyl。另有一美國專利申請案之一請求項亦為相同之化合物，但是 R 為 n-penty（alkyl 之一種）。因而如果把該申請案當作先前技術，則該專利之請求項是被預見的（anticipated）。但是，如果把該專利當作先前技術則該申請案不是預見的。但是，如果 n-penty 是 alkyl 的一顯而易見的選擇，則該專利申請案之請求項就界定了抵觸之標的。

實例 2

　　一美國專利申請案有一請求項為一鍋爐其具有一新穎的安全閥，而另一美國專利有一請求項為該安全閥。先前技術顯示鍋爐具有安全閥是習知的。因而如果把申請案之請求項當作先前技術，則預見了專利之請求項，而如果把專利之請求項當作先前技術，則申請案之請求項為顯而易知，所以該美國專利申請案之請求項界定了抵觸之標的。

實例 3

　　一美國專利申請案有一請求項是使用白金作為觸媒之化學反應。一美國專利有一請求項是相同的化學反應，但是所使用的觸媒乃是擇自白金，鈮及鋯所組成之群組中之一者。由於美國專利所用的觸媒乃由馬庫西方式撰寫，故專利申請案之請求項預見了該專利，而該專利亦預見了專利申請案。因而專利申請案界定了抵觸之標的。

實例 4

　　一美國專利及一美國專利申請案各有一請求項是請求一組合（combination），此組合均包括了結合裝置（means for fastening）。該美國專利之說明書中揭露了用鉚釘作結合裝置，而該美國專利申請案揭露了用粘膠作結合裝置，則雖然此二請求項之用語完全相同，但是該專利申請與該專利並無抵觸情事。除非在此技術領域，鉚釘與粘膠被認為是結構上均等或是彼此是顯而易見的。

實例 5

　　一美國專利有一請求項為一配方(formulation)，包括一界面活性劑：sodium lauryl sulfate。一美國專利申請案有一請求項請求相同之配方，但並未特別指出界面活性劑為 sodium lauryl sulfate，但是在說明書中指出在這種配方中使用 sodium lauryl sulfate 作為界面活性劑是習知的，則此專利與專利申請案之請求項有抵觸之情事。

§25.3　抵觸標的之訂立

　　抵觸標的（count）訂立時需廣到可涵蓋雙方涉入抵觸之請求項之範圍。以下實例說明：

(1)
Party A 之請求項	Party B 之請求項	Count
加熱至 120℃~200℃	加熱至 150℃~240℃	加熱至 120℃~240℃

(2)
Party A 之請求項	Party B 之請求項	Count
化合物擇自 A,B,C 及 D 所組成之族群	化合物擇自 D,E 及 F 所組成之族群	化合物擇至 A,B,C,D,E,F 所組成之族群

(3)
Party A 之請求項	Party B 之請求項	Count
一組合，包括一結合裝置	一組合，包括鉚釘作為結合裝置	一組合，包括一結合裝置

§25.4　抵觸程序之起始

　　一抵觸程序通常是由一專利申請案之專利申請人或審查專利申請案之審查委員建議才會起始。一專利之專利權人如果要建議抵觸程序則需先提出再發證之申請，變成一申請人才能建議起始抵觸程序。由申請人或審查委員建議起始抵觸程序則在抵觸程序宣告之前的程序會有所不同。但是不論由誰提出建議，審查委員需先諮詢抵觸實務專家（Interference Practice Specialists, IPS），IPS 會將所建議之抵觸程序送交專利訴願及抵觸程序委員會（BPAI）。

　　依照 37CFR 41.202，如果申請人要提出抵觸程序之建議，則需提供下列資訊：

(1)　要進行抵觸程序之對方（專利申請案或專利）之足夠資訊（如案號，專利號）以作確認；

(2)　指出申請人要進行抵觸程序之請求項，提出一個或以上之抵觸標的（count）及顯示請求項如何對應於抵觸標的；

(3)　對各抵觸標的（count）提出請求項圖（claim chart），將至少各方之請求項之一與抵觸標的作比較並顯示為何請求項有抵觸；

(4)　詳細解釋為何申請人在發明優先權上會勝出；

(5)　如果加入新的請求項或修正請求項以起始抵觸程序的話，需提供請求項圖以顯示各請求項之在申請人之說明書內有支持的書面敘述；

(6)　對於申請人想要主張利益的各推斷的付諸實施，需提供請求項圖顯示在抵觸之標的之範圍內之揭露中，哪裡提供了推斷的付諸實施。

　　有關上述第(4)項，如果申請人之最早之推斷付諸實施日（例如說明書所揭露之實施例）是早於他方之申請案或專利之表面上的推斷的付諸實施日的話，申請人只要解釋其有權利主張該較早的推斷的付實施日即可。而如果晚於他方的申請案或專利之表面上的推斷的付諸實施日的話，則需(A)證明其發明日期早於他方之專利申請案或專利之推斷的付諸實施日，或(B)證明為何他方無權利主張其書面上的最早推斷的實施日，或(C)提供其他理由證明為何申請人應被認為是較早之發明人。

　　另外需注意的是抵觸的請求項必需是可核准的（allowable），特別是對於支持此抵觸請求項之書面敘述。以往，申請人要起始一抵觸程序時，是將對方的請求項複製（copy）入申請案中。而此實務會產生一問題，就是支持請求項之說明書之揭露之差異可能會使得該請求項對一方是可核准的，但是對另一方卻不是可核准的，或者請求項之語言相同（因為將對方之請求項複製進來），但是在說明書中支持之揭露不同，因而對請求項之解讀會有所不同。故實務上，最好不要完全將對方之請求項複製，而是加入或修正一完全可由自己的說明書支持的請求項，並解釋為何兩方之請求項界定了相同之發明。

　　依照 37CFR 41.202，當審查委員在審查一專利申請案時發現申請人之發明與他人之專利申請案或專利之一請求項所請求之發明相同時，審查委員可要求申請人起始一抵觸程序。如果申請人之申請案已有請求項是請求抵觸之標的時，則不必要求申請人另外加入一請求項，而如果申請人之申請案並無此請求項時，則要求申請人加入或修正請求項。申請人如果在一個月內未提出抵觸程序之建議則被視為對發明優先權讓步（concession of priority），而審查委員就應與抵觸程序專家合作向專利訴願及抵觸委員會建議一抵觸程序。

MPEP 2304.02 Applicant Suggestion [R-4]

＊＊＊＊＊＊＊＊＊＊＊

37 CFR 41.202 Suggesting an interference.

(a) Applicant. An applicant, including a reissue applicant, may suggest an interference with another application or a patent. The suggestion must:

(1) Provide sufficient information to identify the application or patent with which the applicant seeks an interference,

(2) Identify all claims the applicant believes interfere, propose one or more counts, and show how the claims correspond to one or more counts,

(3) For each count, provide a claim chart comparing at least one claim of each party corresponding to the count and show why the claims interfere within the meaning of § 41.203(a),

(4) Explain in detail why the applicant will prevail on priority,

(5) If a claim has been added or amended to provoke an interference, provide a claim chart showing the written description for each claim in the applicant's specification, and

(6) For each constructive reduction to practice for which the applicant wishes to be accorded benefit, provide a chart showing where the disclosure provides a constructive reduction to practice within the scope of the interfering subject matter.

* * * * * * * * * * *

MPEP 2304.02(a) Identifying the Other Application or Patent [R-4]

* * * * * * * * * * *

Usually an applicant seeking an interference will know the application serial number or the patent number of the application or patent, respectively, with which it seeks an interference. If so, providing that number will fully meet the identification requirement of 37 CFR 41.202(a)(1).

* * * * * * * * * * *

MPEP 2304.02(b) Counts and Corresponding Claims [R-4]

The applicant must identify at least one patentable claim from every application or patent that interferes for each count. A count is just a description of the interfering subject matter, which the Board of Patent Appeals and Interferences uses to determine what evidence may be used to prove priority under 35 U.S.C. 102(g)(1).

The examiner must confirm that the applicant has (A) identified at least one patentable count, (B) identified at least one patentable claim from each party for each count, and (C) has provided a claim chart comparing at least one set of claims for each count. The examiner need not agree with the applicant's suggestion. The examiner's role is to confirm that there are otherwise patentable interfering claims and that the formalities of 37 CFR 41.202 are met.<

2304.02(c) Explaining Priority [R-4]

A description in an application that would have anticipated the subject matter of a count is called a constructive reduction-to-practice of the count. One disclosed embodiment is enough to have anticipated the subject matter of the count. If the application is relying on a chain of benefit disclosures under any of 35 U.S.C. 119, 120, 121 and 365, then the anticipating disclosure must be continuously disclosed through the entire benefit chain or no benefit may be accorded.

If the application has an earlier constructive reduction-to-practice than the apparent earliest constructive reduction-to-practice of the other application or patent, then the applicant may simply explain its entitlement to its earlier constructive reduction -to-practice. Otherwise, the applicant must (A) antedate the earliest constructive reduction-to-practice of the other application or patent, (B) demonstrate why the other application or patent is not entitled to its apparent earliest constructive reduction-to-practice, or (C) provide some other reason why the applicant should be considered the prior inventor.

MPEP2304.02(d) Adequate Written Description [R-4]

* * * * * * * * * * * *

The interfering claim must be allowable, particularly with respect to the written description supporting the interfering claim.

Historically, an applicant provoked an interference by copying a claim from its opponent. The problem this practice created was that differences in the underlying disclosures might leave the claim allowable to one party, but not to the other; or despite identical claim language differences in the disclosures might require that the claims be construed differently.

Rather than copy a claim literally, the better practice is to add (or amend to create) a fully supported claim and then explain why, despite any apparent differences, the claims define the same invention. 37 CFR 41.203(a).

MPEP 2304.04 Examiner Suggestion [R-4]

* * * * * * * * * * * *

MPEP 2304.04(a) Interfering Claim Already in Application [R-4]

If the applicant already has a claim to the same subject matter as a claim in the application or patent of another inventor, then there is no need to require the applicant to add a claim to have a basis for an interference.

The examiner may invite the applicant to suggest an interference pursuant to 37 CFR 41.202(a). An applicant may be motivated to do so in order to present its views on how the interference should be declared.

If the applicant does not suggest an interference, then the examiner should work with an Interference Practice Specialist (IPS) to suggest an interference to the Board of Patent Appeals and Interferences (Board). The suggestion should include an explanation of why at least one claim of every application or patent defines the same invention within the meaning of 37 CFR 41.203(a). See MPEP § 2301.03 for a discussion of interfering subject matter. The examiner must also complete Form PTO-850.

* * * * * * * * * * * *

§25.5　抵觸程序之內容

　　抵觸程序之第一步是抵觸之宣告，審查委員會將涉入抵觸程序相關卷宗送至行政專利法官（APJ）處，行政專利法官會與審查委員及抵觸程序專家討論，如果適當，就宣告抵觸程序。抵觸程序宣告後會將抵觸宣告通知書寄交雙方。而雙方就可取得或影印對方之專利或申請案之所有記錄及文件。

　　在抵觸程序宣告之後，雙方可提出初步陳述給行政專利法官，初步陳述主要是提出證據及說明以提供最早的發明日。雙方可同時提出初步訴請（preliminary motion）向美國專利局請求下列事項：①對方的發明不具專利性②沒有抵觸情事③重新定義抵觸標的④取代以不同之申請案⑤請求宣告另外抵觸程序等。

　　在行政專利法官對於初步訴請作出決定後，就會進行證詞階段（testimony Period），此時雙方呈送正式證據以作為最終聽證（final hearing）時專利訴願及抵觸委員會（BPAI）之參考。在 BPAI 最終聽證結束之後，專利訴願及抵觸委員會就會作成最終判決（final decision）。上述程序通常會在二年之內結束。

　　對於最終判決不滿的一方可向地方法院或聯邦巡迴法庭（CAFC）提出上訴，而輸掉抵觸程序的一方，其對應於抵觸標的（count）的請求項將被刪除，但是未對應於抵觸標的的請求項仍會進行審查。而贏的一方其對應於抵觸標的之請求項則仍有效，或者可取得專利。

§25.6　決定發明優先權之法則

　　A，B 表不同之兩方，C 表構想日，RP 表付諸實行日，CP 表關鍵期間（critical period），F 表提出專利申請

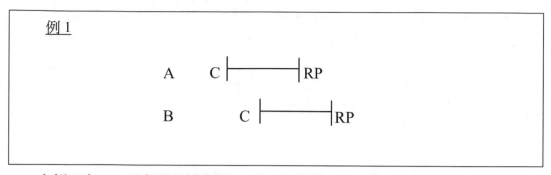

例 1

在例 1 中，A 具有發明優先權，因為 A 先構想且先付諸實施

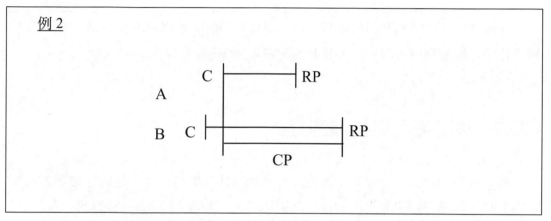

例 2

在例 2 中，如果 B 在關鍵期間（從 A 構想到自己付諸實施之期間）有勤勉度，則 B 具有發明優先權。

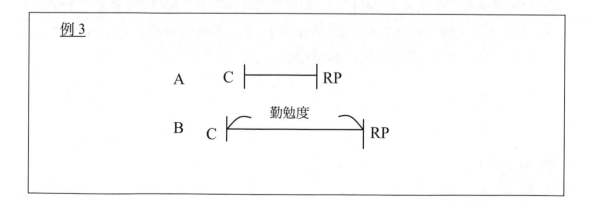

例 3

在例 3 中，B 具有發明優先權，B 雖較晚付諸實施，但構想在先且自構想至付諸實施均有勤勉度

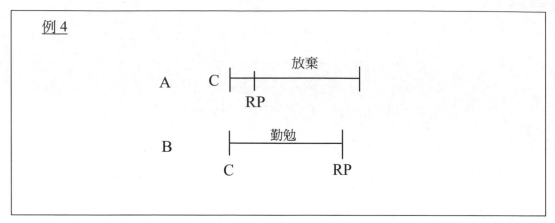

在例 4 中，A 先構想但付諸實施後，放棄一段時間才提出專利申請，但 B 雖在構想在後，但勤勉地付諸實施，故 B 具有發明優先權。

§25.7　共同擁有與抵觸程序

當二專利申請案或一專利申請案及一專利為共同擁有（有一相同之發明人或受讓人）時，通常美國專利局不會宣告抵觸程序。例如兩家不同的公司具有不同之專利申請案均請求相同之發明。之後，兩家公司合併，變成合併後之公司擁有兩個專利申請案，則此新的公司應自行決定發明優先權並告知美國專利局。審查委員亦可要求作選擇。此時申請人（專利權人）需將抵觸之專利申請案請求項刪除，不請求（disclaiming）抵觸之專利請求項或修正申請案之請求項使其不抵觸，或者提出再發證申請案以修正專利之請求項使其不再抵觸。

＊＊＊＊＊＊＊＊＊＊＊＊

MPEP 2304.05 Common Ownership [R-4]

An interference is rarely appropriate between two applications or an application and patent that belong to the same owner. The owner should ordinarily be able to determine priority and is obligated under 37 CFR 1.56 to inform the examiner about which application or patent is entitled to priority. The examiner may require an election of priority between the application and other application or patent. 35 U.S.C. 132(a).

In making the election, the owner must eliminate the commonly claimed subject matter. This may be accomplished by canceling the interfering application claims, disclaiming the interfering patent claims, amending the application claims such that they no longer interfere, or filing a reissue application to amend the patent claims such that they no longer interfere.

第 26 章　揭露義務與 IDS

目　錄

§26.1　揭露義務（Duty of Disclosure）

由於授予專利將影響公眾利益，為了充分保障公眾利益，在審查過程中，審查當局（USPTO）應取得可能影響該申請案獲得專利之相關資訊。

因此，任何與該專利申請案之申請及審查程序有關之以下個人（Individuals）均負有揭露義務：

- 該申請案之任一發明人
- 處理該申請案之美國專利代理人及專利律師
- 與發明人／受讓人／準受讓人一起工作，因而實質涉入該申請案的準備或審查工作之人士（含國外對應案之專利代理人及專利律師）

---

37 CFR 1.56 Duty to disclose information material to patentability

(a) A patent by its very nature is affected with a public interest. The public interest is best served, and the most effective patent examination occurs when, at the time an application is being examined, the Office is aware of and evaluates the teachings of all information material to patentability. Each individual associated with the filing and prosecution of a patent application has a duty of candor and good faith in dealing with the Office, which includes a duty to disclose to the Office all information known to that individual to be material to patentability as defined in this section......

＊＊＊＊＊＊＊＊＊＊＊

(c) Individuals associated with the filing or prosecution of a patent application within the meaning of this section are:

(1) Each inventor named in the application;

(2) Each attorney or agent who prepares or prosecutes the application; and

(3) Every other person who is substantively involved in the preparation or prosecution of the application and who is associated with the inventor, with the assignee or with anyone to whom there is an obligation to assign the application.

＊＊＊＊＊＊＊＊＊＊＊

§26.2　發明人的揭露義務

發明人的揭露義務亦有進一步規定，根據 37CFR 1.63(b)(3)的規定，美國非暫時案之發明人所簽署的發明宣誓書（oath or declaration）中，發明人必須聲明其瞭解其負有 37CFR 1.56 的揭露義務。

---

37 CFR § 1.63 Oath or declaration.

(a) An oath or declaration filed under § 1.51(b)(2) as a part of a nonprovisional application must:

＊＊＊＊＊＊＊＊＊＊＊

(b) In addition to meeting the requirements of paragraph (a) of this section, the oath or declaration must also:

＊＊＊＊＊＊＊＊＊＊＊

(3) State that the person making the oath or declaration acknowledges the duty to disclose to the Office all information known to the person to be material to patentability as defined in § 1.56.

§26.3　需被揭露的資訊

根據 37CFR1.56 規定之相關人等，若所知任何在判定該申請案的專利性時，屬於實質重要的資訊（all information known to that individual to be material to patentability），在得知之時（known），則產生揭露義務。因此並非要求發明人需主動檢索前案。專利性（Patentability）只需考慮得知資訊時之審查中的請求項（pending claim(s)），並非針對說明書揭露的整體發明。但專利性的判定涵蓋所有 35USC102（新穎性）／35USC 103（非顯而易見性）／35USC 112（揭露要件）／double patenting（重複專利）等等之所有各種有關專利性論定之考慮，並非只限於新穎性／非顯而易見性之先前技術。

§26.4 揭露義務之終止

當審查終結或請求項刪除或不再被考慮之後，揭露義務即告終止。

§26.5 實質重要（material to patentability）的認定

根據 37 CFR 1.56(b)，任何可能影響專利性的判斷或賦予專利的資料均屬之。但不能包括與既有之資料重複者（cumulative to information already of record or being made of record in the application）。

例如屬於下述狀況之一：

● 該資訊單獨或與其他文獻合併觀之，會使得請求項推斷上不具專利性者（a *prima facie* case of unpatentability of a claim），不限於先前技術資料（prior art）、若有先使用、公開、銷售、邀約銷售等行為之資料、與其他相關案有重複專利之潛在問題、發明人界定問題等資料均屬之。

● 該資訊與申請人對於下述狀況的陳述相反或不一致者：
回應審定書（Office Action）而提出的不具專利性的答辯
主張具有專利性

---

37 CFR 1.56 Duty to disclose information material to patentability

(a) A patent by its very nature is affected with a public interest. The public interest is best served, and the most effective patent examination occurs when, at the time an application is being examined, the Office is aware of and evaluates the teachings of all information material to patentability. Each individual associated with the filing and prosecution of a patent application has a duty of candor and good faith in dealing with the Office, which includes a duty to disclose to the Office all information known to that individual to be material to patentability as defined in this section. The duty to disclose information exists with respect to each pending claim until the claim is cancelled or withdrawn from consideration, or the application becomes abandoned. Information material to the patentability of a claim that is cancelled or withdrawn from consideration need not be submitted if the information is not material to the patentability of any claim remaining under consideration in the application. There is no duty to submit information which is not material to the patentability of any existing claim. The duty to disclose all information known to be material to patentability is deemed to be satisfied if all information known to be material to patentability of any claim issued in a patent was cited by the Office or submitted to the Office in the manner prescribed by §§ 1.97(b)-(d) and 1.98. However, no patent will be granted on an application in connection with which fraud on the Office was practiced or attempted or the duty of disclosure was violated through bad faith or intentional misconduct. The Office encourages applicants to carefully examine:

(1) Prior art cited in search reports of a foreign patent office in a counterpart application, and

(2) The closest information over which individuals associated with the filing or prosecution of a patent application believe any pending claim patentably defines, to make sure that any material information contained therein is disclosed to the Office.

(b) Under this section, information is material to patentability when it is not cumulative to information already of record or being made of record in the application, and

(1) It establishes, by itself or in combination with other information, a *prima facie* case of unpatentability of a claim; or

(2) It refutes, or is inconsistent with, a position the applicant takes in:

(i) Opposing an argument of unpatentability relied on by the Office, or

(ii) Asserting an argument of patentability.

A *prima facie* case of unpatentability is established when the information compels a conclusion that a claim is unpatentable under the preponderance of evidence, burden-of-proof standard, giving each term in the claim its broadest reasonable construction consistent with the specification, and before any consideration is given to evidence which may be submitted in an attempt to establish a contrary conclusion of patentability.

(c) Individuals associated with the filing or ……

(d) Individuals other than the attorney, agent or inventor may comply with this section by disclosing information to the attorney, agent, or inventor.

(e) In any continuation-in-part application, the duty under this section includes the duty to disclose to the Office all information known to the person to be material to patentability, as defined in paragraph (b) of this section, which became available between the filing date of the prior application and the national or PCT international filing date of the continuation-in-part application.

§26.6　需考慮揭露的資料來源

以下為 USPTO 建議應揭露的幾種資料（但非僅限於此）：

● 該美國專利申請案之對應外國案審查時被引證之資料。或者，基於必要，外國案的審查意見本身。

● 任何在相關之美國共同申請案（co-pending）中所引證之先前技術（理論上，若是延續案，審查委員應自動考慮延續案申請之前，母案中已經列出的所有引證資訊）。

● 如果申請案之標的涉入專利侵權訴訟，所有在訴訟中相關之資料：先前公知公用之資料、有關發明人（inventorship）之資料、先前技術、對於詐欺／不正行為、或違反揭露義務之資料。

§26.7　遞交資訊揭露聲明（IDS）的程序與時間點

要滿足揭露義務，一般利用資訊揭露聲明（Information Disclosure Statement, IDS）程序。填交 PTO/SB/08a Form。

根據提交的時間點，審查委員要考慮的程度，有以下規定

● 第 1 階段：不需另加規費，審查委員需考慮該 IDS：

一般申請案之申請日起 3 個月內或第 1 次審定書發下前；

PCT 申請案進入 U.S.國家階段三個月內；

提出 RCE 後或延續案之第一次 OA 前。

● 第 2 階段：需繳規費，審查委員需考慮該 IDS：

第 1 類別之後，最終審定、審查終結（例如 *Ex parte* Quayle Action）或核准之前不用繳費的例外：

1. 該些 IDS 資料乃是在對應國外申請案之審定書中或檢索報告中第一次被引證，而在此審定書及檢索報告發下三個月內呈送 IDS。

2. 該些 IDS 資料並非在對應國外申請案之審定書或檢索報告中被引證，但是在知曉三個月內呈送 IDS。

- 第 3 階段：依照 37 C.F.R. 1.97(e)作下列之一者之請願，需繳規費，審查委員需考慮該 IDS：

  第 2 階段之後，繳領證費之前，且：

  1. 該些 IDS 資料乃是在對應國外申請案之審定書中或檢索報告中第一次被引證，而此審定書及檢查報告發下三個月內呈送 IDS。

  2. 該些 IDS 資料並非在對應國外申請案之審定書或檢索報告中被引證，但是在知曉三個月內呈送 IDS。

* 若非上述情況，則 USPTO 將不接受，若要該 IDS 被考慮，則需另提 RCE 或延續案

- 第 4 階段：則需提出 RCE 及撤回發證請求（petition）：

繳領證費之後：繳完領證費之後，呈送 IDS 資料並不會被考慮。若希望審查委員考慮，則需另提 RCE 或延續案。

---

§37 C.F.R. 1.97 Filing of information disclosure statement. (或 MPEP609.04(b) Timing Requirements for an Information Disclosure Statement [R-5])

(a) In order for an applicant for a patent or for a reissue of a patent to have an information disclosure statement in compliance with § 1.98 considered by the Office during the pendency of the application, the information disclosure statement must satisfy one of paragraphs (b), (c), or (d) of this section.

(b) An information disclosure statement shall be considered by the Office if filed by the applicant within any one of the following time periods:

(1) Within three months of the filing date of a national application other than a continued prosecution application under § 1.53(d);

(2) Within three months of the date of entry of the national stage as set forth in § 1.491 in an international application;

(3) Before the mailing of a first Office action on the merits; or

(4) Before the mailing of a first Office action after the filing of a request for continued examination under § 1.114.

(c) An information disclosure statement shall be considered by the Office if filed after the period specified in paragraph (b) of this section, provided that the information disclosure statement is filed before the mailing date of any of a final action under § 1.113, a notice of allowance under § 1.311, or an action that otherwise closes prosecution in the application, and it is accompanied by one of:

(1) The statement specified in paragraph (e) of this section; or

(2) The fee set forth in § 1.17(p).

(d) An information disclosure statement shall be considered by the Office if filed by the applicant after the period specified in paragraph (c) of this section, provided that the information disclosure statement is filed on or before payment of the issue fee and is accompanied by:

(1) The statement specified in paragraph (e) of this section; and

(2) The fee set forth in § 1.17(p).

(e) A statement under this section must state either:

(1) That each item of information contained in the information disclosure statement was first cited in any communication from a foreign patent office in a counterpart foreign application not more than three months prior to the filing of the information disclosure statement; or

(2) That no item of information contained in the information disclosure statement was cited in a communication from a foreign patent office in a counterpart foreign application, and, to the knowledge of the person signing the certification after making reasonable inquiry, no item of information contained in the information disclosure statement was known to any individual designated in § 1.56(c) more than three months prior to the filing of the information disclosure statement.

(f) No extensions of time for filing an information disclosure statement are permitted under § 1.136. If a *bona fide* attempt is made to comply with § 1.98, but part of the required content is inadvertently omitted, additional time may be given to enable full compliance.

(g) An information disclosure statement filed in accordance with this section shall not be construed as a representation that a search has been made.

(h) The filing of an information disclosure statement shall not be construed to be an admission that the information cited in the statement is, or is considered to be, material to patentability as defined in § 1.56(b).

(i) If an information disclosure statement does not comply with either this section or § 1.98, it will be placed in the file but will not be considered by the Office.

§26.8　提交 IDS 之資料內容

呈送 IDS 之專利刊物或資料，一般分為美國專利、外國專利文獻及其他文獻三種。

美國專利或美國專利核准前公告文件：僅需填寫專利號／公告號、公告日、第一發明人姓名。不用檢附全文本。

外國專利或外國專利核准前公告文件：需填寫國別、專利號／公告號、公告日、第一發明人姓名，並需檢附清楚全文本與必要之英譯文。

其他文獻：需填寫作者、文章名稱、公開日期、刊物名、刊物出版期數或文本編號、出版商、出版國別等等。並需檢附清楚文本與必要之英譯文。

（非英文之文獻若無附上該非英文文獻與申請案之關聯性之英文簡要說明或其對等英譯，審查委員將不納入考慮。）

---

37 CFR § 1.98 Content of information disclosure statement.

(a) Any information disclosure statement filed under § **1.97** shall include the items listed in paragraphs (a)(1), (a)(2) and (a)(3) of this section.

(1) A list of all patents, publications, applications, or other information submitted for consideration by the Office. U.S. patents and U.S. patent application publications must be listed in a section separately from citations of other documents. Each page of the list must include:

(i) The application number of the application in which the information disclosure statement is being submitted;

(ii) A column that provides a space, next to each document to be considered, for the examiner's initials; and

(iii) A heading that clearly indicates that the list is an information disclosure statement.

(2) A legible copy of:

(i) Each foreign patent;

(ii) Each publication or that portion which caused it to be listed, other than U.S. patents and U.S. patent application publications unless required by the Office;

(iii) For each cited pending unpublished U.S. application, the application specification including the claims, and any drawing of the application, or that portion of the application which caused it to be listed including any claims directed to that portion; and

(iv) All other information or that portion which caused it to be listed.

(3)

(i) A concise explanation of the relevance, as it is presently understood by the individual designated in § 1.56(c) most knowledgeable about the content of the information, of each patent, publication, or other information listed that is not in the English language. The concise explanation may be either separate from applicant's specification or incorporated therein.

(ii) A copy of the translation if a written English-language translation of a non-English-language document, or portion thereof, is within the possession, custody, or control of, or is readily available to any individual designated in § 1.56(c).

(b)

(1) Each U.S. patent listed in an information disclosure statement must be identified by inventor, patent number, and issue date.

(2) Each U.S. patent application publication listed in an information disclosure statement shall be identified by applicant, patent application publication number, and publication date.

(3) Each U.S. application listed in an information disclosure statement must be identified by the inventor, application number, and filing date.

(4) Each foreign patent or published foreign patent application listed in an information disclosure statement must be identified by the country or patent office which issued the patent or published the application, an appropriate document number, and the publication date indicated on the patent or published application.

(5) Each publication listed in an information disclosure statement must be identified by publisher, author (if any), title, relevant pages of the publication, date, and place of publication.

(c) When the disclosures of two or more patents or publications listed in an information disclosure statement are substantively cumulative, a copy of one of the patents or publications as specified in paragraph (a) of this section may be submitted without copies of the other patents or publications, provided that it is stated that these other patents or publications are cumulative.

(d) A copy of any patent, publication, pending U.S. application or other information, as specified in paragraph (a) of this section, listed in an information disclosure statement is required to be provided, even if the patent, publication, pending U.S. application or other information was previously submitted to, or cited by, the Office in an earlier application, unless:

(1) The earlier application is properly identified in the information disclosure statement and is relied on for an earlier effective filing date under 35 U.S.C. 120; and

(2) The information disclosure statement submitted in the earlier application complies with paragraphs (a) through (c) of this section.

§26.9　非相關人提交 IDS

非專利申請案之相關人亦可以提交資料供審查委員參考，但呈送的時間點僅限於為該專案核准前公告日起 2 個月內或核准通知發出前（以較早者為準）。另外，遞件人只能遞交文獻，不能對文獻跟公告案間作任何的評論或解釋。而文獻是否被考慮，由審查委員自行評估。再者，審查委員並無義務一定要考慮。

繳交的文件格式規定類似一般 IDS，外加繳交規費。

---

37 CFR § 1.99 Third-party submission in published application.

(a) A submission by a member of the public of patents or publications relevant to a pending published application may be entered in the application file if the submission complies with the requirements of this section and the application is still pending when the submission and application file are brought before the examiner.

(b) A submission under this section must identify the application to which it is directed by application number and include:

(1) The fee set forth in § 1.17(p);

(2) A list of the patents or publications submitted for consideration by the Office, including the date of publication of each patent or publication;

(3) A copy of each listed patent or publication in written form or at least the pertinent portions; and

(4) An English language translation of all the necessary and pertinent parts of any non-English language patent or publication in written form relied upon.

(c) The submission under this section must be served upon the applicant in accordance with § 1.248.

(d) A submission under this section shall not include any explanation of the patents or publications, or any other information. The Office will not enter such explanation or information if included in a submission under this section. A submission under this section is also limited to ten total patents or publications.

(e) A submission under this section must be filed within two months from the date of publication of the application (§ 1.215(a)) or prior to the mailing of a notice of allowance (§ 1.311), whichever is earlier. Any submission under this section not filed within this period is permitted only when the patents or publications could not have been submitted to the Office earlier, and must also be accompanied by the processing fee set forth in § 1.17(i). A submission by a member of the public to a pending published application that does not comply with the requirements of this section will not be entered.

(f) A member of the public may include a self-addressed postcard with a submission to receive an acknowledgment by the Office that the submission has been received. A member of the public filing a submission under this section will not receive any communications from the Office relating to the submission other than the return of a self-addressed postcard. In the absence of a request by the Office, an applicant has no duty to, and need not, reply to a submission under this section.

§26.10 USPTO 要求相關人提供資料

USPTO 審查委員可以根據需要，要求相關人提交有助於審查之資料或必要之說明。除了先前技術資料外，也可能要求相關人提供跟發明過程、發明之使用或背景技術有關的任何資料。相關人若因故無法提供，則仍應提出回覆說明原因。

---

37 CFR § 1.105 Requirements for information.

(a)

(1) In the course of examining or treating a matter in a pending or abandoned application filed under 35 U.S.C. 111 or 371 (including a reissue application), in a patent, or in a reexamination proceeding, the examiner or other Office employee may require the submission, from individuals identified under § 1.56(c), or any assignee, of such information as may be reasonably necessary to properly examine or treat the matter, for example:

(i) *Commercial databases*: The existence of any particularly relevant commercial database known to any of the inventors that could be searched for a particular aspect of the invention.

(ii) *Search*: Whether a search of the prior art was made, and if so, what was searched.

(iii) *Related information*: A copy of any non-patent literature, published application, or patent (U.S. or foreign), by any of the inventors, that relates to the claimed invention.

(iv) *Information used to draft application*: A copy of any non-patent literature, published application, or patent (U.S. or foreign) that was used to draft the application.

(v) *Information used in invention process*: A copy of any non-patent literature, published application, or patent (U.S. or foreign) that was used in the invention process, such as by designing around or providing a solution to accomplish an invention result.

(vi) *Improvements*: Where the claimed invention is an improvement, identification of what is being improved.

(vii) *In Use:* Identification of any use of the claimed invention known to any of the inventors at the time the application was filed notwithstanding the date of the use.

(viii) *Technical information known to applicant*. Technical information known to applicant concerning the related art, the disclosure, the claimed subject matter, other factual information pertinent to patentability, or concerning the accuracy of the examiner's stated interpretation of such items.

(2) Where an assignee has asserted its right to prosecute pursuant to § 3.71(a) of this chapter, matters such as paragraphs (a)(1)(i), (iii), and (vii) of this section may also be applied to such assignee.

(3) Requirements for factual information known to applicant may be presented in any appropriate manner, for example:

(i) A requirement for factual information;

(ii) Interrogatories in the form of specific questions seeking applicant's factual knowledge; or

(iii) Stipulations as to facts with which the applicant may agree or disagree.

(4) Any reply to a requirement for information pursuant to this section that states either that the information required to be submitted is unknown to or is not readily available to the party or parties from which it was requested may be accepted as a complete reply.

(b) The requirement for information of paragraph (a)(1) of this section may be included in an Office action, or sent separately.

(c) A reply, or a failure to reply, to a requirement for information under this section will be governed by §§ 1.135 and 1.136.

§26.11 違反揭露義務

若被發現有詐欺（fraud）、不正行為（inequitable conduct）而違反揭露義務（violation of duty of disclosure）者，即使是僅對一個請求項造成影響，但會使得所有請求項無效（unpatentable or invalid）（亦即專利無效）。一般專利是否無效，最主要取決於意圖與（intent）與實質性（materiality），亦即是否基於蓄意或惡意，而構成不正行為。

---

37 CFR 1.56 Duty to disclose information material to patentability

＊＊＊＊＊＊＊＊＊＊＊

However, no patent will be granted on an application in connection with which fraud on the Office was practiced or attempted or the duty of disclosure was violated through bad faith or intentional misconduct……

第 27 章　專利權期限之調整與展延

目　錄

§27.1　美國專利權期限

根據 35 USC 154，實用專利與植物專利之專利權期限為申請日起 20 年。

根據 35 USC 173，美國新式樣專利權期限為發證日起 14 年。

（美國實用專利舊制是發證日起 17 年結束。因應烏拉圭協定進行修法後，自烏拉圭協定生效日 1995 年 6 月 8 日後，如果不受任何尾端棄權書（terminal disclaimer）拘束，在烏拉圭協定生效日 1995 年 6 月 8 日或以後仍然生效的實用專利，申請人可任意選用專利申請日起 20 年或發證日起 17 年的期限為專利權結束日。案件申請日已經在 1995 年 6 月 8 日之前，專利權人對於專利權期限可選申請日+20 年或發證日+17 年為長者。申請日在 1995 年 6 月 8 日或之後，以申請日起 20 年的新制計算。）

---

35 U.S.C. 154 Contents and term of patent; provisional rights.

(a) IN GENERAL.

(2) TERM.-Subject to the payment of fees under this title, such grant shall be for a term beginning on the date on which the patent issues and ending 20 years from the date on which the application for the patent was filed in the United States or, if the application contains a specific reference to an earlier filed application or applications under section 120, 121, or 365(c) of this title, from the date on which the earliest such application was filed.

(3) PRIORITY.-Priority under section 119, 365(a), or 365(b) of this title shall not be taken into account in determining the term of a patent.

(c) CONTINUATION.

(1) DETERMINATION.-The term of a patent that is in force on or that results from an application filed before the date that is 6 months after the date of the enactment of the Uruguay Round Agreements Act shall be the greater of the 20-year term as provided in subsection (a), or 17 years from grant, subject to any terminal disclaimers.

(2) REMEDIES.-The remedies of sections 283, 284, and 285 of this title shall not apply to acts which

(A) were commenced or for which substantial investment was made before the date that is 6 months after the date of the enactment of the Uruguay Round Agreements Act; and

(B) became infringing by reason of paragraph (1).

(3) REMUNERATION. The acts referred to in paragraph (2) may be continued only upon the payment of an equitable remuneration to the patentee that is determined in an action brought under chapter 28 and chapter 29 (other than those provisions excluded by paragraph (2)) of this title.

35 U.S.C. 173 Term of design patent.

Patents for designs shall be granted for the term of fourteen years from the date of grant.

§27.2　美國專利權期限調整（Patent Term Adjustment, PTA）

隨著美國專利權期限由發證日 17 年修改為申請日起 20 年，依據美國發明人保護法（AIPA），配套增加了專利權期限調整的規定。確保專利權人若因美國專利局審查上的延滯，其實用專利與植物專利之專利權期限可依法調整。適用於 2000 年 5 月 29 日或以後申請之案件，亦包含該日期以後申請的延續案或者該日期以後申請的 PCT 進入美國之申請案，USPTO 將計算因官方延遲而可以調整的專利權天數。但此條款不適用於新式樣專利，美國新式樣專利仍維持發證日起 14 年。

35 U.S.C. 154 Contents and term of patent; provisional rights.

＊＊＊＊＊＊＊＊＊＊＊

(b) ADJUSTMENT OF PATENT TERM.

(1) PATENT TERM GUARANTEES.

(A) GUARANTEE OF PROMPT PATENT AND TRADEMARK OFFICE RESPONSES. Subject to the limitations under paragraph (2), if the issue of an original patent is delayed due to the failure of the Patent and Trademark Office to

(i) provide at least one of the notifications under section 132 of this title or a notice of allowance under section 151 of this title not later than 14 months after-

the date on which an application was filed under section 111(a) of this title; or

the date on which an international application fulfilled the requirements of section 371 of this title;

(ii) respond to a reply under section 132, or to an appeal taken under section 134, within 4 months after the date on which the reply was filed or the appeal was taken;

(iii) act on an application within 4 months after the date of a decision by the Board of Patent Appeals and Interferences under section 134 or 135 or a decision by a Federal court under section 141, 145, or 146 in a case in which allowable claims remain in the application; or

(iv) issue a patent within 4 months after the date on which the issue fee was paid under section 151 and all outstanding requirements were satisfied, the term of the patent shall be extended 1 day for each day after the end of the period specified in clause (i), (ii), (iii), or (iv), as the case may be, until the action described in such clause is taken.

(B) GUARANTEE OF NO MORE THAN 3-YEAR APPLICATION PENDENCY.- Subject to the limitations under paragraph (2), if the issue of an original patent is delayed due to the failure of the United States Patent and Trademark Office to issue a patent within 3 years after the actual filing date of the application in the United States, not including-

(i) any time consumed by continued examination of the application requested by the applicant under section 132(b);

(ii) any time consumed by a proceeding under section 135(a), any time consumed by the imposition of an order under section 181, or any time consumed by appellate review by the Board of Patent Appeals and Interferences or by a Federal court; or

(iii) any delay in the processing of the application by the United States Patent and Trademark Office requested by the applicant except as permitted by paragraph (3)(C), the term of the patent shall be extended 1 day for each day after the end of that 3-year period until the patent is issued.

(C) GUARANTEE OR ADJUSTMENTS FOR DELAYS DUE TO INTERFERENCES, SECRECY ORDERS, AND APPEALS. Subject to the limitations under paragraph (2), if the issue of an original patent is delayed due to

(i) a proceeding under section 135(a);

(ii) the imposition of an order under section 181; or

(iii) appellate review by the Board of Patent Appeals and Interferences or by a Federal court in a case in which the patent was issued under a decision in the review reversing an adverse determination of patentability, the term of the patent shall be extended 1 day for each day of the pendency of the proceeding, order, or review, as the case may be.

(2) LIMITATIONS.

(A) IN GENERAL. To the extent that periods of delay attributable to grounds specified in paragraph (1) overlap, the period of any adjustment granted under this subsection shall not exceed the actual number of days the issuance of the patent was delayed.

(B) DISCLAIMED TERM. No patent the term of which has been disclaimed beyond a specified date may be adjusted under this section beyond the expiration date specified in the disclaimer.

(C) REDUCTION OF PERIOD OF ADJUSTMENT.

(i) The period of adjustment of the term of a patent under paragraph (1) shall be reduced by a period equal to the period of time during which the applicant failed to engage in reasonable efforts to conclude prosecution of the application.

(ii) With respect to adjustments to patent term made under the authority of paragraph (1)(B), an applicant shall be deemed to have failed to engage in reasonable efforts to conclude processing or examination of an application for the cumulative total of any periods of time in excess of 3 months that are taken to respond to a notice from the Office making any rejection, objection, argument, or other request, measuring such 3-month period from the date the notice was given or mailed to the applicant.

(iii) The Director shall prescribe regulations establishing the circumstances that constitute a failure of an applicant to engage in reasonable efforts to conclude processing or examination of an application.

(3) PROCEDURES FOR PATENT TERM ADJUSTMENT DETERMINATION.

(A) The Director shall prescribe regulations establishing procedures for the application for and determination of patent term adjustments under this subsection.

(B) Under the procedures established under subparagraph (A), the Director shall

(i) make a determination of the period of any patent term adjustment under this subsection, and shall transmit a notice of that determination with the written notice of allowance of the application under section 151; and

(ii) provide the applicant one opportunity to request reconsideration of any patent term adjustment determination made by the Director.

(C) The Director shall reinstate all or part of the cumulative period of time of an adjustment under paragraph (2)(C) if the applicant, prior to the issuance of the patent, makes a showing that, in spite of all due care, the applicant was unable to respond within the 3-month period, but in no case shall more than three additional months for each such response beyond the original 3-month period be reinstated.

(D) The Director shall proceed to grant the patent after completion of the Director's determination of a patent term adjustment under the procedures established under this subsection, notwithstanding any appeal taken by the applicant of such determination.

(4) APPEAL OF PATENT TERM ADJUSTMENT DETERMINATION.

(A) An applicant dissatisfied with a determination made by the Director under paragraph (3) shall have remedy by a civil action against the Director filed in the United States District Court for the District of Columbia within 180 days after the grant of the patent. Chapter 7 of title 5, United States Code, shall apply to such action. Any final judgment resulting in a change to the period of adjustment of the patent term shall be served on the Director, and the Director shall thereafter alter the term of the patent to reflect such change.

(B) The determination of a patent term adjustment under this subsection shall not be subject to appeal or challenge by a third party prior to the grant of the patent.

§27.3　美國專利權期限調整（PTA）的計算

PTA 調整原則為：一般的專利權結束日（自有效申請日起起算 20 年）加上 USPTO 延誤的天數減去申請人延誤的天數。而無論如何，（USPTO 延誤的天數減去申請人延誤的天數）≥0。

USPTO 延誤的天數包括下列四種：

- I：USPTO 超過申請日起 14 個月發出第一次的審定書（Office Action），包括：*Quayle* action、核准通知、限制性選擇或依照 37 CFR §1.105 請求提供資訊。
- II：收到申請人任何回應後 4 個月內沒有回覆或收到領證費及條件齊備時，4 個月內沒有發證公告。
- III：由於牴觸程序／保密命令／訴願（成功）的延誤
- IV：USPTO 沒有在實際申請日起三年內核准公告該申請案

IV 之例外條款：

申請人請求

　因保密命令／牴觸程序／上訴檢視

　由於申請人方面的延遲，例如：

　申請人申請暫停審查達六個月

- 申請 RCE 時以延遲三個月
- 申請人請求暫停審查達三年(Rule 1.103(d)).

註：申請案提出第一次 RCE 後，則自動停止後續任何的 PTA 計算。

申請人的延誤的天數：

● 申請人對前述 I 與第 II 項沒有合理的努力以進行審查，因申請人延遲的天數將需扣減。

● 最常見的延遲扣減：對審定書（OA）答辯申請延期。（除非是因為需額外實驗或天災可請求例外）

● 其他種類的申請人延遲扣減

    (1) 申請人提出暫停審查

    (2) 申請人請求延遲發証

    (3) 放棄或延遲繳交領証費

    (4) 放棄後超過二個月請求復活

    (5) 將暫時申請案轉換成非暫時申時案

    (6) 審定書發下前請求先前修正

    (7) 因疏忽未對審定書答辯

    (8) 非審查委員要求之補充答辯

    (9) 訴願委員會或法院發下決定書後之審定書發下前一個內提出新事証

    (10) 核准通知後提出新事証

    (11) 提出延續案繼續審查

§27.4　PTA 調整天數的通知

第 1 次通知：USPTO 在核發核准通知時，會同時提供一份 PTA 天數的通知書，計算到核准通知發出日之前所累計的 PTA 天數。申請人若認為計算有誤，可以於繳納領證費時一併申請更正。詳細的 PTA 計算天數，領證通知發出後，申請人即可由 USPTO 的公開系統（Public PAIR）查詢。

第 2 次通知：USPTO 的發證通知上會記載預計發證日，以及根據發證日所計算的總 PTA 天數。詳細的 PTA 計算天數，仍然可以藉由 USPTO 的公開系統（Public PAIR）查詢。對於 PTA 計算有疑義者，須於發證日 2 個月內，根據 37 CFR1.705(b)(2)及繳交費用提出正式的

Request for reconsideration of PTA。或者在發證日 180 天內，依據 35USC154(b)(4)向聯邦地方法院對 USPTO 提出民事訴訟。但請求 PTA 重算的權利，僅限於專利權人，第三人無法申請。

37 CFR § 1.705 Patent term adjustment determination.

(a) The notice of allowance will include notification of any patent term adjustment under 35 U.S.C. 154(b).

(b) Any request for reconsideration of the patent term adjustment indicated in the notice of allowance, except as provided in paragraph (d) of this section, and any request for reinstatement of all or part of the term reduced pursuant to § 1.704(b) must be by way of an application for patent term adjustment. An application for patent term adjustment under this section must be filed no later than the payment of the issue fee but may not be filed earlier than the date of mailing of the notice of allowance. An application for patent term adjustment under this section must be accompanied by:

(1) The fee set forth in § 1.18(e); and

(2) A statement of the facts involved, specifying:

(i) The correct patent term adjustment and the basis or bases under § 1.702 for the adjustment;

(ii) The relevant dates as specified in §§ 1.703(a) through (e) for which an adjustment is sought and the adjustment as specified in § 1.703(f) to which the patent is entitled;

(iii) Whether the patent is subject to a terminal disclaimer and any expiration date specified in the terminal disclaimer; and

(iv)

(A) Any circumstances during the prosecution of the application resulting in the patent that constitute a failure to engage in reasonable efforts to conclude processing or examination of such application as set forth in § 1.704; or

(B) That there were no circumstances constituting a failure to engage in reasonable efforts to conclude processing or examination of such application as set forth in § 1.704.

(c) Any application for patent term adjustment under this section that requests reinstatement of all or part of the period of adjustment reduced pursuant to § 1.704(b) for failing to reply to a rejection, objection, argument, or other request within three months of the date of mailing of the Office communication notifying the applicant of the rejection, objection, argument, or other request must also be accompanied by:

(1) The fee set forth in § 1.18(f); and

(2) A showing to the satisfaction of the Director that, in spite of all due care, the applicant was unable to reply to the rejection, objection, argument, or other request within three months of the date of mailing of the Office communication notifying the applicant of the rejection, objection, argument, or other request. The Office shall not grant any request for reinstatement for more than three additional months for each reply beyond three months from the date of mailing of the Office communication notifying the applicant of the rejection, objection, argument, or other request.

(d) If there is a revision to the patent term adjustment indicated in the notice of allowance, the patent will indicate the revised patent term adjustment. If the patent indicates or should have indicated a revised patent term adjustment, any request for reconsideration of the patent term adjustment indicated in the patent must be filed within two months of the date the patent issued and must comply with the requirements of paragraphs (b)(1) and (b)(2) of this section. Any request for reconsideration under this section that raises issues that were raised, or could have been raised, in an application for patent term adjustment under paragraph (b) of this section shall be dismissed as untimely as to those issues.

(e) The periods set forth in this section are not extendable.

(f) No submission or petition on behalf of a third party concerning patent term adjustment under 35 U.S.C. 154(b) will be considered by the Office. Any such submission or petition will be returned to the third party, or otherwise disposed of, at the convenience of the Office.

§27.5　惠氏判例（Wyeth v. Kappos）對 PTA 計算的影響

在 2010 年 1 月的惠氏判例中，主要的爭點為：根據 35 USC 154(b)1(A)：特定審查事項的處理時限。例如：第一次 OA 應於 14 個月內發出、申請人 OA 答辯後應於 4 個月內給予下一次回應、繳納領證費應於 4 個月內發證⋯⋯。超過上述等等期限將計算官方延遲天數。簡稱 A delay。而根據 35 USC 154(b)1(B)：專利審查總期間不超過三年。但因申請人請求 RCE 或因牴觸、保密命令程序⋯⋯者則扣除。超過三年期限將計算官方延遲天數。簡稱 B delay。

　　USPTO 原先的計法：在先發生之 A delay 勢必造成 B delay，因此若 A delay 總天數大於 B delay 天數，則只以 A delay 總天數減去申請人 delay 之天數作為 PTA 之天數。反之，則為 B delay 天數減去申請人 delay 之天數作為 PTA 之天數。但惠氏認為法條並不應如此解讀，因此提出訴訟。最後經法院確認：基於法條解讀，並無 A/B delays 只能擇大者計算之規定，因此只要 A 與 B delay 期間並未重疊，則 A 與 B delay 應加總計算。亦即：

　　A delay 總天數加上 B delay 天數減去申請人 delay 之天數減去 A 與 B 重疊之天數＝PTA 之天數。

　　也因此，USPTO 在此判例發下後，並沒有再上訴，而是決定修改計算方式。在 2010 年 3 月 2 日以後已經更新軟體，此日期以後發證的案件，將會加計上述 B delay。

§27.6　專利權展延（Patent Term Extension）

　　根據 35 USC 156，對於藥品、醫藥裝置、其製造方法或使用方法，若因主管機關，例如，食品藥物管理局（FDA）及美國農業部（USDA）的審查／管制因素延遲上市，則可申請專利權展延，但僅限於發明專利。延展期限最多五年，且必須在專利過期前申請，且必須在獲得主管機關上市許可後的 60 內提出。此展延可與 USPTO 的專利權調整期限（PTA）累加，但根據 Under 35 USC 154(b)，與 FDA/USDA 的延遲累加後，從延期日起最長不能超過 14 年。

---

35 U.S.C. 156 Extension of patent term.

　　(a) The term of a patent which claims a product, a method of using a product, or a method of manufacturing a product shall be extended in accordance with this section from the original expiration date of the patent, which shall include any patent term adjustment granted under section 154(b) if

　　(1) the term of the patent has not expired before an application is submitted under subsection (d)(1) for its extension;

　　(2) the term of the patent has never been extended under subsection (e)(1) of this section;

　　(3) an application for extension is submitted by the owner of record of the patent or its agent and in accordance with the requirements of paragraphs (1) through (4) of subsection (d);

(4) the product has been subject to a regulatory review period before its commercial marketing or use;

(5)

(A) except as provided in subparagraph (B) or (C), the permission for the commercial marketing or use of the product after such regulatory review period is the first permitted commercial marketing or use of the product under the provision of law under which such regulatory review period occurred;

(B) in the case of a patent which claims a method of manufacturing the product which primarily uses recombinant DNA technology in the manufacture of the product, the permission for the commercial marketing or use of the product after such regulatory period is the first permitted commercial marketing or use of a product manufactured under the process claimed in the patent; or

(C) for purposes of subparagraph (A), in the case of a patent which

(i) claims a new animal drug or a veterinary biological product which (I) is not covered by the claims in any other patent which has been extended, and (II) has received permission for the commercial marketing or use in non-food-producing animals and in food-producing animals, and

(ii) was not extended on the basis of the regulatory review period for use in non-food-producing animals, the permission for the commercial marketing or use of the drug or product after the regulatory review period for use in food-producing animals is the first permitted commercial marketing or use of the drug or product for administration to a food-producing animal. The product referred to in paragraphs (4) and (5) is hereinafter in this section referred to as the "approved product."

(b) Except as provided in subsection (d)(5)(F), the rights derived from any patent the term of which is extended under this section shall during the period during which the term of the patent is extended

(1) in the case of a patent which claims a product, be limited to any use approved for the product

(A) before the expiration of the term of the patent

(i) under the provision of law under which the applicable regulatory review occurred, or

(ii) under the provision of law under which any regulatory review described in paragraph (1), (4), or (5) of subsection (g) occurred, and

(B) on or after the expiration of the regulatory review period upon which the extension of the patent was based;

(2) in the case of a patent which claims a method of using a product, be limited to any use claimed by the patent and approved for the product

(A) before the expiration of the term of the patent

(i) under any provision of law under which an applicable regulatory review occurred, and

(ii) under the provision of law under which any regulatory review described in paragraph (1), (4), or (5) of subsection (g) occurred, and

(B) on or after the expiration of the regulatory review period upon which the extension of the patent was based; and

(3) in the case of a patent which claims a method of manufacturing a product, be limited to the method of manufacturing as used to make

(A) the approved product, or

(B) the product if it has been subject to a regulatory review period described in paragraph (1), (4), or (5) of subsection (g). As used in this subsection, the term "product" includes an approved product.

(c) The term of a patent eligible for extension under subsection (a) shall be extended by the time equal to the regulatory review period for the approved product which period occurs after the date the patent is issued, except that

(1) each period of the regulatory review period shall be reduced by any period determined under subsection (d)(2)(B) during which the applicant for the patent extension did not act with due diligence during such period of the regulatory review period;

(2) after any reduction required by paragraph (1), the period of extension shall include only one-half of the time remaining in the periods described in paragraphs (1)(B)(i), (2)(B)(i), (3)(B)(i), (4)(B)(i), and (5)(B)(i) of subsection (g);

(3) if the period remaining in the term of a patent after the date of the approval of the approved product under the provision of law under which such regulatory review occurred when added to the regulatory review period as revised under paragraphs (1) and (2) exceeds fourteen years, the period of extension shall be reduced so that the total of both such periods does not exceed fourteen years, and

(4) in no event shall more than one patent be extended under subsection (e)(i) for the same regulatory review period for any product.

(d)

(1) To obtain an extension of the term of a patent under this section, the owner of record of the patent or its agent shall submit an application to the Director. Except as provided in paragraph

(5), such an application may only be submitted within the sixty-day period beginning on the date the product received permission under the provision of law under which the applicable regulatory review period occurred for commercial marketing or use. The application shall contain

(A) the identity of the approved product and the Federal statute under which regulatory review occurred;

(B) the identity of the patent for which an extension is being sought and the identity of each claim of such patent;

(C) information to enable the Director to determine under subsections (a) and (b) the eligibility of a patent for extension and the rights that will be derived from the extension and information to enable the Director and the Secretary of Health and Human Services or the Secretary of Agriculture to determine the period of the extension under subsection (g);

(D) a brief description of the activities undertaken by the applicant during the applicable regulatory review period with respect to the approved product and the significant dates applicable to such activities; and

(E) such patent or other information as the Director may require.

(2)

(A) Within 60 days of the submittal of an application for extension of the term of a patent under paragraph (1), the Director shall notify

(i) the Secretary of Agriculture if the patent claims a drug product or a method of using or manufacturing a drug product and the drug product is subject to the Virus-Serum-Toxin Act, and

(ii) the Secretary of Health and Human Services if the patent claims any other drug product, a medical device, or a food additive or color additive or a method of using or manufacturing such a product, device, or additive and if the product, device, and additive are subject to the Federal Food, Drug and Cosmetic Act, of the extension application and shall submit to the Secretary who is so notified a copy of the application. Not later than 30 days after the receipt of an application from the Director, the Secretary reviewing the application shall review the dates contained in the application pursuant to paragraph (1)(C) and determine the applicable regulatory review period, shall notify the Director of the determination, and shall publish in the Federal Register a notice of such determination.

(B)

(i) If a petition is submitted to the Secretary making the determination under subparagraph (A), not later than 180 days after the publication of the determination under subparagraph (A), upon which it may reasonably be determined that the applicant did not act with due diligence during the applicable regulatory review period, the Secretary making the determination shall, in accordance with regulations promulgated by the Secretary, determine if the applicant acted with due diligence during the applicable regulatory review period. The Secretary making the determination shall make such determination not later than 90 days after the receipt of such a petition. For a drug product, device, or additive subject to the Federal Food, Drug, and Cosmetic Act or the Public Health Service Act, the Secretary may not delegate the authority to make the determination prescribed by this clause to an office below the Office of the Commissioner of Food and Drugs. For a product subject to the Virus-Serum-Toxin Act, the Secretary of Agriculture may not delegate the authority to make the determination prescribed by this clause to an office below the Office of the Assistant Secretary for Marketing and Inspection Services.

(ii) The Secretary making a determination under clause (i) shall notify the Director of the determination and shall publish in the Federal Register a notice of such determination together with the factual and legal basis for such determination. Any interested person may request, within the 60-day period beginning on the publication of a determination, the Secretary making the determination to hold an informal hearing on the determination. If such a request is made within such period, such Secretary shall hold such hearing not later than 30 days after the date of the request, or at the request of the person making the request, not later than 60 days after such date. The Secretary who is holding the hearing shall provide notice of the hearing to the owner of the patent involved and to any interested person and provide the owner and any interested person an opportunity to participate in the hearing. Within 30 days after the completion of the hearing, such Secretary shall affirm or revise the determination which was the subject of the hearing and notify the Director of any revision of the determination and shall publish any such revision in the Federal Register.

(3) For the purposes of paragraph (2)(B), the term "due diligence" means that degree of attention, continuous directed effort, and timeliness as may reasonably be expected from, and are ordinarily exercised by, a person during a regulatory review period.

(4) An application for the extension of the term of a patent is subject to the disclosure requirements prescribed by the Director.

(5)

(A) If the owner of record of the patent or its agent reasonably expects that the applicable regulatory review period described in paragraphs (1)(B)(ii), (2)(B)(ii), (3)(B)(ii), (4)(B)(ii), or (5)(B)(ii) of subsection (g) that began for a product that is the subject of such patent may extend beyond the expiration of the patent term in effect, the owner or its agent may submit an application to the Director for an interim extension during the period beginning 6 months, and ending 15 days before such term is due to expire. The application shall contain

(i) the identity of the product subject to regulating review and the Federal statute under which such review is occurring;

(ii) the identity of the patent for which interim extension is being sought and the identity of each claim of such patent which claims the product under regulatory review or a method of using or manufacturing the product;

(iii) information to enable the Director to determine under subsection (a)(1), (2), and (3) the eligibility of a patent for extension;

(iv) a brief description of the activities undertaken by the applicant during the applicable regulatory review period to date with respect to the product under review and the significant dates applicable to such activities; and

(v) such patent or other information as the Director may require.

(B) If the Director determines that, except for permission to market or use the product commercially, the patent would be eligible for an extension of the patent term under this section, the Director shall publish in the Federal Register a notice of such determination, including the identity of the product under regulatory review, and shall issue to the applicant a certificate of interim extension for a period of not more than 1 year.

(C) The owner of record of a patent, or its agent, for which an interim extension has been granted under subparagraph (B), may apply for not more than 4 subsequent interim extensions under this paragraph, except that, in the case of a patent subject to subsection (g)(6)(C), the owner of record of the patent, or its agent, may apply for only 1 subsequent interim extension under this paragraph. Each such subsequent application shall be made during the period beginning 60 days before, and ending 30 days before, the expiration of the preceding interim extension.

(D) Each certificate of interim extension under this paragraph shall be recorded in the official file of the patent and shall be considered part of the original patent.

(E) Any interim extension granted under this paragraph shall terminate at the end of the 60-day period beginning on the day on which the product involved receives permission for commercial marketing or use, except that, if within that 60-day period, the applicant notifies the Director of such permission and submits any additional information under paragraph (1) of this subsection not previously contained in the application for interim extension, the patent shall be further extended, in accordance with the provisions of this section-

(i) for not to exceed 5 years from the date of expiration of the original patent term; or

(ii) if the patent is subject to subsection (g)(6)(C), from the date on which the product involved receives approval for commercial marketing or use.

(F) The rights derived from any patent the term of which is extended under this paragraph shall, during the period of interim extension-

(i) in the case of a patent which claims a product, be limited to any use then under regulatory review;

(ii) in the case of a patent which claims a method of using a product, be limited to any use claimed by the patent then under regulatory review; and

(iii) in the case of a patent which claims a method of manufacturing a product, be limited to the method of manufacturing as used to make the product then under regulatory review.

(e)

(1) A determination that a patent is eligible for extension may be made by the Director solely on the basis of the representations contained in the application for the extension. If the Director determines that a patent is eligible for extension under subsection (a) and that the requirements of paragraphs (1) through (4) of subsection (d) have been complied with, the Director shall issue to the applicant for the extension of the term of the patent a certificate of extension, under seal, for the period prescribed by subsection (c). Such certificate shall be recorded in the official file of the patent and shall be considered as part of the original patent.

(2) If the term of a patent for which an application has been submitted under subsection (d)(1) would expire before a certificate of extension is issued or denied under paragraph (1) respecting the application, the Director shall extend, until such determination is made, the term of the patent for periods of up to one year if he determines that the patent is eligible for extension.

(f) For purposes of this section:

(1) The term "product" means:

(A) A drug product.

(B) Any medical device, food additive, or color additive subject to regulation under the Federal Food, Drug, and Cosmetic Act.

(2) The term "drug product" means the active ingredient of

(A) a new drug, antibiotic drug, or human biological product (as those terms are used in the Federal Food, Drug, and Cosmetic Act and the Public Health Service Act) or

(B) a new animal drug or veterinary biological product (as those terms are used in the Federal Food, Drug, and Cosmetic Act and the Virus-Serum-Toxin Act) which is not primarily manufactured using recombinant DNA, recombinant RNA, hybridoma technology, or other processes involving site specific genetic manipulation techniques, including any salt or ester of the active ingredient, as a single entity or in combination with another active ingredient.

(3) The term "major health or environmental effects test" means a test which is reasonably related to the evaluation of the health or environmental effects of a product, which requires at least six months to conduct, and the data from which is submitted to receive permission for commercial marketing or use. Periods of analysis or evaluation of test results are not to be included in determining if the conduct of a test required at least six months.

(4)

(A) Any reference to section 351 is a reference to section 351 of the Public Health Service Act.

(B) Any reference to section 503, 505, 512, or 515 is a reference to section 503, 505, 512, or 515 of the Federal Food, Drug and Cosmetic Act.

(C) Any reference to the Virus-Serum-Toxin Act is a reference to the Act of March 4, 1913 (21 U.S.C. 151 - 158).

(5) The term "informal hearing" has the meaning prescribed for such term by section 201(y) of the Federal Food, Drug and Cosmetic Act.

(6) The term "patent" means a patent issued by the United States Patent and Trademark Office.

(7) The term "date of enactment" as used in this section means September 24, 1984, for human drug product, a medical device, food additive, or color additive.

(8) The term "date of enactment" as used in this section means the date of enactment of the Generic Animal Drug and Patent Term Restoration Act for an animal drug or a veterinary biological product.

(g) For purposes of this section, the term "regulatory review period" has the following meanings:

(1)

(A) In the case of a product which is a new drug, antibiotic drug, or human biological product, the term means the period described in subparagraph (B) to which the limitation described in paragraph (6) applies.

(B) The regulatory review period for a new drug, antibiotic drug, or human biological product is the sum of

(i) the period beginning on the date an exemption under subsection (i) of section 505 or subsection (d) of section 507 became effective for the approved product and ending on the date an application was initially submitted for such drug product under section 351, 505, or 507, and

(ii) the period beginning on the date the application was initially submitted for the approved product under section 351, subsection (b) of section 505, or section 507 and ending on the date such application was approved under such section.

(2)

(A) In the case of a product which is a food additive or color additive, the term means the period described in subparagraph (B) to which the limitation described in paragraph (6) applies.

(B) The regulatory review period for a food or color additive is the sum of

(i) the period beginning on the date a major health or environmental effects test on the additive was initiated and ending on the date a petition was initially submitted with respect to the product under the Federal Food, Drug, and Cosmetic Act requesting the issuance of a regulation for use of the product, and

(ii) the period beginning on the date a petition was initially submitted with respect to the product under the Federal Food, Drug, and Cosmetic Act requesting the issuance of a regulation for use of the product, and ending on the date such regulation became effective or, if objections were filed to such regulation, ending on the date such objections were resolved and commercial marketing was permitted or, if commercial marketing was permitted and later revoked pending further proceedings as a result of such objections, ending on the date such proceedings were finally resolved and commercial marketing was permitted.

(3)

(A) In the case of a product which is a medical device, the term means the period described in subparagraph (B) to which the limitation described in paragraph (6) applies.

(B) The regulatory review period for a medical device is the sum of -

(i) the period beginning on the date a clinical investigation on humans involving the device was begun and ending on the date an application was initially submitted with respect to the device under section 515, and

(ii) the period beginning on the date an application was initially submitted with respect to the device under section 515 and ending on the date such application was approved under such Act or the period beginning on the date a notice of completion of a product development protocol was initially submitted under section 515(f)(5) and ending on the date the protocol was declared completed under section 515(f)(6).

(4)

(A) In the case of a product which is a new animal drug, the term means the period described in subparagraph (B) to which the limitation described in paragraph (6) applies.

(B) The regulatory review period for a new animal drug product is the sum of

(i) the period beginning on the earlier of the date a major health or environmental effects test on the drug was initiated or the date an exemption under subsection (j) of section 512 became effective for the approved new animal drug product and ending on the date an application was initially submitted for such animal drug product under section 512, and

(ii) the period beginning on the date the application was initially submitted for the approved animal drug product under subsection (b) of section 512 and ending on the date such application was approved under such section.

(5)

(A) In the case of a product which is a veterinary biological product, the term means the period described in subparagraph (B) to which the limitation described in paragraph (6) applies.

(B) The regulatory period for a veterinary biological product is the sum of

(i) the period beginning on the date the authority to prepare an experimental biological product under the Virus- Serum-Toxin Act became effective and ending on the date an application for a license was submitted under the Virus-Serum-Toxin Act, and

(ii) the period beginning on the date an application for a license was initially submitted for approval under the Virus-Serum-Toxin Act and ending on the date such license was issued.

(6) A period determined under any of the preceding paragraphs is subject to the following limitations:

(A) If the patent involved was issued after the date of the enactment of this section, the period of extension determined on the basis of the regulatory review period determined under any such paragraph may not exceed five years.

(B) If the patent involved was issued before the date of the enactment of this section and

(i) no request for an exemption described in paragraph (1)(B) or (4)(B) was submitted and no request for the authority described in paragraph (5)(B) was submitted,

(ii) no major health or environment effects test described in paragraph (2)(B) or (4)(B) was initiated and no petition for a regulation or application for registration described in such paragraph was submitted, or

(iii) no clinical investigation described in paragraph (3) was begun or product development protocol described in such paragraph was submitted, before such date for the approved product the period of extension determined on the basis of the regulatory review period determined under any such paragraph may not exceed five years.

(C) If the patent involved was issued before the date of the enactment of this section and if an action described in subparagraph (B) was taken before the date of enactment of this section with respect to the approved product and the commercial marketing or use of the product has not been approved before such date, the period of extension determined on the basis of the regulatory review period determined under such paragraph may not exceed two years or in the case of an approved product which is a new animal drug or veterinary biological product (as those terms are used in the Federal Food, Drug, and Cosmetic Act or the Virus-Serum-Toxin Act), three years.

(h) The Director may establish such fees as the Director determines appropriate to cover the costs to the Office of receiving and acting upon applications under this section.

§27.7　尾端棄權書不拘束專利權展延

　　根據判例，美國法院確認一般因重複專利所提尾端棄權書，不影響專利權人所獲得的專利權展延。也就是說，醫藥品若獲得專利權期限展延，仍可繼續累加於尾端棄權書（terminal disclaimer）之終止日之後。

第 28 章　新式樣專利

目　錄

§28.1 新式樣專利之定義

依據 35USC§171，任何人對製品發明了任何新穎、原創及裝飾之式樣就可能取得新式樣專利。亦即，新式樣專利所請求之標的是具體表現在（embodied in）或施加至（applied to）一製品（或其部份）之式樣。所謂的式樣（design)包括所有種類之式樣，如物品之形狀，表面之裝飾（surface ornamental）及造形（configuration）或其組合。另外，新式樣不能與其具體表現或施加至之物品分開，其不能單獨以一表面裝飾存在，其必需是有固定形體之事物，可複製的，而且不僅僅是一方法之機會結果。

---

35 U.S.C. 171 Patents for designs.

Whoever invents any new, original, and ornamental design for an article of manufacture may obtain a patent therefor, subject to the conditions and requirements of this title.

The provisions of this title relating to patents for inventions shall apply to patents for designs, except as otherwise provided.

MPEP 1502 Definition of a Design [R-2]

In a design patent application, the subject matter which is claimed is the design embodied in or applied to an article of manufacture (or portion thereof) and not the article itself. Ex parte Cady, 1916 C.D. 62, 232 O.G. 621 (Comm'r Pat. 1916). "[35 U.S.C.] 171 refers, not to the design of an article, but to the design for an article, and is inclusive of ornamental designs of all kinds including surface ornamentation as well as configuration of goods." In re Zahn, 617 F.2d 261, 204 USPQ 988 (CCPA 1980).

The design for an article consists of the visual characteristics embodied in or applied to an article.

Since a design is manifested in appearance, the subject matter of a design patent application may relate to the configuration or shape of an article, to the surface ornamentation applied to an article, or to the combination of configuration and surface ornamentation.

Design is inseparable from the article to which it is applied and cannot exist alone merely as a scheme of surface ornamentation. It must be a definite, preconceived thing, capable of reproduction and not merely the chance result of a method.

$$* * * * * * * * * * * *$$

§28.2　新式樣專利說明書應包括的內容

新式樣說明書應包括說明書及圖式。

(A) 說明書

新式樣說明書分為(Ⅰ)前言及名稱(Ⅱ)敘述,及(Ⅲ)請求項三部份。

Ⅰ. 前言及名稱部份

通常包括申請人（發明人）之名字,新式樣之名稱以及新式樣具體表現之物品之本質及用途之簡單敘述。發明之名稱需以大眾習知或使用之名稱訂立。雖然以實線圖表示之所請求之新式樣可能只為物品之一部份,但是新式樣之名稱可敘述整個新式樣。名稱不能只針對較以實線圖表示之所請求之物品為少之部份作敘述。新式樣之名稱不能太廣,否則審查委員不能確定檢索之範圍及分類,大眾亦不能了解具體表現新式樣之物品之性質及用途,如果名稱太籠統,則應於前言部份作物品之性質及用途之簡單敘述。

Ⅱ. 敘述部份

通常只包括圖式之簡單敘述,因為圖式本身就是最佳之敘述。當然,除了圖式之簡單敘述之外,以下之敘述亦是允許的:

(A) 未顯示在圖式中之所請求之新式樣之一部份的外觀敘述;

(B) 未顯示在圖中之物品之放棄專利部份的敘述;

(C) 圖式中虛線部份,例如,不構成請求之新式樣之部份之環境,結構或邊界之敘述;

(D) 所請求之新式樣之環境，用途及本質之敘述；

(E) 新式樣之新穎或非顯而易見之特徵之敘述；

敘述部份通常不能包括以下：

(A) 以實線顯示之所請求的新式樣之任何部份之棄權聲明(disclaimer statement)；

(B) 未顯示在圖中之所請求的新式樣之其他實施例之敘述；

(C) 有關新式樣之功能或其他不相關之敘述。

Ⅲ. 請求項（Claim）

新式樣之說明書只能包括一項請求項（claim），通常之寫法為：The ornamental design for (the article) as shown。在請求項中物品（article）之敘述需與新式樣之名稱一致。當在(Ⅱ)敘述部份包括了新式樣之敘述或者新式樣之改良形式敘述時，新式樣之請求項需寫成"The ornamental design for the article(specifying name)as shown and described"圖式中之實線(full lines)所顯示者才是所請求之新式樣。虛線(broken lines)則通常用作顯示所請求之新式樣之環境等目的。

(B) 圖式

每一新式樣專利申請案必需包括所請求之新式樣之圖式或相片。由於相片或圖式構成了請求項之揭露，故相片或圖式必需清楚且完整，而當各視圖不一致時，審查委員會對圖式作出形式核駁（objection）並要求將各視圖修改成一致。但是當各視圖之不一致相當嚴重而使得新式樣之整體外觀不清楚時，審查委員將以不能據以實施及界定不清楚，發下 35USC112 第 1 及第 2 段之核駁（rejection）。

Ⅰ. 視圖

圖式通常需包括足夠之視圖以揭露所請求之新式樣之完整外觀，例如前視圖，後視圖，上視圖，底視圖及側視圖。最好亦呈送斜視圖（perspective view）以顯示立體的新式樣之外觀。另外，如果左右視圖相同或對稱，亦可只提供一視圖並於敘述中述明，而如果底視圖為平的或並無任何表面裝飾（surface ornamental）則亦可省略。

剖面圖通常是用來顯示內部之構造及功能／機械上之特徵，故通常不必要。如附上剖面圖會被形式核駁並要求刪除。但是，當所請求之新式樣之外表面之造形之精確輪廓（exact contour）無法從各視圖顯示，且並未意圖要顯示內部構造之特微時，圖式中仍可包括有剖面圖。

II. 表面陰影線

表面陰影線（surface shading）在為了要清楚顯示新式樣之立體的外表輪廓時，是必需的。表面陰影線在區別物品之開放及實心的表面時亦為必需。但是，對於以虛線繪製之未請求之標的則不能使用表面陰影線以避免對請求項之範圍產生混淆。

缺乏適當的陰影線可能會使得所請求之新式樣不能據以實施或界定不清楚而遭35USC112 第 1 段及第 2 段之核駁。此外，如果新式樣之表面形狀從揭露不夠明確，在提申後加上表面陰影線亦可能構成新事項（new matter）。通常實線的，黑的表面陰影線是不允許的，除了用來代表黑色或顏色之對比以外。斜的陰影線則可用來顯示透明，半透明，高度打光或反射之表面，如鏡子。材料之對比亦可在一區域使用陰影線而另一區域使用黑點來表示。

III. 虛線

新式樣圖式中使用虛線通常只有兩種用途，一是用來顯示所請求之新式樣期間之環境，一是界定所請求之新式樣之界限。亦即，用虛線表示之部份就不包括在所請求之新式樣的範圍之內。虛線並不能來顯示新式樣之較不重要部份或用來顯示物品之隱藏的平面或表面（hidden planes and surfaces）。

IV.表面處理

一新式樣之裝飾的外觀（ornamental appearance）包括其形狀（shape），造形（configuration），標示物（indicia），對比的顏色或材質（contrasting color or materials），圖案（graphic representations）或其他施加至物品之表面處理（surface treatment）。

在新式樣之圖式或相片中如果揭露了表面處理的話，通常此表面處理就是所請求之新式樣之一部份。而如果將新式樣之平面的表面處理（two-dimensional surface treatment）刪除或改成虛線通常是允許的，如果很清楚地申請人在申請時就擁有該新式樣之造形（形狀）的話。亦即，將一新式樣物品之造形（形狀）上之圖案（表面處理）移除，通常不會造成新事項，如果此表面處理不會使在下面之新式樣不清楚或重疊在新式樣之上的話。但是，如果將立體的表面處理，如圖像，槽，肋移除則會構造新事項而不能被接受。

Ⅴ. 相片及彩色圖式

新式樣申請案之圖式通常是在白紙上以黑色墨汁繪製之圖，只有當①新式樣無法以繪製之圖顯示或②以相片顯示更清楚時，才能用相片代替繪製之圖。

需注意的是，如果相片不夠清楚而使新式樣之細節無法清楚呈現時，審查委員會發下形式核駁（objection），直到申請人補充良好品質之相片之後，新式樣專利申請案才會發證。另外，如果相片無法清楚顯示新式樣之細節、形狀、造形及特徵部份，則審查委員可能發下 35USC112 第 1 段及第 2 段之核駁（rejection）指出無法據以實施及界定不清楚。

另外，新式樣專利申請案不能將圖式與相片併在一起提出。再者，如果以彩色相片或彩色圖式提出新式樣申請，則顏色將被視為所請求之新式樣整體的一部份。而提申之後將顏色去除將被允許，如果申請人在申請時已擁有在顏色以下之物品之形狀，造形之式樣而不包括顏色的話。

MPEP 1503.01 Specification [R-5]

＊＊＊＊＊＊＊＊＊＊＊

I.PREAMBLE AND TITLE

A preamble, if included, should state the name of the applicant, the title of the design, and a brief description of the nature and intended use of the article in which the design is embodied (37 CFR 1.154).

The title of the design identifies the article in which the design is embodied by the name generally known and used by the public but it does not define the scope of the claim. See MPEP § 1504.04, subsection I.A. The title may be directed to the entire article embodying the design while the claimed design shown in full lines in the drawings may be directed to only a portion of the article. However, the title may not be directed to less than the claimed design shown in full lines in the drawings.

＊＊＊＊＊＊＊＊＊＊＊

II.DESCRIPTION

No description of the design in the specification beyond a brief description of the drawing is generally necessary, since as a rule the illustration in the drawing views is its own best description.

* * * * * * * * * * *

In addition to the figure descriptions, the following types of statements are permissible in the specification:

(A) Description of the appearance of portions of the claimed design which are not illustrated in the drawing disclosure.

* * * * * * * * * * *

(B) Description disclaiming portions of the article not shown in the drawing as forming no part of the claimed design.

(C) Statement indicating the purpose of broken lines in the drawing, for example, environmental structure or boundaries that form no part of the design to be patented.

(D) Description denoting the nature and environmental use of the claimed design, if not included in the preamble pursuant to 37 CFR 1.154 and MPEP § 1503.01, subsection I.

* * * * * * * * * * *

(E) A "characteristic features" statement describing a particular feature of the design that is considered by applicant to be a feature of novelty or nonobviousness over the prior art (37 CFR 1.71(c)).

* * * * * * * * * * *

The following types of statements are permissible in the specification:

(A) A disclaimer statement directed to any portion of the claimed design that is shown in solid lines in the drawings is not permitted in the specification of an issued design patent.

* * * * * * * * * * *

(B) Statements which describe or suggest other embodiments of the claimed design which are not illustrated in the drawing disclosure,

* * * * * * * * * * *

(C) Statements describing matters *>that< are directed to function >or are< unrelated to the design.

* * * * * * * * * * *

III.DESIGN CLAIM

* * * * * * * * * * * *

A design patent application may only include a single claim. The single claim should normally be in formal terms to "The ornamental design for (the article which embodies the design or to which it is applied) as shown." The description of the article in the claim should be consistent in terminology with the title of the invention. See MPEP § 1503.01, subsection I.

When the specification includes a proper **>descriptive statement< of the design (see MPEP § 1503.01, subsection II), or a proper showing of modified forms of the design or other descriptive matter has been included in the specification, the words "and described" must be added to the claim following the term "shown"; i.e., the claim must read "The ornamental design for (the article which embodies the design or to which it is applied) as shown and described."

**>Full lines in the drawing show the claimed design. Broken lines are used for numerous purposes.

* * * * * * * * * * * *

MPEP 1503.02 Drawing [R-5]

Every design patent application must include either a drawing or a photograph of the claimed design. As the drawing or photograph constitutes the entire visual disclosure of the claim, it is of utmost importance that the drawing or photograph be clear and complete, and that nothing regarding the design sought to be patented is left to conjecture.

* * * * * * * * * * * *

When inconsistencies are found among the views, the examiner should object to the drawings and request that the views be made consistent. When the inconsistencies are of such magnitude that the overall appearance of the design is unclear, the claim should be rejected under 35 U.S.C.112, first and second paragraphs, as nonenabling and indefinite. See MPEP

* * * * * * * * * * * *

I.VIEWS

The drawings or photographs should contain a sufficient number of views to disclose the complete appearance of the design claimed, which may include the front, rear, top, bottom and sides. Perspective views are suggested and may be submitted to clearly show the appearance of three dimensional designs.

* * * * * * * * * * * *

For example, if the left and right sides of a design are identical or a mirror image, a view should be provided of one side and a statement made in the drawing description that the other side is identical or a mirror image. If the design has a flat bottom, a view of the bottom may be omitted if the specification includes a statement that the bottom is flat and devoid of surface ornamentation.

＊＊＊＊＊＊＊＊＊＊＊

Sectional views presented solely for the purpose of showing the internal construction or functional/ mechanical features are unnecessary and may lead to confusion as to the scope of the claimed design.

Such views should be objected to under 35 U.S.C. 112, second paragraph, and their cancellation should be required. However, where the exact contour or configuration of the exterior surface of a claimed design is not apparent from the views of the drawing, and no attempt is made to illustrate features of internal construction, a sectional view may be included to clarify the shape of said design......

＊＊＊＊＊＊＊＊＊＊＊

II.SURFACE SHADING

While surface shading is not required under 37 CFR 1.152, it may be necessary in particular cases to shade the figures to show clearly the character and contour of all surfaces of any 3-dimensional aspects of the design. Surface shading is also necessary to distinguish between any open and solid areas of the article. However, surface shading should not be used on unclaimed subject matter, shown in broken lines, to avoid confusion as to the scope of the claim.

Lack of appropriate surface shading in the drawing as filed may render the design nonenabling and indefinite under 35 U.S.C. 112, first and second paragraphs. Additionally, if the surface shape is not evident from the disclosure as filed, the addition of surface shading after filing may comprise new matter. Solid black surface shading is not permitted except when used to represent the color black as well as color contrast. Oblique line shading must be used to show transparent, translucent and highly polished or reflective surfaces, such as a mirror. **>Contrast in materials may be shown by using line shading in one area and stippling in another.

＊＊＊＊＊＊＊＊＊＊＊

III.BROKEN LINES

The two most common uses of broken lines are to disclose the environment related to the claimed design and to define the bounds of the claim. Structure that is not part of the claimed design, but is considered necessary to show the environment in which the design is associated, may be represented in the drawing by broken lines.

* * * * * * * * * * *

IV.SURFACE TREATMENT

The ornamental appearance of a design for an article includes its shape and configuration as well as any indicia, contrasting color or materials, graphic representations, or other ornamentation applied to the article ("surface treatment"). Surface treatment must be applied to or embodied in an article of manufacture. Surface treatment, per se (i.e., not applied to or embodied in a specific article of manufacture), is not proper subject matter for a design patent under 35 U.S.C. 171.

* * * * * * * * * * *

A disclosure of surface treatment in a design drawing or photograph will normally be considered as prima facie evidence that the inventor considered the surface treatment shown as an integral part of the claimed design. An amendment canceling two-dimensional surface treatment or reducing it to broken lines will be permitted if it is clear from the application that applicant had possession of the underlying configuration of the basic design without the surface treatment at the time of filing of the application. See In re Daniels, 144 F.3d 1452, 1456-57, 46 USPQ2d 1788, 1790 (Fed. Cir. 1998). Applicant may remove surface treatment shown in a drawing or photograph of a design without such removal being treated as new matter, provided that the surface treatment does not obscure or override the underlying design. The removal of three-dimensional surface treatment that is an integral part of the configuration of the claimed design, for example, removal of beading, grooves, and ribs, will introduce prohibited new matter as the underlying configuration revealed by this amendment would not be apparent in the application as originally filed. See MPEP § 1504.04, subsection II.

V.PHOTOGRAPHS AND COLOR DRAWINGS

Drawings are normally required to be submitted in black ink on white paper. See 37 CFR 1.84(a)(1). Photographs are acceptable only in applications in which the invention is not capable of being illustrated in an ink drawing or where the invention is shown more clearly in a photograph (e.g., photographs of ornamental effects are acceptable). See also 37 CFR 1.81(c) and 1.83(c), and MPEP § 608.02.

Photographs submitted in lieu of ink drawings must comply with 37 CFR 1.84(b). Only one set of black and white photographs is required. Color photographs and color drawings may be submitted in design applications if filed with a petition under 37 CFR 1.84(a)(2).

If the photographs are not of sufficient quality so that all details in the photographs are reproducible, this will form the basis of subsequent objection to the quality of the photographic disclosure. No application will be issued until objections directed to the quality of the photographic disclosure have been resolved and acceptable photographs have been submitted and approved by the examiner. If the details, appearance and shape of all the features and portions of the design are not clearly disclosed in the photographs, this would form the basis of a rejection of the claim under 35 U.S.C. 112,first and second paragraphs, as nonenabling and indefinite.

Photographs and ink drawings must not be combined in a formal submission of the visual disclosure of the claimed design in one application.

＊＊＊＊＊＊＊＊＊＊＊

If color photographs or color drawings are filed with the original application, color will be considered an integral part of the disclosed and claimed design. The omission of color in later filed formal photographs or drawings will be permitted if it is clear from the application that applicant had possession of the underlying configuration of the basic design without the color at the time of filing of the application.

＊＊＊＊＊＊＊＊＊＊＊

§28.3 新式樣專利之法定可予專利標的

依照 35USC171，任何人發明了任何製品（物品）之新穎，原創的，裝飾的式樣可取得新式樣專利。因而，新式樣專利之標的可分為下列三種：

(A) 施加或具體呈現在一製品上之裝飾，印記，印刷或圖案（表面之標示）；

(B) 製品之形狀或造型；及

(C) 上述二種之組合。

需注意，單純的圖案或裝飾並不是新式樣專利之標的。圖案或裝飾必需施加至或具體呈現在製品上才是新式樣專利之標的。

電腦圖標（computer generated icons）如全螢幕之顯示圖像（full screen displays）或及個別之圖標如果單獨存在則為平面之表面裝飾，因此，如果電腦圖標具體呈現在一製品上則是法定可予新式樣專利之標的。亦即，新式樣專利申請案之請求項之標的如果是"顯示在一電腦螢幕、監視器或其他顯示面板或其一部份上之電腦圖標"則符合 35USC171 之規定。

　　35USC171 規定式樣需具體呈現在製品（物品）之上，但是，所請求之式樣可涵蓋複數個物品，或在物品中有複數個部份。如果新式樣涵蓋了複數個物品或在物品中含有複數個部份（零件），則需在新式樣之名稱中註明為一組，一對，一組合，一單位，或一總成（set, pair, combination, unit or an assembly）。而在敘述中亦需聲明請求項是有關於所示之物品之共同的外觀（collective appearance）之新式樣。如果複數個部份顯示在單一視圖之中，則需以括弧括起來表示。如果新式樣是顯示在物品之複數個部份，則物品可用虛線表示，而各不同部份（零件）用實線表示，且不需括弧。

　　新式樣專利的標的必需主要是裝飾性的，而非主要是功能性或機構性的。而在決定新式樣主要是裝飾性或功能性，必需從其整體的外觀來檢視，當然新式樣之特別的元件亦需考慮。

　　某些物品在其被終端使用者使用時可能均是成隱藏之狀態，此時就發生此種使用時被隱藏之新式樣是否具有裝飾性之問題。審查委員在審查此類物品的新式樣專利性時，會考慮具有此式樣之物品在銷售時或展示時是否非在隱藏狀態，因而並非所有使用時被隱藏之新式樣均不符合新式樣之裝飾性之要件。

　　由於新式樣專利中要有原創性，而一模擬已存在之物品或人物之式樣通常被認為是不具原創性的。因此，如此之新式樣會被審查委員以 35U.S.C. §171 稱缺乏原創性而駁回。

　　依照美國專利法 35U.S.C. §171 任何視為對種族、宗教、性別、道德、風俗或國藉有侵犯性之新式樣設計，例如諷刺之創作等均被視為不予新式樣專利之標的。

MPEP 1504.01 Statutory Subject Matter for Designs

The language "new, original and ornamental design for an article of manufacture" set forth in 35 U.S.C. 171 has been interpreted by the case law to include at least three kinds of designs:

(A)　a design for an ornament, impression, print, or picture applied to or embodied in an article of manufacture (surface indicia);

(B)　a design for the shape or configuration of an article of manufacture; and

(C)　a combination of the first two categories.

＊ ＊ ＊ ＊ ＊ ＊ ＊ ＊ ＊ ＊ ＊ ＊

A picture standing alone is not patentable under 35 U.S.C. 171. The factor which distinguishes statutory design subject matter from mere picture or ornamentation, per se (i.e., abstract design), is the embodiment of the design in an article of manufacture.

* * * * * * * * * * * *

MPEP 1504.01(a) Computer-Generated Icons [R-5]

A. General Principle Governing Compliance With the "Article of Manufacture" Requirement

Computer-generated icons, such as full screen displays and individual icons, are 2-dimensional images which alone are surface ornamentation. See, e.g., Ex parte Strijland, 26 USPQ2d 1259 (Bd. Pat. App. & Int. 1992) (computer-generated icon alone is merely surface ornamentation). The USPTO considers designs for computer-generated icons embodied in articles of manufacture to be statutory subject matter eligible for design patent protection under 35 U.S.C. 171. Thus, if an application claims a computer-generated icon shown on a computer screen, monitor, other display panel, or a portion thereof, the claim complies with the "article of manufacture" requirement of 35 U.S.C. 171.

* * * * * * * * * * * *

MPEP 1504.01(b) Design Comprising Multiple Articles or

Multiple Parts Embodied in a Single Article [R-5]

While the claimed design must be embodied in an article of manufacture as required by 35 U.S.C. 171, it may encompass multiple articles or multiple parts within that article. Ex parte Gibson, 20 USPQ 249 (Bd. App. 1933). **>When the design involves multiple articles, the title must identify a single entity of manufacture made up by the parts (e.g., set, pair, combination, unit, assembly). A descriptive statement should be included in the specification making it clear that the claim is directed to the collective appearance of the articles shown. If the separate parts are shown in a single view, the parts must be shown embraced by a bracket "[rcub]". The claim may also involve multiple parts of a single article, where the article is shown in broken lines and various parts are shown in solid lines. In this case, no bracket is needed.< See MPEP §1503.01.

MPEP 1504.01(c) Lack of Ornamentality [R-5] - 1500 Design Patents

I.FUNCTIONALITY VS. ORNAMENTALITY

An ornamental feature or design has been defined as one which was "created for the purpose of ornamenting" and cannot be the result or "merely a by-product" of functional or mechanical considerations.

＊ ＊ ＊ ＊ ＊ ＊ ＊ ＊ ＊ ＊ ＊

To be patentable, a design must be "primarily ornamental." "In determining whether a design is primarily functional or primarily ornamental the claimed design is viewed in its entirety, for the ultimate question is not the functional or decorative aspect of each separate feature, but the overall appearance of the article, in determining whether the claimed design is dictated by the utilitarian purpose of the article."

＊ ＊ ＊ ＊ ＊ ＊ ＊ ＊ ＊ ＊ ＊

Knowledge that the article would be hidden during its end use based on the examiner's experience in a given art or information that may have been submitted in the application itself would not be considered prima facie evidence of the functional nature of the design.

＊ ＊ ＊ ＊ ＊ ＊ ＊ ＊ ＊ ＊ ＊

If there is sufficient evidence to show that a specific design "is clearly intended to be noticed during the process of sale and equally clearly intended to be completely hidden from view in the final use," it is not necessary that a rejection be made under 35 U.S.C. 171.

MPEP 1504.01(d) Simulation

35 U.S.C. 171 requires that a design to be patentable be "original." Clearly, a design which simulates an existing object or person is not original as required by the statute.

Therefore, a claim directed to a design for an article which simulates a well known or naturally occurring object or person should be rejected under 35 U.S.C. 171 as nonstatutory subject matter in that the claimed design lacks originality.

MPEP 1504.01(e) Offensive Subject Matter

Design applications which disclose subject matter which could be deemed offensive to any race, religion, sex, ethnic group, or nationality, such as those which include caricatures or depictions, should be rejected as nonstatutory subject matter under 35 U.S.C. 171

＊ ＊ ＊ ＊ ＊ ＊ ＊ ＊ ＊ ＊ ＊

§28.4 新式樣專利之新穎性

　　35USC§102 之法條亦適用於新式樣專利申請案，但是 35USC§102(d)之期間為 6 個月，而非一年。

　　而在決定新式樣是否為新穎，亦即決定新式樣是否為先前技術所預見（anticipated）時是與實用專利的標準一樣，亦即，先前技術需與新式樣在各方面均需相同。而新式樣專利與先前技術是否相同乃是根據一般觀察者測試（average observer test），亦即一般觀察者如認為新式樣與習知者不同，不是已存在之式樣或不是其改良而已，則具新穎性。但是在判斷是否新穎時，非裝飾（功能）之特徵則不予考慮。亦即審查委委員不必引證先前技術來證明功能之特徵不新穎，或隱藏之特徵不新穎只要證明裝飾之特徵不新穎即可發下缺乏新穎性之核駁。另外，以一般觀察者測試評估新穎性時不要求所請求之新式樣與先前技術為類似之技藝（analogous art）。

　　另外，實驗使用不被視為 35USC§102(b)之公開使用、銷售之例外條款。

　　再者，有關 35USC§102(d)，需注意的是，有許多國家，例如澳洲、瑞士、丹麥、芬蘭、法國、英國、挪威等國家，新式樣在提出申請當天就註冊了，亦即在申請日當天已取得專利（patented）。因而，依照 35USC§102(d)，在這些國家提出申請之六個月之內就需在美國提出新式樣專利申請，否則將違反 35USC§102(d) 之法定不予專利之規定。

MPEP 1504.02 Novelty [R-2]

　　The standard for determining novelty under 35 U.S.C. 102 was set forth by the court in In re Bartlett, 300 F.2d 942, 133 USPQ 204 (CCPA 1962). "The degree of difference [from the prior art] required to establish novelty occurs when the average observer takes the new design for a different, and not a modified, already-existing design." 300 F.2d at 943, 133 USPQ at 205 (quoting Shoemaker, Patents For Designs, page 76). In design patent applications, the factual inquiry in determining anticipation over a prior art reference is the same as in utility patent applications. That is, the reference "must be identical in all material respects." Hupp v. Siroflex of America Inc., 122 F.3d 1456, 43 USPQ2d 1887 (Fed. Cir. 1997).

The "average observer" test does not require that the claimed design and the prior art be from analogous arts when evaluating novelty.

* * * * * * * * * * *

When a claim is rejected under 35 U.S.C. 102 as being unpatentable over prior art, those features of the design which are functional and/or hidden during end use may not be relied upon to support patentability

* * * * * * * * * * *

Further, in a rejection of a claim under 35 U.S.C. 102, mere differences in functional considerations do not negate a finding of anticipation when determining design patentability.

* * * * * * * * * * *

It is not necessary for the examiner to cite or apply prior art to show that functional and/or hidden features are old in the art as long as the examiner has properly relied on evidence to support the prima facie lack of ornamentality of these individual features.

* * * * * * * * * * *

In evaluating a statutory bar based on 35 U.S.C. 102(b), the experimental use exception to a statutory bar for public use or sale (see MPEP § 2133.03(e)) does not usually apply for design patents.

* * * * * * * * * * *

Registration of a design abroad is considered to be equivalent to patenting under 35 U.S.C. 119(a)- (d) and 35 U.S.C. 102(d), whether or not the foreign grant is published.

In order for the filing to be timely for priority purposes and to avoid possible statutory bars, the U.S. design patent application must be made within 6 months of the foreign filing. See also MPEP § 1504.10.

* * * * * * * * * * *

The following table sets forth the dates on which design rights can be enforced in a foreign country (INID Code (24)) and thus, are also useable in a 35 U.S.C. 102(d) rejection as modified by 35 U.S.C. 172. It should be noted that in many countries the date of registration or grant is the filing date.

* * * * * * * * * * *

§28.5 新式樣專利之非顯而易見性

與實用專利之非顯而易見性之審查相同，Graham v. John Deere Co., 之法則亦適用於新式樣專利申請案之非顯而易見性之審查，但需注意以下事項：

(A) 新式樣之先前技術之範圍擴張至所有類似之技藝。當所請求之新式樣與先前技術不同之處只在於物品表面之裝飾時，任何揭示實質上相同之表面裝飾之先前技術均被視為是類似之技藝。

(B) 在決定新式樣之非顯而易見性時需考慮新式樣之整體的外觀。新式樣與先前技術之外觀即使有差異，但此差異是不重要時（de minimis or inconsequential），仍為顯而易見。

(C) 在判斷新式樣專利之非顯而易見性，此技藝人士為"此類物品之設計者"。

(D) 審查委員在判斷新式樣專利之非顯而易見性時仍會考慮第二層考量（Secondary considerations）如商業之成功，被仿冒，業界長期需要等因素。

MPEP 1504.03 Nonobviousness [R-5]

＊＊＊＊＊＊＊＊＊＊＊＊

I.GATHERING THE FACTS

The basic factual inquiries guiding the evaluation of obviousness, as outlined by the Supreme Court in Graham v. John Deere Co., 383 U.S. 1, 148 USPQ 459 (1966), are applicable to the evaluation of design patentability:

(A)Determining the scope and content of the prior art;

(B)Ascertaining the differences between the claimed invention and the prior art;

(C)Resolving the level of ordinary skill in the art; and

(D)Evaluating any objective evidence of nonobviousness (i.e., so-called "secondary considerations").

A. Scope of the Prior Art

The scope of the relevant prior art for purposes of evaluating obviousness under 35 U.S.C. 103(a) extends to all "analogous arts."

While the determination of whether arts are analogous is basically the same for both design and utility inventions (see MPEP § 904.01(c) and § 2141.01(a)),

＊＊＊＊＊＊＊＊＊＊＊

Therefore, where the differences between the claimed design and the prior art are limited to the application of ornamentation to the surface of an article, any prior art reference which discloses substantially the same surface ornamentation would be considered analogous art. Where the differences are in the shape or form of the article, the nature of the articles involved must also be considered.

B. Differences Between the Prior Art and the Claimed Design

In determining patentability under 35 U.S.C. 103(a), it is the overall appearance of the design that must be considered. In re Leslie, 547 F.2d 116, 192 USPQ 427 (CCPA 1977). The mere fact that there are differences between a design and the prior art is not alone sufficient to justify patentability. In re Lamb, 286 F.2d 610, 128 USPQ 539 (CCPA 1961).

All differences between the claimed design and the closest prior art reference should be identified in any rejection of the design claim under 35 U.S.C. 103(a). If any differences are considered de minimis or inconsequential from a design viewpoint, the rejection should so state.

C. Level of Ordinary Skill in the Art

＊＊＊＊＊＊＊＊＊＊＊

"level of ordinary skill in the art" from which obviousness of a design claim must be evaluated under 35 U.S.C. 103(a) has been held by the courts to be the perspective of the "designer of . . . articles of the types presented.

D. Objective Evidence of Nonobviousness (Secondary Considerations)

Secondary considerations, such as commercial success and copying of the design by others, are relevant to the evaluation of obviousness of a design claim. Evidence of nonobviousness may be present at the time a prima facie case of obviousness is evaluated or it may be presented in rebuttal of a prior obviousness rejection.

＊＊＊＊＊＊＊＊＊＊＊

§28.6 新式樣專利申請案如何滿足 35USC112 之要件

由於新式樣專利之請求項只有一項，且敘述為 as shown and described，故新式樣之請求項是否明確界定了新式樣專利保護之標的及圖式（說明書）之敘述是否符合揭露要件（據以實施要件）對新式樣專利來說是同一件事情。如果要以專利保護之新式樣之外觀形狀、造型，因為視覺之揭露（圖式之揭露）不完整而不能確定或了解的話，則請求項就未特別指出及清楚地界定發明之標的。

由於新式樣專利保護者為物品之外觀、形狀、造型、表面裝飾，其以實線表示者，故新式樣之圖式中以實線表示之形狀、造型、表面裝飾必需清楚及正確地繪製才能滿足 35USC112 之第 1 段及第 2 段之要件。

另外，只有在銷售或使用時點可被看見的物品的外觀，表面裝飾，形狀，造型必需充份揭露（清楚及正確地描繪）以符合 35USC112 之要件。在銷售或使用時隱藏起來之部份如果揭露不充份並不違反 35USC112 之要件。

所以，美國專利局之審查委員如果要以圖式揭露不充份而以 35USC112 核駁新式樣專利，亦即，圖式無法滿足據以實施要件及界定不清楚的話，審查委員必需明確指出圖式之缺陷，審查委員不能單單以圖式之品質不良為由，而必需指出圖式的哪些部份不清楚致使不能了解新式樣之形狀或造形或表面裝飾才行。

另外，如果新式樣之圖式之各視圖不一致，且不一致之情況嚴重致使新式樣之整體外觀不清楚時，審查委員可以 35USC112 核駁，指出揭露不能據以實施及界定不清楚。審查委員應將各視圖不一致之情形指出，如果審查委員不能清楚地指出不一致之處，則只能作形式核駁（objection），並請申請人提出修正。

此外，申請人對圖式作修改有可能造成新事項（new matter）。例如改變新式樣之外觀、造型將被視為新事項。移除物品之立體表面處理，例如槽，肋條等均被視為新事項。但是將圖式中之實線改為虛線或將虛線改為實線則非改變新式樣之物品之外觀、造型，則非新事項。

再者，如果新式樣之請求項中使用"或其類似物品"（or similar article"，"or the like，）將被視為不符合 35USC 112 第 2 段之規定。但是在新式樣名稱中使用"或其類似物"則符合 35USC112 第 2 段之規定。當新式樣要被保護之標的從新式樣之揭露

中無法決定時，將被視為不符合 35USC112 第 2 段之規定，例如，新式樣之圖式中
所請求之式樣（以實線表示）與未請求之部份（以虛線表示）之界限不清楚或不易
了解時，雖可能符合 35USC112 第 1 段據以實施之要件，但可能因未清楚界定所要
請求之式樣而被 35USC 112 第 2 段核駁。

MPEP 1504.04 Considerations Under 35 U.S.C. 112 [R-5]

* * * * * * * * * * * *

35 U.S.C. 112, FIRST AND SECOND PARAGRAPHS

Any analysis for compliance with 35 U.S.C. 112 should begin with a determination of whether the claims satisfy the requirements of the second paragraph before moving on to the first paragraph. See In re Moore, 439 F.2d 1232, 169 USPQ 236 (CCPA 1971). Therefore, before any determination can be made as to whether the disclosure meets the requirements of 35 U.S.C. 112, first paragraph, for enablement, a determination of the scope of protection sought by the claim must be made. However, since the drawing disclosure and any narrative description in the specification are incorporated into the claim by the use of the language "as shown and described," any determination of the scope of protection sought by the claim is also a determination of the subject matter that must be enabled by the disclosure. Hence, if the appearance and shape or configuration of the design for which protection is sought cannot be determined or understood due to an inadequate visual disclosure, then the claim, which incorporates the visual disclosure, fails to particularly point out and distinctly claim the subject matter applicant regards as their invention, in violation of the second paragraph of 35 U.S.C. 112.

* * * * * * * * * * * *

However, it should be understood that when a surface or portion of an article is disclosed in full lines in the drawing it is considered part of the claimed design and its shape and appearance must be clearly and accurately depicted in order to satisfy the requirements of the first and second paragraphs of 35 U.S.C. 112.

Only those surfaces of the article that are visible at the point of sale or during use must be disclosed to meet the requirement of 35 U.S.C. 112, first and second paragraphs. "The drawing should illustrate the design as it will appear to purchasers and users, since the appearance is the only thing that lends patentability to it under the design law." Ex parte Kohler, 1905 C.D. 192, 192, 116 O.G. 1185, 1185 (Comm'r Pat. 1905). The lack of disclosure of those surfaces of the article which are hidden during sale or use does not violate the requirements of the first and second paragraphs of 35 U.S.C. 112 because the "patented ornamental design has no use other than its visual appearance......"

＊＊＊＊＊＊＊＊＊＊＊

When a claim is rejected under 35 U.S.C. 112, first and second paragraphs, as nonenabling and indefinite due to an insufficient drawing disclosure, examiners must specifically identify in the Office action what the deficiencies are in the drawing. A mere statement that the claim is nonenabling and indefinite due to the poor quality of the drawing is not a sufficient explanation of the deficiencies in the drawing disclosure. Examiners must specifically point out those portions of the drawing that are insufficient to permit an understanding of the shape and appearance of the design claimed, and, if possible, suggest how the rejection may be overcome. Form paragraphs 15.21 and 15.20.02 may be used.

When inconsistencies between the views of the drawings are so great that the overall appearance of the design is unclear, the claim should be rejected under 35 U.S.C. 112, first and second paragraphs, as nonenabling and indefinite, and the rejection should specifically identify all of the inconsistencies between the views of the drawing. Otherwise, inconsistencies between drawing views will be objected to by the examiner and correction required by the applicant. See MPEP § 1503.02.

＊＊＊＊＊＊＊＊＊＊＊

New Matter

＊＊＊＊＊＊＊＊＊＊＊

A change in the configuration of the claimed design is considered a departure from the original disclosure and introduces prohibited new matter (37 CFR 1.121(f)). See In re Salmon, 705 F.2d 1579, 217 USPQ 981 (Fed. Cir. 1983). This includes the removal of three-dimensional surface treatment that is an integral part of the configuration of the claimed design, for example, beading, grooves, and ribs. The underlying configuration revealed by such an amendment would not be apparent in the application as filed and, therefore, it could not be established that applicant was in possession of this amended configuration at the time the application was filed. However, an amendment that changes the scope of a design by either reducing certain portions of the drawing to broken lines or converting broken line structure to solid lines is not a change in configuration as defined by the court in Salmon. The reason for this is because applicant was in possession of everything disclosed in the drawing at the time the application was filed and the mere reduction of certain portions to broken lines or conversion of broken line structure to solid lines is not a departure from the original disclosure.

＊＊＊＊＊＊＊＊＊＊＊

III.35 U.S.C. 112, SECOND PARAGRAPH

＊＊＊＊＊＊＊＊＊＊＊＊

Use of phrases in the claim such as "or similar article," "or the like," or equivalent terminology has been held to be indefinite. See Ex parte Pappas, 23 USPQ2d 1636 (Bd. Pat. App. & Inter. 1992). However, the use of broadening language such as "or the like," or "or similar article" in the title when directed to the environment of the article embodying the design should not be the basis for a rejection under 35 U.S.C. 112, second paragraph. See MPEP § 1503.01, subsection I.

＊＊＊＊＊＊＊＊＊＊＊

Rejections under 35 U.S.C. 112, second paragraph, should be made when the scope of protection sought by the claim cannot be determined from the disclosure. For instance, a drawing disclosure in which the boundaries between claimed (solid lines) and unclaimed (broken lines) portions of an article are not defined or cannot be understood may be enabling under 35 U.S.C. 112, first paragraph, in that the shape and appearance of the article can be reproduced, but such disclosure fails to particularly point out and distinctly claim the subject matter that applicant regards as the invention.

§28.7　新式樣專利之限制要求

　　當一新式樣專利申請案請求複數個專利性上不同的新式樣時，美國專利局亦會對新式樣之專利申請案作限制要求。而在決定是否專利性上不同時，審查委員必需去比較複數個新式樣之整體外觀。通常如果(A)複數個新式樣之整體外觀基本上具有相同的設計特性（design characteristics），而且(B)複數個新式樣之整體外觀之差異不足以在專利性上區別，則此複數個新式樣不是專利性上不同的新式樣。另外，如果將組合及次組合之新式樣，例如一整體之新式樣及其物品之部份或元件（次組合）之新式樣併在一起申請，通常次組合會有不同之外觀，故在專利性是不同的，則通常美國專利局會發下限制要求。

MPEP 1504.05 Restriction [R-5]

＊＊＊＊＊＊＊＊＊＊＊

Restriction will be required under 35 U.S.C. 121 if a design patent application *>claims< multiple designs that are ** patentably distinct from each other **.

＊＊＊＊＊＊＊＊＊＊＊

II.DISTINCT INVENTIONS

**>In determining patentable distinctness, the examiner must compare the overall appearances of the multiple designs. Each design must be considered as a whole, i.e., the elements of the design are not considered individually as they may be when establishing a prima facie case of obviousness under 35 U.S.C. 103(a). Designs are not distinct inventions if: (A) the multiple designs have overall appearances with basically the same design characteristics; and (B) the differences between the multiple designs are insufficient to patentably distinguish one design from the other.

＊＊＊＊＊＊＊＊＊＊＊

When an application illustrates a component, which is a subcombination of another embodiment, the subcombination often has a distinct overall appearance and a restriction should be required. When an application illustrates only a portion of the design, which is the subject of another embodiment, that portion often has a distinct overall appearance and a restriction should be required.<

＊＊＊＊＊＊＊＊＊＊＊

§28.8 新式樣專利申請案之重複專利核駁

新式樣專利申請案與實用專利申請案一樣有二種重複專利核駁之情形。一是依照35USC171之相同發明型式（same invention type）之重複專利核駁，亦即，一新式樣專利申請案與另一新式樣專利申請案或新式樣專利所請求之式樣為完全相同時，審查委員就會發下重複專利之核駁。另一種重複專利之核駁則並非法定的，而是顯而易見形式（obviousness-type）之重複專利核駁，當兩新式樣專利申請案或一新式樣申請案與另一新式樣專利，請求相同之發明概念，但不同外觀或不同範圍之標的，其在專利性上不能互相區別時，審查委員也會發下重複專利之核駁。

Ⅰ. 相同發明之重複專利核駁

依照 35USC171，對一已取得專利新式樣不能再發給第二個新式樣專利，故審查委員要對新式樣專利申請案作出發明之重複專利之核駁，必需具有相同範圍之相同新式樣被請求兩次時才發下此法定的相同發明之重複專利核駁。

上述法定的相同發明之重複專利核駁只發生在新式樣與新式樣專利申請案之情形。此法定之相同發明之重複專利的核駁不能藉由提出尾端棄權書（terminal disclaimer）來克服。

Ⅱ. 非法定重複專利核駁

非法定重複專利核駁，是當兩新式樣專利申請案請求相同之發明概念，但不同之外觀或不同範圍之標的，其在專利性上不能互相區別時（顯而易見的重複專利）才發生。亦即審查委員要主張兩新式樣是推斷的顯而易見時需(A)兩新式樣之整體外觀基本上具有相同之設計特點，且(B)兩新式樣之差別不足以在專利性上互相區別。所謂差別不足以在專利性上互相區別，即表示對此技藝之設計者是顯而易見的。

此種新式樣專利之顯而易見的重複專利核駁，會發生在一新式樣專利申請案其請求一組合（較窄之請求項）之新式樣，而另一新式樣專利申請案其請求次組合（較廣之請求項）之情形。通常當此二新式樣是專利性上不可區別的，但具有相同的發明概念，審查委員需滿足該窄的請求項（組合之新式樣）是否完全揭露及被涵蓋在較廣之請求項（次組合之新式樣）之中，如果未被揭露及被涵蓋，則不會發下重複專利之核駁。

新式樣專利之顯而易見的重複專利核駁可藉由提出尾端棄權書來克服。

MPEP 1504.06 Double Patenting [R-5]

There are generally two types of double patenting rejections. One is the "same invention" type double patenting rejection based on 35 U.S.C. 171 which states in the singular that an inventor "may obtain a patent." The second is the "nonstatutory-type" double patenting rejection based on a judicially created doctrine grounded in public policy and which is primarily intended to prevent prolongation of the patent term by prohibiting claims in a second patent not patentably distinct from claims in a first patent.

＊＊＊＊＊＊＊＊＊＊＊

I."SAME INVENTION" DOUBLE PATENTING REJECTIONS

A design - design statutory double patenting rejection based on 35 U.S.C. 171 prevents the issuance of a second patent for a design already patented. For this type of double patenting rejection to be proper, identical designs with identical scope must be twice claimed.

＊＊＊＊＊＊＊＊＊＊＊

II.NONSTATUTORY DOUBLE PATENTING REJECTIONS

A rejection based on nonstatutory double patenting is based on a judicially created doctrine grounded in public policy so as to prevent the unjustified or improper timewise extension of the right to exclude granted by a patent. In re Goodman, 11 F.3d 1046, 29 USPQ2d 2010 (Fed. Cir. 1993).

A nonstatutory double patenting rejection of the obviousness-type applies to claims directed to the same inventive concept with different appearances or differing scope which are patentably indistinct from each other. Nonstatutory categories of double patenting rejections which are not the "same invention" type may be overcome by the submission of a terminal disclaimer.

＊＊＊＊＊＊＊＊＊＊＊

To establish a prima facie case of obviousness-type double patenting: (A) the conflicting design claims must have overall appearances with basically the same design characteristics; and (B) the differences between the two designs must be insufficient to patentably distinguish one design from the other. Differences may be considered patentably insufficient when they are de minimis or obvious to a designer of ordinary skill in the art.

＊＊＊＊＊＊＊＊＊＊＊

This kind of obviousness-type double patenting rejection in designs will occur between designs which may be characterized as a combination (narrow claim) and a subcombination/ element thereof (broad claim). See discussion in MPEP § 1504.05, subsection II, B. If the designs are patentably indistinct and are directed to the same inventive concept the examiner must determine whether the subject matter of the narrower claim is fully disclosed in and covered by the broader claim of the reference. If the reference does not fully disclose the narrower claim, then a double patenting rejection should not be made.

＊＊＊＊＊＊＊＊＊＊＊

§28.9　新式樣專利申請案主張國際優先權之規定

35USC172，35USC119(a)-(d)之主張國外優先權之規定亦適用於新式樣專利申請案，但是 35USC102(d)之期限為 6 個月。亦即如要取得較早國外申請日之利益，美國新式樣專利申請案必需在該相同之新式樣在任何外國之最早申請日起 6 個月內提出申請。另外，新式樣專利申請案不能主張暫時申請案之優先權。此外，美國亦承認基於雙邊或多邊協約，例如"Hague Agreement Concerning the International Deposit of Industrial Designs," "Uniform Benelux Act on Designs and Models"，"European Community Design."所提出之新式樣申請之國外優先權。美國新式樣專利申請案如果主張上述基於雙邊或多邊協約申請案之優先權時，除了提供國外申請案之申請日及申請案號之資訊外，尚需提供下列資訊：

(A) 國外申請案所依據提申之協約名稱；

(B) 美國以外之國家之名稱，其中申請案具有效力或者相當於正常之國家申請案者；

(C) 接受國外申請案之國家或國際間政府機構之名稱及地址。

---

35 U.S.C. 172 Right of priority.

The right of priority provided for by subsections (a) through (d) of section 119 of this title and the time specified in section 102(d) shall be six months in the case of designs. The right of priority provided for by section 119(e) of this title shall not apply to designs.

＊＊＊＊＊＊＊＊＊＊＊

MPEP 1504.10 Priority Under 35 U.S.C. 119(a)-(d) [R-5]

＊＊＊＊＊＊＊＊＊＊＊

The provisions of 35 U.S.C. 119(a)-(d) apply to design patent applications. However, in order to obtain the benefit of an earlier foreign filing date, the United States application must be filed within 6 months of the earliest date on which any foreign application for the same design was filed. Design applications may not make a claim for priority of a provisional application under 35 U.S.C. 119(e).

＊＊＊＊＊＊＊＊＊＊＊

The United States will recognize claims for the right of priority under 35 U.S.C. 119(a)-(d) based on applications filed under such bilateral or multilateral treaties as the "Hague Agreement Concerning the International Deposit of Industrial Designs," "Uniform Benelux Act on Designs and Models" and "European Community Design." In filing a claim for priority of a foreign application previously filed under such a treaty, certain information must be supplied to the United States Patent and Trademark Office. In addition to the application number and the date of filing of the foreign application, the following information is required:

(A) the name of the treaty under which the application was filed,

(B) the name of at least one country other than the United States in which the application has the effect of, or is equivalent to, a regular national filing and

(C) the name and location of the national or inter-governmental authority which received the application.

＊＊＊＊＊＊＊＊＊＊＊

§28.10 新式樣專利申請案之延續案

美國專利法 35USC120 亦適用於新式樣專利申請案，亦即當一新式樣專利申請案包括有複數個專利性上不同的新式樣時，申請人可提出分割案並主張母案之申請日之利益。新式樣申請人亦可提出延續案（CPA）及 CIP 申請案。

如果申請人要主張 35USC120 之較早美國申請案之申請日之利益的話，在說明書第 1 行必需包括以下說明。"

"This is a division [continuation] of design Application No.- - - -, filed - - -." 依照 37CFR 1.78(a)(2)，如果在說明書中未（修正）加入上述之說明，則需加入申請資料表（application data sheet）中，否則將被視為放棄主張 35USC120 之優先權之利益。

另外，延續案（CPA）或分割案所請求之新式樣必需揭露在母案之中，否則不能主張先前申請案之申請日之利益。

再者，於 2004 年 9 月 21 日當天或之後申請之新式樣專利申請案如果在申請當天有提出主張先前申請之美國專利申請案之優先權的話，就被視為是參照併入。亦即，如果在母案中有部份資訊因疏忽未加入延續案（分割案）中，就可重新加入而不被視為是新事項（new matter）。

新式樣專利申請案之申請人亦可提出新式樣專利申請之部份延續案（CIP）。但是，如果 CIP 申請案將母案之圖式之中物品之形狀、造型、表面裝飾修改則將被視為是新事項，而不能主張母案之優先權。另外，例如，CIP 申請案中將原母案圖式之一部份由實線改為虛線，則此修改並不會被視為是新式樣之形狀、造形之改變，不是新事項，

另外，如果符合 35USC120 之條件，一新式樣專利申請案亦可主張申請在先之實用專利申請案（utility application）之優先權，亦可主張申請在先之 PCT 申請案之優先權，如果此 PCT 申請案有指定美國的話。

MPEP 1504.20 Benefit Under 35 U.S.C. 120

* * * * * * * * * * *

If applicant is entitled under 35 U.S.C. 120 to the benefit of an earlier U.S. filing date, the statement that, "This is a division [continuation] of design Application No.- - - -, filed - - -." should appear in the first sentence of the specification. As set forth in 37 CFR 1.78(a)(2), the specification must contain or be amended to contain such a reference in the first sentence>(s)< following the title unless the reference is included in an application data sheet (37 CFR 1.76). The failure to timely submit such a reference is considered a waiver of any benefit under 35 U.S.C. 120.

* * * * * * * * * * *

The claimed design in a continuation application and in a divisional application must be disclosed in< the original application. If this condition is not met, the application is not entitled to the benefit of the earlier filing date and the examiner should notify applicant accordingly by specifying the reasons why applicant is not entitled to claim the benefit under 35 U.S.C. 120.

* * * * * * * * * * *

For applications filed on or after September 21, 2004, 37 CFR 1.57(a) provides that a claim under 37 CFR 1.78 for the benefit of a prior-filed application, that was present on the filing date of the application, is considered an incorporation by reference as to inadvertently omitted material. See MPEP § 201.17.<

When the first application is found to be fatally defective under 35 U.S.C. 112 because of insufficient disclosure to support an allowable claim and such position has been made of record by the examiner, a second design patent application filed as an alleged "continuation-in-part" of the first application to supply the deficiency is not entitled to the benefit of the earlier filing date. See Hunt Co. v. Mallinckrodt Chemical Works, 177 F.2d 583, 83 USPQ 277 (F2d Cir. 1949) and cases cited therein. Also, a design application filed as a "continuation-in-part" that changes the shape or configuration of a design disclosed in an earlier application is not entitled to the benefit of the filing date of the earlier application. See In re Salmon, 705 F.2d 1579, 217 USPQ 981 (Fed. Cir. 1983). However, a later filed application that changes the scope of a design claimed in an earlier filed application by reducing certain portions of the drawing to broken lines is not a change in configuration as defined by the court in Salmon. See MPEP §1504.04, subsection II.

§28.11 新式樣專利申請案之加速審查

從 2000 年 9 月 8 日起，新式樣專利申請案之申請人如果自行作了檢索並提出加速審查（Expedited Examination）之請求，美國專利局就會加速審查，加速審查之請求需符合下列要件：

(A) 提出加速審查之請求；

(B) 新式樣專利申請案是完整的，包括符合 35CFR1.84 之圖式；

(C) 提出聲明指出已作了檢索（可提出國外專利事務所之檢索（審查）報告）。此聲明必需包括檢索範圍之列表（包括 US 分類，次分類之專利文獻及非專利文獻）。

(D) IDS 資料；

(E) 基本之新式樣之申請費；

(F) 加速審查之規費。

MPEP 1504.30 Expedited Examination [R-5]

＊＊＊＊＊＊＊＊＊＊＊

37 CFR 1.555 establishes an expedited procedure for design applications. This expedited procedure became effective on September 8, 2000 and is available to all design applicants who first conduct a preliminary examination search and file a request for expedited treatment accompanied by the fee specified in 37 CFR 1.17(k).

＊＊＊＊＊＊＊＊＊＊＊

A design application may qualify for expedited examination provided the following requirements are met:

(A) A request for expedited examination is filed (Form PTO/SB/27 may be used);

(B) The design application is complete and it includes drawings in compliance with 37 CFR 1.84 (see 37 CFR 1.154 and MPEP § 1503 concerning the requirements for a complete design application);

(C) A statement is filed indicating that a preexamination search was conducted (a search made by a foreign patent office satisfies this requirement). The statement must also include a list of the field of search such as by U.S. Class and Subclass (including domestic patent documents, foreign patent documents and nonpatent literature);

(D) An information disclosure statement in compliance with 37 CFR 1.98 is filed;

(E) The basic design application filing fee set forth in 37 CFR 1.16(*>b<) is paid; and

(F) The fee for expedited examination set forth in 37 CFR 1.17(k) is paid.

＊＊＊＊＊＊＊＊＊＊＊

§28.12 新式樣專利之專利權期限及維持費

新式樣專利發證後，專利權為 14 年，自發證日起算，另外，新式樣專利並不需繳交維持費。

MPEP 1505 Allowance and Term of Design Patent

35 U.S.C. 173 Term of design patent.

Patents for designs shall be granted for the term of fourteen years from the date of grant.

實例

發明人 X 於 2004 年 1 月 1 日提出新式樣申請案，該新式樣申請案乃關於一手機之外觀。此新式樣申請案中包括二種手機外觀之實施例，且各包含二組手機之視圖。審查委員於 2005 年 5 月 10 日發下限制要求。發明人選擇第 1 組實施例，並同時於 2005 年 5 月 31 日提出一分割申請案請求第二實施例之手機。原新式樣申請案於 2005 年 12 月 1 日發證，分割案於 2006 年 3 月 12 日發證。請問此二新式樣之專利權之期限為何？

原新式樣之專利期限為自 2005 年 12 月 1 日起至 2019 年 11 月 30 日止，分割案之新式樣之專利權期限自 2006 年 3 月 12 日起至 2020 年 3 月 11 日止。新式樣專利之專利權期間乃自發證日起算 14 年，故分割案之專利權期間並非與母案同時結束，而是自發證日起算 14 年結束。

第 29 章　植物專利

目　錄

§29.1　植物相關發明在美國的保護樣態

植物相關發明在美國可以下方式申請保護。且根據申請標的不同，可以同時申請多種方式的保護。

1. 植物專利（Plant Patent）：無性繁殖的新品種。（35 U.S.C. 161-164, USPTO）
2. 實用專利（Utility Patent）：植物新品種或其方法。（35 U.S.C. 101, USPTO）
3. 植物品種保護法（Plant Variety Protection Act, PVPA）：有性繁殖之新品種。（7 U.S.C. 2331, PVPO, Plant Variety Protection Office, 美國農業部植物品種保護局）

§29.2　植物專利涵蓋的保護標的

凡發明或發現經無性繁殖之任何獨特且新穎的植物新品種，包括經培育而成之芽突變、突變株、雜交株與新發現之苗株，可依法申請取得專利，但不包括塊莖作物或於非栽培地發現之品種。

植物專利的幾個特點：

1. 需屬無性繁殖作物：以無性繁殖方式，諸如：扦插、嫁接、芽接、枝接、壓條、塊根、種球、走莖、單性繁殖或組織培養等方式繁殖。但是馬鈴薯、菊芋等塊莖類作物之繁殖部位因與食用部位相同，因此排除於保護對象範圍。
2. 植物體以無性繁殖方式所得：申請植物專利之植株，申請前先經無性繁殖獲得性狀完全相同之個體群，且申請時需說明其無性繁殖方法，並須證明由其指定之繁殖方法所得之植株個體間性狀表現一致。
3. 提交的新植株個體（突變株）僅限定在栽培地所產生，也就是原存於自然界之野生植物個體並不能直接申請專利保護。

---

35 U.S.C. 161 Patents for plants.

Whoever invents or discovers and asexually reproduces any distinct and new variety of plant, including cultivated sports, mutants, hybrids, and newly found seedlings, other than a tuber propagated plant or a plant found in an uncultivated state, may obtain a patent therefor, subject to the conditions and requirements of this title.

The provisions of this title relating to patents for inventions shall apply to patents for plants, except as otherwise provided.

§29.3　植物專利的發明人

一般來說，發現該植物與達成以無性繁殖該植物之人，均可共列為發明人。

§29.4　植物專利的內容與其請求標的

如一般實用專利，植物專利申請書所揭露的發明內容仍必須符合 35 USC 112 規定。但由於植物無法如同一般工業產品般可以利用文字及圖說完成產品之再製，在 35 USC162 中，特別規定只要申請者已盡可能就揭露內容做詳細說明，審查官不得據 35 USC 112 條款駁回申請案。

植物專利的請求項（claim）只有一項，內容限定於其主張的單一品種的整株植物。

---

35 USC 162 – Description, claim

No plant patent shall be declared invalid for noncompliance with section 112 of this title if the description is as complete as is reasonably possible.

The claim in the specification shall be in formal terms to the plant shown and described.

§29.5　申請植物專利需檢附之文件

申請植物專利需檢附之文件如下：

(1)　植物專利申請表（Plant application transmittal form）

(2)　植物專利申請費用表（Fee transmittal form）：申請費＋檢索費＋審查費

(3) 專利申請資料表（Application data sheet）（see §1.76）

(4) 說明書（Specification）

(5) 圖式（Drawings （in duplicate））

(6) 發明人宣誓書（Executed oath or declaration）（§1.162）

§29.6 植物專利說明書需包含的內容

植物專利說明書的內容需包含下列：

1. 發明名稱

2. 相關申請案之對照參考（若有）

3. 關於聯邦資助研究或發展之陳述（若有）

4. 請求植物之種名與屬名（拉丁名）

5. 變種名

6. 發明之背景

7. 發明之概要

8. 圖式之簡單說明

9. 植物之詳細敘述

10. 單一請求項

11. 揭露之摘要

---

37 CFR §1.163 Specification and arrangement of application elements in a plant application.

(a) The specification must contain as full and complete a disclosure as possible of the plant and the characteristics thereof that distinguish the same over related known varieties, and its antecedents, and must particularly point out where and in what manner the variety of plant has been asexually reproduced. For a newly found plant, the specification must particularly point out the location and character of the area where the plant was discovered.

(b) The elements of the plant application, if applicable, should appear in the following order:

(1) Plant application transmittal form.

(2) Fee transmittal form.

(3) Application data sheet (see § 1.76).

(4) Specification.

(5) Drawings (in duplicate).

(6) Executed oath or declaration (§ 1.162).

(c) The specification should include the following sections in order:

(1) Title of the invention, which may include an introductory portion stating the name, citizenship, and residence of the applicant.

(2) Cross-reference to related applications (unless included in the application data sheet).

(3) Statement regarding federally sponsored research or development.

(4) Latin name of the genus and species of the plant claimed.

(5) Variety denomination.

(6) Background of the invention.

(7) Brief summary of the invention.

(8) Brief description of the drawing.

(9) Detailed botanical description.

(10) A single claim.

(11) Abstract of the disclosure.

(d) The text of the specification or sections defined in paragraph (c) of this section, if applicable, should be preceded by a section heading in upper case, without underlining or bold type.

§29.7 植物專利說明書之圖式

植物專利說明書之圖式必須是清晰完整的。符號或標記並非必要。圖式必須可以顯示該植物一般外觀上的特徵。圖式可以是彩色，若送彩色圖，必須一式兩份。

§ 1.165 Plant Drawings.

(a) Plant patent drawings should be artistically and competently executed and must comply with the requirements of **§ 1.84**. View numbers and reference characters need not be employed unless required by the examiner. The drawing must disclose all the distinctive characteristics of the plant capable of visual representation.

(b) The drawings may be in color. The drawing must be in color if color is a distinguishing characteristic of the new variety. Two copies of color drawings or photographs must be submitted.

§29.8　植物專利的專利要件

需符合新穎性（novelty）及進步性（non-obviousness）要件。然其新穎性一般決定於該植物是否已在國內外為他人熟知或利用（public use or sale）c 或者違反 35 U.S.C. 102d 者，均喪失新穎性。一般的書面公開（照片與一般描述）並不足以為充分的先前揭露，主要是無法達成讓此領域人士據以實施的要求。

性狀審查基準在於與其最接近之現有品種相互比較，當審查官對申請人用以與申請品種比對之最接近對照品種無法認同或對申請品種主要區別性狀有所質疑時，申請人必須負舉證責任，提出證明，必要時更需依審查官意見更換最類似品種，並重新建立性狀比對數據。

進步性審查，主要是基於申請案所申請專利之植物是否運用申請前既有之技術或知識，而為熟習該項技術者所能輕易完成獲預期可得。

植物專利申請案亦可申請 RCE。

§29.9　植物專利的專利權效期與效力

植物專利自專利申請日起 20 年之保護期限內，期間不用繳納維持費。而在美國境內他人不得就已取得專利之植物進行無性繁殖，亦不得販賣、為販賣之要約、及使用以專利所述之方法而繁殖該植物或其任何部分。

不過權利範圍僅及於該育成或發現之原始植物個體及經由繁殖所得與原始植株具相同性狀之植株。另外，植物專利法對以有性繁殖方式（種子繁殖）繁殖受專利保護的植株並不認定構成侵權行為。

35 USC 163 - Grant

In the case of a plant patent the grant shall be of the right to exclude others from asexually reproducing the plant, using, selling, or offering to sell the plant so reproduced, or any of its parts, throughout the United States, or from importing the plant so reproduced, or any parts thereof, into the United States.

20 years (from earliest U.S. filing date); NO maintenance fee; 1 claim; if color distinguishing characteristic, file appln with two copies of color drawings or color photographs + black and white photocopy; printed publication not prior art (not enabling).

第 30 章　異議程序

目　錄

§30.1　異議程序與再審查程序

所謂異議程序（protest），或稱 preissuance opposition（發證前異議），乃是一種法律程序可讓第三者在一專利申請案十八個月早期公告（核准前公告）之前或發下核准通知之前（視哪一日期在先）將與該專利申請案之專利性有關之先前技術文件（prior art document），或該專利申請案已公告，或公開，已放棄，或有發明人適格（inventorship）之問題，或有違反揭露義務（未提交 IDS 資料）等資訊送交美國專利局由審查委員在審查時考慮所送進來之先前技術文件及資訊。

而再審查程序（re-examination）則是一專利發證後由專利權人或第三者提出專利或刊物作為先行技術請求美國專利局對該專利之有效性作再審查。

§30.2　異議程序由何人提起，可作為異議之先前技術及異議程序需於何時提出

任何人，包括個人、公司、政府機關均可依照 37CFR 1.291 提出異議程序。另外，提出異議者亦可以匿名方式由專利律師（專利代理人）或其他代表人提出異議。但是，如果以匿名方式由多數人對一專利申請案提出異議時，需依照 37CFR 1.291(C)(5)，解釋為何第二或之後之異議理由與先前提出之異議不同，及為何此不同之異議理由未在之前一起提出。

可作為異議理由之先前技術（資訊）並不限定於專利及刊物，任何事實或資訊其對一專利申請案專利性不利者均可作為異議理由。通常異議之內容及本質較其是否為書面文件來得重要。美國專利局亦知道作為異議的證據之先前技術不是書面文件時，可能會有是否為真實及證據上（authentication and the probative value）之問題。但美國專利局並不會因為有上述之問題就不去考慮這些證據或者就認為這些證據不能作為異議之理由。

以下是可作為異議之先前技術（資訊）之例子：

(A) 顯示發明在美國於發明日之前已公告或被他人公用，而違反 35USC102 (a) 及／或 103 之資訊；

(B)　顯示發明在美國專利申請案之申請日的一年之前已在美國公開使用及銷售，因而違反 35USC102(b)之資訊；

(C)　申請人已放棄了發明（35USC102(c)）或並非自己發明了要尋求專利之標的（35USC102(f)）之資訊；

(D)　有關 35USC102(g)之發明人適格之資訊；

(E)　有關於揭露是否充份或未揭露最佳模式（35USC112）之資訊；

(F)　任何顯示專利申請案在專利性上違反法律規定之資訊；

(G)　顯示違反 35CFR1.56 之揭露義務或詐欺之資訊。

任何人如果要提出異議，需在專利申請案核准前公告或核准通知寄出之前（視哪一日期為早）提出，除非附有該專利申請案之申請人之書面同意書才能在上述日期之後提出。當然，提出異議需在審查中（pending），尚未放棄之期間提出，異議之資訊及證據才會被考慮。實務上，異議者一旦獲知有先前技術之文獻或資訊就需儘速提出異議以在審查之前就能被審查委員考慮。如果在一專利申請案已發下審定書之後才提出異議，只要該專利申請案仍在審查中，則仍會被考慮。然而如果異議之先前技術文獻（資訊）與最終審定書所引證之先前技術是重複的（cumulative），則審查委員不會重開審查。

MPEP 1901.01 Who Can Protest [R-3].

Any member of the public, including private persons, corporate entities, and government agencies, may file a protest under 37 CFR 1.291. A protest may be filed by an attorney or other representative on behalf of an unnamed ** >real party in interest. 37 CFR 1.291 does not require that the real party in interest be identified. Where a protest is not the first protest by the real party in interest, 37 CFR 1.291(b)(2) requires compliance with 37 CFR 1.291(c)(5). The requirements of 37 CFR 1.291(c)(5) cannot be avoided by multiple protests submitted by different people representing the same real party in interest.<

MPEP 1901.02 Information Which Can Be Relied on in Protest [R-3]

Any information which, in the protestor's opinion, would make the grant of a patent improper can be relied on in a protest under 37 CFR 1.291*. While prior art documents, such as patents and publications, are most often the types of information relied on in protests, 37 CFR 1.291* is not limited to prior art documents. Protests may be based on any facts or information adverse to patentability. The content and substance of the protest are more important than whether prior art documents, or some other form of evidence adverse to patentability, are being relied on. The Office recognizes that when evidence other than prior art documents is relied on, problems may arise as to authentication and the probative value to assign to such evidence. However, the fact that such problems may arise, and have to be resolved, does not preclude the Office from considering such evidence, nor does it mean that such evidence cannot be relied on in a protest under 37 CFR 1.291. Information in a protest should be set forth in the manner required by 37 CFR 1.291(*>c<).

The following are examples of the kinds of information, in addition to prior art documents, which can be relied on in a protest under 37 CFR 1.291*:

(A) Information demonstrating that the inven-tion was publicly "known or used by others in this country...... before the invention thereof by the applicant for patent" and is therefore barred under 35 U.S.C.102(a) and/or 103.

(B) Information that the invention was "in public use or on sale in this country, more than 1 year prior to the date of the application for patent in the United States" (35 U.S.C. 102(b)).

(C) Information that the applicant "has abandoned the invention" (35 U.S.C. 102(c)) or "did not himself invent the subject matter sought to be patented"(35 U.S.C. 102(f)).

(D) Information relating to inventorship under 35 U.S.C. 102(g).

(E) Information relating to sufficiency of disclosure or failure to disclose best mode, under 35 U.S.C. 112.

(F) Any other information demonstrating that the application lacks compliance with the statutory requirements for patentability.

(G) Information indicating "fraud" or "violation of the duty of disclosure" under 37 CFR 1.56 may be the subject of a protest under 37 CFR 1.291*. Protests raising fraud or other inequitable conduct issues will be entered in the application file, generally without comment on those issues. 37 CFR 1.291(*>e<).

＊ ＊ ＊ ＊ ＊ ＊ ＊ ＊ ＊ ＊ ＊

MPEP 1901.04 When Should the Protest Be Submitted [R-3]

* >Except where a protest is accompanied by the written consent of the applicant as provided in 37 CFR 1.291(b)(1), a< protest under 37 CFR 1.291(a) must be submitted prior to the date the application was published >under 37 CFR 1.211< or the mailing of a notice of allowance under 37 CFR 1.311, whichever occurs first, and the application must be pending when the protest and application file are brought before the examiner in order to be ensured of

consideration. As a practical matter, any protest should be submitted as soon as possible after the protestor becomes aware of the existence of the application to which the protest is to be directed. By submitting a protest early in the examination process, i.e., before the Office acts on the application if possible, the protestor ensures that the protest will receive maximum consideration and will be of the most benefit to the Office in its examination of the application. A protest submitted after the mailing of the notice of allowance will not knowingly be ignored if the protest includes prior art documents which clearly anticipate or clearly render obvious one or more claims. However, the likelihood of consideration of a protest decreases as the patent date approaches.

A protest filed after final rejection and complying with >all the requirements of< 37 CFR 1.291* will be considered if the application is still pending when the protest and application are provided to the examiner. However, prosecution will not ordinarily be reopened after final rejection if the prior art cited in the protest is merely cumulative of the prior art cited in the final rejection. **

＊＊＊＊＊＊＊＊＊＊＊＊

§30.3　提出異議需具備之文件

依照 37CFR1.291 提出異議，必需以書面提出，且需註明要提出異議之專利申請案之下列資訊：

(A) 申請人之名稱

(B) 申請案號

(C) 申請日

(D) 發明之名稱

(E) 技藝類別之號碼（art unit number）（如果知道）

(F) 審查委員之名稱（如果知道）

(G) 目前狀態及申請案在何單位（如果知道）

(H) 致該專利申請案之技術中心的審查長

同時需包括據以提出異議之專利刊物及其他資訊一覽表，簡單說明其相關性，及所有相關部份之英文翻譯，以及所有相關先前技術文件之資訊之影本。異議人可用呈送 IDS 之表格來呈送這些文件（資訊）之影本。

　　上述文件除送至美國專利局之該技術中心以外，尚需將一份送交申請人或其代表人，或者如果無法直接送交申請人則送二份至美國專利局。

　　另外，需注意的是提出異議時所有的文件必需齊備，因為美國專利局不給異議者補充資訊，補文件之機會。亦即，異議者提出異議之後，異議者之程序就結束，美國專利局不會對再送進來之文件、資訊考慮。如果異議人要再提供先前技術或文件但是異議之理由不相同時，必需另外提出一異議程序。

MPEP 1901.03 How Protest Is Submitted [R-3]

　　A protest under 37 CFR 1.291* must be submitted in writing, must specifically identify the application to which the protest is directed by application number or serial number and filing date, and must include a listing of all patents, publications, or other information relied on; a concise explanation of the relevance of each listed item; an English language translation of all relevant parts of any non-English language * >patent, publication, or other information relied upon<; and be accompanied by a copy of each patent, publication, or other document relied on. Protestors are encouraged to use form ** >PTO/SB/08A and 08B< "Information Disclosure Statement >By Applicant<" (or an equivalent form) when preparing a protest under 37 CFR 1.291, especially the listing enumerated under 37 CFR 1.291(*>c<)(1). See MPEP § *>609.04(a)<. In addition, the protest and any accompanying papers must either (1) reflect that a copy of the same has been served upon the applicant or upon the applicant's attorney or agent of record; or (2) be filed with the Office in duplicate in the event service is not possible.

　　It is important that any protest against a pending application specifically identify the application to which the protest is directed with the identification being as complete as possible. If possible, the following information should be placed on the protest:

(A)　Name of Applicant(s).

(B)　Application number (mandatory).

(C)　Filing date of application.

(D)　Title of invention.

(E)　Group art unit number (if known).

(F)　Name of examiner to whom the application is assigned (if known).

(G)　Current status and location of application (if known).

(H)　The word "ATTENTION:" followed by the area of the Office to which the protest is directed as set forth below.

INITIAL PROTEST SUBMISSION MUST BE COMPLETE

A protest must be complete and contain a copy of every document relied on by >the< protestor, whether the document is a prior art document, court litigation material, affidavit, or declaration, etc., because a protestor will not be given an opportunity to supplement or complete any protest which is incomplete. Active participation by protestor ends with the filing of the initial protest, as provided in 37 CFR 1.291(*>d<), and no further submission on behalf of protestor will be acknowledged or considered, ** >unless the submission is made pursuant to 37 CFR 1.291(c)(5). 37 CFR 1.291(c)(5) requires that any further submission by the same party in interest must be directed to significantly different issue(s) than those raised in the earlier protest.

＊＊＊＊＊＊＊＊＊＊＊＊

§30.4　美國專利局如何處理第三者提出之異議

　　當有任何人對一特定的專利申請案提出異議時，只要審查委員來得及審理此異議案時，審查委員均會考慮異議者提出之先前技術文件及資訊。但是異議者只會收到美國專利局確收之回郵明信片，異議人並不會被告知被異議之申請案之狀態。亦即除非異議人被准許可檢閱及影印（access）該專利申請案，否則異議人並不能得到任何相關該專利申請案之資訊，包括是否有該專利申請案存在，一直到專利申請案公告為止。亦即，在核准前公告之前，所有資訊均對異議人是保密的。

　　美國專利局之審查委員在對異議理由（證據）作初步之審視之後，如果覺得需要請被異議之申請案的申請人提供意見或針對異議者之特定問題回答時，通常給予一個月之時間回覆。但是審查委員不會要求申請人針對"詐欺"，"不正行為"及"違反揭露義務"部份作意見及回答。

　　審查委員會最後針對異議者所提出之異議證據之重點及爭論仔細考慮並表示審查委員之立場並將結果告知技術中心之審查長。

　　另外，如果有申請人之書面同意，異議人亦可在核准前公告之後的任何時間提出異議，美國專利局亦會接受此異議並考慮其異議理由證據及論點。

MPEP 1901.05 Initial Office Handling and Acknowledgment of Protest [R-3]

I.PROTESTS REFERRED TO EXAMINER

* * * * * * * * * * *

A protest where the application is specifically identified, which is submitted in conformance with 37 CFR 1.291(a)**>, (b), and (c)<, will be considered by the Office >if the protest is matched with the application in time to permit review by the examiner during prosecution<.

* * * * * * * * * * *

II.PROTEST DOES NOT INDICATE SERVICE

If the protest filed in the Office does not, however, indicate service on applicant or applicant's attorney or agent, and is not filed in duplicate, then the Office * >may< undertake to determine whether or not service has been made by contacting applicant or applicant's attorney or agent by telephone or in writing to ascertain if service has been made. If service has not been made and no duplicate has been filed, then the Office may request protestor to file such a duplicate before the protest is referred to the examiner.

* * * * * * * * * * *

III.ACKNOWLEDGMENT OF PROTEST

A protestor in an original or reissue application will not receive any communications from the Office relating to the protest, or to the application, other than the return of a self-addressed postcard which protestor may include with the protest in order to receive an acknowledgment that the protest has been received by the Office.

* * * * * * * * * * *

IV.APPLICATIONS AND STATUS THEREOF MAINTAINED IN CONFIDENCE

The postcard acknowledging receipt of a protest in other than a reissue application will not and must not indicate whether such application in fact exists or the status of any such application.

Thus, unless a protestor has been granted access to an original application, the protestor is not entitled to obtain from the Office any information concerning the same, including the mere fact that such an application exists.

* * * * * * * * * * *

MPEP 1901.06 Examiner Treatment of Protest [R-3]

II.PERIOD FOR COMMENTS BY APPLICANT

If the primary examiner's initial review reveals that the protest is ready for consideration during the examination, the examiner may nevertheless consider it desirable, or necessary, to obtain applicant's comments on the protest before further action. In such situations, the examiner will offer applicant an opportunity to file comments within a set period, usually 1 month, unless circumstances warrant a longer period.

* * * * * * * * * * *

Where necessary or desirable to decide questions raised by the protest, under 37 CFR 1.291(*>f<) the primary examiner can require the applicant to reply to the protest and answer specific questions raised by the protest. The primary examiner cannot require a reply to questions relating to "fraud," "inequitable conduct," or "violation of the duty of disclosure" since those issues are generally not commented on by the Office.

* * * * * * * * * * *

VII.CONSIDERATION OF PROTESTOR'S ARGUMENTS

In view of the value of written protests, the examiner must give careful consideration to the points and arguments made on behalf of the protestor. Any Office action by the examiner treating the merits of a timely submitted protest complying with 37 CFR 1.291(*>c<) must specifically consider and make evident by detailed reasoning the examiner's position as to the major arguments and points raised by the protestor.

* * * * * * * * * * *

VIII. RESULTS OF CONSIDERATION REPORTED TO TECHNOLOGY CENTER (TC) DIRECTOR

After the examiner has considered the protest, the examiner will report the results of such consideration to the TC Director.

§30.5　對再發證申請案提出異議

　　任何人亦可對再發證申請案提出異議（protest）。對再發證申請案提出異議。可在再發證申請案審查中（pending）之期間提出，但是必需在發下核准通知之前。再發證申請案提出之後會在美國專利局之公報上宣告（announcement），但是此宣告（公告）並不排除第三者提出異議。

　　但是，對於再發證申請案提出異議應於其在公報上宣告之日起二個月內提出，超過二個月之後提出異議並不需提出請願（petition），但是如果最終審定書已發下或者審查已結束時，則需同時提出請願（petition）才可。請願時需說明為何需要額外的時間來提出異議及所提異議之本質，同時需將請願書副本一份送交申請人。因此，如果超過了二個月要提出異議，可在二個月到期前提出請願請求以延展此二個月之期間及請求該再發證申請案被延緩審查。

MPEP 1441.01 Protest in Reissue Applications [R-7]

A protest pursuant to 37 CFR 1.291 may be filed throughout the pendency of a reissue application, **>before< the date of mailing of a notice of allowance, subject to the timing constraints of the examination, as set forth in MPEP § 1901.04. While a reissue application is not published under 37 CFR 1.211, the reissue application is published pursuant to 35 U.S.C. 122(b)(1)(A) via an announcement in the Official Gazette (and public availability of the file content) per 37 CFR 1.11(b). Such a publication does not preclude the filing of a protest. 35 U.S.C. 122(c) states:

＊＊＊＊＊＊＊＊＊＊＊

A protest with regard to a reissue application should be filed within the 2-month period following the announcement of the filing of the reissue application in the Official Gazette. If the protest of a reissue application cannot be filed within the 2-month delay period, the protest can be submitted at a later time. Where the protest is submitted after the 2-month period, no petition for entry of the protest under 37CFR 1.182 is needed with respect to the protest being submitted after the 2 months, unless a final rejection has been issued or prosecution on the merits has been otherwise closed for the reissue application.

＊＊＊＊＊＊＊＊＊＊＊

The petition must include an explanation as to why the additional time was necessary and the nature of the protest intended. A copy of the petition must be served upon the applicant in accordance with 37 CFR 1.248. The petition should be directed to the Office of Petitions.

If the protest of a reissue application cannot be filed within the 2-month delay period, the protestor may petition to request (A) an extension of the 2-month period following the announcement in the Official Gazette, and (B) a delay of the examination until the extended period expires.

＊＊＊＊＊＊＊＊＊＊＊

美國專利申請及審查實務

作　　者 / 冠群國際專利商標聯合事務所
責任編輯 / 蕭財政、吳珮琪
圖文排版 / 陳宛鈴
封面設計 / 陳佩蓉

出 版 者 / 冠群國際專利商標聯合事務所
　　　　　台北市信義路四段 279 號 3 樓
　　　　　電話：+886-2-2703-9911　傳真：+886-2-2755-2737
法律顧問 / 毛國樑　律師
印刷製作 / 秀威資訊科技股份有限公司
　　　　　114 台北市內湖區瑞光路 76 巷 65 號 1 樓
　　　　　電話：+886-2-2796-3638　傳真：+886-2-2796-1377
　　　　　http://www.showwe.com.tw
劃撥帳號 / 19563868　戶名：秀威資訊科技股份有限公司
　　　　　讀者服務信箱：service@showwe.com.tw
展售門市 / 國家書店（松江門市）
　　　　　104 台北市中山區松江路 209 號 1 樓
　　　　　電話：+886-2-2518-0207　傳真：+886-2-2518-0778
網路訂購 / 秀威網路書店：http://www.bodbooks.tw
　　　　　國家網路書店：http://www.govbooks.com.tw

2010 年 12 月出版
定價：2000 元　ISBN：978-986-80244-1-0

國家圖書館出版品預行編目

美國專利申請及審查實務 / 蕭財政, 吳珮琪主編.
-- 一版. -- 臺北市 : 冠群事務所, 2010.12
　面 ；　公分.
ISBN 978-986-80244-1-0 (平裝)

1. 專利　2. 美國

440.652　　　　　　　　　　99024784